physiology

The National Medical Series for Independent Study

physiology

EDITORS

John Bullock, M.S., Ph.D.

Associate Professor of Physiology
University of Medicine and Dentistry
 of New Jersey
Newark, New Jersey

Joseph Boyle, III, M.D.

Associate Professor of Physiology
Clinical Assistant Professor of Medicine
University of Medicine and Dentistry
 of New Jersey
Newark, New Jersey

Michael B. Wang, Ph.D.

Associate Professor of Physiology
Temple University School
 of Medicine
Philadelphia, Pennsylvania

ASSOCIATE EDITOR

Robert R. Ajello, M.D.

Attending Staff Physician
Cooper Hospital/University
 Medical Center
Camden, New Jersey

A WILEY MEDICAL PUBLICATION
JOHN WILEY & SONS
New York • Chichester • Brisbane • Toronto • Singapore

Harwal Publishing Company, Media, Pa.

Library of Congress Cataloging in Publication Data
Bullock, John, 1932–
  Physiology.

  (The National medical series for independent study)
(A Wiley medical publication)
  Includes index.
  1. Human physiology—Outlines, syllabi, etc.
I. Boyle, Joseph. II. Wang, Michael B. III. Title.
IV. Series. V. Series: Wiley medical publication.
[DNLM: 1. Physiology—Examination questions. QT
17 P578]
QP41.B95   1984   612   84-6622
ISBN 0-471-09627-X

©1984 by Harwal Publishing Company, Media, Pennsylvania

10 9 8 7 6 5 4

Contents

Preface

Physiology, as part of *The National Medical Series for Independent Study*, represents a unique approach to the teaching of physiologic concepts. The text is designed for students of medicine and other health sciences as well as graduate physicians.

Physiology is a comprehensive and up-to-date treatment of physiologic principles focusing on both normal and pathophysiologic aspects of human physiology. The outline format of *Physiology* provides a coherent, organized framework for study, which saves the reader time in learning or reviewing the conceptual and factual core of this discipline. Therefore, this book is equally useful as a supplementary textbook for a formal course and as a review book for independent study.

The study questions are designed to provide self-evaluation of knowledge derived from the information presented in each chapter of *Physiology*. In addition, the questions grouped in the pretest and post-test allow the student to identify topics requiring further study. These tests are designed to evaluate the student's ability to accumulate factual information, integrate diverse physiologic concepts, and solve scientific and clinical problems. Problem solving is the ultimate skill of the practicing physician, and this ability is tested in *Physiology* using case histories that provide models for the application of physiologic information.

The questions are followed by extensive explanations that discuss the answers in the context of the proposed clinical situations. In this way, the student can practice a skill that he or she must learn in clinical rotations, that is, the ability to develop and support a thesis on the basis of clinical findings.

Physiology was written by professionals with a common interest in the education of students of the health sciences. The authors' collective teaching experience has made them aware of how students grasp physiologic principles and has identified areas that students often find difficult. The authors have attempted to clarify these areas with figures and tables.

John Bullock

Acknowledgments

We express our thanks to the staff at Harwal Publishing Company, particularly Jim Harris, publisher, Debra L. Dreger, project editor, and Jane Edwards, senior editor, for their continued support and patience. The authors also recognize the professional contribution of Wieslawa B. Langenfeld as medical illustrator. Finally, we thank our wives and families for enduring the long hours of isolation generated by the preparation of this book.

Publisher's Note

The objective of the *National Medical Series* is to present an extraordinarily large amount of information in an easily retrievable form. The outline format was selected for this purpose of reducing to the essentials the medical information needed by today's student and practitioner.

While the concept of an outline format was well received by the authors and publisher, the difficulties inherent in working with this style were not initially apparent. That the series has been published and received enthusiastically is a tribute to the authors who worked long and diligently to produce books that are stylistically consistent and comprehensive in content.

The task of producing the *National Medical Series* required more than the efforts of the authors, however, and the missing elements have been supplied by highly competent and dedicated developmental editors and support staff. Editors, compositors, proofreaders, and layout and design staff have all polished the outline to a fine form. It is with deep appreciation that I thank all who have participated, in particular, the staff at Harwal—Debra L. Dreger, Jane Edwards, Gloria Hamilton, Jeanine Kosteski, Wieslawa B. Langenfeld, Keith LaSala, June A. Sangiorgio, Mary Ann C. Sheldon, and Jane Velker.

The Publisher

Introduction

Physiology is one of seven basic science review books in a series entitled *The National Medical Series for Independent Study*. This series has been designed to provide students and house officers, as well as physicians, with a concise but comprehensive instrument for self-evaluation and review within the basic sciences. Although *Physiology* would be most useful for students preparing for the National Board of Medical Examiners examinations (Part I, FLEX, and FMGEMS), it should also be useful for students studying for course examinations. These books are not intended to replace the standard basic science texts but, rather, to complement them.

The books in this series present the core content of each basic science area, using an outline format and featuring a total of 300 study questions. The questions are distributed throughout the book at the end of each chapter and in a pretest and posttest. In addition, each question is accompanied by the correct answer, a paragraph-length explanation of the correct answer, and specific reference to the outline points under which the information necessary to answer the question can be found.

We have chosen an outline format to allow maximal ease in retrieving information, assuming that the time available to the reader is limited. Considerable editorial time has been spent to ensure that the information required by all medical school curricula has been included and that each question parallels the format of the questions on the National Board examinations. We feel that the combination of the outline format and board-type study questions provides a unique teaching device.

We hope you will find this series interesting, relevant, and challenging. The authors, as well as the John Wiley and Harwal staffs, welcome your comments and suggestions.

Pretest

QUESTIONS

Directions: Each question below contains five suggested answers. Choose the **one best** response to each question.

1. Quantitatively, the major anion in the ICF compartment is

(A) Bicarbonate
(B) Chloride
(C) Phosphate
(D) Protein
(E) Sulfate

2. Hypercapnia has its greatest effect on minute ventilation through stimulation of which of the following receptors?

(A) Aortic bodies
(B) Carotid bodies
(C) Baroreceptors
(D) Central chemoreceptors
(E) J receptors

3. Which of the following statements best characterizes taste receptors?

(A) Each of the primary taste qualities is encoded by a specific sensory nerve fiber
(B) The afferent fibers of the tongue all are carried in a single cranial nerve
(C) All areas of the tongue are equally sensitive to each of the primary taste qualities
(D) Taste receptors are very slowly adapting, so taste sensations do not disappear until the stimulus is washed away
(E) Taste receptors live for only a few days and must be regenerated continuously by the taste buds

4. Spironolactone is a substance that is used to

(A) reduce the GFR
(B) increase the renal excretion of Na^+ and decrease the renal excretion of K^+
(C) enhance the action of aldosterone on the distal tubule
(D) antagonize the action of antidiuretic hormone on the collecting duct
(E) maintain Na^+-K^+ balance in adrenalectomized patients

5. A patient who is admitted to the emergency room with hyperpnea has the following data obtained from a laboratory work-up of the serum: $[HCO_3^-]$ = 8 mmol/L; PCO_2 = 20 mm Hg; and pH = 7.22. On the basis of these findings, what is the primary cause of this patient's hyperpnea?

(A) Metabolic acidosis
(B) Respiratory alkalosis
(C) Respiratory acidosis
(D) Hypocapnia
(E) Hyperventilation

6. Which of the following auditory functions is the primary responsibility of the middle ear ossicular chain?

(A) Sound pressure amplification
(B) Auditory signal detection
(C) Sound localization
(D) Pitch discrimination
(E) Auditory adaptation

7. Which of the following conditions produces an increase in the compliance of the lung?

(A) Atelectasis
(B) Surgical removal of one lobe of the lung
(C) Emphysema
(D) Fibrosis of the lung
(E) Abnormal pulmonary surfactant

8. Aldosterone stimulates Na^+ transport mainly in the

(A) ascending limb of the loop of Henle
(B) descending limb of the loop of Henle
(C) proximal convoluted tubule
(D) pars recta of the proximal tubule
(E) distal tubule and collecting duct

9. Which of the following is the major factor that limits expiratory flow rate during most of a maximal forced expiration?

(A) Turbulence in the peripheral airways
(B) Turbulence in the central airways
(C) Maximal velocity of contraction of intercostal muscles
(D) Contraction of the diaphragm
(E) Compression of the airways

10. A thalamic relay is a component of the central pathway for all of the following sensations EXCEPT

(A) vision
(B) olfaction
(C) touch
(D) taste
(E) audition

Questions 11–14

The EKG shown below is from a normal, 35-year-old woman. For each description of events in the cardiac cycle, choose the points on this EKG that designate the appropriate interval.

11. The spread of atrial depolarization occurs between points

(A) 1 and 2
(B) 1 and 3
(C) 1 and 4
(D) 2 and 3
(E) 2 and 4

12. The conduction time through the AV node is measured from point

(A) 1 to 2
(B) 1 to 3
(C) 1 to 4
(D) 2 to 3
(E) 2 to 4

13. The depolarization time through the ventricles is measured from point

(A) 2 to 7
(B) 3 to 10
(C) 3 to 7
(D) 4 to 7
(E) 4 to 10

14. The vulnerable period for the ventricles lies between points

(A) 3 and 5
(B) 3 and 7
(C) 7 and 9
(D) 8 and 9
(E) 9 and 10

(end of group question)

15. The PCO_2 at rest of mixed expired air is approximately

(A) 0.3 mm Hg
(B) 27 mm Hg
(C) 40 mm Hg
(D) 45 mm Hg
(E) 50 mm Hg

16. Which of the following transport processes requires the direct use of energy?

(A) Carrier-mediated transport of Na^+ out of the cell
(B) Facilitated diffusion of glucose into the cell
(C) Na^+-dependent transport of amino acids into the cell
(D) Osmosis of water into the cell
(E) Bulk flow of water out of the capillaries

17. Stimulation of the J receptors initiates which of the following responses?

(A) An increase in PCO_2
(B) Transient respiratory acidosis
(C) Rapid breathing
(D) Stimulation of the apneustic center
(E) Apnea

18. Arterial blood studies in a patient with hyperventilation show a PCO_2 of 25 mm Hg and a $[HCO_3^-]$ of 21.5 mmol/L. The most likely diagnosis is

(A) respiratory acidosis
(B) respiratory alkalosis
(C) metabolic acidosis
(D) metabolic alkalosis
(E) indeterminate because the pH is not given

19. The major determinant of cerebral blood flow rate is

(A) arterial PCO_2
(B) plasma adenosine concentration
(C) aortic blood pressure
(D) neurogenic regulation
(E) body temperature

20. Reciprocal innervation is most accurately described as

(A) inhibition of flexor muscles during an extension
(B) activation of contralateral extensors during a flexion
(C) reduction of Ia fiber activity during a contraction
(D) simultaneous stimulation of alpha and gamma motoneurons
(E) inhibition of alpha motoneurons during a contraction

21. In an animal experiment, a midpontine transection is performed coupled with a bilateral vagotomy. Which of the following responses occurs as a result of this procedure?

(A) Apneustic breathing
(B) Ataxic breathing
(C) Dyspneic breathing
(D) Increased respiratory rate
(E) Apnea

Directions: Each question below contains four suggested answers of which **one or more** is correct. Choose the answer

A   if **1, 2, and 3** are correct
B   if **1 and 3** are correct
C   if **2 and 4** are correct
D   if **4** is correct
E   if **1, 2, 3, and 4** are correct

22. Inhibitory postsynaptic responses are produced by increasing the membrane conductance to

(1) Na^+
(2) K^+
(3) Ca^{2+}
(4) Cl^-

23. The HCO_3^-/CO_2 system is quantitatively the most significant buffer in the

(1) interstitial fluid
(2) intracellular fluid
(3) plasma
(4) erythrocyte

24. Pulse pressure increases as a result of an increase in which of the following measurements?

(1) Heart rate
(2) Stroke volume
(3) Peripheral resistance
(4) Arterial elastance

25. Reactions that occur due to closure of some airways during expiration include

(1) an increase in the work of breathing
(2) an increase in local pulmonary vascular resistance
(3) an uneven distribution of the tidal volume
(4) a decrease in the minute ventilation in the closed areas

SUMMARY OF DIRECTIONS

A	B	C	D	E
1, 2, 3 only	1, 3 only	2, 4 only	4 only	All are correct

26. A patient who is given a physiologic dose of antidiuretic hormone would be expected to show an increased urinary concentration of

(1) Cl^-
(2) Na^+
(3) K^+
(4) urea

27. Statements that accurately describe the state of sleep include

(1) during periods of REM sleep, violent muscular activity is prevented by an inhibition of alpha motoneuron activity
(2) O_2 consumption and blood flow to the brain both decrease during sleep
(3) the progression from light to deep SWS is associated with an increase in the amplitude of the EEG waves
(4) the state of sleep is induced by the periodic reduction in the activity of the ascending reticular activating system

28. Conditions associated with an increase in the closing volume of the lung include

(1) emphysema
(2) old age
(3) airway disease
(4) restrictive lung disease

29. A myopic individual has eyes that

(1) are too long for the refractive power of the lens
(2) require glasses with convex lenses
(3) have a far point that is nearer than 6 m
(4) have a near point that is farther than it is in emmetropic eyes

30. The vital capacity is defined as

(1) the sum of functional residual capacity and inspiratory capacity
(2) the maximal volume expired after a maximal inspiration
(3) the sum of expiratory reserve volume and residual volume
(4) the total lung capacity minus residual volume

31. The receptor cell communicates with its afferent neuron by synaptic transmission in the

(1) Merkle disk
(2) ear
(3) tongue
(4) nose

32. An increased arteriovenous (A-V) PO_2 difference occurs in hypoxia that is

(1) hypokinetic (stagnant)
(2) hypoxic (arterial)
(3) hypemic (anemic)
(4) histotoxic

33. Conditions that are associated with an increase in the total CO_2 content of blood include

(1) metabolic alkalosis
(2) hyperventilation
(3) respiratory acidosis
(4) metabolic acidosis

34. Proteins that play an important role in the contractile activity of smooth muscle include

(1) myosin
(2) actin
(3) ATPase
(4) troponin

35. Mechanisms that decrease airway resistance include

(1) stimulation of sympathetic efferent fibers to the lung
(2) increases in lung volume
(3) increases in the elastance forces of the lung
(4) increases in expiratory effort

36. Statements that accurately describe thermoreceptors include

(1) the skin has more cold fibers than warm fibers
(2) the static firing rate of warm fibers always is higher at warm temperatures than it is at cold temperatures
(3) cold fibers always increase their firing rate when the skin temperature decreases
(4) both warm and cold fibers cease firing at temperatures within the neutral zone

Directions: The groups of questions below consist of lettered choices followed by several numbered items. For each numbered item select the **one** lettered choice with which it is **most** closely associated. Each lettered choice may be used once, more than once, or not at all.

Questions 37–41

The table below shows the arterial blood acid-base data for five individuals who are designated by the letters A–E. For each of the following descriptions of acid-base status, choose the individual with the appropriate acid-base data.

	PCO$_2$ (mm Hg)	[HCO$_3^-$] (mmol/L)	pH	[H$^+$] (nmol/L)
(A)	29	22.0	7.50	31.6
(B)	33	32.0	7.61	24.8
(C)	35	17.5	7.32	48.0
(D)	40	25.0	7.41	38.4
(E)	60	37.5	7.42	38.4

37. Normal

38. Partially compensated metabolic acidosis

39. Fully compensated respiratory acidosis

40. Uncompensated respiratory alkalosis

41. Combined respiratory and metabolic alkalosis

Questions 42–45

For each description of signs of valve dysfunction, choose the valve lesion with which it is most closely associated.

(A) Mitral stenosis
(B) Mitral insufficiency
(C) Aortic stenosis
(D) Aortic insufficiency
(E) None of the above

42. Wide pulse pressure

43. Presystolic murmur

44. Ejection murmur

45. Wide changes in left atrial pressure

Questions 46–50

For each example of solute transport, select the renal tubular segment that is the major site of that transport process.

(A) Proximal convoluted tubule
(B) Distal convoluted tubule
(C) Thin ascending limb of the loop of Henle
(D) Thick ascending limb of the loop of Henle
(E) Collecting duct

46. H$^+$ secretion

47. PAH secretion

48. Glucose reabsorption

49. Hormone-mediated K$^+$ secretion

50. Active Cl$^-$ transport

Questions 51–54

For each condition or substance listed below, select the type of diarrhea most likely to be associated with it.

(A) Defective absorption diarrhea
(B) Motor diarrhea
(C) Osmotic diarrhea
(D) Cotransport diarrhea
(E) Secretory diarrhea

51. Cholera

52. Phenolphthalein

53. Stress

54. Magnesium sulfate

Questions 55–60

Match each of the following events of the ovarian cycle to the appropriate phase.

(A) During the preovulatory phase
(B) During the postovulatory phase
(C) During both the pre- and postovulatory phases
(D) During neither the pre- nor the postovulatory phase

55. Serum estradiol reaches its highest concentration

56. Extraovarian formation of progesterone is greatest

57. Progesterone is secreted in the greatest *amount*

58. Estradiol is secreted

59. LH becomes the dominant pituitary tropic hormone

60. FSH becomes the dominant pituitary tropic hormone

ANSWERS AND EXPLANATIONS

1. The answer is C. *(Chapter 4 II D 3; Table 4-3)* The major intracellular anion actually is a group of organic phosphates, which includes nucleotides (ATP), nucleic acid (DNA and RNA), phospholipids, and phosphoproteins. Thus, most of this intracellular anion is covalently bound to organic compounds. Organic phosphate is quantitatively the most important buffer in the ICF, where its valence as organic phosphate is not known.

2. The answer is D. *(Chapter 3 VIII A 3, B)* The major effect of CO_2 on respiratory drive is through the central chemoreceptors on the surface of the medulla. Although CO_2 does stimulate the aortic and carotid chemoreceptors, its overall effect on these receptors is very small. The J receptors are stimulated by distension of the alveolar walls by fluid and certain drugs, and the baroreceptors are stimulated when the aortic and carotid walls are stretched due to transmural pressure changes.

3. The answer is E. *(Chapter 1 X B 1, 3, C 3, D, E)* The encoding of the primary taste qualities depends on the pattern of sensory nerves activated by a stimulus; a labeled-line mechanism is not used. The front of the tongue is innervated by the facial nerve (cranial nerve VIII), and the back of the tongue is innervated by the glossopharyngeal nerve (cranial nerve IX). Each area of the tongue is particularly sensitive to one of the four primary taste qualities. Taste neurons adapt fairly rapidly so that gustatory sensations disappear quickly. Taste receptors are renewed continuously by cells that differentiate from nonreceptor cells at the base of the taste bud.

4. The answer is B. *(Chapter 4 IX F)* Spironolactone, a synthetic steroid, is a competitive antagonist of the renal action of aldosterone. Therefore, it interferes with the aldosterone-mediated Na^+ reabsorption from the distal tubule and collecting duct of the nephron. It also reduces distal K^+ secretion and excretion. This substance can be used as an effective K^+-sparing diuretic in normal individuals but not in adrenalectomized patients. Thus, spironolactone is effective only when aldosterone is present.

5. The answer is A. *(Chapter 5 VIII D 3; IX B 3, C 1–2)* Metabolic acidosis is characterized by a low arterial pH, a reduced $[HCO_3^-]$, and a compensatory hyperventilation that results in a decreased PCO_2. Metabolic acidosis can be produced by the addition of H^+ or by the loss of HCO_3^-. Acid-base imbalances are defined as metabolic when the $[H^+]$ and the $[HCO_3^-]$ change in opposite directions. In this patient, the variable that showed the greater change is $[HCO_3^-]$, with a 67 percent decrease; the PCO_2 showed a relatively smaller decline of 50 percent. The decrease in the $[HCO_3^-]/S \times PCO_2$ ratio from the normal of 20 to 13 also is consistent with acidosis.

6. The answer is A. *(Chapter 1 VII B 2, 3)* As sound moves from air to water, most of its energy is lost. The middle ear ossicular chain amplifies the sound pressure because the surface area of the tympanic membrane is much larger than that of the oval window. Pitch discrimination is possible because sounds of different frequencies cause vibration in different areas of the basilar membrane (an inner ear component). Auditory detection also is a function of the inner ear. Sound localization and auditory adaptation are carried out by neurons within the brain.

7. The answer is C. *(Chapter 3 III B 3, E)* Emphysema causes an increased compliance of the lung by increasing the size of the air space, which reduces the surface forces, and by destroying the alveolar septa, which reduces the tissue forces. Compliance is a function of the amount of functional lung tissue that is present; thus, lobectomy and atelectasis reduce compliance by eliminating functional tissue. Fibrosis and abnormal surfactant reduce compliance by an increase in the tissue forces and the surface forces, respectively.

8. The answer is E. *(Chapter 4 IX B 1)* Aldosterone stimulates Na^+ transport (reabsorption) from the distal tubule and collecting duct. The total quantity of Na^+ reabsorption dependent on aldosterone is approximately 2 percent of the total filtered Na^+ or about 20 percent of the Na^+ entering the distal tubule.

9. The answer is E. *(Chapter 3 V C 2)* The pleural pressure becomes positive only if expiration is active (i.e., if the expiratory muscles contract forcefully to increase the rate of gas flow). The alveolar pressure always exceeds the pleural pressure, but the pressure in the airways decreases from alveolar to atmospheric pressure along the length of the airways. At some point in the airways, the pleural pressure will equal the pressure inside the airways (equal pressure point, EPP), and downstream from the EPP the airways are narrowed because of the pressure gradient. This narrowing limits the maximal expiratory flow rate so that flow rate is a function of lung volume rather than the effort exerted. This condition is termed effort-independent flow.

10. The answer is B. *(Chapter 1 XI E 4)* All of the sensory systems except the olfactory system project to their cortical receiving areas through a relay in the thalamus. In the olfactory system, the olfactory neurons project directly from the olfactory bulb to the cortex.

11. The answer is A. *(Chapter 2 II E 1 a)* Atrial depolarization occurs during the P wave, which is designated by the segment that extends between points 1 and 2 on this EKG.

12. The answer is D. *(Chapter 2 II C 4 a)* The conduction time for depolarization to proceed through the AV node is measured from the end of the P wave to the beginning of the QRS complex. This interval includes the time for conduction over the bundle branches and the Purkinje fibers; however, conduction speed in these tissues is extremely rapid and so this period would be negligible.

13. The answer is C. *(Chapter 2 III E 1 c)* The ventricles are depolarized during the QRS complex. This interval is designated by the segment that extends from point 3 to point 7 on this EKG.

14. The answer is E. *(Chapter 2 III E 1 d)* During the vulnerable period, the ventricles are in a superexcitable state. This period occurs at about the peak of the T wave and is designated by the segment between points 9 and 10 on this EKG.

15. The answer is B. *(Chapter 3 VI C 2)* Normally, the PCO_2 in alveolar gas is 40 mm Hg; during expiration, however, the PCO_2 in the mixed expired gas is reduced by dilution with the dead space gas. A PCO_2 value of 0.3 mm Hg is much too low because it indicates that dead space ventilation is more than 100 times greater than the alveolar ventilation.

16. The answer is A. *(Chapter 1 I C 1 b, 2, 3)* Na^+ transport by the Na^+-K^+ pump requires the direct hydrolysis of ATP. Carrier-mediated transport of amino acids also requires energy, but the energy is expended to create the Na^+ gradient and is not used directly in the transport process. Although glucose is too large to diffuse through membrane channels, it can be moved across the membrane by facilitated diffusion, a process that does not require the direct use of energy. Osmosis and bulk flow are translocation processes that also do not require the direct use of energy.

17. The answer is C. *(Chapter 3 VIII C 3)* Stimulation of the J receptors results in rapid breathing (tachypnea) due to afferent vagal impulses. These impulses cause an inhibition of the apneustic center to abbreviate inspiration.

18. The answer is B. *(Chapter 4 VIII D 2; IX B 2; XII B 3 b–c)* Respiratory alkalosis is characterized by a decreased $[H^+]$ ($\uparrow$pH), a low PCO_2, and a variable decrement in plasma $[HCO_3^-]$. The pH of the plasma of this patient can be determined with the equation:

$$pH = pK' + \log ([HCO_3^-]/S \times PCO_2), \text{ or}$$
$$pH = 6.1 + \log (21.5/0.75) = 7.56.$$

Respiratory alkalosis must be differentiated from metabolic acidosis, in which the PCO_2 and $[HCO_3^-]$ are also diminished but the pH is decreased ($\uparrow[H^+]$).

The variable that exhibits the greater proportional change is PCO_2, which is decreased by 38 percent in this patient compared to the 10 percent decline in $[HCO_3^-]$. Therefore, the primary cause of this acid-base abnormality is respiratory. Since the $[HCO_3^-]/S \times PCO_2$ ratio is greater than 20, the patient has alkalosis. (Note that it is not necessary to calculate the pH to diagnose the abnormality.)

19. The answer is A. *(Chapter 2 VIII B 2)* The major factor controlling cerebral blood flow is the perivascular PCO_2, which depends largely on the arterial PCO_2. While adenosine is a major dilator in the skeletal muscle and coronary vascular beds, it generally is not recognized as a major factor in the cerebral circulation. The cerebral circulation has a powerful autoregulatory function so that changes in arterial pressure have a minimal effect on blood flow. Sympathetic and parasympathetic nerve fibers do not have a major role in controlling cerebral blood flow.

20. The answer is A. *[Chapter 1 XII A 1 b (2), c, B 2 d, C 1 b (2)]* Reciprocal innervation refers to the inhibition of the antagonist muscle during a contraction. Activation of the contralateral extensors during a withdrawal reflex is called a crossed extensor reflex. Reduction of Ia fiber activity during a contraction is called unloading. Inhibition of alpha motoneurons during a contraction is caused by Ib afferent fibers and is called autogenic inhibition.

21. The answer is A. *(Chapter 3 VIII A 2; Fig. 3-19)* The experimental procedure described removes the inhibitory effects of the vagus nerve and the pneumotaxic center from the apneustic center. This loss of inhibition allows the marked inspiratory tone of the apneustic center to produce inspiratory spasms or apneusis, which reduces the respiratory rate.

22. The answer is C (2, 4). *(Chapter 1 III B 2 b, C 2 a)* Parasympathetic neurons produce an inhibitory effect on the heart by increasing the permeability of the membrane to K^+. In the CNS, **inhibitory** postsynaptic potentials are caused by increasing the membrane conductance to Cl^-; **excitatory** postsynaptic potentials (i.e., membrane depolarizations) usually are produced by increasing the membrane conductance to Na^+ and K^+.

23. The answer is B (1, 3). *(Chapter 5 IV B 1 b, 2, 3)* The HCO_3^-/CO_2 system is the most important buffer of noncarbonic acid in both the plasma and the interstitial fluid (which includes lymph). These two fluid compartments have similar concentrations of HCO_3^-; the $[HCO_3^-]$ of plasma is 25 mmol/L, and the $[HCO_3^-]$ of the interstitial fluid is about 27 mmol/L. Although the HCO_3^-/CO_2 system also is important in the erythrocyte, hemoglobin is quantitatively the more important buffer. The ICF contains large amounts of intracellular protein (i.e., about 60 mmol/L) and organic phosphate compounds, making these systems the quantitatively significant nonbicarbonate buffers in this fluid compartment. The ICF has the capacity to buffer effectively both noncarbonic and carbonic acids as well as bases.

24. The answer is C (2, 4). *(Chapter 2 VI B 2)* Pulse pressure is reduced by increases in heart rate and peripheral resistance, if stroke volume remains unchanged. In both instances, arterial volume increases so that pulse pressure decreases. An increase in stroke volume raises arterial uptake and pulse pressure, whereas an increased elastance (decreased compliance) also increases pulse pressure.

25. The answer is E (all). *(Chapter 3 V B–C)* Closure of airways during respiration indicates that additional work must be done to reopen these airways in the subsequent inspiration. In addition, closure of airways causes a reduction in the compliance of the lung. Closure also reduces the tidal volume in these areas and makes ventilation of the lung uneven. Due to reduced ventilation in the affected areas, there is local hypoxia and hypercapnia, which cause pulmonary vascular constriction and an increase in the vascular resistance.

26. The answer is C (2, 4). *(Chapter 4 VII E 1–3; VIII C)* Physiologic (moderate) amounts of antidiuretic hormone (ADH) increase the urine-to-plasma osmolality ratio (U_{osm}/P_{osm}) via the conservation of water, as reflected in the excretion of a low volume of hyperosmotic urine. ADH decreases free-water clearance, or, stated another way, ADH increases free-water reabsorption. ADH affects solute concentration by increasing the solvent (water) reabsorption from the late distal tubule and collecting duct.

27. The answer is B (1, 3). *(Chapter 1 XIV A 2 c–d, C 1)* Although periodic muscular twitching occurs during REM sleep, more powerful muscular activity is prevented by the generalized paralysis that accompanies REM sleep. As an individual passes from light to deep SWS, there is a decrease in the frequency and an increase in the amplitude of the EEG waves. O_2 consumption by the brain is the same whether an individual is asleep or awake. Sleep is induced by the activity of specific sleep centers within the brain.

28. The answer is A (1, 2, 3). *(Chapter 3 VI B 3)* In restrictive lung disease, the increased elastance forces dilate the airways more than normal and delay the onset of airway closure. The elastance forces are reduced in emphysema and old age so that airway closure begins at higher than normal lung volumes. Early airway closure also occurs with small airway disease because the disease narrows the airways, causing them to close at higher lung volumes.

29. The answer is B (1, 3). *(Chapter 1 VIII F 1)* In myopia, the eye ball is too long for the refractive power of the lens, so the image of an object at the far point (i.e., the focal point) falls in front of the retina. Correction requires the wearing of concave (diverging) lenses so that the focal point is moved back and onto the retina. To see an object clearly, it must be brought closer to the eye than the normal 6 m. Because an object can be seen clearly at this shorter distance without any accommodation, it ultimately can be brought closer than normal and still be seen clearly.

30. The answer is C (2, 4). *(Chapter 3 III C 2 a–d)* Vital capacity (VC) refers to the maximal volume of gas that can be expired after a maximal inspiration. Other expressions of lung capacity include: total lung capacity (TLC), functional residual capacity (FRC), and inspiratory capacity (IC). A lung capacity is a combination of two or more lung volumes. For example:

$$VC = ERV + V_T + IRV,$$
$$TLC = RV + ERV + V_T + IRV, \text{ and}$$
$$FRC = RV + ERV,$$

where ERV = expiratory reserve volume; IRV = inspiratory reserve volume; V_T = tidal volume; and RV = residual volume.

31. The answer is A (1, 2, 3). *[Chapter 1 VI A 3 b; VII B 3 b (1); X C; XI C–D]* The cutaneous receptor cells of Merkle disks are modified epithelial cells that communicate with their primary afferent neuron by synaptic transmission. A similar situation exists in the ear, for the hair cells of the cochlea, and on the tongue, for the taste receptor cells within the papillae. The olfactory receptors are found on the free nerve endings of the olfactory nerve fibers.

32. The answer is B (1, 3). *(Chapter 3 VII C 1–4)* In hypokinetic and anemic hypoxia, O_2 supply is limited by blood flow and arterial O_2 content, respectively. This effect reduces the venous PO_2 and, thus, widens the arteriovenous (A-V) PO_2 difference. By definition, hypoxic hypoxia is associated with a reduced PaO_2, which forces the body to function on the steep portion of the O_2-dissociation curve. This means that adequate O_2 can be withdrawn from the the hemoglobin for a smaller than normal change in PO_2, which narrows the A-V PO_2 difference. The A-V PO_2 difference is reduced by histotoxic hypoxia because the tissues do not remove as much O_2 from the blood.

33. The answer is B (1, 3). *(Chapter 5 V A 2–3; VII A 5)* The term total CO_2 content ([total CO_2]) denotes the sum of the various chemical forms of CO_2 carried in the blood and is represented as:

$$[total \ CO_2] = [HCO_3^-] + [dissolved \ CO_2] + [H_2CO_3].$$

Since the $[H_2CO_3]$ is negligible and [dissolved CO_2] is equal to $0.03 \times PCO_2$, the [total CO_2] exceeds the $[HCO_3^-]$ by 1.2 mmol/L. The major contributor to the [total CO_2] of blood is $[HCO_3^-]$. The physician accepts serum CO_2, total CO_2, and $[HCO_3^-]$ as essentially interchangeable terms.

An increased CO_2 content is associated with the following acid-base conditions: uncompensated metabolic alkalosis, uncompensated respiratory acidosis, respiratory acidosis that is compensated by a metabolic alkalosis, and metabolic alkalosis that is compensated by a respiratory acidosis. Hyperventilation ($\downarrow PCO_2$) and metabolic acidosis ($\downarrow[HCO_3^-]$) are associated with a decreased CO_2 content. .

34. The answer is A (1, 2, 3). *(Chapter 1 IV A 2, C 2)* Myosin, actin, and ATPase all are involved in generating contractile force in smooth muscle. Myosin is on the thick filament, actin is on the thin filament, and ATPase is on the cross-bridge. Troponin, the regulatory protein in striated muscle, is not present in smooth muscle.

35. The answer is A (1, 2, 3). *(Chapter 3 V B 1–2)* An increased expiratory effort causes a compression of the airways and an increase in airway resistance. Sympathetic stimulation causes a bronchodilation secondary to the stimulation of the β_2-adrenergic receptors, while increases in lung volume and elastance forces dilate the airways secondary to an increase in the radial traction exerted by the pulmonary septa.

36. The answer is B (1, 3). *(Chapter 1 VI B 1–3)* The static firing rate of thermoreceptors is a bell-shaped function of temperature. Both the warm and cold fibers fire when the skin temperature is in the neutral range. Cold fibers are defined as those thermoreceptors that increase their firing rate when the skin temperature is cooled.

37–41. The answers are: 37-D, 38-C, 39-E, 40-A, 41-B. *(Chapter 5 VIII D 2; IX B–C; XII; Fig. 5-15)* Metabolic acidosis is characterized by a low arterial pH ($\uparrow[H^+]$), a low plasma $[HCO_3^-]$, and a compensatory hyperventilation leading to a low PCO_2. The individual designated by (**C**) shows a change in the $[H^+]$ that is opposite in direction to the change in the $[HCO_3^-]$; this characteristic denotes a metabolic disturbance. The pH of this individual (7.32) indicates acidosis. The fact that partial respiratory compensation has occurred is evident from the decline in PCO_2.

Respiratory acidosis is characterized by a low arterial pH ($\uparrow[H^+]$), a high PCO_2 (hypercapnia), and a variable increase in the plasma $[HCO_3^-]$. The individual designated by (**E**) shows a compensatory change in the $[HCO_3^-]$ that is in the same direction as the primary change in PCO_2; this is a characteristic of a respiratory disturbance. In this individual, the greater degree of alteration is in the PCO_2. The pH of 7.42 together with the hypercapnia and increased $[HCO_3^-]$ are indicative of complete compensation for a respiratory acidosis. This individual has a $[HCO_3^-]/S \times PCO_2$ ratio that is almost 21:1.

The individual designated by (**A**) has uncompensated respiratory alkalosis as indicated by a 28 percent decrease in PCO_2 (hypocapnia) with only an 8 percent compensatory decrease in the plasma $[HCO_3^-]$.

The condition of respiratory alkalosis with a complicating metabolic alkalosis is characterized by a high pH ($\downarrow[H^+]$), a low PCO_2, and a high $[HCO_3^-]$. In acute respiratory alkalosis, the plasma $[HCO_3^-]$ should fall 2 mmol/L for each 10 mm Hg decrease in PCO_2. The acute decrease of PCO_2 to 33 mm Hg, shown by individual (**B**), should decrease the plasma $[HCO_3^-]$ to 22.6 mmol/L (pH of 7.46) in a case of acute respiratory alkalosis. The high $[HCO_3^-]$ shown by this individual, however, indicates that his or her condition is a combination of respiratory and metabolic alkalosis.

42–45. The answers are: 42-D, 43-A, 44-C, 45-B. *(Chapter 2 V B)* The pulse pressure is the difference between arterial systolic and diastolic pressures; a significant increase in pulse pressure is produced by aortic insufficiency. In this condition, the ventricle must eject the normal stroke volume as well as pump out the regurgitant volume. The pulse in aortic insufficiency frequently is described as a water-hammer pulse because of the large volume ejected and the rapid collapse of the pulse due to the regurgitant volume. A presystolic murmur occurs in mitral stenosis due to turbulence in the blood entering the ventricle during atrial contraction. An early diastolic murmur also occurs in mitral stenosis during the rapid passive filling phase of the cardiac cycle. A diamond-shaped murmur is associated with semilunar valve (aortic) stenosis. The variation in the intensity of the murmur is related to the systolic pressure gradient that occurs between the ventricle and the aorta. With mitral insufficiency, atrial volume is increased greatly during systole by the normal venous return as well as the regurgitant blood from the ventricles. As a result, there are drastic pressure changes in the left atrium.

46–50. The answers are: 46-A, 47-A, 48-A, 49-B, 50-D. *(Chapter 4 VI B 3 a, C 1, 3 c, 4, 5 c)* About 85–90 percent of all H^+ secretion occurs in the proximal convoluted tubule; however, the majority of secreted H^+ combine with filtered HCO_3^- and are reabsorbed into the proximal tubule. In the distal convoluted tubule, most of the remaining H^+ combine with nonbicarbonate buffers, because the small amount of HCO_3^- that did not combine with H^+ in the proximal tubule has been reabsorbed by this point.

PAH is actively secreted by the cells of the proximal convoluted tubule. Glucose reabsorption also occurs in the proximal tubule and normally proceeds to completion. Any glucose that is not reabsorbed proximally is excreted.

K^+ can be both reabsorbed and secreted by the renal tubule. Aldosterone contributes to the regulation of K^+ balance by augmenting the secretion of K^+, an effect that is linked, though not directly, to the aldosterone-mediated increase in Na^+ reabsorption. The cortical collecting duct also appears to be a site of action of aldosterone; however, the regulation of K^+ balance occurs mainly in the distal tubule. Both aldosterone and elevated plasma $[K^+]$ promote active transport of K^+ from the peritubular capillaries into the distal tubular cell, a process that favors K^+ secretion into the distal tubular lumen. Thus, aldosterone stimulates Na^+ transport in the distal tubule and cortical collecting duct.

Active reabsorption of Cl^- occurs in the thick segment of the ascending limb of the loop of Henle, with Na^+ following passively. In both the thin and thick segments of the ascending limb, solute (NaCl) transport into the interstitium occurs without water.

51–54. The answers are: 51-E, 52-E, 53-B, 54-C. *(Chapter 6 IV C 3)* Secretory diarrheas result when secretion of fluid and electrolytes exceeds the colon's absorptive capacity. Examples include use of the laxative phenolphthalein and infection with the enterotoxin-producing organism *Vibrio cholerae*. Motor diarrhea results when hypermotility decreases intestinal transit time, as occurs during the neurogenic diarrhea associated with stress. Osmotic diarrhea occurs in the presence of osmotically active but nonabsorbed substances. These include laxatives such as magnesium sulfate and disaccharides such as lactose, which escape breakdown in the various oligosaccharide deficiencies.

55–60. The answers are: 55-A, 56-A, 57-B, 58-C, 59-B, 60-A. *(Chapter 7 IX F 1 a–c)* Estradiol secretion follows a bimodal pattern, with a major peak occurring in the preovulatory phase and a lesser peak occurring in the postovulatory phase. During the preovulatory (estrogenic) phase, estradiol is secreted by the granulosa cells of the ovarian follicle. During the postovulatory (progestational) phase, estradiol is released by lutein-granulosa cells of the corpus luteum.

Blood progesterone also is derived from the peripheral conversion of adrenal progestogens. This mechanism accounts for the bulk of the circulating levels of progesterone during the preovulatory phase.

LH has two important effects on ovarian function. It forms the corpus luteum and also maintains the functional status of the corpus luteum during the postovulatory phase. There is a negative feedback loop that operates between progesterone and LH.

Plasma FSH levels are high at the beginning of menses, and, therefore, some ovarian follicles begin to develop at this time. The preovulatory phase is dominated by this gonadotropic hormone and even is referred to as the follicular phase.

Neurophysiology

I. CELLULAR HOMEOSTASIS

A. HOMEOSTASIS is the fundamental principle underlying all physiologic functions. It is the process by which an organism maintains a **steady state** condition in which cellular processes can be optimally performed.

1. The **extracellular fluid (ECF)** refers to the blood plasma and interstitial fluid. The composition of the ECF is maintained by the cardiovascular, pulmonary, renal, gastrointestinal, endocrine, and nervous systems acting in a coordinated fashion.

2. The composition of the **intracellular fluid (ICF)** differs from that of the ECF and is maintained by the cell membrane.

3. Table 1-1 lists the major components of the ICF and ECF.

B. CELL MEMBRANE. All animal cells are enveloped by a cell membrane that is composed of a variety of lipids and proteins.

1. The **lipids** (primarily phospholipids and cholesterol) are formed into a **bilayer** (Fig. 1-1).
   a. The polar or hydrophilic ends of the lipid molecules, which contain the phosphate group, line up facing the ICF and ECF.
   b. The hydrophobic ends (hydrocarbon chain) of the lipid molecules face each other in the interior of the bilayer.

2. The **proteins** are inserted into the bilayer. Some proteins span the entire bilayer. Others are contained within the intracellular or extracellular half of the bilayer. The proteins serve a number of important physiologic functions.
   a. Receptors. Cellular proteins act as receptor sites for antibodies and as binding sites for hormones, neurotransmitters, and pharmacologic agents.
   b. Enzymes. A variety of enzymes are bound to the cell membrane. These serve such functions as phosphorylation of metabolic intermediates.
   c. Carriers. Materials can be transported across the cell membrane by carrier proteins contained within the lipid bilayer.
   d. Channels. Membrane channels or **pores**, through which small molecules diffuse, are formed by proteins that span the lipid bilayer.

Table 1-1. Major Ionic Components of the Intracellular and Extracellular Fluids

Ion	Intracellular Concentration		Extracellular Concentration	
	(mOsm/L)	(mEq/L)	(mOsm/L)	(mEq/L)
Na^+	15	15	140	140
K^+	135	135	4	4
Ca^{2+}	10^{-4}	2×10^{-4}	2	4
Mg^{2+}	20	40	1	2
Cl^-	4	4	120	120
HCO_3^-	10	10	24	24
HPO_4^{2-}	10	20	2	4
SO_4^{2-}	2	4	0.5	1
Proteins$^-$, amino acids$^-$, urea, etc.	99	152	0.5	1

Note.—The osmolarity of the intracellular fluid (ICF) is the same as that of the extracellular fluid (ECF). Also, the milliequivalents of cations and anions are equal in the ICF as well as in the ECF.

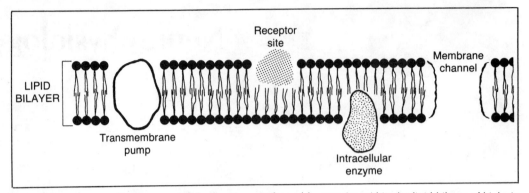

Figure 1-1. Diagram indicating some of the functions performed by proteins within the lipid bilayer of biologic membranes.

C. **TRANSLOCATION OF MATERIAL ACROSS THE CELL MEMBRANE.** Ions, nutrients, and waste products of metabolism are transferred across the cell membrane by a variety of passive and active processes.

1. **Passive transport processes do not require energy.** They are downhill processes.

   a. **Simple diffusion** occurs because all particles in solution are in constant motion.

      (1) Although a single particle moves in an unpredictable (random) direction, it is more likely that the particle will move from an area where it is highly concentrated to an area where its concentration is low than it is for the particle to move in the opposite direction. Thus, a particle moves down its concentration gradient by diffusion.

      (2) Net movement ceases when the concentration of the particle is equal everywhere within the solution. Although random movement of the particle does not cease, the concentration remains the same.

      (3) The rate of diffusion increases as the concentration gradient increases and can be described by **Fick's law of diffusion:**

      $$\text{flux} = \frac{D \bullet A}{t} \bullet (C_{in} - C_{out}),$$

      where flux = the amount of material moved; D = the diffusion coefficient characterizing the material being moved and the substance through which diffusion is taking place; A = the area through which diffusion occurs; t = the thickness of the substance through which the particle is diffusing; and C_{in} and C_{out} = the concentrations of the material on the inside and outside of the substance, respectively.

      (4) Fick's law can be simplified when biologic membranes are involved because the thickness of the membrane is always approximately 10^{-8} m. Dividing D by 10^{-8} yields the permeability (P) of the membrane to the substance, and Fick's law for a unit area of membrane becomes:

      $$\text{flux} = P \bullet (C_{in} - C_{out}).$$

   b. **Facilitated diffusion** requires a carrier. Some particles (e.g., glucose) are too large to diffuse through membrane channels. However, they can be moved across the membrane by a carrier-mediated process that does not require energy.

      (1) The particle binds to a carrier on the outside of the membrane and, while bound to the carrier, is transported through the membrane. The particle dissociates from the carrier when it reaches the ICF. Because the particle binds to the carrier only if it is highly concentrated, the particle can only move **down** its concentration gradient by this process.

      (2) The rate of diffusion increases as the concentration gradient increases until all of the carrier sites are filled. At this point, the rate of diffusion can no longer increase with increasing particle concentration. This is called **saturation kinetics**.

2. **Active transport processes require energy.** They are uphill processes.

   a. **Active transport that has a direct energy source** uses adenosine triphosphate (ATP) directly in the transport process.

      (1) The most common of these active transport systems is the **sodium-potassium (Na^+-K^+) pump**, which uses the membrane-bound ATPase enzyme as a carrier molecule. The Na^+-K^+ pump is responsible for maintaining the high K^+ and low Na^+ concentrations in the ICF.

    (a) Three molecules of Na^+ and one molecule of K^+ bind to the carrier on the inside of the cell.

    (b) Two molecules of K^+ bind to the carrier on the outside of the cell.

    (c) The carrier uses the energy obtained from one molecule of ATP to move two molecules of K^+ into the cell and three molecules of Na^+ out of the cell.

  (2) Other carriers with a direct energy source are available to transport a variety of ions, such as chloride (Cl^-), calcium (Ca^{2+}), and hydrogen (H^+).

b. Active transport that has an indirect energy source uses the energy stored in the Na^+ concentration gradient to transport other particles against a concentration gradient. The system can operate only if the extracellular Na^+ concentration is higher than the intracellular Na^+ concentration. Energy was required to establish the Na^+ concentration gradient, and it is this energy that is used indirectly in the active transport process.

  (1) Glucose and amino acids are transported against their concentration gradients by such a system.

    (a) Na^+ binds to the carrier on the outside of the cell where the Na^+ concentration is high. After Na^+ binds to the carrier, the carrier's affinity for glucose increases, enabling it to bind glucose (Fig. 1-2).

    (b) The carrier then moves both Na^+ and glucose into the cell, where, because of the low Na^+ concentration, Na^+ detaches from the carrier.

    (c) After Na^+ is removed from the carrier, glucose cannot remain attached and so it dissociates from the carrier.

  (2) The energy in the Na^+ concentration gradient also can be used to transport material out of the cell. For example, in **Na^+-Ca^{2+} exchange**, Ca^{2+} is removed from the cell against its electrochemical gradient in exchange for Na^+, which enters the cell down its electrochemical gradient.

3. Osmosis. Water is transported across a cell membrane by a **hydrostatic force** called an **osmotic pressure**. An osmotic pressure difference between the inside and outside of a cell is created when there is a difference in the intracellular and extracellular concentrations of particles not permeable to the membrane.

a. Although it is not known how the particle concentration gradient creates an osmotic pressure, the pressure on either side of the membrane can be calculated using the **van't Hoff equation:**

$$\pi = C \cdot R \cdot T,$$

where π = the osmotic pressure; C = the concentration of the particle in solution; R = the universal gas constant; and T = the temperature (in $^\circ K$).

b. The sum of the intracellular concentration of particles always equals the sum of the extracellular concentration of particles (see Table 1-1) because any change in concentration is eliminated rapidly by the osmotic flow of water across the cell membrane.

  (1) For example, a decrease in the extracellular Na^+ concentration from 140 mmol/L to 135 mmol/L would reduce the total particle concentration from 290 to 285 and the osmotic pressure from 5574 mm Hg to 5478 mm Hg. (These values can be confirmed using the van't Hoff equation, with Na^+ concentrations expressed in mmol/L and the values $T = 310^\circ K$ and $R = 0.062$.)

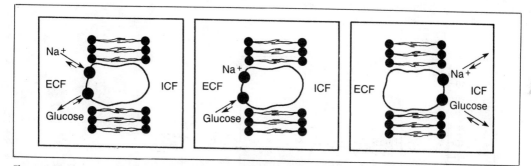

Figure 1-2. Glucose is transported through the membrane by active transport that uses an indirect energy source. As indicated by the length of the *arrows*, the affinity of the carrier for glucose is low in the absence of Na^+. However, after Na^+ binds to the carrier, the affinity for glucose is increased. Glucose and Na^+ diffuse through the membrane together. Once inside the cell, Na^+ is removed from the carrier, the affinity for glucose is reduced, and glucose dissociates from the carrier.

(2) This large osmotic pressure difference of 96 mm Hg causes water to flow into the cell, which dilutes the intracellular particles and equalizes the intracellular and extracellular osmotic pressures.

(3) Note that water moves into the cell, where the osmotic pressure was 5574 mm Hg, from the extracellular space, where the osmotic pressure was 5478 mm Hg. That is, water moves from an area of low osmotic pressure to an area of high osmotic pressure. This paradox is due to the way in which osmotic pressure is calculated and does not imply that water flows from an area of low hydrostatic pressure to one of high hydrostatic pressure.

c. Osmotic pressure is a **colligative property** of solutions; that is, it is related to the number of particles dissolved in the solution.

   (1) Each mmol/L of a particle produces an osmotic pressure of about 19 mm Hg.

   (a) For example, a 1 mmol/L glucose solution produces an osmotic pressure of 19 mm Hg, and a 1 mmol/L sodium chloride (NaCl) solution produces an osmotic pressure of 38 mm Hg (because two particles of solute exist for each molecule of NaCl).

   (b) The osmotic concentration of a substance in solution, therefore, reflects the number of particles dissolved and not the number of molecules. Thus, a 1 mmol/L glucose solution has an osmotic concentration of 1 mOsm/L, and a 1 mmol/L NaCl solution has an osmotic concentration of 2 mOsm/L.

   (2) An osmotic pressure difference is established only if the particle creating the osmotic pressure is not permeable to the cell membrane. For example, because urea easily flows through cell membranes, a change in the extracellular concentration of urea does not cause a net flow of water across the cell membrane.

   (a) The ability of a particle to cause a flow of water across the cell membrane is referred to as its **tonicity**. An extracellular solution that causes water to flow into the cell, making it swell, is called **hypotonic**. An extracellular solution that causes water to flow out of the cell is called **hypertonic**.

   (b) Almost all particles dissolved in blood plasma, except proteins, easily cross the capillary and, thus, do not cause water to flow between the capillary and the interstitial fluid. Plasma proteins have an osmolar concentration of about 1.2 mOsm/L and, thus, create an osmotic pressure of about 23 mm Hg.

   (i) The osmotic pressure produced by the plasma proteins, called the **colloid oncotic pressure**, draws water into the capillary from the interstitial fluid and is counteracted by the hydrostatic pressure of the blood produced by the heart. The movement of water through the capillaries due to hydrostatic or osmotic pressure differences is called **bulk flow**.

   (ii) Whether water flows into or out of the capillaries depends on whether the colloid osmotic pressure is greater or less than the hydrostatic pressure of the blood. When water flows through the capillaries, it carries dissolved particles with it. This is referred to as **solvent drag**.

II. RESTING AND ACTION POTENTIALS.

An electrical potential (voltage) exists between the inside and outside of any cell at rest. This potential difference across the cell membrane, called the **resting potential**, is approximately -80 millivolts (mV) in excitable cells such as nerve and muscle cells. When these cells are stimulated, an **action potential** is produced. During the action potential, the electrical potential inside of the cell changes from -80 mV to approximately $+45$ mV and then returns to the resting potential (Fig. 1-3).

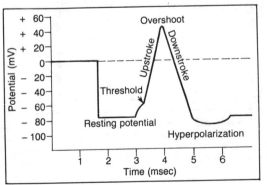

Figure 1-3. When a microelectrode enters a nerve cell, a resting membrane potential is recorded. Stimulation produces an action potential, which also is recorded. The various components of the action potential are indicated on the diagram.

A. **ELECTRICAL RECORDING OF RESTING AND ACTION POTENTIALS.** The electrical activity of excitable cells can be recorded using either intracellular or extracellular electrodes.

1. **Intracellular recordings** of membrane potentials are made with glass microelectrodes that have tip diameters of less than 0.5 μ, allowing them to be inserted through the membrane without causing any damage to the cell.

    a. Figure 1-3 shows the electrical recording made as a microelectrode is inserted into a cell at rest. The electrical potential is 0 mV when the microelectrode is outside the cell and drops to -80 mV as soon as it passes through the membrane into the ICF.

    b. The action potential produced when the cell is stimulated also is shown in Figure 1-3. The components of the action potential are described below.

    (1) The action potential begins when the membrane is depolarized (made less negative) from its resting potential to its **threshold potential** by a stimulus.

    (2) The change in membrane potential observed after threshold is reached is called the **depolarization** or **upstroke** of the action potential. The **overshoot** refers to the positive portion of the action potential.

    (3) The return of the membrane potential toward its resting value is the **repolarization** or **downstroke** of the action potential.

    (4) A small **hyperpolarization** occurs at the end of the action potential. At this stage, the membrane potential is more negative than the resting potential.

2. **Extracellular recordings** usually are made with metal electrodes that are placed on or near the nerve or muscle.

    a. Because these electrodes are outside the cell, they can record only changes in membrane potential; that is, they can record action potentials but not resting potentials. In addition, they cannot record the exact magnitude or time course of the action potentials.

    b. Nonetheless, extracellular recordings are useful in clinical situations when the electrical activity of excitable tissues must be monitored. For example: **electroencephalograms (EEGs)** are used to aid in the diagnosis of brain disease; **electrocardiograms (EKGs)** are used to detect damage to the heart; and **electromyograms** (recordings from nerves and muscles) are used to study neuropathies and myopathies.

B. **THE RESTING POTENTIAL.** Resting and action potentials are produced by the movement of ions across biologic membranes.

1. **The Equilibrium State.** If a cell membrane is permeable to only one ion, an equilibrium state is created. In this condition, the concentration gradient between the inside and outside of the cell does not change because the diffusion of the ion down its concentration gradient is balanced by the membrane potential (i.e., for every ion that moves out of the cell down its concentration gradient another ion is drawn into the cell by the membrane potential).

    a. Figure 1-4A illustrates a cell with a membrane that is permeable only to K^+. Note that the inside of the cell contains much more K^+ than the ECF and that the Na^+ concentration is greater outside the cell than it is inside the cell. These concentration differences are created by the Na^+-K^+ pump discussed in Section I C 2 a (1).

    b. Although the cell membrane shown in Figure 1-4A is permeable to K^+, K^+ cannot flow from the ICF to the ECF unless a negative ion (an anion) travels with it or Na^+ crosses the membrane in the opposite direction. This is due to the law of **electroneutrality**, which states that each positive ion (cation) in solution must be balanced by an anion. Since neither Na^+ nor an anion can flow across the membrane, no net flow of K^+ occurs from the ICF to the ECF.

    c. Nonetheless, some K^+ does cross the membrane. However, it does not enter the ECF. Instead, as diagrammed in Figure 1-4A, these cations stick to the outside of the membrane where they are balanced by anions sticking to the inside of the membrane, thus preserving electroneutrality.

    d. The excess positive charge on the outside of the membrane, or, depending on the vantage point, the excess negative charge on the inside of the membrane, creates an electrical potential difference between the inside and outside of the cell. The magnitude of the potential difference depends on the concentration gradient for K^+. If the concentration gradient is made larger, then more K^+ can cross the membrane and stick to the outside of the cell. This increases the charge separation and, thus, the membrane potential.

    e. The membrane potential can be calculated using the **Nernst equation:**

$$E_m = -\frac{RT}{nF} \cdot \ln \frac{C_{in}}{C_{out}},$$

where E_m = the membrane potential; R = the universal gas constant; T = the temperature (in $^\circ$ K); n = the valence of the ion; F = the Faraday constant, which converts moles of

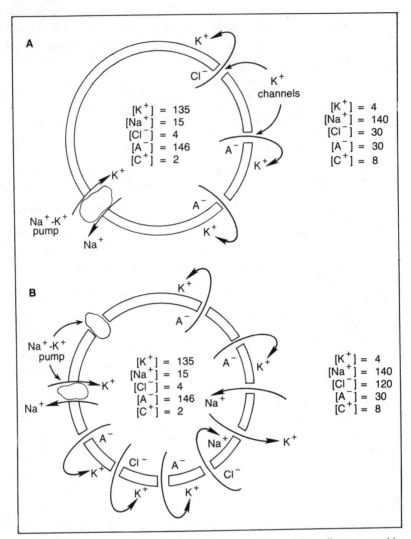

Figure 1-4. *A* The channels contained within the membrane of this cell are permeable to K⁺ only. Equilibrium is achieved when the amount of K⁺ bound to the outside of the membrane is sufficient to produce a membrane potential that can prevent any further increase in K⁺ movement. In this situation, the membrane potential is equal to the Nernst potential for K⁺. *B* The membrane channels in this cell are permeable to both Na⁺ and K⁺. A steady state condition is established by the Na⁺-K⁺ pump. The membrane potential can be calculated using the transference equation. Both the intracellular and extracellular solute concentrations for these cells are expressed in mmol/L.

ions to electrical charge; and C_{in} and C_{out} = the inside and outside concentrations of the ion. Substituting the appropriate values for the constants yields the following equation:

$$E_m = -60 \bullet \log \frac{C_{in}}{C_{out}}.$$

(1) When using this equation, it is essential to include the valence of the ion. If the Nernst potential for an anion (e.g., Cl⁻) is being calculated, the sign in front of the equation becomes positive. If the Nernst potential is being calculated for an ion with a valence of 2 (e.g., Ca²⁺), then the constant becomes 30. Table 1-2 lists the Nernst potentials for some of the common ions found in nerve and muscle cells.

(2) The Nernst potential represents the membrane potential at which the net movement of ions across the membrane is prevented. In the example shown in Figure 1-4*A*, K⁺ diffuses across the membrane and sticks to the outside of the cell, creating a membrane

Table 1-2. Nernst Potential For Ions Commonly Found in Nerve and Muscle Cells

Ion	Concentration (mmol/L) Intracellular	Extracellular	Nernst Potential (mV)
Na^+	15	140	$+58$
K^+	135	4	-92
Ca^{2+}	10^{-4}	2	$+129$
H^+	10^{-4}	40×10^{-6}	-24
Cl^-	4	120	-89
HCO_3^-	10	24	-23

potential. Diffusion stops when the membrane potential becomes large enough to prevent any further movement of K^+. This is an **equilibrium state** because no energy is required to keep the concentration gradient from changing. Since the Nernst potential represents the membrane potential that exists in an equilibrium state, the Nernst potential often is referred to as an **equilibrium potential**.

(3) The Nernst or equilibrium potential has another important meaning. It is the **electrical equivalent of the energy in the concentration gradient** that exists between the inside and outside of the cell.

2. **The Steady State.** The cell shown in Figure 1-4B has a membrane that is permeable to both Na^+ and K^+. Although the intracellular and extracellular concentrations of these ions are the same as in the cell shown in Figure 1-4A, both Na^+ and K^+ can flow across the membrane of the cell in Figure 1-4B without violating the law of electroneutrality. This can occur because every K^+ leaving the cell is replaced by a Na^+ entering the cell from the ECF. Despite the continuous movement of K^+ and Na^+ across the cell membrane, the Na^+-K^+ pump keeps the concentration gradients constant. Because **energy is required** to maintain the concentration gradients, this cell membrane is said to be in a **steady state condition**.

  a. Figure 1-5 is a drawing of an electrical analog of this cell membrane, in which:
    (1) The concentration gradient for each ion is represented by a battery equal in value to its equilibrium potential;
    (2) The channel through which each ion moves across the membrane is represented by a resistor, which has a conductance proportional to the permeability of the membrane for that ion;
    (3) The lipid bilayer portion of the membrane is represented by a capacitor.
  b. The movement of Na^+ and K^+ through the membrane produces a membrane potential, the value of which can be calculated using the **transference equation:**

$$E_m = T_{Na} \cdot E_{Na} + T_K \cdot E_K,$$

where E_m = the membrane potential; T_{Na} and T_K = the **transferences** for Na^+ and K^+; and E_{Na} and E_K = the equilibrium potentials for Na^+ and K^+. The transference for an ion is the **relative conductance** for that ion; that is:

$$T_{Na} = \frac{G_{Na}}{G_{Na} + G_K} \quad \text{and} \quad T_K = \frac{G_K}{G_{Na} + G_K},$$

where G_{Na} and G_K = the conductances for Na^+ and K^+. Conductance (G) is the reciprocal of resistance (R); that is, $G = 1/R$.

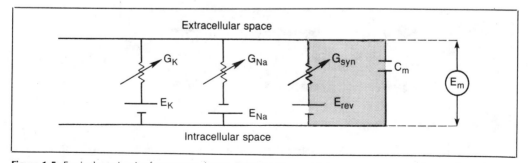

Figure 1-5. Equivalent circuit of nerve membrane. Batteries (E_{Na} and E_K) represent concentration gradients for Na^+ and K^+. Resistors (G_{Na} and G_K) represent channels through which Na^+ and K^+ flow. *Arrows* indicate that the number of open channels can vary. The capacitor (C_m) represents the lipid bilayer. E_{rev} = the synaptic reversal potential; and G_{syn} = the number of channels opened by the synaptic trasmitter.

(1) The transference equation is derived from **Ohm's law**, which states that the flow of current through a resistor equals the conductance of the resistor times the potential difference across the resistor.

(a) Using Ohm's law, the Na^+ current (I_{Na}) and the K^+ current (I_K) are:

$$I_{Na} = G_{Na} \cdot (E_m - E_K) \text{ and } I_K = G_K \cdot (E_m - E_K).$$

(b) In the steady state condition, $I_{Na} = I_K$.

(2) Setting the above expressions for the Na^+ and K^+ currents equal to each other and solving for the membrane potential (E_m) yields the transference equation.

c. The expression, $E_m - E_{ion}$, is the **electrochemical gradient** for an ion.

(1) E_m is the electrical potential acting on the ion, and E_{ion} is the electrical equivalent of the concentration or chemical gradient for the ion.

(2) The expression also is referred to as the **driving force** for the ion. In this regard, it is the net force moving the ion through the membrane.

d. According to the transference equation, the membrane potential depends only on the equilibrium potential and relative conductance (transference) for each ion that can permeate the membrane. However, the Na^+-K^+ pump also contributes to the resting membrane potential.

(1) Although the discussion in Section II B 2 stated that the leakage of K^+ out of the cell was equal to the leakage of Na^+ into the cell, in reality, the Na^+-K^+ pump must move three Na^+ into the cell for every two K^+ it moves out of the cell to maintain the steady state. Consequently, the pump is referred to as **electrogenic** (i.e., it produces a small potential difference across the cell membrane). Adding the **pump potential** (E_{pump}) to the transference equation yields:

$$E_m = T_K E_K + T_{Na} E_{Na} + E_{pump}.$$

(2) Under most physiologic conditions, the value of this pump potential is only a few millivolts and can be disregarded when calculating the membrane potential.

C. THE ACTION POTENTIAL. The action potential illustrated in Figure 1-3 is due to a sequential change in membrane conductance for Na^+ and K^+. The upstroke or depolarization phase of the action potential is due to an increase in Na^+ conductance, and the downstroke or repolarization phase results from an increase in K^+ conductance.

1. Role of Na^+ and K^+ Channels and Gates in Membrane Conductance. Figure 1-6 is a diagram of the Na^+ and K^+ channels found in the membranes of excitable cells. The Na^+ channels are regulated by two gates: the intracellular side of the Na^+ channels contains the **h gate**, and the extracellular side of the channel is covered by the **m gate**. The K^+ channel has a single gate, the **n gate**, located on its extracellular side. The conductance of the membrane depends on the number of channels available for ions to flow through it, and the relative number of Na^+ and K^+ channels that are open during the action potential depends on the behavior of the h, m, and n gates.

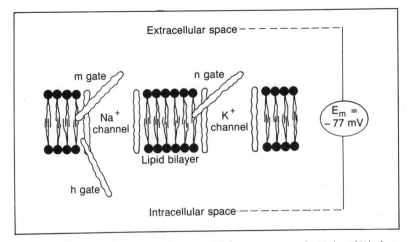

Figure 1-6. A diagrammatic representation of the gates covering the Na^+ and K^+ channels. The negative resting potential tends to keep the m and n gates closed and the h gate opened. E_m = the membrane potential.

a. When the cell is at rest, almost all of the Na^+ channels are closed by the m gate. Although most of the K^+ channels also are closed, there are about nine times as many open K^+ channels as there are open Na^+ channels, meaning that the conductance of the membrane is nine times as great for K^+ as it is for Na^+.

(1) The K^+ and Na^+ transferences at rest are calculated as:

$$T_K = \frac{G_K}{G_K + G_{Na}} = \frac{9}{9 + 1} = 0.9, \text{ and}$$

$$T_{Na} = \frac{G_{Na}}{G_K + G_{Na}} = \frac{1}{9 + 1} = 0.1.$$

(2) Using the values for the equilibrium potentials given in Table 1-2, the membrane potential at rest is calculated as:

$$E_m = E_K \bullet T_K + E_{Na} \bullet T_{Na}$$
$$= -92 \times 0.9 + 58 \times 0.1 = -77 \text{ mV.}$$

b. At the peak of the action potential, the number of open Na^+ channels becomes about 10 times as great as the number of open K^+ channels.

(1) Thus, the transference for K^+ becomes 0.09 and that for Na^+ becomes 0.91. These values can be verified by substituting the relative conductances for K^+ (1) and for Na^+ (10) in the above formulas for determining transference.

(2) The K^+ and Na^+ transferences, can be substituted in the transference equation to show that the membrane potential becomes $+45$ mV at the peak of the action potential.

c. At the end of the downstroke phase of the action potential, the number of open K^+ channels becomes about 15 times as great as the number of open Na^+ channels.

(1) Using the conductances for K^+ (15) and for Na^+ (1), the transferences for K^+ and Na^+ are calculated to be 0.94 and 0.06, respectively.

(2) The K^+ and Na^+ transferences, then, can be substituted in the transference equation to show that at the peak of the hyperpolarization phase of the action potential, the membrane potential is -83 mV.

2. Mechanism of Gating Action. The movement of the m, h, and n gates is controlled by the membrane potential, and the gates, therefore, are described as being **electrically excitable**.

a. When the membrane is at rest, the inside negativity causes the m and n gates to close and the h gates to open (see Fig. 1-6). The m gates are more sensitive to the membrane potential, so more m gates than n gates are closed when the cell is at rest. As a result, the Na^+ transference is lower than the K^+ transference.

b. The m gates open when the cell is depolarized. If the cell is depolarized to threshold, some of the m gates open, causing the Na^+ transference to rise.

(1) The increased transference allows Na^+ to enter the cell, causing it to depolarize further (Fig. 1-7).

(2) The depolarization causes more m gates to open, which further increases the Na^+ transference, and this allows more Na^+ to enter the cell, which further depolarizes the membrane.

(3) As Figure 1-7 illustrates, this is a **positive feedback** or **regenerative mechanism**, in which the response of the membrane to a stimulus (opening of the m gates) causes an effect (membrane depolarization) that produces an even greater response.

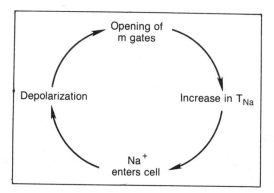

Figure 1-7. When the cell is depolarized to threshold, Na^+ channels open, causing an increase in the transference for Na^+ (T_{Na}). This allows Na^+ to enter the cell, causing further depolarization. The positive feedback system represented by this cycle is responsible for the upstroke of the action potential.

(4) This positive feedback mechanism causes the rapid opening of the m gates and is responsible for the upstroke of the action potential.

c. The upstroke is brought to an end by the h gates, which close or **inactivate** the Na^+ channels, and by the n gates, which open the K^+ channels. Both of these effects are caused by membrane depolarization and occur more slowly than the opening of the m gates.

(1) If the h gates closed the Na^+ channels as rapidly as the m gates opened them, there would be no increase in Na^+ conductance and, thus, no upstroke of the action potential.

(2) Similarly, if the n gates opened along with the m gates, the relative increase in Na^+ conductance (i.e., the increase in Na^+ transference) would not occur, and the upstroke of the action potential would be prevented.

d. Finally, the downstroke occurs because of the inactivation of the Na^+ channels and the opening of the K^+ channels.

3. The "All-or-None" Response.

a. If a stimulus is not strong enough to depolarize the membrane to threshold, the action potential does not occur. If threshold is achieved, a stereotypic action potential is produced. For this reason, the action potential is referred to as an "all-or-none" response.

b. Because of the manner in which the gates respond to a change in membrane potential, the action potential, once initiated by a stimulus, goes through a sequence of membrane conductance and potential changes (Fig. 1-8) in a stereotypic fashion.

(1) At threshold, enough m gates open to initiate the positive feedback mechanism responsible for the upstroke of the action potential.

(2) The upstroke first causes the inactivation of the Na^+ channels (by the closing of the h gates) and then causes the opening of the K^+ channels (by the opening of the n gates) and, thus, is responsible for initiating the downstroke of the action potential.

(3) When the membrane repolarizes, the m and n gates close and the h gates open. Because the K^+ channels close more slowly than the Na^+ channels, the transference for K^+ at the end of the action potential is higher than it is at rest, causing the membrane to hyperpolarize.

4. The Local Response. When a stimulus is not quite strong enough to achieve an action potential, a local response can occur (Fig. 1-9).

a. Under these conditions, the depolarization causes some m gates to open and the membrane begins to depolarize. Because of the marginal strength of the stimulus, not enough m gates open to trigger a regenerative response. Instead, the h gates close and the n gates open, causing the membrane to repolarize. Thus, the action potential cannot begin.

b. The local response varies in size with the strength of the stimulus. Thus, it is a **graded response**, which is localized to the region of membrane where the stimulus is applied and is not propagated along the axon. (See Section II D for a discussion of action potential propagation.)

5. The Passive Response. If a stimulus is too weak to produce a local response, the potential rises and falls in an **exponential** fashion without producing any effect on the electrically excitable gates. The exponential shape of the electrical response to the stimulus is due to the **cable properties** of the membrane.

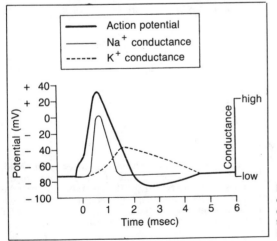

Figure 1-8. The upstroke of the action potential is caused by an increase in Na^+ conductance. At the peak of the action potential, the Na^+ conductance falls due to the closing of the h gates. Repolarization follows the increase in K^+ conductance.

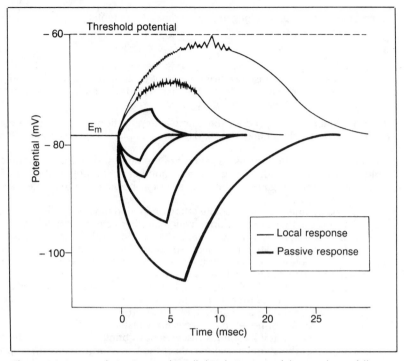

Figure 1-9. Hyperpolarizations and small depolarizations of the membrane follow an exponential time course. The oscillations occur because the depolarization is sufficient to open some m gates. However, not enough open to induce an action potential. This subthreshold response is called a local response. E_m = the membrane potential.

a. The electrical equivalent of the cable properties of a membrane is diagramed in Figure 1-10. Each patch of the membrane contains a resistor representing the conductive pathways through the membrane (R_m); a resistor representing the intracellular pathway for ion flow (R_i); and a capacitor representing the lipid bilayer of the membrane (C_m).

b. Time Constant. When current is applied to a membrane by a stimulus, a charge is added to the capacitor, causing a potential difference to develop across the membrane. The potential difference, V_t, develops at a rate that is determined using the equation:

$$V_t = V_{max} \cdot (1 - e^{-t/R_m C_m}),$$

where V_t = the voltage at time, t; V_{max} = the voltage of the applied stimulus; t = the time after the stimulus is applied; R_m = the membrane resistance; and C_m = the membrane capacitance. The product, $R_m C_m$, is the **time constant** of the membrane and is equal to the time required for the membrane voltage, V_t, to reach two-thirds of V_{max}.

c. Space Constant. The potential produced by the stimulus is spread passively along the

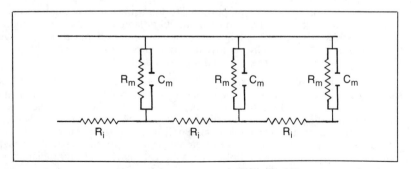

Figure 1-10. Equivalent circuit of an axon. Each patch of membrane is represented by a resistor (R_m) and a membrane capacitor (C_m). The patches are connected by a resistor (R_i) representing the axoplasm.

membrane. At any point along the membrane, the potential, V_x, can be determined using the equation:

$$V_x = V_{max} \bullet e^{-x/\lambda},$$

where V_x = the potential at a point, x, along the membrane; V_{max} = the voltage of the applied stimulus; x = the distance from the point of stimulus application; and λ = the space constant. The **space constant** is the distance from the stimulus to the point at which the voltage falls to about two-thirds of V_{max}.

6. **Refractory Period.** During and for some time after an action potential, the excitability of the membrane is reduced, making it more difficult or impossible to elicit another action potential.
   a. Throughout the entire upstroke and most of the downstroke, the membrane is in the **absolute refractory period**. During this time, another action potential cannot be generated because either the m gates are opening as fast as possible or the h gates have inactivated the Na^+ channels.
   b. For a period of time after the absolute refractory period has concluded, the membrane is in the **relative refractory period**, during which another action potential can be elicited, but the stimulus strength must be increased. This period is due to the increased K^+ conductance and hyperpolarization that follow the downstroke of the action potential.
      (1) The action potential elicited during the relative refractory period differs from the normal action potential in the following ways.
         (a) Its threshold potential is increased; that is, the membrane potential at which the regenerative response occurs is more positive (less negative) than normal.
         (b) The rate of rise and magnitude of the upstroke are less than normal.
      (2) Although the "all-or-none" principle is not violated by these differences, it must be restated to mean that if a stimulus is sufficient to bring the membrane to threshold, the magnitude of the stimulus will not affect the wave form of the action potential.

5. Not all action potentials are like those produced by neurons. For example, the cardiac action potential described in Chapter 2, "Cardiovascular Physiology," Section II B is dependent on not only Na^+ and K^+ channels but on Ca^{2+} channels as well.

D. **PROPAGATION OF THE ACTION POTENTIAL.** Once generated, the action potential must be propagated along the axon in order to transfer information from one location within the nervous system to another. Propagation is possible because the action potential generated at one location on the axon acts as a stimulus for the production of an action potential on the adjacent region of the axon.

1. **Propagation in Unmyelinated Axons.** During the overshoot of the action potential in an unmyelinated axon, the membrane potential becomes about + 40 mV, creating an electrical potential difference between the area of membrane on which the action potential is generated and the adjacent, polarized area. Because of this potential difference, current flows passively between the two areas, causing the polarized region to become depolarized to threshold. This generates an action potential, which becomes the stimulus for generating an action potential in the next area along the axon.
   a. Thus, propagation of the action potential involves the generation of action potentials on contiguous patches of membrane along the axon. Because new action potentials are being generated constantly, the magnitude of the action potential does not change as it is being propagated. This is different from the passive spread of potential, which does decrease in size along the axon.
   b. The speed of propagation is proportional to the **square root** of the fiber diameter.
      (1) To increase the speed of propagation, fiber diameter must be increased.
      (2) However, there is a limit to how large neurons can be if they all are to fit within the nervous system. Generally, unmyelinated neurons are not larger than about 1 μ in diameter and propagate at speeds of less than 1 m/sec.

2. **Propagation in Myelinated Axons.** These axons, which propagate action potentials by **saltatory conduction**, have membranes that are exposed to the ECF only at the **nodes of Ranvier**. The nodes occur at intervals of about 0.1–1 mm, and the membrane area between the nodes is covered by a sheath of **myelin** formed from **Schwann cell** membranes.
   a. Since action potentials can be generated only at the nodes of Ranvier and not at each contiguous patch of membrane, propagation is faster in myelinated axons than in unmyelinated axons. The speed of propagation is proportional to the diameter of the axon and to the internodal distance.
   b. The nervous system uses myelinated fibers when high speeds of conduction are necessary. For example, the axons of the major ascending and descending tracts of the spinal cord,

the sensory axons used for fine tactile discrimination, and the motor axons all are myelinated.

  c. Myelinated fibers range in diameter from 1 to 20 μ. The largest fibers conduct action potentials at speeds of up to 120 m/sec.

III. SYNAPTIC TRANSMISSION

III. SYNAPTIC TRANSMISSION is the process by which nerve cells communicate among themselves and with muscles and glands. The anatomic site of this communication is called a **synapse**. Synaptic transmission is accomplished by a chemical substance called a **neurotransmitter**. The neurotransmitter is released by a neuron (called the **presynaptic cell**) and diffuses to its target cell (called the **postsynaptic cell**). The presynaptic cell activates the postsynaptic cell, producing one of the following responses: an **action potential** if the target is another neuron, **contraction** if the target is a muscle, or **secretion** if the target is a gland. The transmitter also can be inhibitory, in which case it decreases the activity of the postsynaptic cell. The space between the presynaptic and postsynaptic cells is the **synaptic cleft**.

  A. NEUROMUSCULAR TRANSMISSION refers to synaptic communication between alpha motoneurons and skeletal muscle fibers. Neuromuscular transmission also occurs between autonomic efferent fibers and both smooth and cardiac muscle (see Section III B, "Autonomic Synaptic Transmission").

    1. Physiologic Anatomy (Fig. 1-11).
      a. Figure 1-11A is a drawing of the neuromuscular synapse as viewed with a light microscope.
        (1) The alpha motoneuron branches as it approaches the muscle, sending axon terminals to a number of skeletal muscle fibers. The number of skeletal muscle fibers innervated by an alpha motoneuron depends on the type of muscle fiber involved.
          (a) Alpha motoneurons going to large muscles used primarily for strength and postural control innervate hundreds to thousands of skeletal muscle fibers.
          (b) Alpha motoneurons going to small muscles used for precision movements innervate just a few skeletal muscle fibers.
        (2) Each skeletal muscle fiber receives only one axon terminal.
        (3) The axon terminal lies in a groove called the **synaptic trough**, which is formed by an invagination of the skeletal muscle fiber.
        (4) Synaptic transmission occurs in the **end-plate** region of the skeletal muscle fiber.
      b. The details of the presynaptic and postsynaptic (end-plate) membranes can be observed in the electron microscopic view of the synaptic junction shown in Figure 1-11B.
        (1) **Synaptic vesicles** that contain the neurotransmitter, **acetylcholine (ACh)**, are found in the presynaptic terminal concentrated around specialized membrane structures called **dense bars**.
        (2) The postsynaptic membrane contains receptor sites to which the neurotransmitter binds.
        (3) The synaptic cleft is filled with an amorphous structure to which the enzyme, **acetylcholinesterase**, is bound. Acetylcholinesterase is responsible for degrading the ACh after the neurotransmitter produces its effect on the end-plate membrane of the skeletal muscle fiber.

    2. Synthesis, Storage, and Release of ACh.
      a. ACh is synthesized in the nerve terminal from choline and acetyl coenzyme A (acetyl-CoA) by the enzyme, **choline acetyltransferase**.
        (1) Choline acetyltransferase is synthesized in the nerve cell body and is transported to the nerve terminal.
        (2) Acetyl-CoA is supplied by normal cellular metabolism.
        (3) Choline is derived from dietary sources and is transported to the nerve terminal by a Na^+-dependent cotransport process, which is described in Section I C 2 b.
      b. The newly synthesized ACh is stored within the synaptic vesicles, with each vesicle containing 10,000–50,000 molecules of ACh.
      c. Spontaneous release of ACh occurs by **exocytosis** (Fig. 1-12) whenever a vesicle comes into contact with **attachment sites** on the dense bars.
        (1) In exocytosis, the vesicles fuse with the presynaptic membrane, exposing their contents to the ECF.
        (2) ACh diffuses out of the vesicle into the synaptic cleft, and the vesicle merges with the presynaptic membrane.
        (3) Later in time, new vesicles are formed from invaginations of the presynaptic membrane. These vesicles are refilled and used again to discharge ACh into the synaptic cleft.

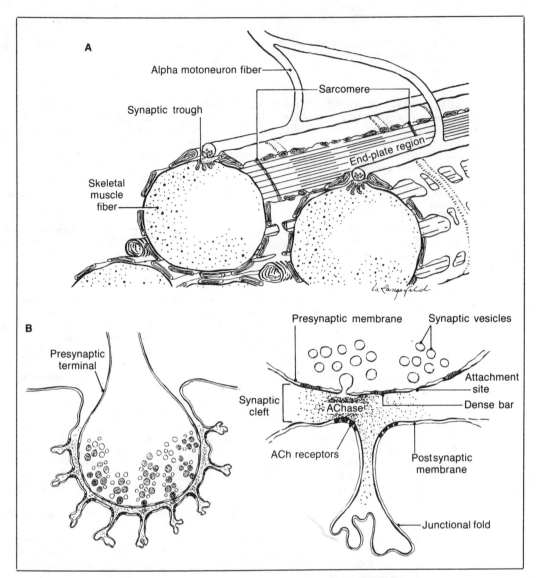

Figure 1-11. *A* Light microscopic view of a neuromuscular junction. The alpha motoneuron branches as it reaches the muscle, and each branch forms a synapse with a single muscle fiber at the end-plate region of the muscle. The axon terminal lies in the synaptic trough. Junctional folds increase the surface area of the postsynaptic membrane. *B* Electron microscopic view, showing the synaptic vesicles within the presynaptic terminal, the acetylcholinesterase (*AChase*) within the synaptic cleft, and the postsynaptic receptor sites. *ACh* = acetylcholine.

3. Events in Synaptic Transmission.

   a. Synaptic transmission begins when an action potential is propagated into the nerve terminal.

   (1) Depolarization of the nerve terminal by the action potential causes Ca^{2+} channels to open, allowing Ca^{2+} to enter the cell down its electrochemical gradient.

   (2) The increase in intracellular Ca^{2+} concentration causes about 200–300 vesicles to bind to attachment sites and release their contents into the synaptic cleft.

   b. The ACh released into the synaptic cleft, either spontaneously or as a consequence of an alpha motoneuron action potential, binds to ACh receptors on the end-plate membrane, where it causes a postsynaptic response called the **end-plate potential (EPP)**.

   (1) When ACh binds to the postsynaptic receptor, it causes the opening of a channel that is permeable to both Na^+ and K^+.

   (a) The ACh-activated channel is **chemically excited**. Unlike the Na^+ and K^+ channels found on electrically excitable membranes, the ACh-activated channel is opened by the binding of the neurotransmitter to the receptor and not by membrane depolarization.

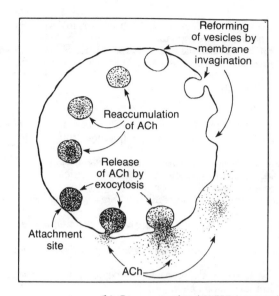

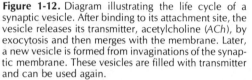

Figure 1-12. Diagram illustrating the life cycle of a synaptic vesicle. After binding to its attachment site, the vesicle releases its transmitter, acetylcholine (*ACh*), by exocytosis and then merges with the membrane. Later, a new vesicle is formed from invaginations of the synaptic membrane. These vesicles are filled with transmitter and can be used again.

   (b) Consequently, the EPP is a graded response (i.e., its magnitude is proportional to the number of open ACh channels), and it is not propagated.
(2) Na$^+$ enters the cell down its electrochemical gradient, while K$^+$ leaves the cell down its electrochemical gradient. Since the electrochemical gradient for Na$^+$ is greater than that for K$^+$, the amount of Na$^+$ entering the cell exceeds the amount of K$^+$ leaving the cell, and the cell depolarizes. The Na$^+$ current (I_{Na}) and K$^+$ current (I_K) through the synaptic channel can be calculated using the equations:

$$I_{Na} = G_{Na} \bullet (E_m - E_{Na}), \text{ and}$$

$$I_K = -G_K \bullet (E_m - E_K).$$

The minus sign indicates that the current is flowing out of the cell.
   (a) These currents become equal to and opposite each other when the membrane potential becomes -17 mV. (This can be verified by substituting the appropriate values, $E_{Na} = +58$ mV and $E_K = -92$ mV, in the above equations and solving for E_m when $I_{Na} = I_K$. Since $G_{Na} = G_K$, the conductance terms cancel out.)
   (b) The membrane potential at which the net current flowing through the synaptic channel is zero is called the **reversal potential**. This value is analogous to the equilibrium (Nernst) potential and can be used as the voltage of the battery representing the synaptic channel in the equivalent circuit of the end-plate membrane (see Fig. 1-5).
   (c) For the EPP, net current through the synaptic channel is zero when the membrane potential is -17 mV. Thus, the reversal potential is -17 mV.
(3) In the equivalent circuit of the muscle fiber, a battery representing the reversal potential of the synapse and a resistor representing the conductance (i.e., the number of open synaptic channels) is added in parallel to the Na$^+$ and K$^+$ batteries and conductances (see Fig. 1-5).
   (a) The conductances, G_{Na} and G_K, represent the nonsynaptic Na$^+$ and K$^+$ channels found on the muscle membrane adjacent to the end-plate.
   (b) The value of the membrane potential (E_m) can be calculated using the transference equation:

$$E_m = E_K T_K + E_{Na} T_K + E_{rev} T_{syn},$$

where E_{rev} = the reversal potential and T_{syn} = the relative number of open synaptic channels.
 c. The contents of a single synaptic vesicle are referred to as a **quantum** of ACh. When a single vesicle spontaneously releases its quantum of ACh into the synaptic cleft, it opens several thousand ACh channels, producing an EPP of about 1 mV. This is called a **miniature end-plate potential (MEPP)**.
(1) Its magnitude can be calculated by substituting the following values: $T_K = 0.89$, $T_{Na} = 0.06$, and $T_{syn} = 0.05$ in the transference equation, which yields a membrane potential (E_m) of:

$$E_m = -92 \bullet 0.89 + 58 \bullet 0.06 + -17 \bullet 0.05 = -79 \text{ mV.}$$

Thus, the synaptic channels that are opened by a quantum of ACh increase the synaptic conductance to a value that changes the membrane potential from -80 mV to -79 mV.

(2) MEPPs occur on an average of about one per second.

d. When an action potential invades the nerve terminal, it causes the release of several hundred quanta of ACh, producing an EPP of about -40 mV. This can be calculated by substituting the following values: $T_K = 0.33$, $T_{Na} = 0.02$, and $T_{syn} = 0.65$ in the transference equation, which yields a membrane potential (E_m) of:

$$E_m = -92 \bullet 0.33 + 58 \bullet 0.02 + -17 \bullet 0.65 = -40 \text{ mV.}$$

Thus, the synaptic channels that are opened by the amount of ACh released when an action potential invades the nerve terminal change the membrane potential from -80 mV to -40 mV.

e. The EPP acts as a stimulus to generate an action potential on the electrically excitable membrane adjacent to the end-plate. (Because the end-plate lacks electrically excitable gates, an action potential cannot be generated directly on it.)

(1) The EPP spreads passively into the adjacent region of the membrane, depolarizing it to threshold.

(2) An action potential is generated and propagated along the muscle membrane, initiating a muscle contraction.

(3) Each time an alpha motoneuron fires an action potential it causes all of the muscle fibers it innervates to contract.

f. After binding to the ACh receptor, the ACh dissociates from the receptor and is hydrolyzed by the acetylcholinesterase found in the synaptic cleft.

(1) Degradation of ACh is necessary to prevent it from causing multiple muscle contractions.

(2) Enzymatic destruction is a unique method of inactivating the transmitter, which occurs only at ACh synapses. At all other synapses, the transmitter action is terminated when the transmitter diffuses out of the synaptic region or is actively transported back into the nerve terminal.

B. AUTONOMIC SYNAPTIC TRANSMISSION. The autonomic nervous system is divided into the parasympathetic and sympathetic divisions. Each division has a preganglionic neuron in the central nervous system (CNS) and a postganglionic neuron in the peripheral nervous system. In both divisions of the autonomic nervous system, ACh is the transmitter used to communicate between the pre- and postganglionic fibers. ACh also is used by the parasympathetic postganglionic fibers, whereas norepinephrine is the transmitter used by the sympathetic postganglionic fibers. A large number of postganglionic fibers, particularly those within the ganglia of the gastrointestinal tract, use neither ACh nor norepinephrine as the transmitter substance.

1. Overview. Ganglionic transmission within the sympathetic and parasympathetic divisions of the autonomic nervous system is essentially the same as that at the neuromuscular junction. ACh is released from the preganglionic (presynaptic) fiber, diffuses across the synaptic cleft, and binds to receptors on the postganglionic fiber, causing it to depolarize.

a. In almost all respects, this process is identical to that described for skeletal muscle neurotransmission. However, a single preganglionic fiber does not release enough neurotransmitter to depolarize the postganglionic fiber to threshold. The postganglionic fiber can discharge an action potential only when there is a summation of the postsynaptic response.

b. The ACh receptors on the postganglionic fibers are called **nicotinic receptors** because they also are activated by nicotine.

(1) Nicotine is extremely toxic; it causes vomiting, diarrhea, sweating, and high blood pressure.

(2) Similar to the receptors on skeletal muscle fibers, the ACh receptors on postganglionic fibers cause the membrane to depolarize by opening channels that are permeable to K^+ and Na^+.

2. Effects of Parasympathetic Postganglionic Fibers. These fibers can have either an **excitatory** or an **inhibitory** effect depending on the postsynaptic receptor that is activated.

a. Excitatory effects are produced on a variety of smooth muscles (e.g., stomach, intestine, bladder, and bronchi) and on glands.

(1) The action of ACh on these target organs is similar to its action on skeletal muscle and postganglionic neurons; that is, ACh opens membane channels that are permeable to K^+ and Na^+, and the cell depolarizes.

(2) However, the nature of the receptor is different. ACh receptors innervated by parasympathetic fibers are called **muscarinic** because they respond to the drug, muscarine, and not to nicotine.

(3) Muscarinic ACh receptors also are blocked by the drug, **atropine**, which does not block the nicotinic receptors on postganglionic and skeletal muscle fibers. These two types of ACh receptors are blocked by different drugs, which indicates that they are not identical.

    (a) Skeletal muscle receptors are blocked by the drug **curare**.

    (b) Receptors on the autonomic postganglionic fibers are blocked by **hexamethonium**.

b. Inhibitory effects of parasympathetic fibers are produced on the heart, primarily on the pacemaker regions of the atrium.

    (1) The ACh receptor is classified as muscarinic because it is activated by muscarine and is blocked by atropine.

    (2) This receptor produces its inhibitory effect by increasing the permeability of the membrane to K^+.

        (a) The channel opened by the muscarinic ACh receptor allows K^+ to enter the cell down its electrochemical gradient, causing the cell to hyperpolarize.

        (b) As a result, the membrane potential is driven further from threshold, making it more difficult to excite the cell.

        (c) The magnitude of the hyperpolarization can be determined using the transference equation and the values: $T_K = 0.75$, $T_{Na} = 0.05$, and $T_{syn} = 0.2$, which yields a membrane potential (E_m) of:

$$E_m = -92 \bullet 0.75 + 58 \bullet 0.05 + -92 \bullet 0.2 = -85 \text{ mV}.$$

Thus, the membrane hyperpolarizes by 5 mV.

        (d) The reversal potential for the channel opened by ACh is -92 mV, because K^+ is the only ion able to move through it.

3. Effects of Sympathetic Postganglionic Fibers. These fibers release norepinephrine and can have an **excitatory** or **inhibitory** effect, depending on the type and location of the receptor activated.

  a. Alpha and Beta Receptors.

    (1) Receptors activated by norepinephrine are divided into alpha (α) and beta (β) types based on the drugs that activate and inhibit them.

        (a) Alpha receptors are activated preferentially by **epinephrine** and are blocked by **phenoxybenzamine**. They produce an excitatory effect on most of the smooth muscles that contain them (e.g., the blood vessels, vas deferens, and uterus). However, alpha receptors produce an inhibitory effect on the smooth muscle of the gastrointestinal tract.

        (b) Beta receptors are activated preferentially by **isoproterenol** and are blocked by **propranolol**. They produce an inhibitory effect on all smooth muscles that contain them, most importantly, the smooth muscle of the bronchioles. However, beta receptors produce an excitatory effect on the heart.

    (2) The mechanism by which alpha receptors produce their effects is not well understood.

        (a) Most likely, alpha receptors produce excitation (depolarization) of smooth muscle by opening a channel through which Na^+ and K^+ can flow. It is possible that Ca^{2+} also may flow through the channel. Another possibility is that Ca^{2+} enters the cell through the receptor-activated channel and, in turn, causes the opening of a channel permeable to K^+ and Na^+.

        (b) Alpha receptors most likely produce inhibition (hyperpolarization) of gastrointestinal smooth muscle by opening channels through which K^+ can flow.

    (3) Beta-receptor inhibition of smooth muscle is brought about by an increase in the cyclic adenosine $3',5'$-monophosphate (cAMP) concentration of the cell.

        (a) The beta receptor is the enzyme, adenyl cyclase, which converts ATP to cAMP when norepinephrine binds to it.

        (b) Transport proteins are activated by cAMP to pump Ca^{2+} out of the cell, resulting in a decrease in the contractile force of the muscle. (See Section IV C 2 for a discussion of the role of Ca^{2+} in smooth muscle contraction.)

        (c) Cyclic AMP is referred to as a **second messenger** because it is produced by the action of the first messenger, the synaptic transmitter. Second-messenger mechanisms are common within the nervous system and can produce effects on their target cells without affecting membrane potential. This is called **pharmacomechanical coupling**.

    (4) Beta receptors cause an increase in the contractile force of cardiac muscle by increasing the amount of Ca^{2+} entering the cell during the cardiac action potential.

  b. Synthesis, Storage, and Release of Norepinephrine.

    (1) Biosynthesis of norepinephrine occurs in the sympathetic nerve terminals.

        (a) The first step is the hydroxylation of **tyrosine** to **dihydroxyphenylalanine (dopa)** by the enzyme, **tyrosine hydroxylase**.

(b) This is followed by the decarboxylation of dopa to **dopamine** by the enzyme, **decarboxylase**, and the oxidation of dopamine to **norepinephrine** by the enzyme, **dopamine-β-hydroxylase**.

(2) The newly synthesized norepinephrine is stored in vesicles that have a dense core when viewed in electron micrographs. This is in contrast to the synaptic vesicles containing ACh, which appear clear in electron micrographs.

(3) Norepinephrine is released by exocytosis in a manner identical to that of ACh release.

(4) Inactivation of norepinephrine occurs by active transport into the nerve terminal and diffusion out of the synaptic cleft. Enzymatic degradation is not used as a mechanism of inactivating norepinephrine, as it is for ACh.

4. The **enteric nervous system** is an independent component of the autonomic nervous system. It is composed of the ganglia found within the walls of the gastrointestinal tract and is responsible for coordinating the activity of the gastrointestinal smooth muscle. (Similar neuronal networks probably exist to control the activity of other visceral organs of the body.)

a. The enteric nervous system receives synaptic input from the sympathetic and parasympathetic postganglionic fibers.

b. All of the neurotransmitters used by the enteric nervous system neurons have not been identified. However, known neurotransmitters are similar to those used by the CNS and include: serotonin, the enkephalins and endorphins, somatostatin, vasoactive intestinal peptide (VIP), and the purines, ATP and adenosine.

C. THE CNS

1. Synaptic Mechanisms. An enormous variety of synaptic connections occur within the CNS. In a few locations (e.g., within the retina and olfactory bulb), some synaptic transmission is accomplished by passive spread of electric current between two cells. However, all important synapses within the CNS use chemical transmitter mechanisms.

a. Electrical synaptic transmission occurs at specialized junctions between two cells called **gap junctions** (Fig. 1-13).

(1) These junctions are located at points where the pre- and postsynaptic membranes are separated by only 20 Å. Gap junctions are formed by membrane bridges that contain aqueous channels through which small molecules and ions can pass from one cell to another, establishing cytoplasmic continuity.

(a) When an action potential propagating along the membrane in one cell reaches the gap junction, electric current passively flows through the gap into the other cell, causing it to depolarize. This cannot occur at chemical synapses, because the pre- and postsynaptic cells are too far apart and are not connected by channels.

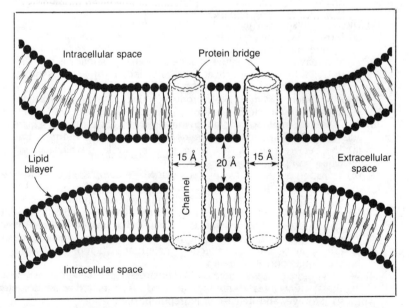

Figure 1-13. A gap junction forms a low-resistance electrical pathway between two cells. Although the two cells remain separate, they maintain cytoplasmic continuity via a channel within a protein bridge.

        (b) Electric current can pass through the gap in both directions, allowing either cell to serve as the pre- or postsynaptic cell.

    (2) Although gap junctions are not important in CNS synaptic transmission, they play an important role in coordinating muscle contraction in the heart and viscera. Gap junctions rapidly transmit an action potential that is generated in one cell to all the other cells within the organ.

  b. Chemical synaptic transmission within the CNS occurs by a mechanism similar to that described previously.

    (1) The synaptic transmitter is synthesized in the nerve terminal, stored in vesicles, and released by exocytosis when an action potential invades the nerve terminal.

    (2) After being released from the presynaptic terminal, the transmitter diffuses across the synaptic cleft, binds to a postsynaptic receptor, and causes the opening of channels through which ions can flow. Both excitatory and inhibitory receptors exist on the postsynaptic cell.

    (3) The transmitter is inactivated in one of three ways.

      (a) It diffuses out of the synaptic cleft.

      (b) It is actively transported into the presynaptic terminal.

      (c) It is enzymatically degraded (if the transmitter is ACh).

2. Summation. The postsynaptic cell integrates the information it receives from thousands of presynaptic terminals. This is illustrated in Figure 1-14, using the alpha motoneuron found in the ventral horn of the spinal cord as an example. (Although synaptic junctions cover most of the dendritic and somatic membrane, only a few of these are shown in the figure.)

  a. Because the synaptic junctions are only about 1–2 μ in diameter, only a few synaptic vesicles release their contents each time an action potential invades the nerve terminal. As a result, the postsynaptic response is only a few millivolts in amplitude.

  b. In order for an **excitatory postsynaptic potential (EPSP)** produced by the excitatory nerve terminals to depolarize the alpha motoneuron to threshold, the EPSPs must summate.

    (1) Although the synaptic channel is open for only a few milliseconds, the membrane capacitance causes the EPSP to decay more slowly. As a result, the duration of the EPSP is about 15 msec.

    (2) If another action potential invades the nerve terminal before the first EPSP has disappeared, the second EPSP will add to the first, producing a larger response. This is called **temporal summation**. If the presynaptic nerve terminal fires frequently enough, the EPSPs can sum sufficiently to depolarize the alpha motoneuron to threshold.

    (3) Summation also can occur if several nerve terminals fire synchronously. This is called **spatial summation**.

  c. The inhibitory synaptic junctions produce an **inhibitory postsynaptic potential (IPSP)**, which, like the EPSPs, can summate to produce a larger response.

  d. The alpha motoneuron membrane potential at any given moment is determined by the combination of EPSPs and IPSPs.

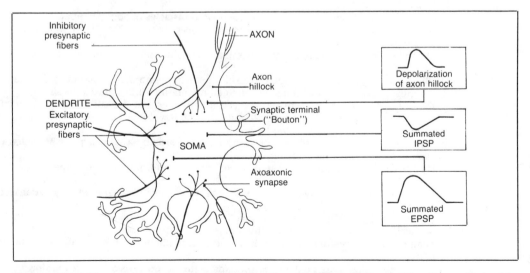

Figure 1-14. Inhibitory and excitatory synapses are formed on an alpha motoneuron. When the amplitude of the summated excitatory postsynaptic potentials (*EPSP*) exceeds the amplitude of the summated inhibitory postsynaptic potentials (*IPSP*), the axon hillock is depolarized to threshold and an action potential is generated.

(1) The channel opened by the excitatory neurotransmitter allows Na^+ and K^+ to move through it and, like the EPP, has a reversal potential of -17 mV.
(2) The channel opened by the inhibitory neurotransmitter allows Cl^- to flow through it and has a reversal potential of -89 mV.
(3) The membrane potential can be calculated as in Section III A 3 b (3), using the transference equation. In this case, both the excitatory and inhibitory channels are included.

$$E_m = E_K T_K + E_{Na} T_{Na} + \underset{(EX)}{E_{rev}} \underset{(EX)}{T_{syn}} + \underset{(INH)}{E_{rev}} \underset{(INH)}{T_{syn}}.$$

  c. The EPSPs and IPSPs produced all along the cell membrane spread passively to the **axon hillock** (see Fig. 1-14), where, if threshold is reached, an action potential will be generated.

  3. Presynaptic inhibition, the final type of synaptic process to be considered, is carried out by the axoaxonic synapse illustrated in Figure 1-14.
    a. The number of vesicles releasing their contents into the synaptic cleft depends on the amount of Ca^{2+} entering the presynaptic terminal, which, in turn, depends on the magnitude of the action potential invading the nerve terminal.
      (1) The axoaxonic synapse increases the transference of the nerve terminal to Cl^-, causing the transference of Na^+ to diminish and, consequently, the size of the action potential to decrease.
      (2) Since the action potential is smaller, the amount of transmitter released from the nerve terminal is reduced, and the size of the EPSP is decreased.
    b. In presynaptic inhibition, the excitability of the postsynaptic cell is not diminished, whereas an IPSP reduces the effectiveness of all excitatory input to a cell. By using presynaptic inhibition, a particular excitatory input can be inhibited without affecting the ability of other excitatory synapses to fire the cell.

IV. MUSCLE CONTRACTION. Muscle fibers are divided into two types based on their appearance in light micrographs. **Striated muscle**, which includes **skeletal** and **cardiac** muscle, is characterized by alternating light and dark bands, whereas **smooth muscle** has no distinguishing surface features. Although all muscle types function in a similar way, several important differences exist. In the following discussion, the structural and contractile properties of skeletal muscle are noted first, with the major differences in cardiac and smooth muscle noted afterward.

A. SKELETAL MUSCLE
  1. Structure (Fig. 1-15). Skeletal muscle fibers vary in diameter from about 10 to 100 μ and can be several centimeters in length.
    a. Fascicles. Skeletal muscle fibers are grouped into **fascicles** of about 20 fibers by a connective tissue sheath called the **perimysium**, which is continuous with the connective tissue surrounding the entire muscle.
      (1) The perimysium is continuous with the **endomysium** surrounding each individual muscle fiber.
      (2) The endomysium is continuous with the **sarcolemma**, a glycoprotein-containing sheath that closely envelops the true cell membrane of the muscle fiber.
      (3) The tight connection between the cell membranes and the surrounding connective tissue structures makes it possible for the force developed by the muscle fibers to be transmitted effectively to the tendons.
    b. Myofibrils. Each muscle fiber is divided into **myofibrils** by a tubular network called the **sarcoplasmic reticulum** (see Fig. 1-15).
      (1) The myofibrils are about 1 μ in diameter and extend from one end of the muscle fiber to the other.
      (2) The myofibrils are divided into functional units, called **sarcomeres**, by a transverse sheet of protein called the **Z disk**.
      (3) The Z disks from neighboring myofibrils are lined up with each other so that in histologic slides a **Z line** spans the entire width of the fiber.
    c. Filaments. The myofibrils contain **thick** and **thin filaments** composed of contractile proteins.
      (1) The **thin filaments** contain the proteins **actin**, **tropomyosin**, and **troponin**, and are about 50 Å in diameter and 1 μ in length.
        (a) One end of the thin filament is attached to the Z disk, so that filaments from oppos-

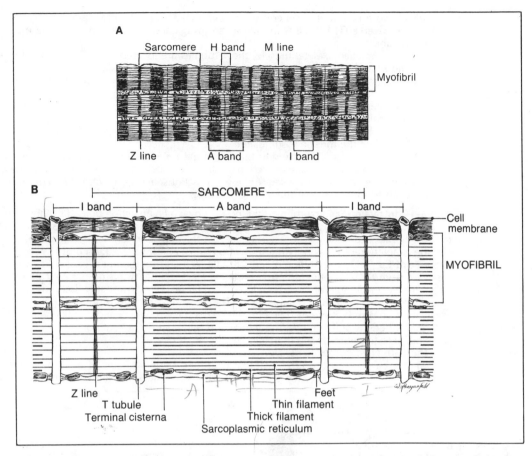

Figure 1-15. High- and low-power light microscopic view of human skeletal muscle. *A* Under low power, the alternating A and I bands can be seen. The functional unit of the muscle fiber, the sarcomere, is bounded by Z lines. *B* Under high power, the T tubules can be seen at the junction of the A band and I band. The terminal cisternae of the sarcoplasmic reticulum form specialized junctions with the T tubules called junctional feet.

ing Z disks extend longitudinally into the center of the sarcomere toward each other.

    (b) The thin filaments appear to be held in place by a thin protein that connects their free ends.

  (2) The **thick filaments** contain the protein, **myosin**, and are about 110 A in diameter and 1.6 μ in length.

    (a) The thick filaments are interspersed between the thin filaments at the center of the sarcomere. They are not attached to the Z disk.

    (b) Projections from the thick filaments, called **cross-bridges**, extend toward the thin filaments. Cross-bridges play a fundamental role in muscle contraction.

  (3) The interdigitating thick and thin filaments create the pattern of light and dark bands that characterizes light microscopic views of skeletal muscle (see Fig. 1-15).

    (a) The dark areas in the center of the sarcomere are called **A bands**. These contain the thick filaments.

    (b) The light areas on either side of the Z disk are called **I bands**. They contain the thin filaments.

    (c) The thick and thin filaments overlap to some extent in the A band. The area of the A band without any thin filaments is called the **H band**. At the center of the H band is the **M line**. This is the region of the thick filaments that does not contain any cross-bridges.

    (d) The amount of overlap between thick and thin filaments varies with sarcomere length and determines how much force skeletal muscle will develop when it is stimulated. When the muscle is stretched or shortened, the thick and thin filaments slide past each other, and the I band increases or decreases in size.

    d. Tubules. Two tubular networks course through skeletal muscle fibers (see Fig. 1-15).

      (1) The **transverse (T) tubule** is formed as an invagination of the surface of the muscle membrane.

        (a) In mammalian skeletal muscle, the T tubules are located at the junction of the A and I bands. (In frog muscle, the T tubules are at the Z disk.)

        (b) An action potential spreading over the surface of the muscle membrane is propagated into the T tubular network.

      (2) The T tubule forms specialized contacts with the **sarcoplasmic reticulum (SR)**, the internal tubular structure that runs between the myofibrils.

        (a) The SR has a high concentration of Ca^{2+}, which is used to initiate muscle contraction when the muscle is stimulated.

        (b) The ends of the SR expand to form **terminal cisternae (TC)**, which make contact with the T tubule. Small projections, called **feet**, span the 200 Å separating the two tubular membranes. These are not gap junctions, and no electrical continuity appears to exist between the SR and T tubule.

  2. Excitation-contraction (EC) coupling is the process by which an action potential initiates the contractile process. EC coupling involves four steps: the propagation of the action potential into the T tubule and release of Ca^{2+} from the TC; the activation of the muscle proteins by Ca^{2+}; the generation of tension by the muscle proteins; and the relaxation of the muscle.

    a. Release of Ca^{2+}. Depolarization of the T tubule by the action potential causes Ca^{2+} to be released from the TC of the SR. Low-resistance electrical pathways do not exist between the T tubule and TC membrane; therefore, the release of Ca^{2+} probably does not involve the opening of electrically excitable gates. The actual mechanism is not yet known.

    b. Activation of Muscle Proteins. For a muscle to contract, the thick and thin filaments must interact. When the cell is at rest, this interaction is inhibited. Ca^{2+} removes this inhibition by binding to troponin on the thin filament.

      (1) The Thin Filament Proteins (Fig. 1-16*A*).

        (a) The backbone of the thin filament is formed by two chains of **actin** molecules, which wind around each other. On each actin molecule is a myosin binding site.

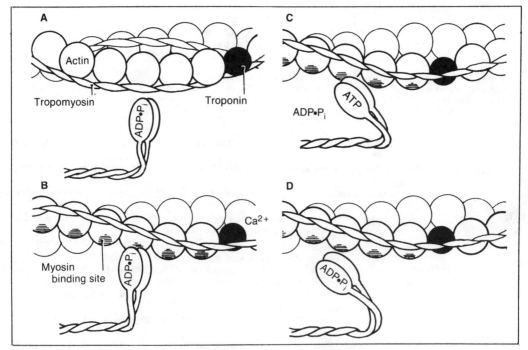

Figure 1-16. *A* The thin filament is composed of three proteins. Actin forms the backbone of the thin filament. At rest, the tropomyosin is covering the myosin binding site on actin. When Ca^{2+} binds to troponin, the troponin undergoes a conformational change, pulling tropomyosin away from the myosin binding site and allowing the cross-bridge cycle to begin. *B* The first step of the cycle is the binding of myosin to actin. *C* The second step is the bending of the cross-bridge and the sliding of the thin filament over the thick filament. After the products of ATP hydrolysis are released from the cross-bridge, a new molecule of ATP binds to the cross-bridge. *D* In the third step, the cross-bridge detaches from the thin filament. *A* The cross-bridge then stands up, and a new cycle begins. Cycling continues for as long as intracellular Ca^{2+} concentration remains high.

(b) When not stimulated, the myosin binding site is covered by another thin filament protein, **tropomyosin**. This long molecule coils around the actin chain.

(c) The position of tropomyosin on the thin filament is controlled by another protein, **troponin**. The troponin molecule contains a binding site for Ca^{2+}. When the binding site is occupied, troponin undergoes a conformational change that causes tropomyosin to move away from its position covering the myosin binding site on actin.

(d) Once uncovered, the binding site on actin combines with the cross-bridges from the thick filament, and contraction begins.

(2) The Thick Filament Protein. About 300 **myosin** molecules coalesce to form the thick filament.

(a) Myosin is a complex protein consisting of a globular head, the cross-bridge, and a long tail.

(b) Rotation can occur at two points in the myosin molecule. One, located in the tail region, is used to rotate the molecule outward when the spacing between the thick and thin filaments changes. The other, located at the junction of the head and tail, is the site where the cross-bridge bends when generating tension.

(c) Located on the cross-bridge is the enzyme, ATPase, which is used to hydrolyze ATP and, thus, provide the energy for muscle contraction.

c. Generation of Tension. Tension is generated by the cycling of the cross-bridges, which occurs after they bind to the thin filament.

(1) The first step in the cross-bridge cycle is the binding of actin and myosin (see Fig. 1-16*B*). This occurs spontaneously, after Ca^{2+} binds to troponin and tropomyosin moves away from its position blocking the myosin binding site on actin.

(a) At rest, the intracellular Ca^{2+} concentration is less than 10^{-7} mol/L.

(b) When stimulated, enough Ca^{2+} is released from the TC to raise the intracellular Ca^{2+} concentration to 10^{-5} mol/L. At this concentration, all of the muscle protein is activated.

(2) The second step is the bending of the cross-bridge and the sliding of the thin filament across the thick filament (see Fig. 1-16*C*).

(a) The energy used to bend the cross-bridge and generate tension is obtained from ATP.

(b) The ATP molecule and its hydrolyzing enzyme, ATPase, both are attached to the cross-bridge.

(3) The third step is the detachment of the cross-bridge from the thin filament. This occurs after the cross-bridge has bent (see Fig. 1-16*D*).

(a) For detachment to occur, the products of ATP hydrolysis (i.e., ADP and inorganic phosphate) must be removed from the cross-bridge and a new molecule of ATP put in its place.

(b) If no ATP is available, the thick and thin filaments cannot be separated. This is the cause of rigor mortis.

(c) As soon as myosin separates (or, perhaps, while it is separating) from actin, the initial steps of ATP hydrolysis occur. The product of these metabolic reactions is a high-energy ATP intermediate, myosin • ADP • P_i.

(d) Complete dissociation of the phosphate from the ADP does not occur until after myosin has bound to actin and completed its bending cycle.

(4) In the final step, the cross-bridge returns to its original upright position. Once there, it can participate in another cycle. Cycling continues for as long as Ca^{2+} is attached to troponin.

d. Relaxation occurs when the Ca^{2+} is removed from the cytoplasm by Ca^{2+} pumps located on the SR membrane. When the intracellular Ca^{2+} concentration falls below 10^{-7} mol/L, troponin returns to its original conformational state, tropomyosin moves back to cover the myosin binding site on actin, and cross-bridge cycling stops.

3. Mechanical Properties. The force developed by the bending of the cross-bridge is transmitted through the thin filament to the Z disk and then through the sarcolemma and tendonous insertions of the muscle to the bones. The contractile properties of the muscle can be studied in two types of mechanical conditions, **isometric** and **isotonic**.

a. In **isometric contractions**, the muscle length remains constant during the contractile event, and the force developed during the contraction is measured.

(1) Muscle Twitch. Figure 1-17 illustrates the intracellular Ca^{2+} concentration and the force developed by a muscle during a single **twitch**.

(a) A single electrical stimulus is used to depolarize the muscle membrane to threshold and produce an action potential.

(b) The action potential causes the release of Ca^{2+} from the TC.

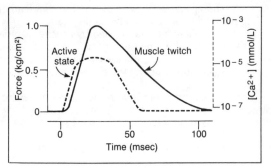

Figure 1-17. When intracellular Ca^{2+} concentration (*right ordinate*) rises above 10^{-7} mmol/L, cross-bridge cycling causes an increase in muscle force (*left ordinate*). The duration of cross-bridge cycling is the active state, and the resulting force development is the muscle twitch.

    (c) Ca^{2+} initiates the cross-bridge cycle, which lasts until the Ca^{2+} is resequestered by the Ca^{2+} pumps on the SR membrane.
    (d) The period during which the intracellular Ca^{2+} concentration is above resting values and the cross-bridges are cycling is called the **active state**.
  (2) An increased strength of electrical stimulation causes an increase in the isometric force developed by the muscle.
    (a) The increased force is the result of more muscle fibers being activated. Although this is a predictable result, it is important to note because it is one of the two major methods used by the motor control system to increase the force of skeletal muscle contraction.
    (b) If the initial conditions are the same, then the force developed by each muscle fiber during a twitch remains constant. This is because the amount of Ca^{2+} released from the TC activates all of the muscle protein, and the time for the SR Ca^{2+} pump to resequester the Ca^{2+} remains the same. However, the force developed by the muscle can be changed by varying the length of the sarcomeres and the frequency of stimulation.
  (3) Figure 1-18 illustrates the relationship between muscle length and force development during a muscle twitch. This is called the **length-tension relationship**. The figure shows both the actual muscle length and the length of the individual sarcomeres. The sarcomere length plays an important role in the length-tension relationship.
    (a) The initial length of the muscle, called the **preload**, is set by a load (or force) applied to the muscle before it is stimulated.
      (i) Maximal force is obtained when the preload is set at a sarcomere length of 2.2 μ. At this length, the overlap between thick and thin filaments is optimal, since every cross-bridge from the thick filament is opposite an actin molecule on the thin filament. When contraction is initiated, each of the cross-bridges binds to the thin filament and contributes to the contractile force.
      (ii) Increasing the preload causes a decrease in the force developed by the muscle fiber. At preloads greater than 2.2 μ, the overlap between thick and thin

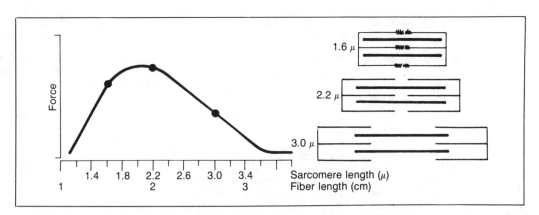

Figure 1-18. The length-tension relationship results from the overlap between thick and thin filaments. At a sarcomere length of 2.2 μ, overlap is optimal and force development is maximal. At lengths greater than 2.2 μ, force decreases because cross-bridge overlap is lessened. At lengths less than 2.2 μ, force is less because the thin filaments meet at the center of the sarcomere, causing an increased resistance to shortening.

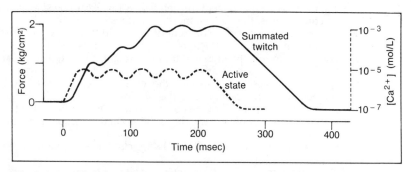

Figure 1-19. When the duration of the active state is increased by repetitive firing of the muscle, force development increases because there is sufficient time for the series elastic component to be stretched completely.

filaments is decreased. Thus, some cross-bridges do not have actin filaments to combine with, and the force developed by the muscle is diminished.

(iii) It is not as obvious why decreasing the preload below 2.2 μ causes a decrease in force development, since every cross-bridge has an actin site with which to combine. However, at short sarcomere lengths the thin filaments bump into each other, making it more difficult for the muscle to develop force. Also, it is possible that the amount of Ca^{2+} released by the TC is reduced as a function of sarcomere length.

(b) Varying the preload is not an important method of varying the contractile force of the skeletal muscle. Often the muscle length is determined by the particular motor task being performed. However, if muscle length is not constrained by the motor activity, more force can be obtained by holding the muscle at its optimal length (i.e., where the sarcomere length is 2.2 μ).

(4) **An increased frequency of electrical stimulation** also increases the force of muscle contraction.

(a) Varying the frequency of muscle stimulation is the other major method used by the motor control system to vary the force of muscle contraction. This is illustrated in Figure 1-19. Note that the Ca^{2+} concentration does not increase above 10^{-5} mol/L but that the duration of the active state is increased due to the repetitive release of Ca^{2+} from the TC.

(b) The increase in muscle force due to repetitive stimulation is called **summation**. When the frequency of stimulation is rapid enough to allow the force to rise smoothly to a maximum, the response is called a **tetanus**. Figure 1-20 is a simple mechanical analog of the contractile properties of the muscle, which can be used to explain the phenomena of summation and tetanus.

(i) The thick and thin filaments are represented on Figure 1-20 by the **contractile component (CC)**, and the tendons and other compliant structures of the muscle are represented by a **series elastic component (SEC)**.

(ii) In order for the force developed by the cross-bridges to be transmitted to the bones, the SEC must be stretched. The more the SEC is stretched, the more force is transmitted.

(iii) Since each cross-bridge is only 100 Å long, its rotation will move the thin filament about 75 Å. At least 30 cross-bridge cycles are required to stretch the SEC sufficiently to transmit all the force of the attached cross-bridges to the bone.

(iv) Since each cycle lasts about 1 msec, the cross-bridges must cycle for at least 30 msec. This cannot happen in a single twitch, because the Ca^{2+} is resequestered too rapidly. By repeatedly stimulating the muscle, however, the Ca^{2+} remains

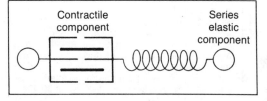

Figure 1-20. The contractile properties of a muscle can be explained using a mechanical analog consisting of a force generating contractile component in series with an elastic element.

in the cytoplasm for a longer time and allows the SEC to be stretched sufficiently.
b. In **isotonic contractions**, the skeletal muscle shortens.
   (1) Figure 1-21 illustrates the development of force and the change in muscle length that occur during an isotonic contraction.
      (a) In order to shorten, the muscle must lift a weight, called the **afterload**, which is applied after the muscle begins to contract.
      (b) When the muscle is stimulated, the cross-bridges begin to cycle, the SEC is stretched, and force is applied to the afterload.
         (i) During the initial part of the contraction, the force is less than the afterload and shortening does not occur. This is the isometric portion of the isotonic contraction.
         (ii) After the SEC has been sufficiently stretched, the force equals the afterload and the muscle begins to shorten.
         (iii) While the muscle is shortening, the length of the SEC remains constant and the force remains equal to the afterload, thus the term, **isotonic** contraction.
         (iv) The velocity of shortening also remains constant during the contraction.
   (2) As shown in Figure 1-21, the magnitude of the afterload affects the characteristics of the contraction.
      (a) As the afterload increases, the duration of the isometric portion of the contraction increases. This occurs because the SEC must stretch more to transmit the force required to lift the greater afterload.
      (b) The amount of shortening that can occur decreases as the afterload increases due to the length-tension relationship. As the sarcomere length is reduced below 2.2 μ, the force that can be developed decreases. At some length, the maximal force that can be developed by the muscle becomes less than the afterload, and further shortening is impossible. The greater the afterload, the sooner this length is reached.
      (c) Finally, the velocity of shortening decreases with an increase in the afterload. This **load-velocity** relationship is an important characteristic of muscle because it indicates that the greatest velocity of shortening is generated when the afterload on the muscle is zero. The peak velocity of shortening is an indication of the cross-bridge cycling speed. (The effect of norepinephrine on the load-velocity relationship in cardiac muscle is illustrated in Chapter 2, "Cardiovascular Physiology," Figure 2-16.)

B. CARDIAC MUSCLE, like skeletal muscle, is striated. Although it is similar to skeletal muscle in structure and behavior, cardiac muscle displays some important differences.

   1. Structure.
      a. Cardiac muscles are smaller than skeletal muscles. Each fiber is about 15–20 μ wide, about 100 μ long, and only about 5 μ thick, making its shape more ribbon-like than cylindrical.
      b. The striation pattern of a cardiac muscle fiber is similar to that described for a skeletal muscle fiber.

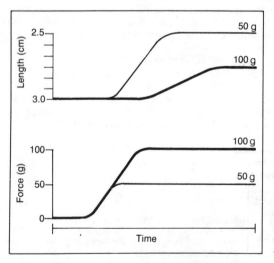

Figure 1-21. During an isotonic contraction, sufficient force must be generated in order for the muscle to shorten against the afterload on the muscle. Once sufficient force is developed, the muscle shortens at a constant velocity. Increasing the afterload increases the amount of force that must be developed before shortening can begin and decreases the velocity and extent of shortening. *Heavy lines* represent an afterload of 100 g, and *light lines* represent an afterload of 50 g.

     c. Cardiac muscle also has both T tubular and SR systems. The T tubule is larger than in skeletal muscle and located at the Z disk rather than at the junction of the A and I bands. The SR makes contact with the T tubule and with the cell membrane.

  2. **EC coupling** of cardiac muscle also differs in significant ways from that of skeletal muscle. Most importantly, the amount of Ca^{2+} in the cytoplasm is never enough to activate all of the muscle protein. As a result, the force of contraction can vary with the amount of Ca^{2+} entering the cell. Several mechanisms may be involved in regulating Ca^{2+} entry during the action potential.

     a. The cardiac action potential has a prolonged plateau, during which Ca^{2+} can enter the cell. This Ca^{2+} can add to the Ca^{2+} released from the TC and be used to initiate contraction.

     b. As in skeletal muscle, Ca^{2+} is released from the TC when the cell membrane is depolarized. In addition, Ca^{2+} present in the cytoplasm can trigger the release of additional Ca^{2+} from the TC. This is called **Ca^{2+}-induced Ca^{2+} release**.

     c. The Na^+-Ca^{2+} exchange mechanism is an important regulator of intracellular Ca^{2+} concentration. Under resting conditions, this exchange mechanism removes Ca^{2+} from the cell. However, when the cell depolarizes, the amount of Ca^{2+} removed is reduced, or the exchange mechanism reverses direction and actually adds Ca^{2+} to the cytoplasm.

  3. The **mechanical properties** of cardiac muscle differ from those of skeletal muscle because of the duration of the action potential and the ability of cardiac muscle to regulate the amount of Ca^{2+} entering the cell.

     a. The action potential and the period of Ca^{2+} removal from the cytoplasm are nearly equal in duration, and so summation and tetanus are not possible. This is not a physiologic disadvantage for the heart, which must relax after each beat so blood can reenter it.

     b. The sarcomere length in a cardiac muscle prior to contraction (the preload) depends on how much blood has entered the heart. Because this amount is under physiologic control, it is an important regulator of the force of cardiac muscle contraction.

     c. The force of contraction can vary at a given sarcomere length if the amount of Ca^{2+} entering the cell is changed. This also is under physiologic control and, thus, is an important regulator of cardiac muscle contractile force.

C. **SMOOTH MUSCLE** differs greatly from striated muscle in appearance and in its mechanism of EC coupling.

  1. **Structure.**

     a. Smooth muscles are thinner than cardiac muscles. They are about 5–10 μ wide and 10–500 μ long.

     b. Smooth muscles are not divided into sarcomeres with interdigitating thick and thin filaments, so they lack striations.

       (1) Smooth muscle proteins are contained in thick and thin filaments, but these filaments are not organized into sarcomeres. Instead they are dispersed throughout the cell.

       (2) The thin filaments are attached to **dense bodies**. Some of these bodies are anchored to the cell membrane, but most are floating within the cytoplasm.

       (3) Cross-bridges extend from the thick filaments, and the cross-bridge cycle appears similar to that of striated muscle.

     c. T tubules are absent and unnecessary in smooth muscle because the cell is small enough for a surface action potential to be effective in activating the contractile machinery.

     d. SR exists in all smooth muscle, but it is not as important in visceral smooth muscle as it is in vascular smooth muscle.

  2. **EC Coupling.** Smooth muscle differs most from striated muscle in its mechanism of activation. In striated muscle, the cross-bridge is prevented from binding to actin by the troponin-tropomyosin regulatory complex. In smooth muscle, there is no troponin-tropomyosin complex inhibiting cross-bridge cycling. However, the myosin cross-bridges cannot bind to actin until they are activated, and this activation requires that one of the light-chain proteins on the myosin head be phosphorylated.

     a. Myosin phosphorylation is catalyzed by **myosin light-chain kinase (MLCK)**, which is inactive at rest. MLCK is activated by combining with another enzyme, **calmodulin**, but this cannot occur until calmodulin has combined with Ca^{2+}. Thus, the role of Ca^{2+} in smooth muscle is to initiate a sequence of reactions ending in the phosphorylation of myosin.

     b. Another enzyme, **myosin phosphatase**, removes the phosphate from the myosin. The amount of myosin active at any time depends on the relative activity of the MLCK and the phosphatase. Relaxation occurs when the intracellular Ca^{2+} concentration falls and the MLCK becomes inactive. At this point, the phosphatase removes phosphate from the myosin light-chain protein, and cross-bridge cycling ceases.

V. GENERAL SENSORY MECHANISMS

A. CLASSIFICATION OF RECEPTORS. The function of receptors is to transform or **transduce** stimulus energy into electrical energy and to transmit information about the nature of the stimulus to the CNS. Receptors can be classified on the basis of the following criteria.

1. **Source of Stimulus.**
   a. Exteroceptors, such as the eye, ear, and taste receptors, receive stimuli from outside of the body.
   b. Enteroceptors, such as the chemoreceptors for measuring blood gases and the baroreceptors for measuring blood pressure, receive stimuli from within the body.

2. **Type of Stimulus Energy.**
   a. Mechanical receptors detect skin deformation and sounds.
   b. Thermal receptors detect environmental temperatures.
   c. Photoreceptors detect light.
   d. Chemoreceptors detect odors and sapid substances. *(Pleasantly flavorful)*

3. **Type of Sensation Produced.** Receptors are classified as touch, heat, cold, pain, light, sound, taste, and olfactory receptors.

4. **Rate of Adaptation.**
   a. Slowly adapting receptors (also called **tonic** or **static** receptors) fire action potentials continuously while a stimulus is being applied.
   b. Rapidly adapting receptors (also called **phasic** or **dynamic** receptors) fire action potentials at a decreasing rate during stimulus application.

5. **Neuron Type.** *(Nerve fibers)*
   a. Neurons are classified on the basis of **size** and **propagation velocity.**
      (1) Neurons divide into four different size groups designated I–IV.
         (a) Group I fibers are 12–20 μ in diameter, group II fibers are 6–12 μ, group III fibers are 1–6 μ, and group IV fibers are less than 1 μ in diameter. All but group IV fibers are myelinated.
         (b) Figure 1-22A is a histogram showing the size distribution of the neurons within a nerve trunk. The peaks in the distribution correspond to each of the four groups, indicating that they are not arbitrary divisions.
      (2) Figure 1-22B illustrates the recording of a **compound action potential** from a nerve trunk. The compound action potential is the summed activity of all the fibers within the nerve bundle. The action potentials are generated at one end of the nerve trunk and propagated into the other end, where the recording electrodes are placed.

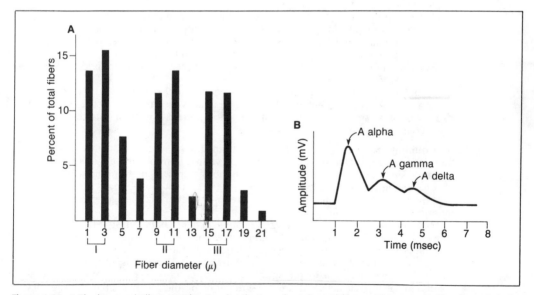

Figure 1-22. *A* The bar graph illustrates the size distribution of myelinated fibers within a nerve. The three peaks in the distribution correspond to the three functional groups of fibers, I, II, and III. *B* The compound action potential recorded from a nerve bundle has three peaks, which correspond to the functional grouping of myelinated nerve fibers according to conduction velocity (i.e., A alpha, A gamma, and A delta fibers).

(a) The fibers with the fastest conduction velocities reach the recording electrodes first and are responsible for the first wave of the compound action potential. Subsequent peaks occur when the fibers with slower conduction velocities reach the recording electrodes.

(b) The fibers responsible for generating the first peak are called **A alpha fibers**, those producing the second peak are called **A gamma fibers**, and those producing the final peak are called **A delta fibers**. All are myelinated fibers. The slowest conducting fibers, called **C fibers**, are unmyelinated and are not shown in Figure 1-22*B*.

  b. Certain important sensory and motor fibers are identified with particular neuron types.

  (1) Muscle spindle afferent neurons belong to the group I and group II fibers.

  (2) Pain and temperature afferent neurons belong to the A delta fibers, if myelinated, and to the C fibers, if unmyelinated.

  (3) Alpha motoneurons are so-called because they are part of the A alpha group of fibers.

  (4) Similarly, gamma motoneurons are part of the A gamma fiber group.

B. TRANSDUCTION OF STIMULUS ENERGY BY RECEPTORS. Each receptor is specialized to receive a particular type of stimulus called an **adequate** or **appropriate** stimulus. The receptor is exquisitely sensitive to its adequate stimulus. For example, the retina can detect the presence of a single photon of light. However, the receptor can respond to other forms of stimulus energy if the stimulus is strong enough. For example, a mechanical stimulus such as rubbing the eyes can produce a sensation of light. The functional components of transduction in a simple mechanical receptor are illustrated in Figure 1-23 and described below.

1. In this case, the **transducer region** is at the end of the nerve terminal. In contrast to this, in the eye, ear, and tongue, the transducer region is on a receptor cell that is separate from the afferent neuron, and the two cells communicate by synaptic transmission.

  a. The transducer region of the receptor is responsible for converting the stimulus energy into an electrical signal called the **receptor** or **generator potential**.

  b. Each receptor has a particular mechanism for producing a receptor potential. These mechanisms will be described later for each receptor as it is discussed. In the mechanical receptor, the stimulus produces a receptor potential by deforming the nerve terminal.

  (1) The deformation opens channels that are permeable to Na^+ and K^+ and causes the membrane to depolarize.

  (2) As illustrated in Figure 1-23, the receptor potential follows the time course of the mechanical stimulus.

2. The **spike generator region** of the receptor is responsible for converting the receptor potential into a train of action potentials.

  a. Figure 1-23 illustrates the repetitive discharge of action potentials by the receptor axon in response to a receptor potential.

  b. The receptor potential passively spreads from the transducer region to the spike generator region. If the spike generator is depolarized to threshold, an action potential is generated.

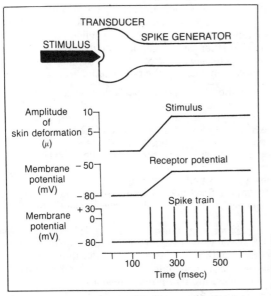

Figure 1-23. When a stimulus is applied to a sensory nerve ending, a generator potential results. The magnitude of the generator potential and the frequency of action potential discharge are proportional to the magnitude of the stimulus.

c. At the end of the action potential, the receptor potential again causes the spike generator membrane to depolarize toward threshold. If threshold is reached again, another action potential is generated.

d. In a tonic receptor, action potentials are generated continuously during stimulus application. In a phasic receptor, adaptation causes the discharge frequency to decrease.

C. ADAPTATION. The most obvious function of adaptation is to decrease the amount of sensory information reaching the brain. However, the brain stem mechanisms responsible for consciousness and attention are capable of keeping the information flow to the brain within tolerable limits so that sensory adaptation is not required for this purpose. The major role of sensory adaptation is to permit the receptor to encode information about the rate of stimulus application.

1. Adaptation occurs by two major mechanisms.
   a. In one, the transducer mechanism fails to maintain a receptor potential despite continued stimulus application.
   b. In the other, the spike generator fails to sustain a train of action potentials despite the presence of a receptor potential. This occurs because the excitability of the spike generator membrane decreases. Although the mechanism of this decreased excitability is not known, it may be due to an increase in:
   (1) The membrane conductance to K^+
   (2) The activity of an electrogenic Na^+-K^+ pump
   (3) The inactivation of Na^+ channels

2. The discharge rate in a phasic receptor increases as the rate of stimulus application increases.
   a. When a stimulus is applied too slowly, adaptation of the spike generator occurs before an action potential can be generated.
   b. When a stimulus is applied rapidly enough, excitation exceeds adaptation, and a steady discharge of action potentials occurs.

D. SENSORY CODING. The intensity, location, and quality of a stimulus are encoded by the sensory system.

1. Intensity is encoded by the firing frequency of a sensory neuron.
   a. As the intensity of a stimulus is increased, the magnitude of the receptor potential is increased. This relationship is expressed by the **Steven's power law function**:

$$V = k \bullet I^n,$$

   where V = the magnitude of the receptor potential; k = a constant; I = the intensity of the stimulus; and n = a constant. Figure 1-24 is a plot of this relationship, in which the maximal value of the receptor potential is about 50 mV.
   (1) Assuming that the sensory system requires a receptor potential of at least 1 mV to detect the presence of a stimulus, the maximal stimulus that can be detected is 50 times as great as the minimally detectable stimulus.
   (2) The power law relationship between stimulus intensity and receptor potential magnitude enables this relatively small range of receptor potential amplitudes (50 to1) to encode a range of stimulus intensities that is greater than 1 million to 1.
   b. The graph in Figure 1-24 also illustrates that sensory perception and the firing frequency of a sensory neuron are related to stimulus intensity by the Steven's power law. This substantiates the concept that stimulus intensity is encoded by the rate of sensory nerve discharge.

2. Stimulus location is encoded primarily by the location of the sensory projection to the cerebral cortex. This mechanism of encoding, called **topographic representation**, is used by the visual and somatosensory systems to localize the point of stimulus application. The auditory system uses a different mechanism, which is discussed in Section VII D 2 c.
   a. Each sensory neuron receives information from a particular sensory area called its **receptive field**. The smaller the receptive field, the more precise the encoding of stimulus localization.
   (1) The receptive fields of the fovea are the smallest within the eye.
   (2) Similarly, the receptive fields within the fingertips are much smaller than those on the hands and back, where it is difficult to localize precisely the point of a mechanical stimulus.
   b. Stimulus localization can be made more precise by **lateral inhibition.** This is demonstrated by the drawing in Figure 1-25, which shows three overlapping receptive fields in the skin.
   (1) Two stimuli are applied to the skin, and the sensory task is to indicate whether one or two stimuli are present. The closest that two stimuli can be to each other and still be distinguished as separate stimuli is called the **two-point threshold**.

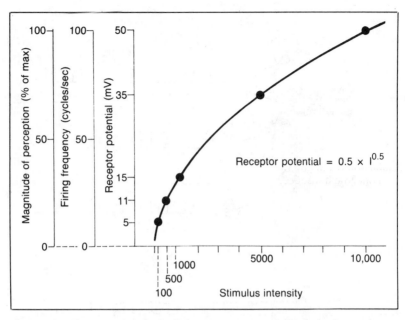

Figure 1-24. The proportionality between generator potential and stimulus can be expressed as a power function. This same relationship can be used to describe the proportionality between stimulus magnitude and firing frequency and between the magnitude of the stimulus and its perceived intensity. *I* = intensity.

(2) Without lateral inhibition (see Fig. 1-25*A*), the two stimuli are recognized as separate only if they are placed in receptive fields that are separated from each other by a nonstimulated receptive field.

(3) Lateral inhibition (see Fig. 1-25*B*) can decrease the two-point threshold by reducing the discharge of the neuron innervating the receptive field in the center and, thus, making it apparent to the CNS that two stimuli are present.

3. Stimulus quality is encoded by a variety of mechanisms.
   a. The simplest mechanism uses a **labeled line**, in which the stimulus is encoded by the particular neural pathway that is stimulated.
      (1) The basic sensory modalities are encoded in this way. Thus, the sensation of touch is elicited whether the receptors on the skin are excited by mechanical deformation or by electrical stimulation. Similarly, light always is evoked no matter how the retina is activated, and sound always results from stimulation of the cochlea.
      (2) The same type of sensation results no matter where along the sensory pathway the stimulus is applied. For example, stimulating electrodes placed on the visual cortex evoke the sensation of light, while seizures within the olfactory cortex produce sensations of smell.
   b. A more complex mechanism of coding uses the pattern of activity within the neural pathway that is carrying information to the brain.
      (1) In **temporal pattern coding**, the same neuron can carry two different types of sensory information depending on its pattern of activity. For example, cutaneous cold receptors indicate temperatures below 30° C by firing in bursts and temperatures above 30° C by firing without bursts.
      (2) In **spatial pattern coding**, the activity of several neurons is required to elicit a sensation. For example, three neurons may be required to encode different taste sensations. A sour taste may result if all three of them are activated, while a salty taste may result if only two of them fire.
   c. The most sophisticated mechanism of sensory coding uses **feature detectors**. These are neurons within the brain, which integrate information from a variety of sensory fibers and fire to indicate the presence of a complex stimulus.
      (1) For example, the location of an object in space can be encoded by cortical cells receiving information from a single eye. However, special feature detectors receiving information from both eyes are required to specify the depth of an object in space.

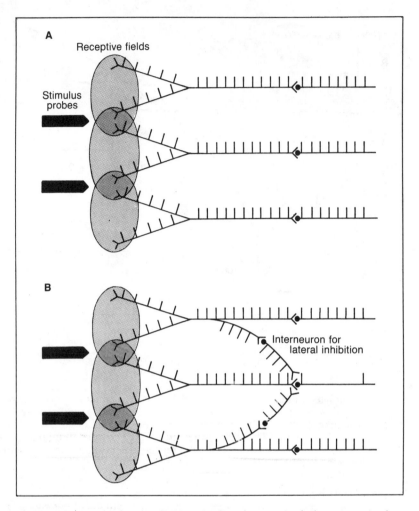

Figure 1-25. In order for two distinct stimuli to be perceived, they must stimulate receptive fields separated by an unstimulated receptive field. *A* Without lateral inhibition, the stimuli produce equal amounts of discharge in all three neurons. *B* In the presence of lateral inhibition, however, the neuron with the receptive field in the center is presynaptically inhibited by collaterals from the neurons with receptive fields located laterally. As a result, the receptive field in the center does not fire and two stimuli are perceived.

　　　　(2) Similarly, the location of a sound in space requires integration of information from both ears by feature detectors within the brain stem.

VI. CUTANEOUS SENSATION. About 1 million sensory nerve fibers innervate the skin. Most of these are unmyelinated nerve fibers that are responsible for crude somatosensory mechanical sensation. Although far fewer in number, the large myelinated (group II) sensory fibers encode the important sensory qualities of touch, vibration, and pressure. The sensations of temperature and pain are encoded by small myelinated (group III, A delta) fibers and unmyelinated (group IV, C) fibers.

　　A. MECHANICAL SENSATION. The sensation produced by mechanical stimulation of the skin can be divided into three **submodalities**: vibration, touch, and pressure. Each of these sensations is encoded by a different set of mechanical receptors.

　　　　1. Vibration is encoded by the **pacinian corpuscle** (Fig. 1-26*A*).
　　　　　a. The pacinian corpuscle is a rapidly adapting **encapsulated receptor**.
　　　　　　(1) Its capsule is an onion-like lamellar structure, which is about 1 mm in diameter.

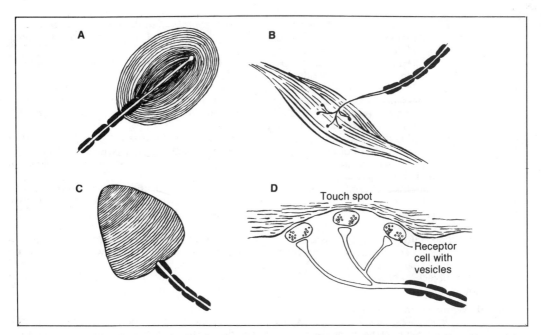

Figure 1-26. The encapsulated organs of the skin all are innervated by a single, large afferent fiber. The pacinian corpuscle (A) is a rapidly adapting receptor that encodes vibration. The Ruffini ending (B) is a slowly adapting receptor that encodes pressure. The Meissner   corpuscle (C) is a rapidly adapting receptor responsible for detecting the speed of stimulus application. The Merkel disk (D) is a slowly adapting receptor capable of encoding the location of a stimulus. Its receptor cells communicate with the primary afferent neurons by synaptic transmission.

  (2) A single, large (group II) sensory axon enters the capsule.
 b. When a mechanical stimulus deforms the outer lamellae of the capsule, the deformation is transmitted through the capsule to the nerve terminal.
  (1) The deformation of the nerve terminal increases the membrane permeability to Na^+ and K^+, producing a depolarizing receptor potential.
  (2) The size of the receptor potential increases in proportion to the magnitude of the deformation. However, because the pacinian corpuscle is a very rapidly adapting receptor, only a few action potentials will be generated, regardless of stimulus intensity.
 c. Adaptation occurs because the deformation of the nerve terminal is not maintained during continuous stimulus application.
  (1) Only rapid deformations are transmitted effectively to the core of the capsule.
  (2) If the stimulus is applied slowly or left in place, the inner lamellae become rearranged so that they no longer deform the nerve terminal.
  (3) This is an example of adaptation caused by a failure to maintain the receptor potential.
 d. A steady discharge of the pacinian corpuscle can be generated by a vibratory stimulus.
  (1) Each time the stimulus is removed and reapplied, the pacinian corpuscle discharges another action potential.
  (2) The frequency of discharge by the pacinian corpuscle equals the frequency of a vibratory stimulus in the range of 50 to 500 hertz (Hz or cycles/sec).
   (a) This is about the same range of frequencies identifiable by humans, indicating that the frequency of firing is encoding the frequency, not the intensity, of vibration.
   (b) Intensity is encoded by the number of action potentials generated by each deformation (limited to two or three) and the number of pacinian corpuscles responding to the vibration.

 2. Pressure is encoded by the **Ruffini nerve ending** (see Fig. 1-26B).
  a. The Ruffini ending also is encapsulated. Its capsule is a liquid-filled collagen structure, in which a single group II sensory axon terminates. Collagen strands within the capsule make contact with the nerve fiber and the overlying skin. Thus, any deformation or stretch of the skin causes the nerve terminal to depolarize and generate action potentials.
  b. The Ruffini ending is a slowly adapting receptor. The train of action potentials continues for as long as the stimulus is present.
  c. The magnitude of the stimulus is encoded by the firing frequency of the Ruffini ending.

3. Both the **Meissner corpuscle** and the **Merkle disk** contribute to the encoding of **touch**.
   a. The **Meissner corpuscle** is a rapidly adapting encapsulated receptor receiving a single group II afferent nerve fiber (see Fig. 1-26C).
      (1) Its frequency of firing is proportional to the rate of stimulus application; thus, the Meissner corpuscle encodes the velocity of the stimulus application.
      (2) Rapid deformation occurs when the skin is jabbed quickly with a probe and when the fingers move over a rough object. This use of velocity information (i.e., identification of the contour of objects palpated by the fingers and hands) is especially important to blind individuals who read braille letters.
   b. The **Merkel disk** is a unique type of cutaneous receptor because the transducer is not on the nerve terminal but on the epithelial cells that make up the disk (see Fig. 1-26D).
      (1) The epithelial sensory cells form synaptic connections with branches of a single group II afferent fiber.
      (2) The disk is about 0.25 mm in diameter and can be stimulated only if the stimulus is applied directly to the disk. The small receptive field of the Merkel disk makes it an ideal receptor to encode information about the location of the stimulus.

B. TEMPERATURE SENSATION is encoded by thermoreceptors located on the free endings of small myelinated (A delta) and unmyelinated (C) fibers. Separate receptors with discrete receptive fields exist for the encoding of warm and cold sensations. Warm fibers respond only when a warm object is applied to them, and cold fibers respond only to cool objects. Warm fibers are not as numerous as cold fibers and almost exclusively are unmyelinated.

1. Warm Fibers.
   a. Figure 1-27 shows the static firing rate of a typical warm fiber as a function of skin temperature. (The firing rates are measured after the skin has adapted to the ambient temperature.) Warm fibers begin to respond at a skin temperature of 30° C, reach a maximal response at about 45° C, and cease activity at about 47° C.
   b. A warm fiber always increases its firing rate phasically when the skin is warmed and phasically decreases it when the skin is cooled. Soon after the initial response to the change in temperature, the fiber adapts its firing rate to the steady level indicated on the graph in Figure 1-27.

2. Cold Fibers. The response of cold fibers is analogous to that of the warm fibers (see Fig. 1-27).
   a. Cold fibers begin to display a static discharge at skin temperatures of about 15° C, reach a maximal response at about 25° C, and cease firing at about 42° C.
   b. If the skin temperature is raised above 45° C, the cold fibers begin to fire again. This is the threshold for activating pain fibers. Thus, if the skin is heated to a temperature above 45° C, the sensation of pain is accompanied by a feeling of coolness. This is referred to as **paradoxical cold**.
   c. Cold fibers always increase their rate of firing transiently when the skin temperature is cooled and decrease it when the skin temperature is warmed. Also like warm fibers, cold fibers quickly adapt their firing rate to the static levels indicated in Figure 1-27.

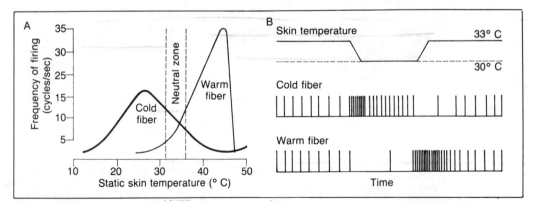

Figure 1-27. *A* Graph illustrating the tonic level of firing in both warm and cold fibers as a function of temperature. *B* Spike trains illustrating the dynamic responses of warm and cold fibers to a change in temperature. When the temperature decreases, cold fibers increase their rate of firing and then adapt to the firing rate indicated by the graph. Similarly, when the temperature increases, warm fibers phasically increase their firing rate before adapting to the rate indicated by the graph.

3. Sensory coding by thermoreceptors is complicated by the fact that the hypothalamic thermoreceptors that monitor the body's core temperature also contribute to the perception of the thermal environment.

    a. At skin temperatures within the **neutral** or **comfort zone** (31°–36° C), complete perceptual adaptation occurs (i.e., awareness of temperature disappears). Persistent warmth and coolness are felt at temperatures above and below this zone. Temperatures below 25° C are described as unpleasantly cold, while those below 15° C and above 45° C are painful. This is due to stimulation of pain receptors and not to the stimulation of thermoreceptors.

    b. The skin is most sensitive to temperature changes when the skin temperature is 31°–36° C (the neutral zone). At these temperatures, changes of less than 0.5° C are detected easily.

        (1) Transient temperature changes can be encoded by labeled-line mechanisms because cold fibers always increase their firing rate when the skin is cooled, and warm fibers always increase their firing rate when the skin is warmed.

        (2) Coding is more difficult under static conditions, however because cold fibers have the same firing rate at temperatures above and below 30° C, and warm fibers have the same firing rate at temperatures above and below 37° C. Temperatures that produce the same firing rates can be discriminated by two coding mechanisms.

            (a) **Spatial coding** can be used since at temperatures between 30° C and 42° C both cold and warm fibers are firing, at temperatures below 30° C only cold fibers are firing, and at temperatures above 42° C only warm fibers are active.

            (b) **Temporal coding** can be used by cold fibers because action potentials fire in bursts at temperatures below 30° C but not at temperatures above 30° C.

C. PAIN SENSATION is different from other sensations because its purpose is not to inform the brain about the quality of a stimulus but rather to indicate that the stimulus is physically damaging to the individual. Although unpleasant, pain is a useful sensation if it leads to activities that remove the damaging stimulus. However, the sensation of pain can endure long after the stimulus is removed and the injury is healed. This sort of **chronic pain** is an extremely debilitating disease that is very difficult to treat. Much more information is required about the physiologic and psychologic mechanisms of pain before its elimination becomes a routine part of medical practice.

    1. The **peripheral mechanisms** of pain sensation are reasonably well understood.

        a. The receptors for pain, which are called **nociceptors** to indicate that they respond to noxious stimuli, are on the free nerve endings of small myelinated (A delta) and unmyelinated (C) fibers.

            (1) Nociceptors are specific for painful stimuli, responding to damaging or potentially damaging mechanical, chemical, and thermal stimuli. Cutaneous receptors that respond to nonpainful levels of these stimuli do not elicit pain sensations no matter how intense the stimulus.

            (2) Although the adequate stimulus for nociceptors is not known, it is assumed that a chemical such as histamine or bradykinin is released from cells damaged by the pain stimulus and that the chemical substance activates the nociceptors.

        b. Two types of pain sensation result from application of a strong, noxious stimulus to the skin.

            (1) **Fast** or **initial pain** is a discrete, well-localized, pin-prick sensation that results from activating the nociceptors on the A delta fibers.

            (2) **Slow** or **delayed pain** is a poorly localized, dull, burning sensation that results from activating the nociceptors on the C fibers.

        c. The two types of pain sensation use different pathways to reach the centers of consciousness in the brain.

            (1) Action potentials that are propagated by the fast-pain fibers travel faster and, thus, reach the brain before those conducted by the slow-pain fibers. The sensory fibers for fast pain have small receptive fields, travel to the cortex through the spinothalamic tract, and are topographically represented on the cortex—all factors that account for the ability of these fibers to encode the location of the stimulus producing the fast pain.

            (2) Slow pain, which has a more diffuse pathway, travels to the brain through the spino-reticulo-thalamic system. Collaterals of this system pass through the reticular formation to activate fiber tracts that produce the emotional perceptions accompanying pain sensations. These pathways account for the intense unpleasantness associated with slow pain.

        d. The two types of pain sensation elicit different reflexes.

            (1) Fast pain evokes a withdrawal reflex (see Section XII A) and a sympathetic response including an increase in blood pressure and a mobilization of body energy supplies.

            (2) Slow pain produces nausea, profuse sweating, a lowering of blood pressure, and a generalized reduction in skeletal muscle tone. (Pain sensations originating in the muscles, blood vessels, and viscera produce similar reflexes.)

2. The **central mechanisms** of pain sensation are not well-known.
   a. Generally it is assumed that pain sensation is conveyed to the CNS through the anterolateral quadrant; however, surgical section of the cord through this area is not very successful in relieving chronic pain.
      (1) Failure to relieve pain with this procedure may result from the existence of parallel pain pathways outside the anterolateral tracts.
      (2) Alternatively, chronic pain may result from the spontaneous activity of pain centers within the CNS.
         (a) Reverberating circuits that develop because of continuous pain input may fail to stop firing when the input is removed.
         (b) Denervation supersensitivity may develop in pain centers subsequent to the removal of pain fiber input. (Denervation supersensitivity is an increased sensitivity to circulating neurotransmitters, which results when the normal synaptic input is removed from a neuron).
   b. Treatment of chronic pain is based on attempts to remove the pain area within the brain by surgery or to reduce the activity of the pain pathways by activating inhibitory pathways projecting to the pain areas.
      (1) **Transcutaneous electrical stimulation** of peripheral sensory neurons has been somewhat successful in alleviating chronic pain. Presumably, the success of acupuncture in treating pain is based on its stimulation of peripheral nerve fibers.
      (2) Electrical stimulators also have been placed on the dorsal columns of the spinal cord to stimulate the ascending sensory nerve tracts.
      (3) Another treatment is to stimulate electrically the pathway that originates in the brain stem and passes through the periaqueductal gray matter. This pathway also can be activated by morphine and the endogenous morphine agonists, the endorphins and enkephalins. Morphine's ability to reduce pain may derive from its stimulation of this pathway.

3. **Referred Pain.**
   a. Pain originating in visceral organs is **referred** to sites on the skin. These sites are innervated by nerves arising from the same spinal segment as the nerves innervating the visceral organs. Because of this anatomic relationship, diagnosis of visceral disease can be made based on the location of the referred pain. For example, ischemic heart pain is referred to the chest and the inside of the arm.
   b. Most likely, referred pain occurs because the visceral and somatic pain fibers share a common pathway to the brain. Since the skin is topographically mapped and the viscera are not, the pain is identified as originating on the skin and not within the viscera.

4. Referred pain should not be confused with **projected pain**, which occurs as a result of directly stimulating fibers within a pain pathway.
   a. Because a labeled-line mechanism is used to encode the location of the pain, stimulation anywhere along the pathway will result in the same perception. For example, striking the elbow causes pain to be projected to the hand.
   b. Amputees often have sensations that appear to come from the severed limb. This is known as **phantom limb sensation** and presumably results from activation of the sensory pathway either at the site of amputation or within the CNS. Occasionally, the phantom sensation is one of pain, but with no obvious source of stimulation it is difficult to eliminate the pain.

VII. AUDITION

A. PHYSICS OF SOUND

1. **Sound waves** are produced by vibrating objects that cause alternating phases of compression and rarefaction in the medium through which the sound is transmitted.
   a. In air, sound travels at about 330 m/sec (1100 ft/sec or 700 miles/hr).
   b. In water, sound travels much faster, at about 1500 m/sec.

2. Sound waves are characterized by their **frequency** and **amplitude**.
   a. The frequency of sound is measured in hertz (Hz).
      (1) The range for human hearing is about 20–20,000 Hz.
      (2) The human voice produces sounds of about 1000–3000 Hz when speaking.
   b. The amplitude or **intensity** of sound is measured using a logarithmic scale. The unit of intensity is the **decibel**.

(1) Decibels are measured using the formula:

$$db = 20 \log \frac{P_{sound}}{P_{SPL}},$$

where db = the number of decibels; P_{sound} = the pressure of the sound stimulus; and P_{SPL} = the sound pressure level (SPL) at the threshold for human hearing.
  (a) Thus, sound intensities are measured on a ratio scale using a subjective intensity (threshold) as a base.
  (b) The actual SPL intensity is 0.0002 dynes/cm².
(2) According to the formula for calculating sound pressures in decibels, a stimulus that has a pressure 10 times as great as another will be 20 db greater in intensity than the other stimulus.
  (a) The sound pressures used during normal conversation are about 1000 times as great as threshold or 60 db. (This value can be confirmed using the above equation.)
  (b) An airplane produces a sound pressure of about 100,000 times that of threshold or 100 db.
  (c) Sound pressures above 140 db (10^7 times threshold) are painful and damaging to the auditory receptors.

B. STRUCTURE OF THE EAR. The ear consists of three divisions: the external ear, the middle ear, and the internal ear (Fig. 1-28).

1. **The external ear** consists of two parts that function to direct sound waves into the auditory apparatus.
   a. **Pinna or Auricle.** In lower animals, the cartilagenous pinna can be moved by muscular action in the direction of a sound source to collect sound waves for the receptor organ. In humans, these muscles have little action. By transforming the sound field, however, the convoluted pinna does help identify sound sources. For example, the pinna aids in distinguishing a sound source placed in front of the head from one placed behind the head.
   b. The **external auditory canal** extends from the pinna to the tympanic membrane of the middle ear. Its outer portion is cartilagenous and its inner portion is osseous.

2. The **middle ear** or **tympanic cavity** is an air-filled cavity within the temporal bone. It contains the **tympanic membrane** (eardrum) and the **auditory ossicles** (ossicular chain). Sound stimuli pass through the pinna and strike the tympanic membrane, causing it to vibrate. The major

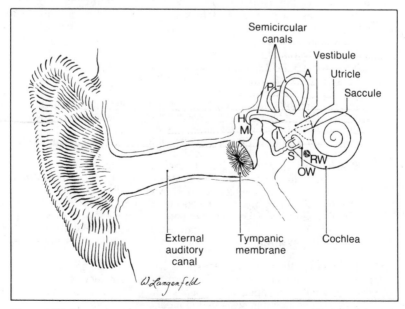

Figure 1-28. The major anatomic components of the human ear. The bony labyrinth consists of the sac-like otolith organs (i.e., the saccule and utricle) and three semi-circular canals: the horizontal (*H*), anterior (*A*), and posterior (*P*) canals. *M* = malleus; *I* = incus; *S* = stapes; *OW* = oval window; and *RW* = round window.

function of the ossicular chain is to convey the sound stimulus from the tympanic membrane to the inner ear. Its presence is essential for normal hearing.

  a. Components of the Ossicular Chain.

  (1) **Malleus.** This small bone resembles a mallet. The **manubrium** (handle) of the malleus is connected to the inner surface of the tympanic membrane.

  (2) **Incus.** This bone, said to resemble an anvil, articulates with the head of the malleus.

  (3) **Stapes.** This ossicle resembles a stirrup. The head of the stapes articulates with the incus, and the oval **footplate** contacts the membrane of the **oval window** of the cochlea.

  b. Impedance Matching. To reach the auditory receptors, the sound must be transferred from the air-filled middle ear to the fluid-filled inner ear.

  (1) Effective transfer of sound energy from air to water is difficult because most of the sound is reflected. This is due to the different mechanical (i.e., elastic, resistive, and inertial) properties of the two media and is described as an impedance mismatching.

  (2) The middle ear functions as an impedance matching device primarily by amplifying the sound pressure.

  (a) Most of this amplification occurs because the area of the tympanic membrane (55 mm²) is about 17 times as great as the stapes-oval window surface area. This concentrates the same force over a smaller area and, thus, amplifies the pressure.

  (b) A small additional amount of amplification is obtained by the mechanical advantage that is due to the leverage of the middle ear bones.

  (c) Together these effects increase the sound pressure by 22 times (27 db).

  c. Middle Ear Muscles.

  (1) **Structure and Innervation.**

  (a) **Tensor Tympani.** This elongated muscle inserts on the manubrium of the malleus and is innervated by the trigeminal nerve (cranial nerve V).

  (b) **Stapedius.** This small muscle inserts on the neck of the stapes and is innervated by the facial nerve (cranial nerve VII).

  (2) **Action.** These muscles contract reflexly in response to intense sounds. They act on the mobility and transmission properties of the ossicular chain to reduce the pressure of sounds reaching the inner ear.

  (3) **Functional Importance.** Although the middle ear muscles may act to prevent receptors from being damaged by high-intensity sounds, contraction occurs too long after the stimulus to provide much protection. Muscle contraction also occurs just prior to vocalization and chewing, which suggests that the middle ear muscles may act to reduce the intensity of the sounds produced by these activities.

3. The inner ear consists of two components. The **bony labyrinth** is a network of cavities within the petrous part of the temporal bone. The bony labyrinth consists of three divisions: the semicircular canals, the cochlea, and the vestibule. Within the bony labyrinth is a series of interconnected membranous ducts called the **membranous labyrinth**. The membranous labyrinth is surrounded by a fluid called **perilymph**, which is rich in Na^+. The fluid within the membranous ducts is **endolymph**, which has a high concentration of K^+.

  a. The **cochlea** is a spirally arranged tube (two and one-half turns in humans) that functions as the auditory analyzer of the ear. Figure 1-29 depicts the cochlea as an uncoiled, linear structure. The cochlea contains the following components.

  (1) The **cochlear duct** is an elongated endolymph-filled stocking that lies within the coiled length of the cochlea. The open space within this membranous duct is the **scala media**. Four membranes enclose the scala media.

  (a) The **spiral ligament** is a thickening of the endosteum of the cochlea, which forms the peripheral edge of the cochlear duct.

  (b) **Reissner's membrane** is an extremely delicate membrane that forms the roof of the cochlear duct. It separates the scala media from the **scala vestibuli**.

  (c) The **limbus** forms the central, narrow edge of the cochlear duct.

  (d) The **basilar membrane** represents the floor of the cochlear duct and separates the scala media from the **scala tympani**. It also is the seat of the organ of Corti. The basilar membrane is broader at the apex (near the **helicotrema**) than it is at the base (near the **oval window**), and the membrane is 100 times more stiff at the base than it is at the apex.

  (2) The **scala vestibuli** refers to the space above the cochlear duct.

  (3) The **scala tympani**, the space below the cochlear duct, is continuous with the scala vestibuli at the very apex of the cochlear duct. This junction is termed the **helicotrema**. The scala vestibuli and scala tympani contain perilymph.

  (4) The **oval window** is the membrane-covered opening of the scala vestibuli, which joins the footplate of the stapes.

  (5) The **round window** is the membrane-covered opening of the scala tympani.

  b. The **organ of Corti** is the specialized region that contains the sensory receptor cells of the

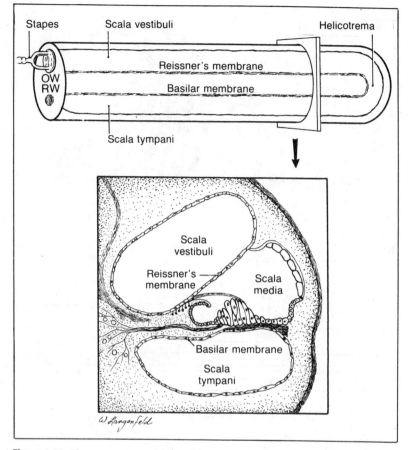

Figure 1-29. The components of the cochlea, shown as it would appear if uncoiled (*upper diagram*) and in cross section (*lower diagram*). *OW* = oval window; and *RW* = round window.

ear. It lies on the basilar membrane, adjacent to the limbus, and contains the following structures (Fig. 1-30).

(1) Hair Cells. The essential excitatory event in auditory transduction is the deflection of the cilia protruding from the surface of hair cells. When the cilia are deflected, transmitter substance is released from the hair cells.

(2) Tectorial Membrane. This flap of tissue rests on the tips of the cilia of the hair cells and is attached to the limbus at one end. Any movement of the basilar membrane causes the free end of the tectorial membrane to drag or push the cilia of the hair cells.

(3) Nerve Fibers. The peripheral endings of the afferent fibers form synapses with the hair cells, and the cell bodies form the **spiral ganglion**. Axons of these neurons join the vestibulocochlear nerve (cranial nerve VIII).

C. AUDITORY TRANSDUCTION

1. **Transmission of Sound Waves into the Cochlea.** When the stapes vibrates as a consequence of sound stimulation, **traveling waves** progress through the cochlea toward the helicotrema. The fluid-filled cochlea tends to dampen the waves as they progress down its length, so that the waves dissipate before reaching the helicotrema. However, the traveling waves cause the basilar membrane to vibrate. Waves transmitted to the underlying scala tympani are dissipated by the elastic membrane covering the round window.

2. **Oscillation of the Basilar Membrane.** The pressure waves in the scala vestibuli cause the basilar membrane to vibrate. The pattern of vibration depends on the frequency of the traveling wave in the perilymph.

   a. **Low-frequency waves** cause the entire basilar membrane to vibrate, and maximal excursion is at the **apical end**.

   b. **High-Frequency Waves.** As the frequency of the imposed wave increases, the broad, com-

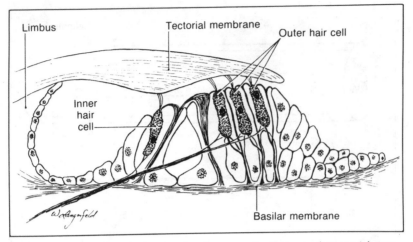

Figure 1-30. The organ of Corti, showing the connection between the tectorial membrane and the cilia of the hair cells.

pliant apical end of the basilar membrane becomes unable to follow the rapid pressure waves in the cochlea, and the point of maximal displacement moves progressively closer to the stiff, basal end of the membrane. Thus, sounds of very high frequency cause maximal excursion at the **basal end** of the basilar membrane.

c. **Frequency Coding.** Tones of different frequencies are distinguished in the cochlea because they produce maximal stimulation at different positions along the basilar membrane. This is called the **place principle** of frequency discrimination. A shift in the position of maximal stimulation engages a different set of hair cells. Thus, tones of different frequencies activate different afferent neurons, since each set of hair cells is innervated by a specific nerve fiber.

d. **Intensity Coding.** If the intensity of a pure tone is increased, the point of maximal stimulation along the basilar membrane is unchanged. However, the amplitude of vibration is increased, and a large portion of the membrane vibrates. The CNS interprets a larger and broader deflection of the basilar membrane as an increase in loudness.

3. **Discharge Pattern of Afferent Nerve Fibers.** The nerve fiber is activated by a neurotransmitter that is released by the hair cell. Although a single nerve fiber responds to tones of different frequencies, it responds best (i.e., has the lowest intensity threshold) to a particular sound frequency. The best frequency for a particular neuron depends on the location of the hair cells it innervates. A fiber innervating hair cells near the oval window responds best to high-frequency tones. A neuron innervating a region closer to the helicotrema, at the apical end of the basilar membrane, responds best to low-frequency tones.

D. CENTRAL AUDITORY PROCESSING

1. **Auditory Pathway.**
   a. **General Considerations.** The auditory pathway is more complex than the other sensory pathways due to the following features.
      (1) **Series-Parallel Circuits.** In ascending from one nucleus to the next, some fibers bypass one auditory center to terminate at a higher point in the pathway. Thus, while some neurons progress in a series along the main pathway, other neurons skip to the next auditory station without synapsing.
      (2) **Decussation.** At several points in the auditory pathway, fibers arising from a nucleus cross the midline to terminate in a nucleus of the contralateral side.
   b. **The Principal Pathway.** The following are the major nuclei and tracts of the auditory pathway.
      (1) **Vestibulocochlear Nerve.** The auditory portion of this nerve is composed of the centrally directed branches of the afferent fibers whose cell bodies lie in the spiral ganglion. The peripheral branches innervate the hair cells of the organ of Corti.
      (2) **Cochlear Nuclei.** The dorsal and ventral cochlear nuclei, lying at the junction of the medulla and pons, receive input from the afferent fibers.
      (3) **Superior Olivary Nucleus.** Fibers exiting the cochlear nuclei pass, via the **trapezoid body**, to the contralateral superior olive.
      (4) **Inferior Colliculus.** The output of the superior olive passes, via the **lateral lemniscus**, to the inferior colliculus of the midbrain tectum.

    (5) Medial Geniculate Body. Fibers of the inferior colliculus traverse the brachium of the inferior colliculus to synapse in the medial geniculate body of the thalamus.

    (6) Auditory Cortex (areas 41 and 42). Thalamic fibers project to the superior temporal gyrus of the cortex. The **gyri of Heschl** represent the primary auditory cortex.

  c. Alternate Pathways. Some neurons of the principal pathway terminate in the nucleus of the trapezoid body or in the nucleus of the lateral lemniscus. Decussating fibers arise from the superior olive to ascend in the contralateral lateral lemniscus, and the commissure of the inferior colliculus interconnects the tecta of the two sides. In addition, some fibers of the cochlear nuclei ascend without crossing.

  d. Reflex Pathways. Collaterals of the auditory pathway terminate in the motor nucleus of the facial nerve and the motor nucleus of the trigeminal nerve. These branches serve as input for control of the stapedius and tensor tympani muscles. Descending fibers from the inferior colliculus pass to motor centers for coordination of acoustic reflexes (e.g., turning the head).

  e. Afferents to the Reticular Formation. Collaterals of the auditory pathway enter the ascending reticular activating system. Sound stimulation is effective in producing arousal and attentiveness.

  f. Descending Pathways. Coextensive with the principal ascending sensory pathway is a parallel descending system. The descending system is **inhibitory** and gates the ascending input so as to modulate and refine the information proceeding to the cortex. The most significant component of the descending pathway is the **olivocochlear bundle**, which originates in the olivary complex and terminates in the ipsilateral and contralateral organ of Corti. The inhibitory input at the junction of the hair cell and afferent neuron plays a role in filtering out sounds, particularly when attention is directed toward other stimuli in the environment.

2. Physiology of Central Auditory Neurons.

  a. Tonotopic Organization. Generally, any neuron of the auditory pathway can be tested with tones of different frequencies to determine a best frequency for that cell. Neurons responding best to low-frequency tones will be located at one end of a nucleus, while neurons responding best to high-frequency tones will be represented at the opposite end of the nucleus. This orderly arrangement of frequency sensitivity, termed tonotopic organization, resembles the retinotopic organization of the visual system and the somatotopic organization of the somatosensory system. The tonotopic map reflects the methodical arrangement of frequency sensitivity along the length of the basilar membrane from base to apex. Tonotopic organization is prominent in the cochlear nuclei but becomes less precise in more rostral structures of the auditory pathway.

  b. Feature Detection. Higher auditory centers respond to particular features of sound stimuli. For example, cortical neurons may respond specifically to a shift from high- to low-frequency notes, which is why lesions of the auditory cortex may not impair the ability to discriminate frequency. Instead, lesions of the auditory cortex cause a loss of ability to recognize a patterned sequence of sounds. In addition, the ability to identify the position of a sound source is impaired.

  c. Localization of Sound in Space. Detection of the position of a sound source depends on the ability of the CNS to compare intensity differences and phase differences. A sound source located behind the head, closer to one ear than the other, produces a slightly more intense sound in the near ear than in the remote ear. Sound absorption by the tissues of the head attenuates sound intensity in the remote ear. In addition, at low frequencies there may be phase differences in the sound waves striking the two ears. These minute phase and intensity differences between the two ears are detected and discriminated by the auditory system. The extensive decussation and complex circuitry of the auditory pathway provide the CNS with the information necessary to identify a sound source.

E. HEARING IMPAIRMENTS

1. Tinnitus is a ringing sensation in the ears caused by irritative stimulation of either the inner ear or the vestibulocochlear nerve.

2. Deafness. Two classes of deafness are distinguished based on the location of the abnormality or lesion.

  a. Conduction deafness is caused by interference with the transmission of sound to the sensory mechanism of the inner ear. This form of deafness can result from wax accumulation in the external ear, perforation of the tympanic membrane, or otosclerosis (loss of mobility of the ossicular chain due to excess bone growth). Conduction deafness represents a defect of the external or middle ear.

  b. Nerve deafness results from defects of either the inner ear or the vestibulocochlear nerve. Prolonged exposure to high-intensity sound may cause this form of hearing impairment.

Localized damage to hair cells at a specific region along the basilar membrane may result from prolonged exposure at a certain frequency range (e.g., due to occupational requirements). The resulting hearing loss then is restricted to that frequency range. Deafness due to a lesion in the CNS structures generally is not found clinically because of the redundancy and bilateral nature of the central auditory pathways.

F. DIAGNOSTIC TESTS OF AUDITORY FUNCTION

1. **Audiometry.** With this procedure, the threshold intensity for detecting sounds across the frequency spectrum is determined in an individual and compared to normal. Since the sound is applied through earphones, each ear can be tested separately. Audiometry permits the audiologist to define (in decibels) the severity of the hearing loss at each frequency.

2. **Bone Conduction.** By applying a vibratory stimulus to the head, sound is conducted directly to the cochlea through the bones of the skull without passing through the middle ear. In patients suffering from conduction deafness, air-conducted sounds are not perceived as well as bone-conducted sounds.

3. **Auditory-Evoked Potentials.** Auditory click stimuli produce a characteristic multiple-wave potential in recording electrodes placed on the scalp of normal individuals. Because of the small amplitude of the evoked potential (especially in relation to the ongoing EEG waves recorded through the electrodes), computer-averaging techniques must be used to enhance the signal-to-noise ratio. Each wave of the evoked potential can be correlated with the discharge of neurons in the principal stations of the ascending auditory pathway. Changes in waveform can be interpreted to localize brain lesions to specific levels of the CNS.

VIII. OPTICS

A. **NATURE AND PROPERTIES OF LIGHT.** **Visible light** represents only a small part of the electromagnetic spectrum. Light waves have wavelengths of 380 nm–760 nm (3800 A–7600 A).

1. **Color.** The color perceived depends on the **wavelength** of the light rays striking the retina (Table 1-3). The human eye can detect the difference in color of sources that differ in wavelength by 1 nm.

2. **Speed of Light.** The speed of light in a vacuum is a constant denoted by **c**, where $c = 3 \times 10^8$ m/sec. The speed of light is reduced when it travels through any medium. The speed of light in a medium is denoted by c_n.

3. **Index of Refraction.** The ratio of the speed of light in a vacuum to the speed in a medium is called the index of refraction (**n**) of that medium. Thus, $n = c/c_n$. The index of refraction is a measure of the optical density of the medium.

4. **Refraction** is the bending of light rays as they pass from one medium into another medium.
   a. When a light ray passes from one medium to another at an angle that is **perpendicular** or **normal** to the plane of the interface between the two media, the ray of light is not bent but may be slowed in velocity.
   b. When a light ray enters another medium at an **angle** that is not normal to the surface, the light is **refracted** (Fig. 1-31).
   c. When light enters a medium of higher optical density it is refracted **toward** the normal. Conversely, when light leaves a medium of higher optical density to enter one with a lower index of refraction, it is bent **away** from the normal.

B. LENSES

1. **Convex Lens.** A light ray entering a lens with a spherical surface is bent toward the normal. Upon leaving the lens, the ray enters the less optically dense air medium and is bent away

Table 1-3. The color perceived by the human eye for different wavelengths of light

Wavelength (nm)	Color
4000–4240	violet
4240–4912	blue
4912–5750	green
5750–5850	yellow
5850–6470	orange
6470–7000	red

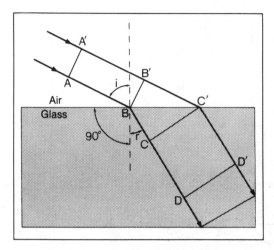

Figure 1-31. Refraction of light at an air-glass interface. The wavefront (*AA′*) is shown approaching the surface at an angle to the normal. As the wave travels through air from B′ to C′, it travels at a slower velocity in glass from B to C (BC < B′C′). The new wavefront (CC′ and DD′) is rotated relative to AA′. The wave is deviated toward the normal; that is, the angle of incidence (*i*) is smaller than the angle of refraction (*r*).

from the normal. Both refractions bend the ray further from its original direction (Fig. 1-32). Parallel rays striking the lens are focused onto a single point due to the curvature of the lens, which causes each ray to strike the surface at an angle slightly different from neighboring rays. The ray of light passing through the **axis** of the lens is not refracted because it strikes the lens normal to the surface. However, a ray that strikes near the periphery of the lens makes a considerable angle with the normal to the surface at that point (Fig. 1-33).

    a. Principal Focus. After passing through a convex lens, rays that are parallel to the axis converge to meet at a point called the principal focus **(F)** or focal point.

    b. Focal Plane. Rays that are parallel to each other but not to the axis of the lens converge at a point on the focal plane.

    c. Focal length, denoted by **f**, is the distance from the lens to the focal point.

2. Concave Lens. Refraction at the surfaces of a double concave lens causes parallel rays to diverge; therefore, no real image is formed by a concave lens (Fig. 1-34).

3. Parallel-Faced Slab. A ray passing through two parallel surfaces, at an angle to the surfaces, passes through without change of direction but with some displacement of the ray (Fig. 1-35). In thin lenses this displacement is so small that it can be ignored.

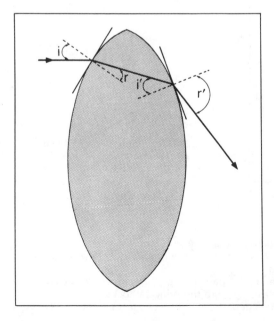

Figure 1-32. Horizontal ray entering a convex lens. The ray is bent toward the normal upon entering the glass (i > r) and away from the normal upon leaving (i′ < r′).

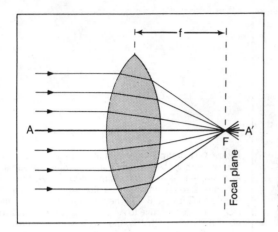

Figure 1-33. Principal focus of a convex lens. The principal axis (*line AA′*) passes through the centers of curvature of the lens. Rays parallel to the axis converge at the principal focus (*F*). The distance between F and the lens is the focal length, *f*.

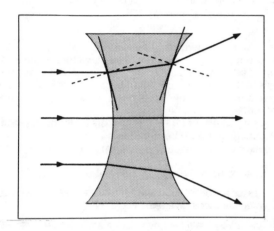

Figure 1-34. Horizontal rays entering a double concave lens. Refraction at each surface of the lens causes parallel rays to diverge. The focal length is a negative quantity, hence the lens is also called a negative lens.

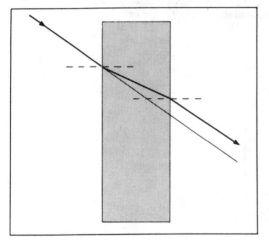

Figure 1-35. Effect of a parallel-faced slab. The direction of a ray is unchanged if it passes through a planar lens. The displacement of the ray in thin lenses is negligible.

C. IMAGE FORMATION

1. Ray Tracing. To determine the properties of an image formed by a convex lens, three rays from representative points on an object are traced, although only two are needed. The following rays are used (Fig. 1-36).

    a. Ray 1 travels parallel to the axis and must pass through the principal focus (F). Recall that by definition, rays parallel to the axis travel to the principal focus.

    b. Ray 2 passes through the optical center of the lens. The two faces of the lens encountered

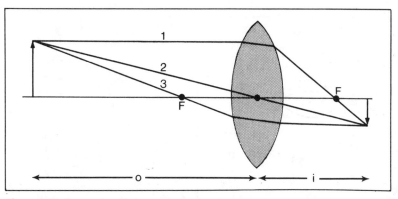

Figure 1-36. Ray tracing. To locate the image of a point when the position of an object is known, three rays should be graphed. *F* = the principal focus; *o* = the object distance; and *i* = the image distance.

by this ray are parallel and act as a parallel-faced slab. Recall that in thin lenses, rays passing through such a slab are unchanged in direction.

c. Ray 3 passes through the principal focus on the object side of the lens and emerges from the lens as a ray parallel to the axis.

2. **Lens Equation.** The relation between object distance **(o)**, image distance **(i)**, and focal length **(f)** is: $1/f = 1/o + 1/i$.

   a. From this equation it follows that if the object is at infinity, $f = i$ (i.e., the image falls at the principal focal length of the lens). Rays emanating from infinity approach parallel to the axis. A point source of light farther than 20 feet (6 m) from the lens produces rays that diverge so little they may be considered parallel rays.

   b. A point source of light that is less than 6 m from the lens, but is beyond the principal focus (as in Figure. 1-36), produces an image that is focused at a point called the **conjugate focus**. The lens equation is used to determine the distance between the conjugate focus and the lens. This distance is i.

   c. If the point source is placed at the principal focus or closer, no real image is produced. There is not enough refraction to converge the rays onto a conjugate focus, and the rays remain divergent.

3. **Magnification.** It can be demonstrated that:

$$\frac{\text{image size}}{\text{object size}} = \frac{i}{o}.$$

The ratio i/o is called the magnification.

4. **Refractive Power.** The effect of a lens on the curvature of a wave front is called the refractive power of the lens. Measured in **diopters**, the refractive power of a lens equals the reciprocal of its focal length or **1/f**, when f is measured in meters. A lens with a focal length of 20 cm, therefore, has a refractive power of + 5 diopters. The refractive power of most eyeglass lenses ranges from + 5 to − 5 diopters. (A negative number denotes a diverging lens.) The human eye has a refractive power of about + 60 diopters.

   a. Convex Lens. A wave front of parallel rays first contacts the lens at the center, where the lens is thickest. By the time the wave front has passed through the thick portion of the glass, peripheral portions of the front have proceeded beyond the lens, because they traversed a smaller thickness of glass. Consequently, the wave front must curve and converge to a point, the principal focus. Beyond this point, the wave front again diverges. Any lens that is thicker at the center than at the edges is a **converging** or **positive lens**, and its refractive power is expressed as a positive number. The thicker the central portion of a lens relative to the periphery, the greater the refraction of the wave and the smaller the focal length. Thus, a thick convex lens has a high refractive power.

   b. Concave Lens. The periphery of a concave lens is thicker than the center. The resulting curvature of the wave front causes the wave front to diverge. Any lens that is thicker at the periphery than at the center is a **diverging** or **negative lens**, and its refractive power is expressed as a negative number.

D. REFRACTIVE MEDIA OF THE EYE. A horizontal section through the human eyeball is shown in Figure 1-37. To reach the retina, light must pass through four refractive interfaces: the anterior

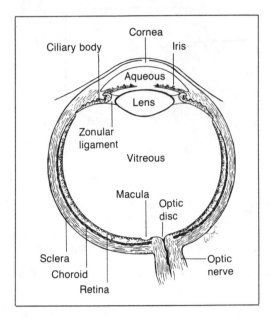

Figure 1-37. Horizontal section of the eyeball, as seen from above.

surface of the cornea, the posterior surface of the cornea, the anterior surface of the lens, and the posterior surface of the lens.

1. **Cornea.**
   a. **Transparency.** The absence of blood vessels in the cornea allows light to pass through unhindered. The cornea, consequently, must receive oxygen and nutrients via diffusion through the aqueous humor. Elevated intraocular pressure, a condition termed **glaucoma**, can compromise the transparency of the cornea. Intraocular pressure is measured clinically by a tonometer, which records the resistance to a calibrated indentation of the cornea.
   b. **Refractive Power.** The greatest amount of refraction of the eye occurs at the anterior surface of the cornea, because the difference between the indices of refraction of air and of the cornea is larger than the differences between the indices of refraction of the other ocular media. The cornea contributes more than +40 diopters of refractive power. The amount of refraction contributed by the cornea is fixed.

2. **Lens.**
   a. **Transparency.** The lens, like the cornea, is avascular. It consists of a capsule, a layer of epithelial cells, and a system of transparent fibers. New fibers are continually laid down. As an individual ages, the lens becomes harder and less malleable. **Cataracts** are opacities of the lens which reduce transparency. One cause of cataracts is an elevated glucose level in the aqueous humor, as occurs in uncontrolled diabetes.
   b. **Refractive Power.** The refractive power of the lens, unlike that of the cornea, is adjustable. The lens is suspended by the tough **zonular fibers**, which stretch the lens and flatten it. The tension on the zonular fibers is released by contraction of the **ciliary muscle**. Contraction results in the constriction of this ring of muscle fibers, which lies anterior to the attachment of the zonular fibers. The release of tension on the zonular fibers results in an increase in the curvature of the anterior surface of the lens. The increased convexity of the lens increases its refractive power. The lens can add, at most, +12 diopters of power to the ocular apparatus.

E. OPTICAL PROPERTIES OF THE EYE

1. **Image Formation in the Normal Eye.** To simplify the task of ray tracing through the refractive surfaces of the eye, a model known as the **reduced eye** is used to approximate the optical actions of the four refractive interfaces of the eye. The basic assumption of this model is that all refraction occurs at one interface with an index of refraction of 1.33 (Fig. 1-38).
   a. **Focal Distance.** The interior focal distance of the reduced eye is 17 mm. This simulates the distance between the cornea and retina in the normal, or **emmetropic**, eye.
   b. **Refractive Power.** The refractive power of the reduced eye is 1/0.017 or 59 diopters. The cornea contributes about three-fourths of this power.

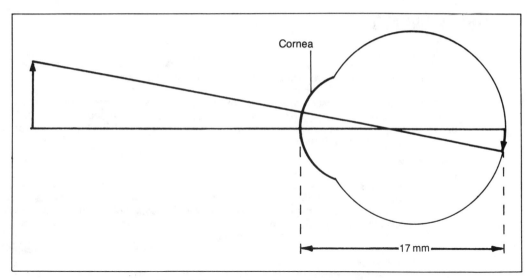

Figure 1-38. Image formation in the reduced eye.

    c. Magnification. The image of an object on the retina is **inverted** and **reduced** in size. An object 40 m high at a distance of 2000 m is 0.3 mm long on the retina.

2. Accommodation. Objects located more than 6 m from the eye are focused on the retina of the normal eye at rest. At this distance, the rays coming from the object are considered parallel, and the retina is at the focal length of the reduced eye. The image will, therefore, be focused on the retina. If objects are less than 6 m from the eye, the rays diverge. Unless the eye adjusts, the rays focus at a point behind the retina. At the point where the rays actually strike the retina they are not in focus and appear blurred.

    a. Accommodation Reflex. To bring near objects in focus on the retina, a three-part reflex is invoked.

        (1) Bulging of Lens. Contraction of the ciliary muscle releases tension on the zonular fibers, increasing the convexity and dioptric power of the lens.

        (2) Pupillary constriction reduces spherical and chromatic aberration and increases depth of field. The result is an improved image quality.

        (3) Convergence. The eyes converge by simultaneous contraction of the medial recti muscles.

    b. Near Point. As an object is brought closer to the eye, a point is reached where even the strongest contraction of the ciliary muscle does not produce a clear image because the rays are too divergent. The nearest point at which an object is clearly seen with full accommodation is known as the near point.

    c. Far Point. A distant object that is in focus without accommodation is said to be at, or past, the far point of the eye. In the emmetropic eye the far point is 6 m.

3. Function of the Pupil. The iris diaphragm controls the diameter of its central opening, the pupil. The iris contains two muscles—the **sphincter** and **dilator** muscles. Activation of the sphincter muscle constricts the pupil; this is called **miosis**. Dilatation of the pupil, called **mydriasis**, results from the action of the dilator muscle. The pupil has three major functions.

    a. Reduction of Optical Aberrations. A reduction in pupil diameter decreases chromatic and spherical abberation.

        (1) Chromatic Aberration. In any lens, light of different wavelengths is refracted differently. For example, the violet component of white light, which is refracted more strongly, may be focused in front of the image; the red component, which is refracted less strongly, may fall somewhat behind the image. The resulting image may have a color fringe.

        (2) Spherical Aberration. The focal length of rays entering a lens at the periphery may not coincide perfectly with the focal length of rays entering near the center of the lens. Cutting off the outer zone with a diaphragm produces a sharper image.

    b. Provision of an Increased Depth of Field. By reducing the pupil diameter, acceptable focusing is achieved for a wide range of object distances.

    c. Protection Against Excessive Illumination. Reduction in the pupil diameter reduces the amount of light striking the retina.

F. OPTICAL DEFECTS

1. Myopia
  a. **Description.**
   (1) In this condition, the eye is too long for the refractive power of the accessory tissues, and parallel rays are focused at a point in front of the retina.
   (2) The far point is closer than 6 m and may be only a few centimeters.
   (3) The near point is closer than in the emmetropic eye. The myopic eye can focus on very near objects. Thus, the term **nearsightedness** is used for this defect.
  b. **Treatment.** Concave lenses are prescribed. Such lenses cause sufficient divergence of the rays to bring distant objects to focus on the retina.

2. Hyperopia.
  a. **Description.**
   (1) In this condition, the eyeball is too short for the refractive power of the accessory tissues. Parallel rays of light, therefore, are focused behind the retina. Distant objects can be seen only if the hyperope accommodates. The action of the lens causes the distant object to focus on the retina. Thus, the hyperope is continuously accommodating and as a result suffers eyestrain and headache.
   (2) The near point is more distant than in the emmetropic eye, which is why this optical defect is termed **farsightedness**.
  b. **Treatment.** Convex lenses are prescribed to bring the rays to focus on the retina.

3. Presbyopia.
  a. **Description.**
   (1) As an individual ages, the lens becomes less flexible, and there is a consequent decline in the amplitude of accommodation. The lessened capacity of the lens to bulge causes increasing difficulty in seeing near objects. Although the decline in the plasticity of the lens is a continuous process, individuals between the ages of 40 and 50 years begin to notice a decreasing ability to read.
   (2) The far point is unchanged in this defect.
   (3) The near point recedes further from the eye as a result of aging.
  b. **Treatment.** Convex lenses, usually bifocals, are prescribed for work and reading.

4. Astigmatism.
  a. **Description.** Ordinarily, the surface of the cornea is spherical, with equal curvature in all meridians. In the astigmatic eye, the corneal surface is not perfectly spherical and the curvature along one meridian is greater than the curvature along another meridian. Consequently, rays emanating from a luminous point do not form a point-image on the retina.
  b. **Treatment.** Astigmatism is corrected with a lens that has different curvatures in the appropriate meridians to compensate for the anomalies in the cornea.

IX. VISION.
The **retina** is the receptor mechanism of the eye and the ultimate tool of vision. Light and color-sensitive elements of the retina communicate their messages, via **neural pathways**, to the **visual cortex** of the brain.

A. RETINA

1. Morphology and Function of Retinal Components.
  a. **Layers, Fovea, and Optic Disk.** The retina is divided into 10 histologically defined **layers** (Fig. 1-39). In the center of the retina is a yellowish region termed the **macula lutea**. The **fovea** is a central depression in the macula lutea, about 0.3 mm in diameter, where some of the retinal layers are displaced and the retina, consequently, is thinned. The fovea is the visual center of the eye and the area of highest resolution. About 3 mm to the nasal side of the fovea is the **optic disk**, where the ganglion cell axons turn and pierce the retina to form the optic nerve. Receptor elements are absent at the optic disk, which creates a **blind spot** on the retinal field.
  b. **Cells.** The retina is composed of the following cells.
   (1) **Pigment cells** contain melanin and are arranged in a single lamina. Their major functions include:
    (a) Absorption of stray light
    (b) Phagocytosis of disks that have been sloughed from the outer segments of receptor cells
   (2) **Receptor cells** divide into two distinct types: **rods** and **cones.**
    (a) The important **morphologic features of** rods and cones are illustrated in Figure 1-40 and are discussed below.
     (i) The **outer segment** of rods and cones invaginates to form membrane disks. In

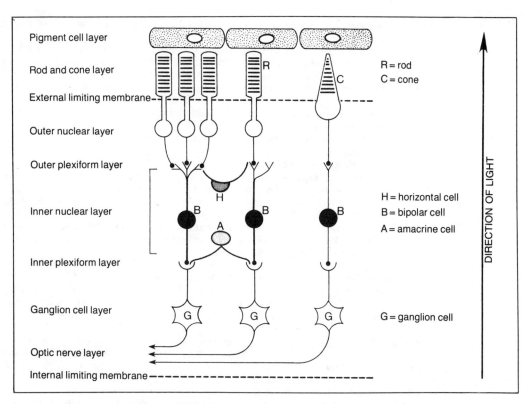

Figure 1-39. Cytoarchitecture of the retina. The three-neuron circuit of the rod system differs from that of the cone system. Convergence of receptors onto bipolar cells is seen only in the rod system. Horizontal cells and amacrine cells link adjacent retinal circuits.

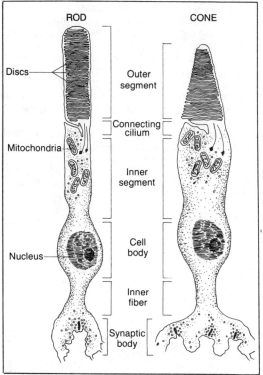

Figure 1-40. Morphology of rod and cone receptor cells. Cones, which are responsible for color perception and high visual acuity, are found in the fovea. Rods, which are responsible for night vision, are located in the peripheral retina.

rods, but not in cones, these pinch off and become independent of the cell membrane.

In rods, the disks undergo constant renewal. Disks are formed initially at the basal pole of the outer segment and migrate through the outer segment until they reach the apical pole. Here the disintegrating disks are phagocytosed by the pigment cells. Cones show a more diffuse pattern of disintegration.

The photopigment molecules are embedded in the disk membranes.

    (ii) The **inner segment** contains numerous mitochondria as well as the nucleus of the rod or cone.

    (iii) The **receptor terminal** contains synaptic vesicles.

  (b) The **functions** of rods and cones divide as follows.

    (i) **Color Vision.** Only cones are involved with color vision.

    (ii) **Sensitivity to Light.** Cones function at high light intensities and, therefore, are responsible for day **(photopic)** vision. Cones are not as sensitive as rods to low levels of illumination. Rods function at low light intensities and are responsible for night **(scotopic)** vision.

    (iii) **Visual Acuity.** Cones have a high level of visual acuity and are present in the fovea. Rods have a much lower level of acuity; they are lacking in the fovea and are confined to the peripheral retina.

    (iv) **Dark Adaptation.** When an individual moves into a darkened environment, the eyes slowly adapt to lower levels of illumination. The threshold for sensation of low-intensity light stimuli falls progressively, as shown in Figure 1-41. The dark adaptation curve has two limbs, representing adaptation of cones followed by that of rods. Rod function is suppressed during day vision and comes to the fore only when background illumination is low.

(3) **Müller cells** are glial elements that have cell bodies in the **inner nuclear layer**. Their processes extend to form the **internal limiting membrane** as well as the **external limiting membrane**. The processes also fill the spaces between receptor cells and serve as a connecting and supporting tissue.

(4) **Bipolar Cells and Ganglion Cells.**

  (a) **Bipolar cells** synapse with the receptor terminals of rods or cones in the **outer plexiform layer**. Bipolar cells in turn terminate on ganglion cells in synapses located in the **inner plexiform layer**.

  (b) **Ganglion Cells.** The dendrites of these neurons receive their input from the bipolar cells. Their axons course along the retina to the optic disk and form the **optic nerve**.

  (c) **Function in Light Sensitivity and Visual Acuity.** The low sensitivity but high acuity of the cone system results from the connection of single cones to individual bipolar cells and subsequently to single ganglion cells projecting to the brain (see Fig. 1-39). The rod system has a higher sensitivity because a large number of these receptors converge onto an individual bipolar cell, and a photon striking any one of these rods is effective in activating the bipolar cell. The bipolar cell of the cone is less sensitive, since a photon must strike a particular receptor location to be effective. Because stimuli at several adjacent rod locations all can result in an identical response in the second-order neuron, acuity is sacrificed in the rod system.

(5) **Local Circuit Neurons.**

  (a) **Horizontal Cells.** The processes of these cells are confined to the **outer plexiform layer**. They form a triad synapse at the junction of the receptor terminals and the bipolar cells.

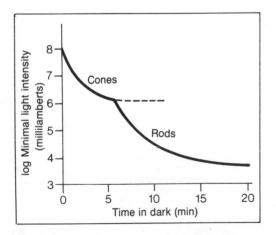

Figure 1-41. Dark adaptation. The eye becomes more sensitive to low-intensity light stimuli with increasing time in the dark. The *ordinate* indicates threshold light intensities.

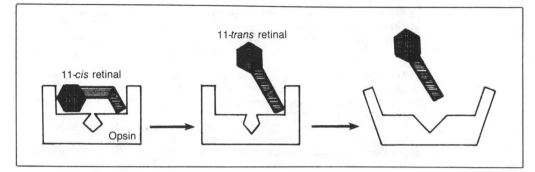

11-trans retinal

11-cis retinal

Opsin

Figure 1-42. Action of light on the rhodopsin molecule. Initially, the chromophore moiety, 11-cis retinal, fits closely into the opsin moiety. Light causes the conformation of the 11-cis retinal to change to the elongated all-trans form, and the chromophore no longer fits into the opsin portion of the molecule. Release of 11-trans retinal causes the opsin configuration to change.

> **(b) Amacrine cells** form complex synapses in the **inner plexiform layer** at the junction of the bipolar cells and the dendrites of the ganglion cells.

2. Photochemistry of Receptor Cells.
  a. Bleaching of Rhodopsin. When light strikes the eye it is absorbed by the photopigments of the retina, causing a photochemical transformation to occur (Fig. 1-42). During this process, the color of the photopigment changes from purple to pink, and so this process is called **bleaching**.
   (1) The effect of light is to cause a photoisomerization of 11-cis **retinal** into an intermediate form that, after going through a series of spontaneous changes, finally is converted into all-trans retinal.
   (2) The all-trans retinal dissociates from the opsin and then is converted into all-trans **retinol** (vitamin A).
  b. Regeneration of Rhodopsin. All-trans retinol is enzymatically isomerized to 11-cis retinol and then enzymatically oxidized to 11-cis retinal. The 11-cis retinal then spontaneously combines with opsin, reforming rhodopsin.
   (1) The conversion of the retinol to retinal is dependent on the pigment epithelium.
   (2) Deficiency of vitamin A prevents the regeneration of rhodopsin. For this reason, lack of vitamin A leads to **night blindness**.
  c. Absorption Spectrum of Visual Pigments.
   (1) Although rhodopsin absorbs light over most of the visible spectrum, it displays a maximal absorption at about 500 nm (the wavelength of green light).
   (2) There are three types of cones, each of which contains a photopigment that absorbs light best at a particular wavelength (Fig. 1-43).

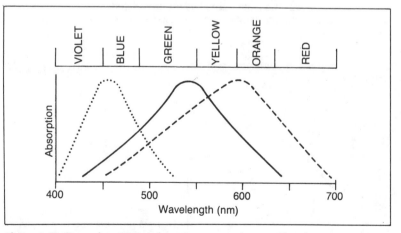

Figure 1-43. Spectral sensitivity of the three types of cones. The photopigment that is present in the outer segment of each cone type is responsible for the variance in the maximal light absorption and sensitivity in these cells. The *dotted line* indicates cones with blue-sensitive pigment; the *solid line* indicates cones with green-sensitive pigment; and the *dashed line* indicates cones with orange-sensitive pigment.

(a) **Red cones** have a maximal absorption at a wavelength of 570 nm.

(b) **Green cones** have a maximal absorption at a wavelength of 535 nm.

(c) **Blue cones** have a maximal absorption at a wavelength of 445 nm.

(3) **Color blindness** results from the inability to produce one or more of the cone photopigments.

(a) **Monochromats** have no cone photopigments and must rely entirely on rods for their vision. They are unable to see any colors.

(b) **Protonopes** lack the red pigment and, thus, make color confusions in the red-green portion of the spectrum.

(c) **Deuteranopes** lack green pigment and, like the protonopes, are unable to distinguish colors in the red-green portion of the spectrum.

(d) **Tritanopes** lack blue pigment and, thus, have difficulty distinguishing colors in the blue portion of the spectrum.

(e) Protonopes and deuteranopes are called red-green color blind because of their inability to see colors in this region of the spectrum normally. However, they are not truly blind in that they are able to see colors. Their deficiency is that they do not see them normally. Similarly, loss of the blue pigment does not prevent the seeing of blue light because light in the blue end of the spectrum is absorbed to some extent by the red and green pigments.

(4) **Color Coding.** Over most of the visible spectrum, light is absorbed by all three cone pigments (see Fig. 1-43). Thus, the encoding of color depends on the relative amount of activity in each of the three cones. For example, if the red cones are stimulated most strongly, the CNS interprets this activity as red light. If the red and green cones are stimulated equally, then a sensation of yellow is generated by the CNS.

3. Electrophysiology of the Retina.

a. Receptor Cell Potential.

(1) Receptor cells have a resting membrane potential of about -40 mV due to a high membrane conductance for Na^+.

(2) When receptor cells are stimulated by light, Ca^{2+} is released from storage sites within the cell.

(3) Ca^{2+} causes the Na^+ channels to close, producing a hyperpolarizing receptor potential. This is unlike other vertebrate sensory receptors, which exhibit depolarization in response to stimulation.

(4) Increasing the light intensity increases the receptor hyperpolarization (Fig. 1-44). The amplitude of the hyperpolarizing potential varies with the logarithm of light intensity.

b. Organization of Ganglion Cell Receptive Fields (Fig. 1-45).

(1) **Central Circuit.** Each ganglion cell receives synaptic input from many **bipolar cells**, which, in turn, receive input from a large number of receptor cells. All of the receptor cells that connect to a ganglion cell through bipolar cells form the receptive field **center** for that ganglion cell.

(a) In the dark, the receptor cells are depolarized and, thus, release synaptic transmitter. Because the transmitter is inhibitory, it keeps the bipolar cell hyperpolarized. As a result, the ganglion cell is not stimulated.

(b) When light strikes the receptor cells, it causes them to hyperpolarize. As a result,

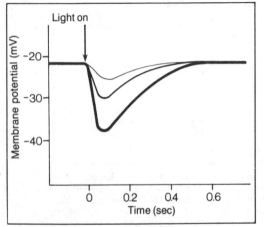

Figure 1-44. Response of a receptor cell to light. Responses to three brief light flashes of increasing intensity are shown. The receptor response is a graded hyperpolarization that increases with increasing light intensity. The *heaviest line* indicates the light flash of highest intensity.

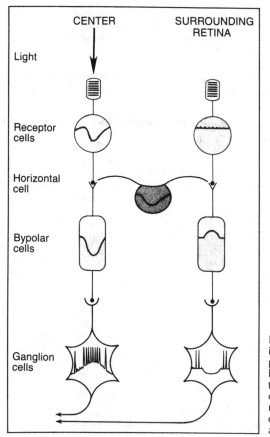

CENTER   SURROUNDING
                   RETINA

Light

Receptor
cells

Horizontal
cell

Bypolar
cells

Ganglion
cells

Figure 1-45. Organization of retinal circuits. Light strik-ing one spot on the retina (*center circuit*) causes hyper-polarizing responses in the receptor and bipolar cells but depolarization and spikes in the ganglion cells of this circuit. By the action of horizontal cells, bipolar cells in the surrounding retinal circuits are depolarized, resulting in the hyperpolarization of the ganglion cells of these circuits. Note that of the cell types shown, true action potentials are seen in only the ganglion cells.

receptor cell transmitter release is decreased, the bipolar cell depolarizes, and the ganglion cell is stimulated.

    (c) Thus, when any of the cells in the receptive field center of a ganglion cell are stimulated by light, it causes the ganglion cell to fire.

(2) Surrounding Circuit. The ganglion cell also is influenced by receptors that connect to the bipolar cells through **horizontal cells**. Receptor cells lying within an annulus sur-rounding the receptive field of the ganglion cell form the receptive field **surround**. When light strikes any of the receptors in the receptive field surround, the ganglion cell is inhibited through the horizontal to bipolar to ganglion cell pathway.

(3) This organization of the ganglion cell receptive field is called an **on-center, off-surround** organization to indicate that light hitting the center of the field causes ganglion cell excitation, while light hitting the surround of the field causes ganglion cell inhibition. There is a complementary type of organization that has **off-center, on-surround** characteristics. Ganglion cells with this type of receptive field are inhibited by light hitting the center of the field and are excited when light is shined on the cells within the area surrounding the center.

    (a) Because of the center-surround organization of the receptive fields, the retina only transmits information about changes in levels of illumination within the visual field.

    (b) For example, a bright spot surrounded by darkness will cause a large increase in the firing of an on-center ganglion cell, whereas a dark area surrounded by light will cause the ganglion cell to stop firing. If both the center and the surround are bathed by light or placed in darkness, then the ganglion cell firing rate will not change.

B. NEURAL PATHWAYS OF VISION

    1. Optic Tract. The axons of the ganglion cells are rearranged at the **optic chiasm**. Axons from the right half of each retina project to the **right lateral geniculate**, and axons from the left half of each retina pass to the **left lateral geniculate**. This means that fibers of the **temporal** retina on each side pass through the chiasm without crossing, while fibers of the **nasal** half of each retina decussate at the chiasm.

2. Lateral Geniculate Nucleus. The inputs from each eye remain segregated in the lateral geniculate; uncrossed fibers go to layers 2, 3, and 5, and crossed fibers go to layers 1, 4, and 6. Neurons in the lateral geniculate also retain the on-center, off-surround (or the off-center, on-surround) organization of the ganglion cells.

C. **PRIMARY VISUAL CORTEX.** Neurons of the cortex are responsible for visual perception. This is accomplished by cells whose input configuration allows for **feature detection** (i.e., sensitivity to stimuli of specific shape and orientation).

1. **Feature detectors** divide into three types: **simple cells**, **complex cells**, and **hypercomplex cells.**
   a. **Simple Cells.**
      (1) **Response Characteristics.** Like ganglion and lateral geniculate cells, simple cortical cells have antagonistic center-surround organizations. However, rather than circular receptive fields, simple cells have elongated receptive fields. The best simple cell response is obtained with light or dark bars. The long axis of the bar must fall on a specific place on the retina and be oriented at a particular angle.
      (2) **Synaptic Input.** The response characteristics of simple cells arise from the input they receive. Several adjacent ganglion cells, arrayed in a straight line at a particular orientation on the retina, project (via the lateral geniculate) onto a single simple cell of the visual cortex. When the ganglion cell array is activated with a bar of light, the cortical neuron is activated simultaneously. The simple cortical cell receiving input from this ganglion cell array can only be activated by (i.e., recognize) a bar of light that is at the proper retinal position and orientation.
   b. **Complex Cells.**
      (1) **Response Characteristics.** Like simple cells, complex cortical cells respond best to light lines or dark bars of a particular orientation. The complex cell, however, responds even if the oriented stimulus does not fall on a particular retinal position. The stimulus can be delivered to a larger area of the visual field and still evoke a response.
      (2) **Synaptic Input.** The response of a complex cell can be interpreted as a response to a **series of simple cells**, representing adjacent portions of the retina. The antagonistic center-surround organization is less prominent in complex cells.
   c. **Hypercomplex Cells.** These units respond only to light lines of particular length or to edges. They receive and **integrate input from a variety of complex cells**. Hypercomplex cells are more prominent in the **prestriate cortex** (areas 18 and 19) than in the primary visual cortex (area 17).

2. **Cortical Columns.** Cortical cells respond to lines, bars, and edges. The orientation of the linear stimulus is critical in determining whether a cell in the visual cortex will respond. Cells that respond to stimuli of a particular orientation are grouped together in an **orientation column**, which is perpendicular to the cortical surface.

D. **VISUAL PATHWAYS BEYOND THE PRIMARY VISUAL CORTEX.** Information concerning visual stimuli is analyzed further by cortical areas surrounding the primary visual cortex. Information is passed from the primary visual cortex to the prestriate cortex and from there onto the inferotemporal cortex (middle and inferior temporal gyri).

1. **Connections.** The prestriate cortex receives input from:
   a. The striate cortex
   b. Commissural fibers from the contralateral hemisphere
   c. The pulvinar thalami

2. **Function.** The prestriate cortex is involved in the recognition and identification of objects and patterns.

3. **Lesions** of the prestriate cortex result in an inability to use visually presented information. Lesions of the inferotemporal cortex result in visual learning deficits.

E. **RETINAL PROJECTIONS TO THE MIDBRAIN.** In lower vertebrates, the tectum is the highest visual processing center of the brain. In more advanced animals, the superior colliculus and pretectal area subserve the following basic functions:

1. Visual reflexes, including reflex shifts of the eyes to maintain fixation of the fovea on a visual target and the pupillary light reflex (pretectal area)

2. Voluntary shifts of gaze

3. Orientation of the head and body toward visual stimuli

F. LESIONS OF THE VISUAL PATHWAY (Fig. 1-46)

1. **Optic nerve lesions** cause a loss of vision in one eye but do not affect the vision in the other eye.

2. **Optic chiasm lesions** characteristically cause a loss of vision in the temporal fields of both eyes, a defect termed **bitemporal hemianopia**. This is a **heteronymous** hemianopia because the affected visual fields view disparate targets. Tumors of the pituitary gland frequently compress the optic chiasm to produce this defect.

3. **Optic tract lesions** produce **homonymous hemianopia**. A lesion of the right optic tract, for example, cuts the uncrossed fibers from the temporal retina of the right eye (causing a loss of vision in the **right nasal visual field**) and the crossed fibers from the nasal retina of the left eye (causing a loss of vision in the **left temporal visual field**). These two hemifields view the same target; thus, the defect is described as homonymous.

4. **Optic radiation lesions** affect the axons of the lateral geniculate neurons that project to the primary visual cortex. Fibers representing the inferior portion of the retina (superior visual field) sweep down into the temporal lobe (Meyer's loop). Consequently, lesions of the optic radiations result in **homonymous quadrantanopia** involving loss of vision in superior or inferior portions of the homonymous hemifields.

5. **Primary visual cortex lesions** cause homonymous hemianopia marked by **macular sparing**. This is due to the macular portion of the retina being heavily represented on the occipital lobe.

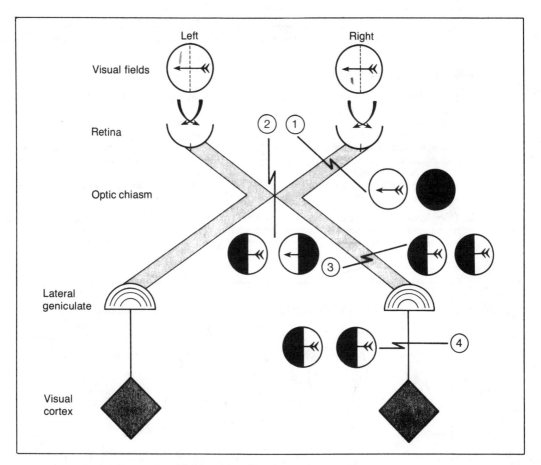

Figure 1-46. Abnormalities in visual fields produced by discrete lesions of nerve fibers. *Blackened areas* represent portions of the visual field where vision is absent. A lesion of the right optic nerve (*1*) causes blindness in one eye. A lesion of the optic chiasm (*2*) causes heteronymous hemianopia. A lesion of the optic tract (*3*) causes homonymous hemianopia. A lesion of the optic radiations (*4*) causes homonymous quadrantanopia.

X. TASTE

A. DEFINITION. Taste, or gustation, is the sensory modality mediated by the chemoreceptors of the tongue, mouth, and pharynx. Taste should be distinguished from flavor, which includes the olfactory, tactile, and thermal attributes of food in addition to taste. The taste chemoreceptors are sensitive to dissolved chemicals, both organic and inorganic.

B. PSYCHOPHYSICS

1. Taste Qualities. There are four primary taste qualities; all taste sensations are assumed to result from various combinations of these four primaries.
   a. Sweet sensation is produced by various classes of organic molecules including sugars, glycols, and aldehydes. The tip of the tongue is the area most sensitive to sweet stimuli.
   b. Bitter sensation is produced by alkaloids, such as quinine and caffeine. The back of the tongue is the area most sensitive to these stimuli.
   c. Salt sensation is produced by the anions of ionizable salts. The greatest sensitivity to salty tastes occurs in the front half of each side of the tongue.
   d. Sour sensation is produced by acids; this sensation relates, to some degree, to the pH of applied stimulus solutions. The greatest sensitivity to these stimuli occurs in the posterior half of each side of the tongue.

2. Threshold. The lowest concentration of a sapid substance that can be discriminated is the threshold concentration of that substance. Increasing the area of the tongue exposed to the solution reduces the threshold concentration. Individual differences in sensitivity are genetically determined. For example, some individuals cannot detect phenylthiocarbamide unless it exists in very high concentrations.

3. Adaptation. The intensity of a sensation resulting from continuous application of a taste stimulus declines with time until the solution becomes tasteless. The firing rate of afferent nerve fibers from the tongue similarly declines.

C. PERIPHERAL RECEPTORS

1. Lingual Papillae. These specialized protuberances on the surface of the tongue are of four morphological types (Fig. 1-47).
   a. Filiform papillae are mechanical, nongustatory structures.
   b. Fungiform papillae have 8–10 taste buds on each papilla.
   c. Circumvallate papillae are arranged in a V-shaped row of 7–12 on the posterior part of the tongue. Each papilla has approximately 200 taste buds. The taste buds are located on the sides of these large structures.
   d. Foliate papillae, located on the lateral border of the tongue anterior to the circumvallate papillae, have numerous taste buds.

2. Taste Buds. Each taste bud represents a cluster of 40–60 taste cells (Fig. 1-48) as well as supporting cells. Taste buds are located on the tongue papillae, hard palate, soft palate, epiglottis, and in the pharynx.

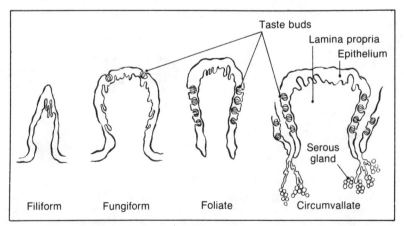

Figure 1-47. Structure of the four types of tongue papillae. Taste buds are located, as shown, on all but the filiform type of papillae.

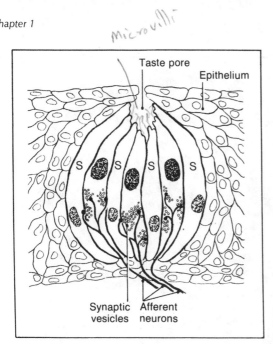

Microvilli

Figure 1-48. Structure of a taste bud. Taste receptor cells contain synaptic vesicles and receive afferent nerve terminals. Supporting cells (*S*) are not innervated.

3. Taste Cells.

   a. Structure. These chemosensitive receptor cells are elongated cells with microvilli at the apical surface. Dissolved substances enter the taste pore of the taste bud to contact the microvilli. The taste cells are innervated by afferent neurons.

   b. Turnover. Each taste cell has a life cycle of only a few days. The degenerating taste cell is replaced by a cell that arises from the supporting epithelial cells. Contact with the afferent neuron converts an epithelial cell into a taste cell. Conversely, nerve transection causes the disappearance of taste cells.

D. INNERVATION. Taste cells are innervated by branches of the facial, glossopharyngeal, and vagus nerves (cranial nerves VII, IX, and X, respectively). Nerve fibers terminate on receptor cells to form synapses. [The tactile and temperature receptors of the mouth, tongue, and pharynx are innervated by the trigeminal nerve (cranial nerve V).]

   1. The taste buds in the **anterior two-thirds of the tongue** are innervated by lingual branches of the **facial nerve**; the lingual nerve, which branches from the chorda tympani, is part of the facial nerve. The cell bodies are located in the geniculate ganglion, and the nerve terminals end in the **nucleus solitarius** of the medulla.

   2. The taste buds in the **posterior third of the tongue** are innervated by the **glossopharnygeal nerve**. The cell bodies lie in the superior and inferior ganglia of this nerve. The fibers relating to taste sensation terminate in the nucleus solitarius.

   3. Taste receptors in the **pharyngeal aspect of tongue** and on the hard palate, soft palate, and epiglottis are innervated by fibers of the **vagus nerve**. The cell bodies are located in the superior and inferior ganglia of the vagus and terminate in the nucleus solitarius.

E. TASTE CODING. Each nerve fiber in the gustatory nerves responds to more than one taste stimulus. However, each fiber responds best to one of the four primary taste qualities. Thus, the coding of a gustatory sensation is not a simple, labeled-line, chemical sensory system. Taste sensation depends on the **pattern** of nerve fibers activated by a particular stimulus.

F. CENTRAL GUSTATORY PATHWAYS (Fig. 1-49)

   1. Medulla. All taste information, whether carried on the facial, glossopharyngeal, or vagus nerve, is consigned to the anterior part of the nucleus solitarius. Second-order fibers pass to either the pons or thalamus. The medulla receives both ipsilateral and contralateral input.

   a. Pons. Fibers originating in the nucleus solitarius may terminate in the parabrachial nucleus, a cellular area just ventral to the brachium conjunctivum (superior cerebellar peduncle). Third-order fibers from this area then pass to the thalamus or to regions of the brain related to feeding behavior (i.e., the lateral hypothalamus and amygdala).

   b. Thalamus. The taste receiving area is located in the nucleus ventralis posteromedialis of the thalamus, a region that receives other afferent input relating to the orofacial region.

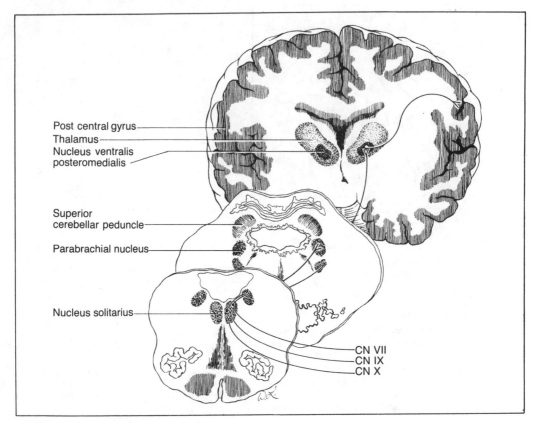

Figure 1-49. The taste pathways. Diagrammatic sections of medulla, pons, and forebrain show principal synaptic terminations of the gustatory system. *CN* = cranial nerve.

    2. Cortex. From the thalamus, third-order fibers pass to the cortex. The taste receiving area is located on the postcentral gyrus on the lateral convexity of the cerebral cortex. This area is associated with the face area of the somatosensory cortex. The cortex receives ipsilateral and contralateral input.

XI. OLFACTION

    A. DEFINITION. Olfaction, like taste, is a chemical sense involving receptors that are sensitive to chemicals in solution. For olfaction, the chemicals initially are airborne; however, they must dissolve in the mucous layer lining the nose before they can come in contact with olfactory receptors.

    B. PSYCHOPHYSICS

        1. Characteristics of Odorants. To be an effective odorant, a substance must be:
            a. Volatile, since the olfactory receptors respond to chemicals transported by air into the nose
            b. Water soluble (to some degree) in order to penetrate the watery mucous layer lining the nasal epithelium to reach the receptor cell membrane
            c. Lipid soluble (to some degree) in order to penetrate the cell membranes of the olfactory receptor cells to stimulate those cells

        2. Threshold. The olfactory receptors have varying sensitivity to substances. For many substances, the sensitivity is so high that a few molecules interacting with a receptor are sufficient to produce excitation.

        3. Odor Discrimination. The olfactory system has the capacity to discriminate various chemical compounds. This ability depends on the receptor sites on the membranes of receptor cells, which vary in their capability to accept odorant molecules of complementary conformation. In some instances, the olfactory system can discriminate dextro- and levorotatory forms as well as cis- and trans- conformations of a molecule. In contrast to gustatory sensation, no primary olfactory qualities can be identified.

4. Dynamic Range. The olfactory system has limited ability to discriminate differences in odorant concentration in ambient air. Perceived odor intensity conforms to a power function with an exponent of less than 1 (see Section V D 1).

5. Adaptation. Olfactory sensation decreases very rapidly with continued exposure to an odorant.

C. OLFACTORY MUCOSA. The olfactory mucosa refers to the specialized part of the nasal mucosa that contains the olfactory receptor cells. It is distinguished from the surrounding respiratory mucosa by the presence of tubular **Bowman's glands**, the absence of the rhythmic ciliary beating that characterizes the respiratory mucosa, and by a distinctive yellow-brown pigment. A mucous layer covers the entire epithelium.

1. Location. The olfactory mucosa is located on the superior nasal concha, adjacent to the nasal septum. Because of its superior position in the nasal cavity, the olfactory mucosa is not directly exposed to the flow of inspired air entering the nose. Odorant molecules come in contact with the olfactory mucosa by sniffing (i.e., by short, forceful inspirations). Sniffing produces turbulence in the air flow and thereby transports molecules to the receptor cells.

2. Innervation of the olfactory mucosa is by the olfactory nerves (cranial nerve I) and some branches of the trigeminal nerve (cranial nerve V). The irritative character of some odorants is due to stimulation of the free nerve endings of the trigeminal nerve.

3. Cells. The olfactory mucosa is composed of three cellular elements (Fig. 1-50).

    a. Receptor cells are bipolar neurons with a single dendrite that extends to the surface and terminates in a knob. Projecting from the knob of each sensory neuron are cilia that have the familiar microtubule configuration of two central microtubules surrounded by a ring of nine microtubules. The cilia reach the outer surface of the mucous layer. The axon that leaves the basal end of the receptor cell is unmyelinated. Receptor cells degenerate and have a life span of about 60 days.

    b. Supporting cells have a columnar shape. Microvilli extend from the surface of these cells into the mucous layer covering the nasal mucosa.

    c. Basal cells are stem cells from which new receptor cells are formed. There is a continuous replacement of receptor cells by mitosis of basal cells.

D. RECEPTOR CELL ELECTROPHYSIOLOGY

1. Receptor Potential. The adsorption of odorant molecules to the plasma membrane of the cilia of the receptor cells generates a receptor potential in the receptor cell.

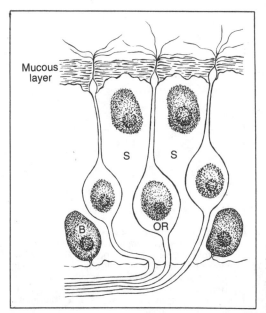

Figure 1-50. The olfactory epithelium. Olfactory receptor cells (*OR*) are situated among supporting cells (*S*). The cilia lie within the mucous layer covering the epithelium. New receptors are generated from basal cells (*B*). Bowman's glands, which contribute to the mucus secretion, are not shown.

2. **Electro-olfactogram (EOG).** The EOG is a tracing of the slow monophasic negative potential detected by an electrode placed on the olfactory mucosa. It reflects the local, graded receptor potentials evoked in the population of olfactory receptors by an odorous stimulus. The amplitude of the EOG increases with increases in the intensity of the stimulus.

3. **Impulse Initiation.** Neural impulses are generated by the receptor potential. Spikes originate at the point in the olfactory receptor where the cell narrows to form the **olfactory fiber** or **axon**.

4. **Spike Activity.** The resting activity in olfactory fibers is extremely low. Odorous stimulation causes excitation of most fibers, but a reduction in firing rate, or inhibition, is noted in some cases. Conduction velocity is slow because the fibers are unmyelinated and small in diameter.

5. **Sensory Coding.** A specific olfactory receptor does not respond to a particular compound or category of compounds; instead, an individual receptor responds to many odors. Furthermore, no two receptor cells have identical responses to a series of stimuli. Sensory perception, therefore, is based on the pattern of receptors activated by the stimulus.

E. CENTRAL PATHWAYS

1. **Fila Olfactoria.** Axons of receptor cells join upon leaving the olfactory mucosa and are sheathed by a **Schwann cell**. Contiguous axon bundles unite to form the fila olfactoria (olfactory nerves). The fila olfactoria penetrate the cribriform plate of the ethmoid bone to enter the olfactory bulb of the brain.

2. **Olfactory Bulb** (Fig. 1-51).
   a. **Lamination.** The olfactory bulb is organized in concentric layers. From outermost to innermost, these layers are:
   (1) Olfactory nerve layer
   (2) Glomerular layer (the olfactory nerve axons terminate within these synaptic complexes)

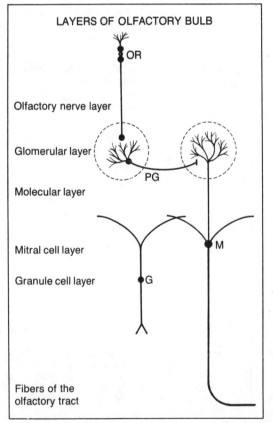

Figure 1-51. Synaptic organization of the olfactory bulb. Input arises from olfactory receptors (*OR*) in the olfactory mucosa. *PG* = periglomerular cells; *M* = mitral cells; and *G* = granule cells. Input to the granule cells arising from the anterior olfactory nucleus, the contralateral olfactory bulb, and cortical structures is not shown. Tufted cells, which resemble mitral cells, also are not depicted.

(3) Molecular layer
(4) Mitral cell layer
(5) Granule cell layer
(6) Fibers of the olfactory tract

b. **Neuronal Elements.**
   (1) **Projection neurons** receive input from the olfactory nerves and project their axons down the olfactory tract. These neurons exist in two forms.
      (a) **Mitral cells** are the most important second-order neurons of the olfactory bulb. These neurons have the following types of dendrites:
         (i) Apical dendrites, which arborize to form the olfactory glomeruli
         (ii) Basal dendrites
      (b) **Tufted cells** are small cells that resemble mitral cells.
   (2) **Intrinsic neurons** do not project axons out of the olfactory bulb. Two types of intrinsic neurons exist in the olfactory bulb.
      (a) **Periglomerular cells** interconnect neighboring glomeruli.
      (b) **Granule cells** interconnect the basal dendrites of neighboring mitral cells. These small, axonless neurons form complex local circuits (including dendrodendritic synapses) with mitral cell basal dendrites.

c. **Synaptic Organization.**
   (1) **Central Circuit.** Axons of receptor cells converge to form synapses with mitral cell dendrites in the glomeruli. Afferent volleys in the olfactory nerves cause excitation of mitral cells. Impulses generated in mitral cells travel to subsequent stations of the olfactory system.
   (2) **Surrounding Circuits.**
      (a) Dendrites of periglomerular cells, located in the glomeruli, also are excited by olfactory nerve afferents. Impulses along the axons of these cells cause inhibitory postsynaptic potentials in mitral cell dendrites of neighboring glomeruli
      (b) Granule cells cause inhibitory postsynaptic potentials in basal dendrites of neighboring mitral cells. Granule cells receive input from mitral cells, the contralateral olfactory bulb, and corticofugal fibers from higher olfactory centers.

3. **Olfactory Tract.**
   a. **Projection fibers** arising from mitral and tufted cells form the outflow pathway of the olfactory bulb. Caudally, the olfactory tract divides into the **lateral**, **intermediate**, and **medial olfactory striae.**
   b. **Anterior olfactory nucleus** is the name given to those cells scattered in the olfactory tract that receive collaterals of the projection fibers. Neurons of the anterior olfactory nucleus project to the contralateral olfactory bulb through the **anterior commissure.**

4. **Olfactory Cortex.** Olfactory tract axons project directly to the primary olfactory cortex. This contrasts with other sensory systems, in which a thalamic relay intervenes in the projection pathway to the cortex. The following areas of the cortex receive the input of the olfactory tract.
   a. **Prepiriform Cortex and Periamygdaloid Cortex.** These cortical regions are located within the uncus of the temporal lobe. They receive their input via the lateral olfactory stria and represent the primary olfactory cortex.
   b. **Olfactory Tubercle.** This cortical region lies within the anterior perforated substance and receives its input via the intermediate olfactory stria.
   c. **Subcallosal Gyrus.** This cortical area receives its input from the medial olfactory stria.

XII. SPINAL REFLEXES. A reflex is an automatic response to a stimulus, carried out by a relatively simple neuronal network consisting of a receptor, an afferent pathway, and an effector organ. Spinal cord reflexes are classified as originating from cutaneous receptors or from muscle receptors.

A. **CUTANEOUS REFLEXES.** The most important of the cutaneous reflexes is the **withdrawal, flexor**, or **pain reflex**, by which a body part is withdrawn from the site of a painful stimulus. Several reflex pathways within the spinal cord act together to coordinate the activity of all of the muscles necessary to produce a smooth movement.

1. The receptors for the withdrawal reflex are the nociceptors located on the free nerve endings of A delta and C fibers. Upon entering the spinal cord, the pain fibers synapse on many interneurons. Some of these convey information to the CNS. Others form several reflex pathways that coordinate the withdrawal of the limb.
   a. The major afferent pathway travels through several interneurons before synapsing on the alpha motoneurons innervating the muscles used to withdraw the limb from the painful stimulus. Because several interneurons are interposed between the afferent and efferent

neurons, the reflex is described as **multisynaptic** or **polysynaptic**. The interneurons form several pathways of different lengths, through which the pain stimulus can reach the alpha motoneuron.

   (1) **Reverberating circuits** are a consistent feature of the withdrawal reflex pathways. In these circuits, a branch from the axon of an interneuron in the reflex pathway feeds back onto previously excited interneurons, causing them to become reexcited. Reverberating circuits cause activity in the reflex pathway to continue even after the sensory receptor has stopped firing. This is called **afterdischarge**.

   (2) Under most circumstances, only the minimal number of muscles necessary to withdraw the limb are activated. This is called **local sign**. For example, if the hand accidently touches a hot stove, just the fingers may move away.

   (3) When the stimulus is very strong, however, the withdrawal reflex may require the involvement of more muscle groups. This is called **irradiation**. Thus, if the hand picks up a hot coal, not only will the fingers open to drop it, but the entire arm will withdraw and the individual may leap away from the site of painful stimulation.

   b. Another important pathway formed by the interneurons within the spinal cord terminates on the antagonistic motoneurons. This is an **inhibitory** pathway.

   (1) By inhibiting the motoneurons that innervate muscles antagonistic to those withdrawing the limb, this pathway ensures that the flexion movement is not impeded by contraction of the extensors.

   (2) This type of neuronal organization, in which the reflex pathway activating one group of alpha motoneurons also inhibits its antagonistic motoneuron, is quite common within the spinal cord and is called **reciprocal innervation**.

   c. Finally, the interneurons form pathways that cross the spinal cord to innervate the extensor motoneurons on the contralateral side. This is called a **crossed extensor reflex**.

2. Integration of the withdrawal reflex occurs on the alpha motoneuron, the **final common pathway** through which all the afferent fibers act. If the excitatory pathways dominate, the alpha motoneuron discharges a train of action potentials; if the inhibitory pathways dominate, the neuron does not fire.

3. The effector organs are the skeletal muscles that cause withdrawal of the limb. Although they are called flexors, these muscles are flexors in the physiologic, not the anatomic, sense. For example, the muscles that cause the fingers to open in order to drop a hot coal, although anatomically referred to as extensors, are considered to be flexors because they are involved in the withdrawal reflex.

4. The anatomic organization of the withdrawal reflex produces the following characteristic behavior.

   a. The withdrawal reflex has a relatively long latency because the afferent pathway uses small, slowly conducting fibers and involves many synapses.

   b. The afterdischarge that results from the parallel pathways and reverberating circuits causes the response to outlast the stimulus. This keeps the affected limb away from the painful stimulus while the brain determines where to put it.

   c. The crossed extensor reflex produces a patterned response in which the affected limb flexes while the contralateral limb extends. In the lower limbs, this allows the contralateral limb to support the body while the other limb is raised off the ground.

B. MUSCLE REFLEXES. Two important reflexes originate in the muscles: the **stretch reflex** and the **lengthening reaction**. The stretch reflex causes the reflex contraction of a muscle that is stretched. For example, when the patella tendon is tapped by a reflex hammer, it causes the quadraceps muscle to be stretched. The stretched muscle contracts reflexly, causing the leg to be elevated (called the knee jerk reflex). The lengthening reaction causes inhibition of alpha motoneurons innervating muscles that are under tension, allowing them to lengthen.

1. **The Stretch Reflex.**

   a. The receptor for the stretch reflex is the muscle spindle. This complex, spindle-shaped, encapsulated receptor contains muscle fibers that have both sensory and motor innervation (Fig. 1-52).

   (1) The number of spindles in each muscle depends on the task performed by the muscle. Muscles involved in precision movements contain many more spindles than muscles used to maintain posture. For example, hand muscles have about 80 spindles, which is about 20 percent of the number of spindles contained in back muscles weighing 100 times as much.

   (2) The muscle fibers within the spindle are called **intrafusal muscle fibers** in contrast to the **extrafusal muscle fibers** responsible for generating tension. There are two types of intrafusal muscle fibers.

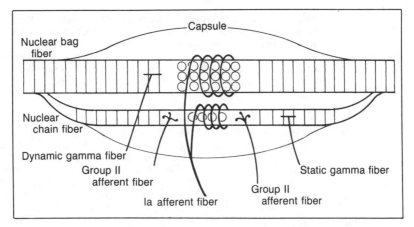

Figure 1-52. Diagram of an intrafusal muscle fiber, showing its nuclear bag and nuclear chain fibers. The afferent innervation (Ia and II fibers) and efferent innervation (gamma dynamic and gamma static fibers) of the intrafusal muscle fiber also are illustrated.

       (a) The **nuclear bag fibers** are 30 μ in diameter and 7 mm in length and are so-called because their nuclei appear to be bunched up in the center of the cell as if in a bag. About 2–5 nuclear bag fibers exist in a typical spindle.

       (b) The **nuclear chain fibers** are 15 μ in diameter and 4 mm in length and are so-called because their nuclei are lined up in a single file in the center of the fiber. About 6–10 nuclear chain fibers exist in each spindle.

       (c) The connective tissue surrounding the intrafusal fibers is continuous with that of the extrafusal fibers. As a result, when the extrafusal fiber contracts the intrafusal fiber is shortened, and when the extrafusal fiber stretches the intrafusal fiber is lengthened. The muscle spindle, thus, is **in parallel** with the extrafusal muscle fibers.

  (3) Two sensory neurons emerge from the muscle spindle.

       (a) A single large fiber, called a **group Ia fiber** or **primary ending**, sends branches to every intrafusal fiber within the muscle spindle.

       (b) Several smaller neurons, called **group II fibers** or **secondary endings**, innervate the nuclear chain fibers.

       (c) The primary endings surround the center of the intrafusal muscle fiber, while the secondary endings terminate on either side of the primary endings.

  (4) The efferent fibers to the muscle spindle are called **gamma fibers** because their axons belong to the A gamma group of fibers. There are two types of gamma fibers.

       (a) **Static gamma fibers** primarily innervate nuclear chain intrafusal muscle fibers and produce tonic activity in the Ia afferent fibers.

       (b) **Dynamic gamma fibers** primarily innervate nuclear bag intrafusal muscle fibers and generate phasic activity in the Ia afferent fibers.

  b. The Ia fiber enters the spinal cord through the dorsal root and sends branches to every alpha motoneuron that goes to the muscle from which the Ia originated. Although some branches go to synergistic muscles, these are not physiologically important.

    (1) In contrast to the withdrawal reflex, the stretch reflex has a rapidly conducting afferent fiber, is monosynaptic, has a short latency, and does not exhibit afterdischarge and irradiation.

    (2) In the stretch reflex, as in the withdrawal reflex, the alpha motoneuron is the final common pathway, serving as both an integrating center and efferent pathway.

    (3) Also like the withdrawal reflex, the stretch reflex is characterized by reciprocal innervation. Thus, when a stretch reflex is elicited, the muscle antagonistic to the stretched muscle is inhibited, allowing the agonistic muscle to contract without interference.

  c. Both extensor and flexor muscles exhibit stretch reflexes. These are elicited routinely during a neurologic examination to test for damage to either the spinal cord or the sensory or motor neurons. The efferent fibers control the sensitivity of the receptors to stretch. Their role is discussed in Section XII C 2.

2. The Lengthening Reaction.

  a. The receptors for the lengthening reaction are the **Golgi tendon organs**.

    (1) These small (0.5 mm long), encapsulated receptors are located in the tendons, between the muscles and tendon insertions.

(2) The Golgi tendon organs are stretched whenever the muscle contracts and, thus, in contrast to the muscle spindle, are **in series** with the extrafusal muscle fibers.

(3) The Golgi tendon organs have neither muscle fibers nor an efferent innervation.

b. The afferent fiber innervating the Golgi tendon organ is called a **group Ib fiber**. It enters the dorsal root and forms a disynaptic pathway, which ends on the alpha motoneurons that send axons to the muscle from which the Ib fiber originated.

(1) The disynaptic pathway of the Golgi tendon organ is inhibitory to the alpha motoneuron.

(2) The Ib fiber, like all sensory fibers, releases an excitatory transmitter. To produce inhibition, an inhibitory interneuron must be activated.

c. Like the stretch reflex, the lengthening reaction occurs in both flexor and extensor muscles, lacks afterdischarge and irradiation, and displays reciprocal innervation.

d. Historically, the lengthening reaction has been described as a protective reflex in which a strong and potentially damaging muscle force reflexly causes the muscle to be inhibited, and, as a result, the muscle lengthens instead of trying to maintain the force and risk being damaged. Although this is true, it is now clear that the reflex plays a more important role in regulating tension during normal muscle activity. The lengthening reaction is described as **autogenic inhibition**, which indicates that the force generated when the muscle contracts is the stimulus for its own relaxation.

C. ROLE OF STRETCH REFLEX IN THE CONTROL OF MOVEMENT. The stretch reflex is used by the motor control system to aid in the performance of a movement. During activity generated by the motor command system, the Ia fibers from the muscle spindle inform the motor control system about the changes in muscle length and provide the alpha motoneuron with a source of excitatory input in addition to that coming from higher centers.

1. Figure 1-53 illustrates the effect of muscle stretch and contraction on Ia discharge when the gamma motoneurons are not activated.

a. **Stretching** the extrafusal muscle fiber causes the intrafusal muscle fiber to be stretched. Most of the stretch occurs in the central, most compliant region of the intrafusal fiber, which lacks sarcomeres.

(1) When the intrafusal muscle fiber is stretched, the Ia fiber terminal is deformed. The

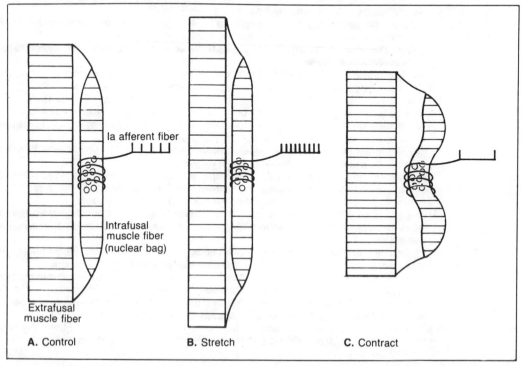

Figure 1-53. The firing rate of the Ia afferent fiber increases when the extrafusal muscle fiber is stretched because the intrafusal muscle fiber also is stretched. Most of the stretch occurs at the center region of the intrafusal muscle fiber. When the extrafusal muscle fiber contracts, the Ia afferent discharge decreases because the intrafusal muscle fiber slackens (i.e., the intrafusal muscle fiber is unloaded).

deformation causes a receptor potential, which, in turn, generates a train of action potentials.

(2) The action potentials initially discharge at a frequency proportional to the velocity of stretch and then adapt to fire tonically at a freqency proportional to the amount of muscle stretch.

b. Shortening the extrafusal fiber causes the central region of the intrafusal muscle fiber to be compressed.

(1) This reduces the deformation of the Ia fiber terminal, causing the Ia fiber to cease firing.

(2) The reduction of firing that occurs during muscle contraction, called **unloading**, is functionally disadvantageous because the CNS stops receiving information about the rate and extent of muscle shortening.

2. Unloading can be prevented by the activity of the gamma motoneurons, as shown in Figure 1-54. The gamma motoneurons cause the intrafusal muscle fiber sarcomeres to shorten along with the shortening of the extrafusal muscle fiber. As a result, the central region of the intrafusal muscle fiber remains stretched during muscle contraction and unloading does not occur.

   a. When the dynamic gamma motoneurons are fired, only the nuclear bag intrafusal fibers shorten. Since the nuclear bag fibers are responsible for producing the phasic (i.e., velocity-sensitive) portion of the Ia response to stretch, stimulation of the dynamic gamma fibers causes an increase in phasic activity without affecting the tonic activity.

   b. When the static gamma motoneurons are fired, only the nuclear chain fibers shorten. Since nuclear chain fibers are responsible for the tonic component of the Ia response, stimulation of the static gamma fibers causes an increase in the tonic (length-sensitive) level of Ia firing without affecting the phasic level.

3. During a normal movement, the motor command system **coactivates** both the alpha and gamma motoneurons. This diminishes the amount of unloading that occurs during muscle contraction, allowing the CNS to determine whether its motor commands are being carried out.

   a. For example, when the motor control system issues a command to lift a weight, the alpha and gamma motoneurons are coactivated.

   (1) As the extrafusal muscle fiber shortens, the intrafusal muscle fiber sarcomeres also shorten.

   (2) If the two muscle fibers shorten at the same rate, then the central region of the intrafusal fiber is neither compressed nor lengthened, thus keeping Ia activity at a constant level.

   (3) The constant level of Ia input to the CNS during a movement indicates that the motor command is being carried out.

   b. If the weight to be lifted is underestimated by the CNS, the motor command system does not activate a sufficient number of alpha motoneurons to lift the weight. Thus, the extrafusal muscle fibers do not shorten.

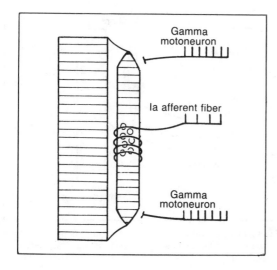

Figure 1-54. The firing rate of the Ia afferent fiber does not decrease if the gamma motoneuron discharge accompanies contraction of the extrafusal muscle fiber. Unloading is prevented because the intrafusal muscle fiber also shortens. Since the center region of the extrafusal muscle fiber does not contain sarcomeres, it does not shorten and, in fact, may lengthen if the gamma motoneuron causes sufficient shortening of the intrafusal muscle fiber sarcomeres.

(1) However, the intrafusal muscle fibers do shorten. But because the tendon ends of the muscle cannot move, the central portion of the intrafusal fiber lengthens.

(2) Stretching the central region of the fiber causes Ia activity to increase, indicating that the motor command is not being carried out. The CNS uses this information to readjust its command to the spinal cord.

(3) However, even before the CNS responds to the information provided by the Ia fibers, the Ia activity is used at the spinal cord level to adjust the alpha motoneuron activity to meet the unexpectedly high load.

   (a) Since the Ia fiber synapses on the alpha motoneuron, its activity increases the excitability of the alpha motoneuron.

   (b) The increase in excitability leads to an increase in the frequency of action potential generation and an increase in muscle force development.

4. Gamma Loop. The CNS can initiate movements directly by stimulating just the gamma motoneurons, using a pathway called the **gamma loop** (Fig. 1-55).

   a. Increasing gamma motoneuron activity causes the intrafusal muscle fiber sarcomeres to shorten, which in turn leads to a stretch of the central portion of the intrafusal fiber and activation of the Ia fiber. Firing the Ia fibers causes alpha motoneuron activity to increase, which results in an increased amount and force of skeletal muscle activity.

   b. Although the gamma loop can elicit movement on its own, it normally does not do so. However, because of coactivation, the gamma loop is activated during all movements and thus contributes to the excitability and firing rate of the alpha motoneurons.

XIII. THE MOTOR CONTROL SYSTEM

A. INTRODUCTION. Figure 1-56 is a block diagram illustrating the major components of the motor control system. This system is a complex and highly integrated network in which all parts work together to produce a movement. The specific tasks assigned to each component can be analyzed as a basis for understanding how movements are coordinated.

1. The **idea** for a movement is generated within the cortical association areas of the parietal lobe.

2. The idea then is transferred to the motor areas of the frontal lobe, where it is organized into a **motor command**.

3. The command then is sent to the spinal cord for execution. The basal ganglia, cerebellum, brain stem, and spinal cord all participate in the production of a coordinated movement by modifying the motor command.

   a. The basal ganglia provide the motor patterns necessary to maintain the postural support required for motor commands to be carried out properly.

   b. The cerebellum receives information from the motor cortex about the nature of the in-

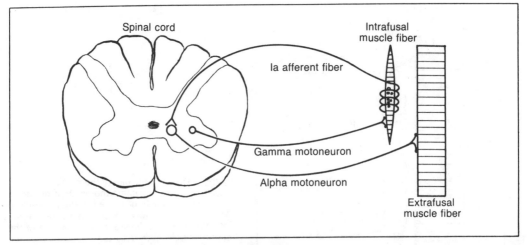

Figure 1-55. The gamma loop increases the firing of the alpha motoneuron during muscle contraction. The loop begins with the gamma motoneuron, which discharges to cause intrafusal muscle fiber contraction. This leads to an increase in Ia activity, which in turn causes increased alpha motoneuron discharge via a monosynaptic reflex.

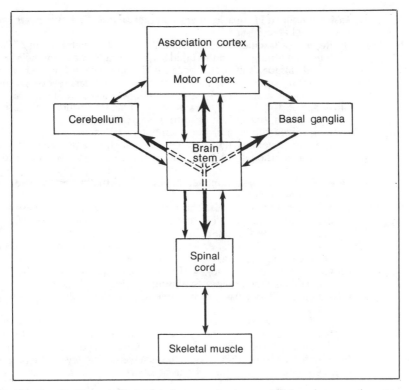

Figure 1-56. Diagram illustrating the extensive interconnections between the components of the motor control system. Note that all of the descending pathways except for the pyramidal tract communicate with the spinal cord through the brain stem.

tended movement and from the spinal cord about how well it is being performed. This information is used to adjust the motor command so that the intended movement is executed smoothly.

   c. The brain stem is the major relay station for all motor commands except those requiring the greatest precision, which are transferred directly to the spinal cord. The brain stem also is responsible for maintaining normal body posture during motor activities.

   d. The spinal cord contains the final common pathways through which a movement is executed. By selecting the proper motoneurons for a particular task and by reflexly adjusting the amount of motoneuron activity, the spinal cord plays an important role in the coordination of motor activity.

B. THE SPINAL CORD

   1. Physiologic Anatomy.

   a. The alpha and gamma motoneurons within the spinal cord that innervate a particular muscle are collected together into a **motoneuron pool**.

   (1) The alpha and gamma motoneurons are randomly distributed within the pool and overlap to some extent with cells from other motoneuron pools.

   (2) Usually, the motoneurons leave the spinal cord in several contiguous ventral roots and then combine into a single motor nerve containing the alpha and gamma motoneurons as well as the Ia, Ib, and II afferent fibers from the muscle.

   b. Each motoneuron and all the muscle fibers it innervates form a **motor unit**. The motor units within a muscle vary in size from a few muscle fibers to several thousands of fibers. Muscles that perform precise movements have smaller motor units than those responsible for large body movements and for maintenance of posture.

   (1) Whenever an alpha motoneuron fires, all of the muscle fibers in its motor unit are activated.

   (2) The muscle fibers belonging to a single motor unit are dispersed throughout the muscle so that the force they produce is distributed evenly.

   (3) All of the muscle fibers in a motor unit are of the same physiologic type and are categorized according to their histochemical and contractile characteristics.

(a) **Fast fatigable (FF)** muscle fibers contract quickly, fatique easily, and have the following characteristics.

    (i) **Rapid Contractile Speeds.** This results from the high myosin ATPase activity of their cross-bridges and the rapid sequestering of Ca^{2+} by their SR.

    (ii) **Rapid Fatigue.** This occurs because FF fibers have few mitochondria and, thus, cannot make use of oxidative metabolism. FF fibers rely on glycolysis for their ATP supply and fatigue when their glucose stores are depleted.

    (iii) **A Sparse Capillary Supply.** Because FF fibers do not make use of oxidative metabolism, the growth of surrounding capillaries is limited.

    (iv) **Large Size.** Although FF fibers cannot sustain activity for long periods of time, they can generate large contractile forces. Their large size is not a disadvantage from a diffusional point of view because they do not make use of oxidative metabolism.

(b) **Slow (S)** muscle fibers contract slowly, are virtually nonfatigable, and have the following characteristics.

    (i) **Slow Contractile Speeds.** This is due to the low myosin ATPase activity of their cross-bridges and the slow sequestering of Ca^{2+} by their SR. These characteristics reduce the amount of ATP used by S fibers and, thus, contribute to their resistance to fatigue. Their long contraction times make summation and tetanus possible at low frequencies of stimulation.

    (ii) **Great Resistance to Fatigue.** This is a consequence of the ability of S fibers to use oxidative metabolism as a primary source of ATP.

    (iii) **A rich capillary supply** is available to provide for the oxygen (O_2) needs of S fibers.

    (iv) **Small Size.** Although S fibers cannot produce a large amount of force, they can sustain force for a long time. Their small size is necessary for O_2 to diffuse into the center of the fiber and for waste products to diffuse out of the fiber.

(c) **Fast fatigue-resistant (FR)** muscle fibers have characteristics that are intermediate between those of the FF and S muscle fibers.

2. **Task Performance.** During the performance of a motor task, the small motor units, because they are more excitable, are recruited before the large ones. This **size principle** of motor unit selection has significant physiologic advantages.

  a. To perform a precision movement requiring small amounts of force, it is advantageous to use small motor units. When more force is required, larger motor units must be activated.

    (1) The proper motor units are selected automatically because the motoneuron pool is organized according to the size principle.

      (a) When a small amount of force must be applied, the motor cortex provides a minimal amount of input to the motoneuron pool, activating only the smallest motor units.

      (b) If more force is required, the motor cortex increases its input to the motoneuron pool and larger motor units are recruited.

    (2) Because of the size principle, the motor cortex does not need to specify the particular motoneuron to be recruited during a movement. This simplifies the organization of the motor command structure by reducing the number of cortical neurons that are involved in generating a movement.

  b. To perform a task requiring endurance, the fatigue-resistant motor units must be recruited. When power is required, the larger motor units must be recruited.

    (1) Again, motor unit selection is accomplished automatically because of the size principle.

      (a) When an endurance movement must be performed, the motor cortex provides a minimal input to the spinal cord, and the smallest, most fatigue-resistant motor units are recruited.

      (b) When the amount of force being generated by the muscle is not sufficient to execute the movement, the motor cortex increases its input and recruits more motor units. In all cases, the most fatigue-resistant fibers are recruited first without requiring that the motor cortex determine which motoneuron to activate.

    (2) The relation between fatigue resistance and motoneuron size also is a consequence of the size principle. A muscle fiber's fatigability can be altered by its activity.

      (a) When a muscle fiber is contracted repeatedly, its oxidative capacity and blood supply increase, making the fiber more fatigue resistant. Conversely, when a muscle fiber is required to contract against large afterloads, the fiber increases in size.

      (b) Small motor units are recruited first. Thus, they are involved in all movements and, consequently, develop fatigue resistance.

C. THE BRAIN STEM contains the medulla, pons, midbrain, and parts of the diencephalon. Neuronal circuits within these areas control many physiologic functions including blood pressure, respiration, body temperature, sleep, and wakefulness. In addition, the **reticular formation** and **vestibular nuclei** contribute importantly to the motor control system.

1. The **reticular formation** is the relay station for all descending motor commands except those traveling directly to the spinal cord through the medullary pyramids.
   a. **Normal Function.** The reticular formation receives and modifies motor commands to the proximal and axial muscles of the body and is involved in the performance of all motor activities except those fine movements performed by the distal muscles of the fingers and hands. The reticular formation also is responsible for maintaining normal postural tone.
      (1) Neurons within the pontine reticular formation send axons to the spinal cord in the medial reticulospinal tract and are excitatory to the alpha and gamma motoneurons innervating the extensor **antigravity muscles**.
      (2) These neurons are prevented from firing too rapidly by inhibitory input derived from the cerebral and cerebellar components of the motor control system. The amount of inhibition is increased to reduce postural tone and is decreased to enhance postural tone. The withdrawal of inhibition, called **release of inhibition**, frequently is used to increase activity within the CNS.
   b. **Lesions.**
      (1) If the brain stem is severed from the spinal cord, the syndrome of **spinal shock** results.
         (a) The initial result of removing the spinal cord from the control of the brain stem is the complete loss of reflex activity, which can last for days in cats and for months in humans.
         (b) When reflexes return, they no longer are under the influence of the brain stem and, so, do not follow their normal patterns. For example, the local sign that characterizes the withdrawal reflex disappears. Instead, even a light touch on the foot can cause activation of all the flexor muscles in the body.
         (c) The excitability of the motoneurons ultimately becomes greater than normal, causing particular groups of muscles to contract continuously.
      (2) If the cerebral and cerebellar inhibitory inputs to the reticular formation are removed by severing the brain stem above the pontine reticular formation, the condition of **decerebrate rigidity** results.
         (a) Without the inhibitory input from these higher centers, the pontine reticular formation fires uncontrollably, subjecting both the alpha and gamma antigravity motoneurons to intense excitation.
            (i) The excessive firing of the alpha motoneurons causes the antigravity muscles (i.e., the leg extensors and the arm flexors) to contract continuously.
            (ii) The firing of the gamma motoneurons activates the gamma loop, causing increased discharge of the Ia afferent fibers. These reflexly add to the alpha motoneuron excitation produced by the reticulospinal tract.
         (b) Decerebrate rigidity is analogous to the condition of **spasticity**.
            (i) Spasticity normally is caused by lesions of the descending pathways from the motor cortex to the reticular formation.
            (ii) Spasticity is characterized by an increased amount of antigravity muscle activity. This activity is increased further when the affected muscle is stretched, due to the increased activity of the gamma motoneurons. Because these fibers are firing at higher than normal rates, the muscle spindles become more sensitive to stretch.
            (iii) Cutting the dorsal roots reduces the amount of Ia input to the spinal cord and reduces the amount of alpha motoneuron activity. This reduces the spasticity.

2. **The Vestibular System.** Vestibular system reflexes are responsible for maintaining tone in antigravity muscles and for coordinating the adjustments made by the limbs and eyes in response to changes in body position.
   a. **Vestibular Receptors.** The receptors that initiate the vestibular reflexes are located within the **labyrinth** (see Fig. 1-28)—a system of fluid-filled, membrane-bound canals and sacs that are continuous with the cochlea. The labyrinth consists of the sac-like **otolith organs** (i.e., the **saccule** and **utricle**) and three **semicircular canals**. The saccule communicates directly with the cochlea, which is beneath it, and the utricle, which is above it. Both ends of all three semicircular canals emerge from the utricle. Each canal has an expanded end, called the **ampulla**, which contains the receptor cells.
      (1) **Orientation.**
         (a) The semicircular canals are responsible for detecting angular accelerations of the head. The **horizontal canal** lies in a plane that is approximately parallel to the earth when the head is held in a normal upright position. The vertical canals lie in vertical planes. The plane of the **anterior vertical canal** is oriented along a line from the center of the head toward the eye, while the plane of the **posterior vertical canal** is oriented along a line from the center of the head toward the ear. The orientation of

the posterior canal on one side of the head is roughly parallel to the orientation of the anterior canal on the other side of the head.

(b) The otolith organs are responsible for detecting linear acceleration and the static position of the head. The receptors within the utricle are oriented in a horizontal plane, while those in the saccule are oriented in a vertical plane.

(2) Receptor Cells. The receptor cells of the vestibular system are similar to those of the cochlea. However, unlike the auditory receptor cells, the cilia of the vestibular hair cells are polarized (Fig. 1-57). A large cilium, called the **kinocilium**, is located at one end of the cell. All of the others are called **stereocilia**. Whenever the stereocilia are bent toward the kinocilium, the cell is depolarized. When the stereocilia are bent away from the kinocilium, the cell is hyperpolarized.

(a) The hair cells of the semicircular canals are located on a mass of tissue (the **crista**) within the ampulla.

(i) The cilia are embedded in a gelatinous structure called the **cupula**, which completely fills the ampullar space.

(ii) When the head begins to move, the fluid within the semicircular canals lags behind and pushes the cupula backward. This causes the cilia of the hair cells to bend, producing either a depolarization or hyperpolarization depending on whether the stereocilia are pushed toward or away from the kinocilium.

(iii) After 15–20 seconds of continuous movement at a constant velocity (e.g., the movement of a twirling dancer), the velocity of fluid movement catches up to that of the head, and the cupula returns to its resting position. This causes the cilia to return to their upright position and the hair cell to return to its resting membrane potential. Thus, the semicircular canals signal changes in motion (acceleration) and are insensitive to movements at a constant angular velocity.

(iv) When the head stops moving, the fluid within the semicircular canals continues to move, pushing the cupula forward. This causes the cilia to bend in the opposite direction. Thus, if the original movement caused the hair cell to depolarize, the hair cell hyperpolarizes when the movement ceases.

(v) The hair cells of the horizontal canals are oriented with the kinocilium located closest to the utricle. Head movements that bend the stereocilia toward the utricle (utriculopedal movements) cause the hair cells to depolarize, while movements that bend the stereocilia away from the utricle (utriculofugal movements) cause hyperpolarization. For example, when the head is rotated toward the left, the endolymph in the left horizontal semicircular canal pushes the cupula and stereocilia toward the utricle (i.e., the fluid lags behind the head movement and causes the cupula to move toward the right), causing the hair cell to depolarize (Fig. 1-58). At the same time, the hair cells within the right horizontal canal are pushed away from the utricle and hyperpolarize. When the head stops rotating, the fluid continues to move, pushing the cupula in the opposite direction. This causes the hair cells in the left horizontal canal to hyperpolarize and those in the right semicircular canal to depolarize.

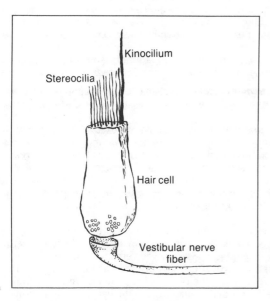

Kinocilium

Stereocilia

Hair cell

Vestibular nerve fiber

Figure 1-57. Diagram illustrating the polarized hair cells of the vestibular system. The large cilium is called the kinocilium. The smaller cilia are called stereocilia. When the stereocilia are bent toward the kinocilium, the cell depolarizes. When the stereocilia are bent away, the cell hyperpolarizes.

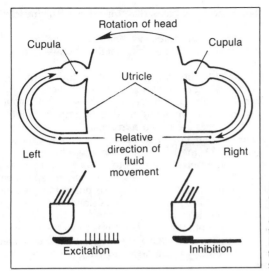

Figure 1-58. Diagram illustrating the movement of the endolymph within the semicircular canal when the head rotates to the left. The fluid lags behind the head and pushes the cupula toward the right. As indicated, the stereocilia are bent toward the kinocilium on the left and away from the kinocilium on the right. Thus, the hair cells within the left horizontal canal are stimulated, and those on the right are inhibited.

    (vi) The vertical canals also work in pairs: when the anterior vertical canal on one side is stimulated, the posterior vertical canal on the other side is inhibited. In the vertical canals, the kinocilium is located on the side of the hair cell away from the utricle.

    (vii) Because each of the three canals is oriented in a different plane, movement of the head in any direction causes a unique pattern of activity to be generated by the semicircular canals. This information is used by the CNS to interpret the speed and direction of head movement and to make the appropriate adjustments in posture and eye position.

  (b) The hair cells of the utricle and saccule are located on a mass of tissue called the **macula**.

    (i) The cilia are enmeshed in a gelatinous structure filled with small calcium carbonate crystals called **otoconia**.

    (ii) Because the otoconia are heavier than the fluid of the otolith organs, they tend to sink downward and, thus, can bend the cilia of the hair cells.

    (iii) When the head deviates from the horizontal position, the hair cells bend. The kinocilium of each hair cell is oriented in a different plane so that, regardless of the direction in which the head is tilted, some of the hair cells are maximally stimulated while others are maximally inhibited.

    (iv) Linear acceleration, which occurs when jumping down stairs or accelerating forward in a car, also causes the otoconia to be displaced and to stimulate the otolith organs.

 b. Neuronal Pathways. The hair cells of the labyrinth synapse on the vestibulocochlear nerve (cranial nerve VIII), which in turn projects to the vestibular nuclei within the brain stem. In addition, the vestibular nuclei receive inhibitory input from the cerebrum and cerebellum.

  (1) The vestibular nuclei send their axons into the spinal cord through a number of vestibulospinal tracts. This input is excitatory to antigravity alpha motoneurons.

  (2) If the inhibitory input from the cerebrum and cerebellum is removed, the vestibular nuclei greatly increase their firing rate and contribute to the syndrome of decerebrate rigidity.

    (a) The rigidity produced by the vestibular nuclei differs from that produced by the reticular formation in that the vestibular nuclei primarily affect the alpha motoneurons rather than both alpha and gamma motoneurons.

    (b) Spasticity produced by the vestibular nuclei, therefore, is not reduced greatly by cutting the dorsal roots and eliminating the Ia input. This type of spasticity is called alpha rigidity to distinguish it from the type of spasticity caused by the reticular formation, which is called gamma rigidity.

    (c) When discussing the effects of the brain stem on antigravity muscles, the terms spasticity and rigidity are used interchangeably. Clinically, however, spasticity and rigidity are not alike. Spasticity refers to the condition in which the stretch reflexes of the antigravity muscles are increased due to increased activity of the alpha or gamma motoneurons. Rigidity refers to the condition seen in Parkinson's disease (see Section XIII D 2 d), in which there is increased activity of all of the muscles at a joint.

c. Vestibular Reflexes. The vestibular system participates in a number of reflexes.

(1) The most important of these reflexes is responsible for maintaining visual fixation during movement of the head.

(a) For example, if the head is rotated to the left, the eyes move toward the right in order to prevent an image from moving off the fovea. When the eyes have rotated as far as they can, they are rapidly returned to the center of the socket. If rotation of the head continues, the eyes once again move in the direction opposite the head rotation. These movements of the eyes are called **nystagmus**.

(b) The slow movement of the eyes to maintain visual fixation is initiated by the activity within the semicircular canals. When the head rotates to the left, the activity of receptors in the left horizontal canal causes the eyes to move toward the right.

(c) After the body has been rotated and then movement ceases, the receptors within the right horizontal canal are stimulated, and these cause **postrotary** eye movements to occur in the opposite direction. That is, the eyes slowly move to the left until they reach the end of the socket, at which point they return quickly to the center. These movements continue until the cupula returns to its resting position.

(d) Lesions within the vestibular pathways can cause nystagmus to occur spontaneously, while damage to the vestibular receptors can prevent nystagmus from occurring during rotation of the head.

(2) Another important reflex is initiated by receptors within the otolith organs and occurs when an individual walks down stairs or jumps from a platform.

(a) When making such a descent, the muscles in the leg begin to contract before the feet reach the ground in order to cushion the force of impact. The otolith receptors responsible for this reflex are stimulated by the linear acceleration of the head that occurs during the descent.

(b) Individuals lacking otolith reflexes are prone to leg injuries because of the large contact forces that occur when descending from a height (e.g., stepping off of a bus).

D. THE BASAL GANGLIA are interconnected nuclei within the cerebrum and do not make direct connections with the spinal cord. It appears that the major function of the basal ganglia is to aid the motor cortex in generating commands concerned with controlling proximal muscle groups during a movement. For example, when the hand is used to write on a blackboard, the large muscles of the arm and shoulders are used to hold the hand in its proper position for writing. The coordination of these muscle groups is believed to be under the control of the basal ganglia.

1. Physiologic Anatomy. The basal ganglia consist of: the **striatum** (which is composed of the **caudate nucleus** and **putamen**), the **globus pallidus** (also called the **pallidum**), the **subthalamic nucleus**, and the **substantia nigra**.

a. These nuclei have complex interconnections.

(1) The globus pallidus is in the center of three major feedback loops.

(a) One pathway leaves the globus pallidus and travels through the thalamus, cortex, and striatum before returning to the globus pallidus.

(b) Another pathway follows the same route from the globus pallidus to the thalamus but travels to the striatum and back to the globus pallidus without going through the cortex.

(c) The third pathway connects the globus pallidus with the subthalamic nucleus.

(2) Another important feedback loop is established between the striatum and the substantia nigra.

b. These pathways between the nuclei are described in a highly diagrammatic way because not enough is known about how the basal ganglia function to allow a more realistic and more detailed description.

2. Lesions in the basal ganglia produce characteristic deficits in motor behavior.

a. Lesions in the globus pallidus result in an **inability to maintain postural support** of the trunk muscles. The head bends foward so that the chin touches the chest, and the body bends at the waist.

(1) The motor deficits are not due to muscular weakness or failure of voluntary control because individuals with these lesions can stand upright when requested to do so.

(2) Since the globus pallidus is the major outflow tract of the basal ganglia, it is possible that the motor deficits occur because the cortex is deprived of information it needs for automatic control of the trunk muscles.

b. Lesions in the subthalamic nucleus cause spontaneous, wild, flinging, ballistic movements of the limbs. This syndrome is called **hemiballismus**.

(1) The movements, which are caused by a release of inhibition, appear on the side opposite the lesion.

(2) Because the movements appear to be like those performed when an individual is thrown off balance, the subthalamic nucleus is believed to be involved with controlling the centers that issue the motor commands for balance.

    (a) Normally, the subthalamic nucleus responds to the need for initiating the balancing movement by momentarily withdrawing its inhibition from these centers.

    (b) When there is a lesion in the subthalamic nucleus, these centers no longer are under inhibitory control and, so, the movements are generated spontaneously.

  c. Lesions within the striatum produce a variety of motor syndromes related to a release of inhibition.

    (1) **Huntington's chorea** is characterized by continuous, uncontrollable, quick movements of the limbs.

    (2) In **athetosis** the limbs, fingers, and hands perform continuous, slow, irregular, twisting motions.

    (3) **Dystonia** is typified by, twisting, tonic-type movements of the head and trunk.

  d. Lesions within the substantia nigra produce **Parkinson's disease**.

    (1) This disease is characterized by **rigidity, hypokinesia** (reduction in voluntary movement), and **tremor**.

      (a) The rigidity in Parkinson's disease involves all of the muscles at a joint and, thus, is different from spasticity that is associated with cortical lesions. It has been described as **lead-pipe rigidity** because the rigid limb, when moved, remains where it is placed. The rigidity seen in Parkinson's disease also has been described as **cogwheel rigidity**. When an examiner tries to move the limb, the limb periodically gives way and then reestablishes its resistance to movement like cogs on a wheel.

      (b) The hypokinesia is not due to a loss of muscle strength or power because normal movements occur under certain conditions. It also is not caused by rigidity because some cases of hypokinesia occur in the absence of any rigidity. The hypokinesia reduces the movement patterns normally associated with motor activity. For example, the arms do not swing when walking, and facial expressions do not change during conversation.

      (c) The tremor associated with Parkinson's disease occurs at a frequency of about 4–7 cycles/sec. This is slower than the normal physiologic tremor, which has a frequency of about 10 cycles/sec. The tremor in Parkinson's disease occurs at rest and usually disappears during voluntary activity.

    (2) The lesion of Parkinson's disease involves a pathway that uses **dopamine** as its neurotransmitter. Some success has been acheived in treating Parkinson's disease with L-**dopa**, a precursor of dopamine that can cross the blood-brain barrier. Dopamine inhibits the striatum sufficiently to reduce some of the clinical signs of the disease.

E. THE CEREBELLUM is intimately associated with all aspects of motor control. Its removal produces no deficits in emotional or intellectual function but causes profound disturbances in the ability to produce smooth, coordinated movements. The functions of the cerebellum appear to be related to the control of the timing, duration, and strength of a movement.

  1. Anatomy

    a. Two transverse fissures divide the cerebellum into three lobes: the **anterior, posterior**, and **flocculonodular lobes**. These lobes have developed at different times during evolution.

      (1) The oldest lobe, the flocculonodular lobe, is called the archicerebellum.

      (2) Next to evolve was the anterior lobe, which is called the paleocerebellum.

      (3) The last lobe to evolve was the posterior lobe, which is named the neocerebellum.

    b. Another description of the cerebellum is based on the connections it makes with other components of the motor control system.

      (1) The entire anterior lobe and those parts of the posterior lobe that receive information from the spinal cord are called the **spinocerebellum**.

      (2) The remainder of the posterior lobe receives its input from the cerebral cortex and, thus, is called the **cerebrocerebellum**.

      (3) The flocculonodular lobe is functionally related to the vestibular apparatus and so is also called the **vestibulocerebellum**.

  2. Lesions in the various divisions of the cerebellum produce characteristic motor deficits.

    a. Lesions in the vestibulocerebellum cause deficits related to the loss of vestibular function, principally, a loss of equilibrium and an ataxic gate. Individuals with this type of lesion are unable to maintain their balance and, so, tend to fall over when standing. When walking, these individuals keep their legs far apart and stagger from place to place.

b. Lesions in the spinocerebellum have no obvious effects in humans, most likely because spinocerebellar functions can be assumed by the neocerebellum. In cats, lesions in the anterior lobe cause an increase in the tone of the antigravity muscles.

c. Lesions in the cerebrocerebellum cause small motor deficits unless a wide area of the cerebellar cortex is affected. If the outflow pathways are damaged, however, the ability to produce smooth, coordinated movements is lost.

   (1) A major sign of neocerebellar disease is **decomposition of movement.** Instead of all the muscles acting in a coordinated way to produce a movement, the muscles are used one at a time. For example, when reaching for an object, the arm first extends at the shoulder, followed by extension at the elbow, and finally by the movement of the hand.

   (2) Dysmetria (the inability to stop a movement at the appropriate time or to direct it in the appropriate direction) is another sign of neocerebellar disease. These effects result from the loss of the neuronal circuitry required to control the duration and strength of a movement.

   (3) The **intention tremor** that results from neocerebellar lesions also is related to the inability to time and sequence movements properly. The intention tremor is different from the resting tremor of Parkinson's disease and appears to occur because an entire movement cannot be directed by a single motor command. Instead, the movement is directed partway to the target and then halted. Several other motor commands, each directing the movement closer to the target, are required before the movement is completed.

   (4) Adiadochokinesia is the inability to make rapidly alternating movements such as turning the hands back and forth. This, too, appears related to the inability to time the duration of a movement, which occurs as a consequence of lesions within the neocerebellum.

F. THE CEREBRAL CORTEX contains the neuronal circuits responsible for the conception and generation of motor commands. Two parallel systems of descending pathways originate in the cerebral cortex: the **pyramidal** and **extrapyramidal** systems. Clinically, these systems are considered together because lesions within the cortex almost always involve both of them. These systems are functionally different, however, and should be considered separately.

1. Physiologic Anatomy.

   a. Several areas within the cortex are responsible for generating motor commands.

   (1) The primary motor area occupies the frontal cortex, just in front of the central sulcus. It is **topographically organized**, with the cortical areas that control the muscles of the hands and face occupying more space than the cortical areas that control the muscles of the limbs and trunk.

   (2) The supplementary motor area is located on the medial surface of the cortex, and it, too, has a complete representation of the entire body.

   (3) The secondary motor area occupies part of the parietal lobe, just across the central sulcus from the region of the primary motor area that controls the facial muscles. The neurons in this area affect movements on both sides of the body.

   b. All of these cortical areas send projections to both of the descending motor systems.

   (1) The axons of the pyramidal system travel directly from their origin in the cortex to their destination in the spinal cord.

   (a) However, they also send collateral fibers to all areas of the motor control system and, thus, communicate their motor command to the basal ganglia, cerebellum, and brain stem.

   (b) The cortical cells that synapse directly on cranial nerves controlling facial muscles perform the same function as pyramidal tract neurons and, thus, are considered to be part of the pyramidal system.

   (c) The pyramidal tract system is responsible for controlling muscles that make precision movements. These include the muscles that move the fingers and hands and the muscles that produce speech.

   (2) The axons of the extrapyramidal system end on relay neurons within the brain stem. These neurons influence the spinal cord through the reticulospinal tracts.

2. Lesions.

   a. Lesions in the pyramidal system cause relatively small motor deficits considering the importance of this system and the large number of neurons involved (over 1 million in each pyramid). The major deficit resulting from pyramidal tract lesions is weakness and a loss of precision in the muscles controlling fine movements of the fingers.

   b. Lesions within the extrapyramidal system produce more profound conditions. These lesions cause the release of cerebral inhibition of the reticular formation, which leads to the production of spasticity [see Section XIII C 1 b (2)].

XIV. SLEEP AND CONSCIOUSNESS. Many of the body's regulatory mechanisms vary in their activity during the day. For example, body temperature is about 1° C higher during the early evening than it is at dawn, and adrenocortical hormones are secreted at levels that are higher in the morning than they are at night. The cycle of these **diurnal** or **circadian rhythms** is roughly 24 hours. However, if an individual is isolated from the environmental stimuli that indicate the normal day-night periods, the cycle time lengthens, demonstrating that the circadian rhythms are not rigidly linked to the rotation of the earth but can be driven by an individual's internal **biological clock**. It is necessary to understand these normal variations in physiologic activities when evaluating pathologic functions. For example, it would be misleading to compare fevers or blood cortisol levels obtained at different times of the day. The most obvious, and probably most important, diurnal rhythm is the sleep-wake cycle.

A. ASSESSMENT OF SLEEP STATES. When awake, an individual is able to perform all activities that are required for individual and species survival (e.g., eating, drinking, learning, and procreating). When asleep, an individual is not aware of the environment and is unable to perform activities that require consciousness. The presence of sleep can be assessed by behavioral analysis. An individual who does not move and does not respond when spoken to or touched probably is asleep. However, a more accurate assessment of sleep can be obtained from an **EEG**.

1. Obtaining the EEG.
    a. The EEG is obtained by placing electrodes on the scalp. The location of the electrodes and the amplification and paper speed of the polygraph used for recording the EEG are standardized. (The same is true for the EKG, which measures cardiac electrical activity; see Chapter 2, "Cardiovascular Physiology," Section III E.)
    b. The brain waves recorded by an EEG represent the summated activity of millions of cortical neurons. The IPSPs, EPSPs, and the passive spread of electrical activity into the dendrites of these neurons, rather than their action potentials, form the basis for the EEG.

2. Variations in the EEG During Sleep and Wakefulness. The two states of sleep—slow-wave and fast-wave sleep—have characteristic EEG patterns.
    a. When an individual is awake, the electrical activity recorded from the brain is asynchronous and of low amplitude (Fig. 1-59). This type of brain electrical recording is called a **beta wave**.
    b. If an individual sits quietly for a while, the brain waves gradually become larger and highly synchronized. The typical resting EEG pattern has a frequency of 8–13 cycles/sec and is called an **alpha wave**. When the eyes open or when conscious mental activity is initiated, the EEG shifts from an alpha to a beta pattern. This is called **alpha blocking**.
    c. As consciousness is reduced still further, an individual enters a state of sleep called **slow-wave sleep (SWS)**. SWS progresses in an orderly way from light to deep sleep (see Fig. 1-59).
        (1) Light sleep is characterized by an EEG that shows high-amplitude waves of about 12–15 cycles/sec, called **sleep spindles**, which periodically interrupt the alpha rhythm.
        (2) During moderate sleep, the EEG displays slower and larger waves called **theta waves**.
        (3) The deepest stage of SWS produces an EEG pattern with very slow (4–7 cycles/sec) and large waves called **delta waves**.
        (4) Behaviorally, SWS is characterized by a progressive reduction in consciousness and an

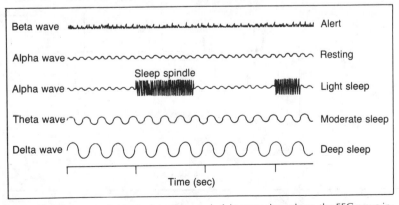

Figure 1-59. As an individual passes from wakefulness to deep sleep, the EEG wave increases in amplitude and decreases in frequency. Sleep spindles indicate the presence of light sleep.

increasing resistance to being awakened. Muscle tone is reduced, the heart and respiratory rates are decreased, and, in general, body metabolism is slowed.
  d. The other type of sleep, called **fast-wave** or **desynchronized sleep**, is characterized by the same high-frequency and low-amplitude EEG pattern that is seen in the waking state. In this case, however, the individual clearly is unresponsive to environmental stimuli and, thus, is asleep. For this reason, fast-wave sleep also is called **paradoxical sleep**.
    (1) Because this state of sleep is characterized by the presence of rapid eye movements, it also is called **rapid eye movement (REM)** sleep.
    (2) Dreaming also occurs during REM sleep, and so the name **dream sleep** also is applied to it.
    (3) Behaviorally, REM sleep is quite different from SWS.
      (a) It is as difficult to arouse an individual from REM sleep as it is from deep sleep. However, when awakened from REM sleep, the individual is immediately alert and aware of the environment.
      (b) The eyes are not the only organ that is active during REM sleep. The middle ear muscles are active, penile erection occurs, heart rate and respiration become irregular, and there are occasional twitches of the limb musculature. Fortunately, muscle tone also is reduced tremendously during REM sleep so the frequency and intensity of muscle twitching do not produce injuries or awaken the individual.

B. THE SLEEP CYCLE. There is an orderly progression of sleep stages and states during a typical sleep period (Fig. 1-60).

1. When an individual falls asleep, the light stage of SWS is entered first. During the next hour or so, the individual passes into progressively deeper stages of sleep until deep sleep is reached. After about 15 minutes of deep sleep, the depth of sleep starts to decrease and continues to do so until the individual reenters the light stage of sleep (about 90 minutes after the start of the first sleep cycle). At this point, the individual passes from SWS to REM sleep.

2. This cycle repeats itself about five times during the night. However, as can be observed in Figure 1-60, after the second cycle, the intervals between periods of REM sleep become shorter and the duration of each period of REM sleep becomes longer. Also, as morning is approached, an individual spends less time in the deeper stages of SWS and periodically is awakened.

3. The sleep cycle indicated in Figure 1-60 is typical of an adult. The cycle varies greatly with age.
  a. During infancy, about 16 hours of every day are spent asleep. This drops to 10 hours during childhood and to 7 hours during adulthood. Elderly individuals spend less than 6 hours of each day sleeping.
  b. It is interesting to note that prematurely born infants spend about 80 percent of their sleep time in REM sleep, whereas full-term infants spend about 50 percent (8 hours) in REM sleep. The total time spent in REM sleep is reduced to about 1.5–2 hours by puberty and remains unchanged thereafter.
  c. During infancy and childhood, therefore, the reduction in sleep time from 16 to 10 hours

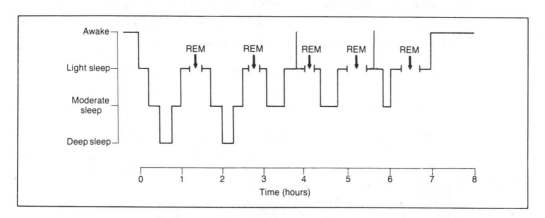

Figure 1-60. Diagram indicating the pattern of sleep during one sleep cycle. As the evening progresses, the depth of slow wave sleep decreases and the duration and frequency of rapid eye movement (*REM*) sleep episodes increase. Note that occasional periods of wakefulness occur during the night.

occurs almost entirely by a reduction of the amount of time spent in REM sleep. In adulthood, the reduction in sleep time is caused by a reduction in the time spent in the deep stages of SWS.

C. PHYSIOLOGIC BASIS FOR SLEEP. Areas throughout the entire brain participate in the sleep-wake cycle.

1. **The waking state** is maintained by the ascending reticular activating system, which is a diffuse collection of neurons within the medulla, pons, midbrain, and diencephalon. Electrical stimulation anywhere within this area causes the EEG pattern to change abruptly from that of the sleep state to that of the waking state. This is called a **cortical alerting** or **arousal** response.
   a. A lesion blocking the connection between the ascending reticular activating system and the thalamus produces a permanent state of sleep or **coma**. In this situation, stimulation of sensory pathways can cause a momentary desynchronization of the EEG but does not produce any behavioral signs of arousal.
      (1) Coma is not simply a deep sleep state. It is characterized by a loss of consciousness from which arousal cannot be elicited.
      (2) Brain O_2 consumption is reduced during coma. This is in marked contrast to normal sleep, in which there is no change in brain O_2 consumption from the waking state.
   b. A transient pathologic loss of consciousness is called **syncope** (fainting). More persistent losses of consciousness are called **stupor**, from which arousal can be obtained.
   c. **Brain death** occurs when the brain no longer can achieve consciousness. Because of the desire to obtain organs for transplant operations and the desire to remove heroic life support systems, an objective standard for determining the presence of brain death has been developed.
      (1) Brain death is said to occur when a loss of consciousness is accompanied by a flat EEG (i.e., an EEG with no brain waves) and a loss of all brain stem regulatory systems (e.g., those systems that control respiration and blood pressure).
      (2) Moreover, these clinical signs must be due to traumatic or ischemic anoxia and not to hypothermia or metabolic poisons, from which later recovery is possible.
      (3) Finally, the criteria for brain death must be present for 6–12 hours.

2. **The sleep state** does not result from the passive withdrawal of arousal. Two sleep centers exist in the brain stem; one is responsible for producing SWS, and the other produces REM sleep.
   a. The SWS center is located in a midline area of the medulla containing the **raphe nuclei**.
      (1) The neurons within these nuclei use serotonin (5-hydroxytryptamine) as a neurotransmitter.
      (2) Administration of serotonin directly into the cerebral ventricles of experimental animals induces a state of SWS, whereas lesions in this region induce a permanent state of insomnia.
   b. The REM sleep center is located in specific nuclei of the pontine reticular formation, including the **locus ceruleus**, which uses norepinephrine as a neurotransmitter. Lesions within this area eliminate the electrophysiologic and behavioral signs of REM sleep.

3. **Sleep Disorders.** As noted previously, there is a cycling between SWS and REM sleep during a normal sleep period.
   a. In **narcolepsy**, REM sleep is entered directly from the waking state.
      (1) Individuals suffering from narcolepsy often report an intense feeling of sleepiness just prior to an attack, although sleep sometimes occurs without warning.
      (2) In some narcoleptics, the profound reduction in muscle tone characteristic of REM sleep can occur without loss of consciousness. During such an attack, called **cataplexy**, the individual suddenly becomes paralyzed, falls to the ground, and is unable to move.
      (3) Another symptom associated with narcolepsy is the presence of a dream-like state during wakefulness, which narcoleptics describe as a hallucination.
   b. Most of the other symptoms of sleep disorders are associated with SWS. These include sleepwalking (**somnambulism**), bed-wetting (**nocturnal enuresis**), and nightmares (**pavor nocturnus**), all of which occur during stages of SWS.
      (1) During a nightmare that occurs in SWS sleep, the individual wakes up screaming and appears terrified. However, no reason for the acute anxiety is recalled.
      (2) By contrast, terrifying dreams that occur during REM sleep are graphically remembered.

STUDY QUESTIONS

Directions: Each question below contains five suggested answers. Choose the **one best** response to each question.

1. All of the following characteristics of neuromuscular synaptic transmission are also true of sympathetic postganglionic synaptic transmission EXCEPT

(A) the release of neurotransmitter requires the entry of Ca^{2+} into the presynaptic terminal
(B) the release of neurotransmitter from the presynaptic terminal is accomplished by exocytosis
(C) the magnitude of the postsynaptic response is proportional to the amount of neurotransmitter released
(D) the channels activated by the neurotransmitter are not activated by changes in membrane potential
(E) the termination of the postsynaptic response is accomplished by enzymatic degradation of the neurotransmitter

2. Which of the following statements best characterizes osmosis?

(A) Water moves from an area of high osmotic pressure to an area of low osmotic pressure
(B) A 1 mmol/L glucose solution produces the same osmotic pressure as a 1 mmol/L sodium chloride solution
(C) Solutions that have the same osmolarity also have the same tonicity
(D) The colloid oncotic pressure causes the osmotic flow of water from the interstitial fluid into the capillaries
(E) A saline solution is hypotonic if an erythrocyte shrinks when placed in it

3. Which of the following receptor properties is related most to the detection of pressure?

(A) High sensitivity
(B) Slow rate of adaptation
(C) Small receptive field
(D) High conduction velocity
(E) Complex encapsulation

4. The central nervous system and the neuromuscular junction have in common which of the following characteristics of synaptic transmission?

(A) The postsynaptic response can be reduced by presynaptic inhibition
(B) Summation is required to drive the postsynaptic cell to threshold
(C) Excitatory postsynaptic potentials are generated by opening channels through which both Na^+ and K^+ can flow
(D) Inhibitory postsynaptic potentials are generated by opening channels through which Cl^- can flow
(E) The synaptic transmitter agent is removed from the synaptic cleft by being actively transported into the presynaptic terminal

5. An eye with a refractive power of 50 diopters has a focal length of

(A) 100 mm
(B) 50 mm
(C) 20 mm
(D) 5 mm
(E) 2 mm

6. SWS is associated with

(A) an asynchronous EEG pattern
(B) nightmares
(C) rapid eye movements
(D) occasional twitches of the limb musculature
(E) an irregular heart beat

7. The stretch reflex and the withdrawal reflex have in common which of the following characteristic features?

(A) Irradiation
(B) Afterdischarge
(C) Reciprocal innervation
(D) Several synapses within the spinal cord
(E) Small afferent fibers

8. Increasing the frequency of stimulation causes the force of skeletal muscle contraction to be increased due to an increase in the

(A) intracellular Ca^{2+} concentration
(B) number of cross-bridges binding to actin
(C) duration of the active state
(D) activity of myosin ATPase
(E) number of muscle fibers contracting

9. Movement disorders caused by the release of inhibition are produced by lesions in all of the following parts of the motor control system EXCEPT

(A) the striatum
(B) the globus pallidus
(C) the subthalamic nucleus
(D) the extrapyramidal system
(E) the substantia nigra

Questions 10 and 11

The diagram below shows the electrical recording made by an electrode placed in a cell that has been stimulated to produce an action potential. Points A-E on the diagram represent the phases of the action potential.

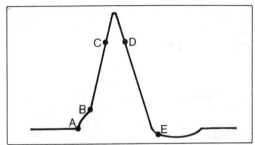

10. At which point on the action potential illustrated above are the greatest number of Na^+ channels open?

(A) A
(B) B
(C) C
(D) D
(E) E

11. At which point on the action potential illustrated above are the greatest number of K^+ channels open?

(A) A
(B) B
(C) C
(D) D
(E) E

Directions: Each question below contains four suggested answers of which **one or more** is correct. Choose the answer

  A if **1, 2, and 3** are correct
  B if **1 and 3** are correct
  C if **2 and 4** are correct
  D if **4** is correct
  E if **1, 2, 3, and 4** are correct

12. The protein components of the cell membrane function as

(1) synaptic receptors
(2) phosphorylation enzymes
(3) carriers in Na^+ transport
(4) membrane channels

13. A decrease in the diameter of the pupil causes an increase in

(1) spherical aberration
(2) visual acuity
(3) sensitivity to light
(4) depth of field

14. The intracellular and extracellular concentrations of four ions are listed in the table below. Which of the ions are in equilibrium when the membrane potential is -60 mV?

Ion	Concentration (mmol/L)	
	Intracellular	**Extracellular**
Cl^-	15	150
Mg^{2+}	15	1.5
K^+	150	15
Na^+	15	150

(1) Cl^-
(2) Mg^{2+}
(3) K^+
(4) Na^+

SUMMARY OF DIRECTIONS

A	B	C	D	E
1,2,3 only	1,3 only	2,4 only	4 only	All are correct

15. An increase in the afterload of an isotonically contracting muscle causes an increase in the

(1) speed of shortening
(2) force of muscle contraction
(3) amount of shortening
(4) interval between stimulation and the initiation of shortening

16. Statements correctly describing the generation of pain sensation include

(1) stimulation of unmyelinated pain fibers causes a dull, burning, poorly localized pain sensation
(2) stimulation of nonpain fibers can reduce the sensation of pain
(3) stimulation of pain fibers within the viscera causes painful sensations that appear to originate on the skin
(4) stimulation of thermoreceptors can cause the sensation of pain

17. True statements concerning auditory function include which of the following?

(1) Most of the sound energy entering the cochlea travels through the helicotrema to the scala tympani
(2) All frequencies of sound in the audible range (20–20,000 Hz) are heard equally well
(3) The middle ear reflexes are essential to protect the ear from damage due to loud noises
(4) High-frequency sounds cause maximal displacement of the basal end of the basilar membrane.

18. In comparison to motor units that fire later, motor units that fire at the beginning of a movement are

(1) able to be tetanized at a lower frequency of stimulation
(2) able to generate greater amounts of force
(3) more dependent on oxidative metabolism
(4) more easily fatigued

19. Motor signs attributable to cerebellar lesions include the inability to

(1) perform rapidly alternating hand movements
(2) coordinate the activity of the shoulder and arm muscles when reaching for an object
(3) stop the arm from moving after it reaches its target
(4) keep the arm from moving spontaneously

20. Lesions in the cortical auditory area are likely to cause serious impairment of the ability to

(1) identify sound patterns
(2) discriminate between pure tones of different frequencies
(3) localize sounds in space
(4) detect loud sounds

21. The neurons of the gamma loop include the

(1) gamma motoneuron
(2) alpha motoneuron
(3) Ia afferent fiber
(4) Ib afferent fiber

22. The accommodation reflex is invoked to bring near objects into focus. Components of this reflex include

(1) relaxation of the ciliary muscles
(2) dilation of the pupils
(3) increase of the focal length of the eyes
(4) convergence of the eyes

ANSWERS AND EXPLANATIONS

1. The answer is E. (*III A, B*) The mechanism of transmitter release is the same at all synapses. Also at all synapses, the postsynaptic membrane is chemically, not electrically, excitable, and the magnitude of the membrane response is proportional to the amount of transmitter binding to the postsynaptic receptors. Acetylcholine, the neuromuscular transmitter, is the only transmitter that is enzymatically degraded. The sympathetic transmitter, norepinephrine, is inactivated by removal from the synaptic cleft.

2. The answer is D. (*I B 3*) Water moves from an area of low osmotic pressure to an area of high osmotic pressure. Osmotic pressure depends on the number of dissolved particles. Since a molecule of sodium chloride dissolved into two particles, it has twice the osmotic pressure of glucose. Tonicity refers to the ability of a dissolved particle to cause the net movement of water by osmosis, whereas osmolarity depends on only the number of particles dissolved. The colloid oncotic pressure (i.e., the osmotic pressure produced by the plasma proteins) drives water from the interstitial fluid into the capillaries.

3. The answer is B. (*VI A 1–3*) In order to detect pressure, a receptor must be able to encode the strength of a stimulus. Receptors that adapt slowly can do this. Rapidly adapting receptors stop firing to steady levels of stimulation and, thus, are better able to detect the velocity with which a stimulus is applied. All of the other properties listed are not necessary for pressure detection.

4. The answer is C. (*III A, C*) Presynaptic inhibition occurs only within the central nervous system (CNS); there are no inhibitory potentials (pre- or postsynaptic) generated in skeletal muscle fibers. The only response is an end-plate potential (called an excitatory postsynaptic potential in the CNS), which is generated by opening channels through which Na^+ and K^+ can move. In the CNS, the amount of transmitter released by a single action potential is not sufficient to drive the postsynaptic cell to threshold, so either temporal or spatial summation is required. At the neuromuscular junction, each time the presynaptic neuron fires, enough transmitter is released to bring the skeletal muscle fiber to threshold.

5. The answer is C. (*VIII C 4*) By definition, the power of a lens in diopters is equal to the reciprocal of the focal length in meters. Substituting 50 into the formula yields a focal length of 0.02 m or 20 mm.

6. The answer is B. (*XIV A 2, C 3*) When awakened from a nightmare that occurs during SWS, an individual has a tremendous feeling of dread but no memory of what caused the emotional response. Vivid, terrifying dreams also can occur during REM sleep, but when awakened from these, an individual can accurately recall the subject matter of the dream. REM sleep is so-named because of the rapid-eye movements that occur during this sleep state. In addition, the muscles occasionally twitch and the heart beats irregularly during REM sleep.

7. The answer is C. (*XII A 1 a–b, B 1 a–b*) Reciprocal innervation refers to the organization of reflex pathways by which the antagonist of an activated muscle is inhibited. Reciprocal innervation is a general feature of most reflexes, including the stretch and withdrawal reflexes. Only the withdrawal reflex is polysynaptic and exhibits afterdischarge and irradiation. The Ia afferent fibers moderating the stretch reflex are much larger than the A delta and C fibers moderating the withdrawal reflex.

8. The answer is C. (*IV A 2 a–b, 3 a–b*) When skeletal muscle is activated, the intracellular concentration of Ca^{2+} becomes 10^{-5} mol/L, which is sufficient to activate all of the muscle protein. The number of cross-bridges binding to actin depends on the preload and is not changed by the frequency of stimulation. The activity of the myosin ATPase and the number of muscle fibers contracting also are not affected by the frequency of stimulation. However, by increasing the frequency of stimulation, Ca^{2+} remains in the cytoplasm longer, increasing the duration of the active state and, thus, increasing the amount of stretch on the series elastic component.

9. The answer is B. (*XIII D 2, F 2 b*) The globus pallidus is one of five nuclei that comprise the basal ganglia; it also is the major outflow tract of this component of the motor control system. Lesions within the globus pallidus result in a reduction of activity in the trunk musculature responsible for maintaining posture. Lesions in all of the other nuclei as well as lesions in the extrapyramidal system cause excessive activity in motoneurons due to a loss of inhibition.

10 and 11. The answers are: 10-C and 11-D. (*II C; Figs. 1-3 and 1-8*) The m gates, which cover the extracellular sides of the Na^+ channels, begin to open at threshold (point A). Most of the m gates are opened during the upstroke (point B to point C). At the peak of the action potential, the h gates, which cover the intracellular sides of the Na$^+$ channels, are beginning to close the Na$^+$ channels. The K$^+$ channels are opened by the n gates near the peak of the action potential. Even though the K$^+$ transference is greatest during the hyperpolarization phase (point 5), more K$^+$ channels are opened during the downstroke of the action potential (point 4).

12. The answer is E (all). (*I B 2*) The proteins contained in the cell membrane perform many physiologic functions. Although proteins that act as receptors and enzymes can be confined to the extracellular half of the membrane bilayer, other membrane proteins can span the entire cell membrane.

13. The answer is C (2, 4). (*VIII E 3*). When the diameter of the pupil is reduced, spherical aberration is decreased because light enters the eye only at the center of the lens, where it is most free from optical defects. Thus, visual acuity and depth of field are increased. Because less light enters through the smaller pupil, sensitivity to light is decreased.

14. The answer is B (1, 3). (*II B 1*) The Nernst potential for each ion can be calculated using the given concentrations. Solving the Nernst equation for each of the ions yields the following values: Cl$^-$ = -60 mV; Mg^{2+} = -30 mV; K$^+$ = -60 mV; and Na$^+$ = $+60$ mV. When the membrane potential equals the Nernst potential for an ion, no energy is required to keep the concentration gradient from changing, and the ion is in an equilibrium state. When calculating the Nernst potential, it is essential to include the valence of the ion.

15. The answer is C (2, 4). (*IV A 3 b*) Both the speed and the amount of shortening decrease when the load on a muscle is increased. However, in order to lift the load, the force generated by the muscle must increase. Since more time is needed for the muscle to achieve this force, the interval between stimulation and shortening increases.

16. The answer is A (1, 2, 3). (*VI B 3 a, C 1 a–b, 2 b, 3*) Stimulation of C fibers produces a poorly localized, burning sensation because of the large receptive field and diffuse pathways of these fibers. Stimulation of large sensory fibers through acupuncture, transcutaneous electrical stimulation, or by electrodes placed on the dorsal columns can reduce pain by activating the same pathways activated by morphine and its endogenous agonists. Visceral pain is referred to skin locations, and, therefore, it is important to know these locations when diagnosing visceral disease. Thermoreceptors encode sensation through labeled lines and, thus, cannot produce painful sensations.

17. The answer is D (4). (*VII A 2, C 1–3*) Most of the sound energy entering the cochlea is damped out before it reaches the helicotrema. The energy is transferred directly to the scala tympani through the basilar membrane. The basilar membrane is narrowest and stiffest at the basal end (the end closest to the middle ear) and, thus, vibrates maximally to high tones in this region. Not all tones are amplified equally well by the middle ear, and so some, particularly those in the range of 1000 to 3000 Hz, are heard better than others.

18. The answer is B (1, 3). (*XIII B 1–2*) According to the size principle, the smallest motoneurons are activated before the larger ones. The smaller motoneurons are part of small motor units and innervate muscle fibers that are small, dependent on oxidative metabolism, and resistant to fatigue. Because these muscle fibers also contract slowly, they can be tetanized at lower frequencies of stimulation.

19. The answer is A (1, 2, 3). (*XIII E 2*) The cerebellum is responsible for coordinating the strength, duration, and sequence of a motor act. Thus, lesions to the cerebellum result in deficits in the performance of motor behaviors that require coordination. Although an intention tremor occurs during a movement, no spontaneous movements are associated with cerebellar disease.

20. The answer is B (1, 3). (*VII D 2*) Although all auditory functions are impaired by cortical lesions in the auditory area, the most complex functions suffer the greatest deficits. These include the ability to identify patterns of sound and to localize of the point in space from which a sound originates. Simple functions such as detection of loudness and pure tone discrimination still can be accomplished.

21. The answer is A (1, 2, 3). (*XII C 4*) The gamma loop prevents or reduces the amount of unloading during a muscle contraction. Activation of the gamma motoneuron causes the intrafusal fiber to contract. This stimulates the Ia afferent fiber, which leads to an increase in alpha motoneuron excitability.

22. The answer is D (4). (*VIII E 2 a*) The accommodation reflex has three components. The ciliary muscles contract to increase the refractive power of the lens (and decrease its focal length). The pupils constrict to reduce chromatic and spherical aberration. Finally, the eyes converge by simultaneous contraction of the medial recti muscles.

2

Cardiovascular Physiology

I. THE VESSELS. The major functions of the cardiovascular system are distribution of metabolites and oxygen (O_2) to all body cells and collection of waste products and carbon dioxide (CO_2) for excretion. The cardiovascular system also is involved in thermoregulation and in distribution of hormones to distant sites. The **heart** provides the driving force for this system; the **arteries** serve as distributing channels to the organs; the **veins** collect blood and serve as blood reservoirs; and the **capillaries** are the exchange area of the system.

A. ELASTIC ARTERIES

1. **Structure.** The **aorta** and its large branches—the carotid, iliac, and axillary arteries—are composed of a thick medial layer containing large amounts of **elastin** and some **smooth muscle cells**. The **distensibility** of these vessels enables them to take up the volume of blood ejected from the heart with only moderate increases in pressure.

2. **Function.**
   a. **Compliance.** Elastic arteries serve as a Windkessel (i.e., they accommodate the stroke volume of the heart because they are distensible). Figure 2-1 shows pressure-volume relationships for the aorta in different age groups. These curves are relatively linear in the middle range, where pressure and volume coordinates are normally. Distensibility or **compliance** is reduced as an individual ages (i.e., the vessels stiffen). Thus, for any volume increment into the aorta, there is an increase in the pressure fluctuation. Clinically, this causes an increased pulse pressure in older individuals.
   b. **Elastic Recoil.** During diastole, the elastic recoil provides potential energy to maintain blood flow during this phase of the cardiac cycle.

B. MUSCULAR ARTERIES. Most of the named arteries in the body are muscular arteries.

1. **Structure.** As these vessels extend farther from the heart, their medial layer contains more and more smooth muscle.

2. **Function.** These vessels are distributing channels to the body organs, and they have a relatively large lumen-to-wall thickness ratio. The large lumen minimizes the pressure drop resulting from resistance losses (Fig. 2-2).

3. **Cross-Sectional Area and Velocity of Flow.** The cross-sectional area of the arterial tree increases as it proceeds toward the periphery (see Fig. 2-2). Over time, the same volume of

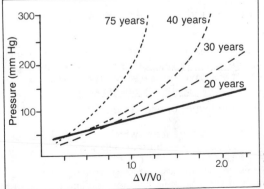

Figure 2-1. Average volume/pressure relationships of arterial systems from 20-, 30-, 40-, and 75-year old individuals. Note the increasing elastance (decreased compliance) that occurs with advancing age. ΔV = volume change; and V_0 = unstressed volume.

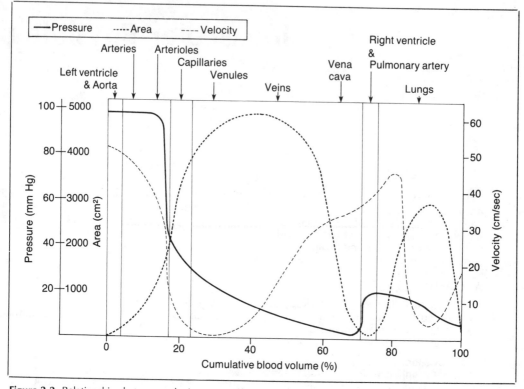

Figure 2-2. Relationships between velocity, area, volume, and pressure in various segments of the cardiovascular system. The same volume of blood must pass through each segment of the system per unit time; therefore, the velocity and area are inversely related. The resistance of the large vessels is minimal, so pressure loss also is minimal until the small arterioles are reached.

blood traverses each class of vessels, making the flow velocity an inverse function of the cross-sectional area.

C. **ARTERIOLES,** the "stopcocks" of the circulation, control the volume of blood flowing to the various vessel beds. The major pressure drop in the cardiovascular system occurs in the arterioles (see Fig. 2-2).

1. **Structure.** Arterioles have a thick layer of **smooth muscle** within the medial coat and a relatively narrow lumen.

2. **Functions.**
   a. **Control of Flow.** Arterioles markedly alter their diameter by contraction or relaxation of the smooth muscle in their walls. The smooth muscle tone varies with the activity of the sympathetic nerves and the local concentrations of metabolites and certain hormones. Variation in the lumen diameter allows fine control of the distribution of cardiac output to different organs and tissues.
   b. **Dampening of Pulsations.** Arterioles convert the pulsatile flow in the arteries to a steady flow in the capillaries.

D. **THE MICROCIRCULATION** consists of vessels smaller than 100 μ in diameter; it includes **metarterioles, arterioles, capillaries,** and **postcapillary venules.** Short, low-resistance connections between arterioles and veins are called **arteriovenous (A-V) shunts**; these occur in some tissues and commonly in the skin, where they function in thermoregulation.

1. **Organization.** The microcirculation is a meshwork of fine vessels that distribute blood to every cell in the body.
   a. **Capillaries** arise directly from arterioles or metarterioles, which are high-resistance conduits between arterioles and veins. A cuff of smooth muscle cells surrounds the origin of many capillaries and is termed the **precapillary sphincter.**
   b. **Vasomotion.** Not all capillaries in a tissue are functional at any moment, and the capillary blood flow is controlled by the precapillary sphincter. The opening and closing of these

sphincters and the resultant alteration in blood flow is termed vasomotion. The sphincters contract in response to high internal vascular pressure and relax in the presence of high CO_2 tension (PCO_2) and low O_2 tension (PO_2).

2. **Function.** Exchange between the **blood** and the **interstitial fluids** occurs in the microcirculation, and the interstitium, in turn, equilibrates with the body cells. Many capillaries arise from each arteriole and metarteriole, providing a total cross-sectional area of 0.4–0.5 m². This large area results in an average blood flow velocity of 0.3–0.4 mm/sec in the capillaries. Within short periods of time in the same capillary, however, the velocity varies widely (i.e., from 0–1.0 mm/sec) due to vasomotion.

3. **Endothelial Structure.** The microcirculatory endothelium varies depending on the function of an organ.
   a. **Fenestrations.** In the renal glomeruli, gastrointestinal tract, and in some glands there are large (200–1000 Å in diameter) transcellular fenestrations through the endothelial cells. These fenestrations are coupled with an incomplete basement membrane. A thin membrane may or may not cover the fenestrations. In the brain, the microcirculatory endothelium has no fenestrations, and a complete basement membrane is present. These structures are morphologic evidence of the blood-brain barrier that retards or prevents the transfer of many substances.
   b. **Gap Junctions.** Large gaps occur between the endothelial cells in the bone marrow, liver, and spleen. The postcapillary venules (20–60 μ in diameter) are the most permeable portion of the microcirculation. Large molecules are likely to leak from this site after the administration of such substances as histamine and bradykinin.
   c. **Tight Junctions.** In most body tissues the clefts between endothelial cells appear fused, but evidence indicates there are pores of about 40 Å in diameter between endothelial capillary cells. In cerebral capillaries these pores are absent, which is further evidence of a blood-brain barrier.

4. **Mechanisms of Exchange.** The microcirculation provides a total surface area of about 700 m² for exchange between the circulatory system and the interstitial compartment.
   a. **Diffusion**, the principal mechanism of microvascular exchange, is the kinetic movement of molecules from an area of high concentration to an area of low concentration.
      (1) **Rate of Diffusion.** Diffusion rate depends on the **solubility** of the substance in the tissues, the **temperature**, and the **surface area** available; it is inversely related to **molecular size** and the **distance** over which diffusion occurs. Molecules such as O_2, CO_2, water, and glucose rapidly equilibrate across the microvascular endothelium. Larger molecules such as albumin cross the endothelial barrier very slowly or are impeded completely due to the fact that the average pore size is smaller than albumin.
      (2) **Diffusion Path.** There is controversy over how most substances are transported. Transportation may be through gap junctions, fenestrae, intracellular vesicles (cytopemphis), or transcellular routes.
   b. **Bulk Flow.** A hydrostatic pressure difference across the endothelium results in the bulk flow of water and solutes from vascular to tissue space, although the retention of large molecules (**proteins**) within the vascular bed is a retarding force (**oncotic pressure**). The factors determining the balance between filtration and reabsorption were originally formulated by Ernst Starling, and the concept is termed **Starling's hypothesis** (Fig. 2-3). Bulk flow of water (Q_w) is expressed as:

$$Q_w = K [(P_c - P_i) - (p_c - p_i)],$$

   where K = the filtration coefficient, which varies with the permeability of the tissues and the available exchange surface area; P_c = the hydrostatic pressure in the capillaries; P_i = the hydrostatic pressure in the interstitium; p_c = the oncotic pressure in the capillary; and p_i = the oncotic pressure in the interstitium.
      (1) The **filtration coefficient** (K) varies primarily with changes in the surface area of the functional microcirculation at any one time. Dilation of arterioles, and especially dilation of precapillary sphincters, can increase the number of functional capillaries (**recruitment**). Normally, only a fraction of capillaries exhibit blood flow at any one time. Conditions such as increased PCO_2, decreased PO_2, increased hydrogen ion (H^+) concentration, and hyperkalemia as well as such substances as histamine, bradykinin, and adenosine all can cause vasodilation and an increase in the number of perfused capillaries.
      (2) The **capillary hydrostatic pressure** (P_c) represents a force that pushes fluid out into the interstitium. P_c depends on the arterial blood pressure, the arteriolar resistance, the state of the precapillary sphincters, and especially on the venous pressure. P_c varies widely but ranges from 30–45 mm Hg in most tissues. If blood flow is present, the P_c must exceed the venous pressure. A low arteriolar resistance allows a greater propor-

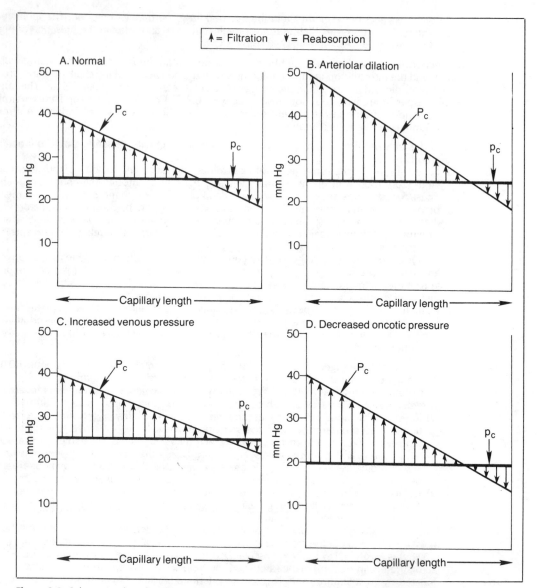

Figure 2-3. Schematic plot of Starling's hypothesis in idealized capillaries. P_c = capillary pressure; and p_c = oncotic pressure of plasma proteins. (A) Normally, filtration slightly exceeds reabsorption, and the difference is removed by the lymphatics. (B) Arteriolar dilation leads to increases in both capillary pressure and filtration forces. It is assumed that the arterial pressure is maintained by vasoconstriction in other vascular beds. (C) An increased venous pressure is totally reflected by an increased capillary pressure. (D) The loss of plasma proteins leads to an excess of filtration and edema formation.

tion of the arterial pressure to be transmitted to the capillaries. For example, in the kidneys, the glomerular P_c is about 50–60 mm Hg, which provides a large force for glomerular filtration. Contraction of the precapillary sphincter can occlude the vascular lumen so that P_c equilibrates with the local venous pressure; under these conditions fluid may be reabsorbed along the entire length of the capillary.

(3) The **interstitial pressure** (P_i) has been shown to be either slightly positive or slightly negative, depending on experimental technique. P_i increases as the interstitial fluid volume increases **(edema)**, which hinders further fluid filtration from the capillary.

(4) The **oncotic pressure of blood** (p_c) is due to the osmotic pressure of blood proteins, which is only a small fraction (0.5 percent) of the total osmotic pressure of blood. The normal p_c is 25–27 mm Hg, a force tending to cause reabsorption of fluid into the capillaries.

(5) Oncotic Pressure of the Interstitium (p_i). Significant amounts of plasma proteins cross the endothelial barrier through either intercellular pores or fenestrae. Protein in the interstitial space produces a force causing outward filtration of fluid from the vascular system. The effective p_i in the interstitium generally is estimated as 5–10 mm Hg.

  c. Cytopemphis. Some substances are transported across the endothelium by incorporation into **cytoplasmic vesicles** that are discharged into the interstitium. Transport of large, lipid-insoluble molecules by cytopemphis has been proposed, but the transport of water and small solutes would be negligible by this mechanism.

E. LYMPHATICS start as closed endothelial tubes that are permeable to fluid and high molecular weight compounds.

  1. Function. The lymphatic system removes from the interstitium albumin and other macromolecules, which have escaped from the microcirculation. Also, lymphatics drain fluid from the interstitium to maintain a gel state.

  2. Mechanism. Lymph is mobilized by the contraction of large lymphatics and contiguous skeletal muscles. The lymphatics have an extensive system of one-way valves to maintain the flow of lymph to the heart.

  3. Flow Rate. Lymphatic flow rate varies in different organs with tissue activity and capillary permeability. Normal lymphatic flow is about 2 L/day for the entire body; this volume contains approximately 200 g of protein that has been lost from the microcirculation. Most lymph comes from the gastrointestinal tract and the liver.

  4. Significance. The lymphatic system represents the only mechanism of returning interstitial proteins to the circulatory system. The concentration of albumin in the blood makes back-diffusion of albumin into the capillaries impossible. Lymph collected from the peripheral tissues is returned to the cardiovascular system via the thoracic duct, which empties into the left subclavian vein near its junction with the left internal jugular vein.

F. VEINS

  1. Structure. Veins are thin-walled structures that contain small amounts of **elastic tissue** and **smooth muscle**. Veins provide a larger cross-sectional area than arteries at equivalent distances from the heart; therefore, the resistance to flow and the velocity of flow are less in veins than in arteries.

  2. Functions.
    a. Conduits. The systemic veins carry blood from the tissues to the right atrium; from the heart, blood passes into the lungs for oxygenation. The pulmonary veins carry the effluent blood from the lungs to the left atrium.
    b. Reservoir. Veins serve as a fluid reservoir; at any one time, 65–75 percent of the circulating blood volume is in the veins.
    c. Venoconstriction. The smooth muscle of veins constricts in response to **sympathetic stimulation**. This constriction displaces blood toward the heart and raises the ventricular filling pressure, increasing cardiac output by raising ventricular end-diastolic volume and ventricular stroke volume.

II. THE HEART

A. MYOCARDIAL SYNCYTIUM. The myocardium is composed of discrete cardiac muscle cells in contact at the intercalated discs. The intercalated discs are low-resistance pathways between the cells; depolarization of one cell is transmitted to adjacent cells; thus, the myocardium is a functional syncytium. Actually, the heart is two syncytia; the atria comprise one and the ventricles comprise the other. These two syncytia are separated by the fibrous atrioventricular (AV) valve ring, which insulates the electrical events of the atria from those of the ventricles. Normally, the only functional connection between atria and ventricles is the AV node.

B. ELECTRICAL ACTIVITY

  1. Resting Membrane Potential (RMP).
    a. Magnitude. A voltage difference of 60–90 mV exists between the inside and the outside of a myocardial cell; inside is negatively charged with respect to ground.
    b. Mechanism.
      (1) RMP is caused by:
        (a) The difference in **membrane permeability** to sodium ion (Na^+) and potassium ion (K^+)

(b) The differences in the **concentration** of these ions, which exist between the inside and the outside of the cell

(2) The concentration differences are maintained by an electrogenic ion pump involving the membrane-bound, magnesium ion (Mg^{2+})-activated Na^+-K^+ adenosine triphosphatase (Na^+-K^+-ATPase).

(3) The resting cell can be considered a K^+ battery; the RMP of a resting cell is 90 mV, and the K^+ equilibrium potential is 96 mV. The difference between the actual membrane potential and the K^+ equilibrium potential is due to the permeability of the membrane to other ions, primarily Na^+.

2. Threshold Potential. Excitable cells, such as muscle and nerve cells, undergo rapid depolarization if the membrane potential is reduced to a critical level, called the **threshold potential**. Once the threshold potential is reached, the remainder of the depolarization is spontaneous.

3. Action Potential (AP). Myocardial cells are divided into slow and fast fibers, depending on the configuration and the propagation velocity of the AP (Fig. 2-4).

a. Slow fibers normally are present only in the sinoatrial (SA) and AV nodes, but fast fibers can be converted into slow fibers by the effects of hypoxia **(ischemia)** or by tetrodotoxin, which blocks Na^+ channels.

(1) The **RMP** of slow fibers is about 50–60 mV.

(2) The **AP** of slow fibers results from the inward ionic conductance through slow Na^+ and calcium ion (Ca^{2+}) channels. The upstroke of the AP occurs in approximately 100 msec in slow fibers, compared to 1 msec in fast fibers.

(3) **Conduction velocity** is a function of the RMP and the rate of depolarization. The AP of slow fibers has a conduction velocity of 0.02–0.1 m/sec.

(4) The **relative refractory period** may last for several seconds in slow fibers. A stimulus applied during this period must be stronger than normal to elicit an AP. This AP has an even slower conduction velocity and a lower amplitude than normal, which frequently results in conduction block within the AV node so that all atrial depolarizations do not reach the ventricles. The **absolute refractory period** may continue even after the RMP has stabilized.

b. Fast fibers include the normal atrial and ventricular myocardial cells as well as the specialized conducting tissues in the heart.

(1) The **RMP** of fast fibers is 80–90 mV.

(2) The **AP** of fast fibers rises rapidly at a rate directly related to the RMP. The duration of the AP varies in different fast fibers and is longest in the Purkinje and bundle of His fibers. This property provides the heart with some protection against reentry phenomena and ventricular fibrillation [see Section II C 5 c (1)].

(3) **Conduction velocity** varies from 0.3–1.0 m/sec in myocardial cells to approximately 4 m/sec in Purkinje fibers. The fast conduction velocity ensures that the entire myocardium is depolarized nearly instantaneously, which improves the efficiency of myocardial contraction.

(4) The **absolute refractory period** in fast fibers lasts until the repolarization has reached a membrane potential of 50–60 mV. The **relative refractory period** ends when the RMP is achieved.

c. Phases of the AP. The AP is divided into five phases (see Fig. 2-4).

(1) Phase 0. The resting membrane is relatively impermeable to Na^+. Theoretically, the cell membrane contains specific channels for certain ions. The RMP controls the Na^+

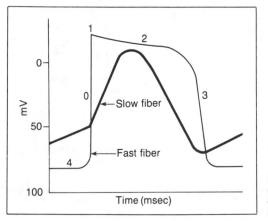

Figure 2-4. Schematic plot of the action potential from a slow and a fast fiber, with the different phases of the action potential represented. Note the diastolic depolarization and the slow phase 0 in the slow fiber.

channels; when the RMP is reduced to threshold, the Na$^+$ channels open and allow Na$^+$ to diffuse rapidly into the cell along its electrochemical gradient. This process is self-regenerative (i.e., inward movement of Na$^+$ reduces the membrane potential, opening more Na$^+$ channels and so allowing more Na$^+$ to enter the cell). The movement of Na$^+$ is a current-carrying process and results in rapid depolarization (phase 0) in fast fibers. In slow fibers, phase 0 coincides with an increase in the Ca^{2+} conductance, so that Ca^{2+} is the current-carrying ion in these fibers.

(2) Phase 1. During phase 1, the Na$^+$ conductance rapidly decreases due to an inactivation of the Na$^+$ channels, and the membrane conductance for both Ca^{2+} and K$^+$ increases. The overall effect is a small change in the membrane potential toward repolarization, which prevents the membrane potential from reaching the Na$^+$ equilibrium potential of +40 to +60 mV.

(3) Phase 2. The **plateau** of the AP in fast fibers coincides with an increased membrane conductance for Ca^{2+}. The movement of Ca^{2+}, a divalent cation, into cells prevents a further repolarization and is important in the excitation-contraction process. The Purkinje fibers exhibit the longest plateau (phase 2) of any cardiac tissues, which also provides them with the longest refractory period.

(4) Phase 3. The phase of **rapid repolarization** is due to a decrease in the inward Na$^+$ and Ca^{2+} currents and a large increase in the outward K$^+$ current.

(5) Phase 4. Nonpacemaker cells (i.e., normal atrial and ventricular myocardium) exhibit a constant membrane potential during phase 4. In pacemaker tissues, such as SA and AV nodes and Purkinje fibers, a slow **diastolic depolarization** during phase 4 is indicative of automaticity. The diastolic depolarization, also termed a **pacemaker potential**, brings the membrane potential toward threshold. The cell whose membrane potential first reaches threshold is the pacemaker of the heart, and the depolarization is conducted to the remainder of the myocardium. The diastolic depolarization in SA nodal cells is caused by decreased K$^+$ permeability, but pacemaker potentials in Purkinje fibers are due to a slow inward current carried by Ca^{2+}. Normally, the SA node is the pacemaker of the heart, because its cells have the highest automaticity (i.e., steepest slope of phase 4). Suppression of SA nodal activity or an increased slope of phase 4 in other areas of the heart produces an **ectopic pacemaker** (i.e., a pacemaker outside of the SA node).

C. CONDUCTION PATHWAYS

1. The **SA node**, located near the junction of the superior vena cava and the right atrium, is the analog of the sinus venosus in lower vertebrates.

   a. Structure. The SA node consists of two types of cells.

   (1) Pacemaker or **P cells** are ovoid and pale-staining. These cells are thought to perform the actual pacemaker function of the node.

   (2) Transitional or **T cells** are elongated and lie between the P cells and the typical atrial muscle cells.

   b. Determinants of Heart Rate. The normal SA node has an intrinsic firing rate of 90–120 beats/min; the intrinsic rate is approximately 120 beats/min in young individuals and declines with advancing age.

   (1) Temperature. The intrinsic firing rate of the SA node is reduced by cooling and increased by heating.

   (2) Ions. The major effects of ions on automaticity are caused by changes in the concentrations of K$^+$ and Ca^{2+}.

   (a) K$^+$. Hyperkalemia reduces the RMP by decreasing the ratio of K$^+$in/K$^+$out. Hyperkalemia in excess of 8 mEq/L inactivates the membrane Na$^+$ channels, which decreases the AP rate of rise and slows the conduction rate. These changes lead to conduction disturbances (e.g., atrial standstill, AV nodal block, reentry phenomena, and increased ectopic activity), which can lead to ventricular fibrillation. **Hypokalemia** causes a decreased membrane conductance for K$^+$, which reduces RMP; this alters automaticity in the same manner as in cases of hyperkalemia.

   (b) Ca^{2+}. Hypercalcemia slows heart rate and decreases the excitability of cardiac tissues while it increases the tension developed by cardiac muscle. Marked hypercalcemia results in cardiac arrest during systole. **Hypocalcemia** increases the slope of phase 4 of the AP and reduces RMP; both effects increase heart rate.

   (3) Vagus Nerve.

   (a) Effect on Heart Rate. The right vagus nerve densely innervates the SA node and liberates **acetylcholine** (ACh) from its terminals. Normal vagal activity **(vagal tone)** slows heart rate from the intrinsic SA rate of 90–120 beats/min to the actual rate of about 70 beats/min. Stimulation of the left vagus nerve in some species produces AV nodal block.

  (i) Vagal stimulation or the application of ACh results in an increased (i.e., more negative) RMP and a decreased slope of phase 4 of the AP. Both effects are due to an increased membrane permeability for K^+ and lead to a decrease in heart rate.

  (ii) Strong vagal stimulation (e.g, during carotid sinus massage) produces sinus arrest and cardiac standstill, which persist until an ectopic pacemaker begins to function.

(b) Inotropic Effect. ACh increases K^+ conductance in atrial muscle, which shortens phase 2 of the atrial AP. This decreases the amount of Ca^{2+} influx and thus the force of atrial contraction **(negative inotropic effect)**. The vagi do not innervate the ventricles, and injections of ACh do not alter ventricular electrical and mechanical properties.

(4) Sympathetic Nerves.

  (a) Effect on Heart Rate. Stimulation of cardiac sympathetic nerves or the injection of **sympathomimetic drugs** increases heart rate and the automaticity of ectopic sites. These effects are caused by an increase in the slow inward current carried by the Ca^{2+}, which increases the slope of phase 4 depolarization.

  (b) Effect on Conduction Velocity. Sympathetic stimulation increases the amplitude and the slope of phase 0 of the AP; both effects increase the conduction velocity.

  (c) Effect on AP Duration. Sympathetic stimulation causes an increased rate of **K^+ efflux** during phase 2, which shortens the duration of the AP. Abbreviation of electrical systole also causes a shortening of mechanical systole, which is important in maintaining an adequate diastolic filling period for the ventricles during increased heart rates.

  (d) Inotropic Effect. Stimulation of the cardiac sympathetic nerves causes a marked **positive inotropic effect** partly due to the increased Ca^{2+} influx.

2. Interatrial Tract (Bachmann's Bundle). This band of specialized muscle fibers runs from the SA node to the left atrium and causes a more synchronous contraction of both atria, because the conduction velocity through this tract is faster than through the regular atrial muscle fibers.

3. Internodal Tracts. The **anterior, middle**, and **posterior** internodal tracts are bundles of specialized cells that connect the SA and AV nodes. They are more resistant than the atrial myocardium to the effects of hypoxia and hyperkalemia (e.g., hyperkalemia produces atrial standstill, but SA nodal impulses are carried over these tracts to initiate ventricular depolarization).

4. The **AV node** is located just beneath the endocardium on the right side of the interatrial septum near the tricuspid valve. Normally, the AV node is the only pathway for **excitation** to proceed from the atria to the ventricles. Occasionally, an alternate pathway occurs between atria and ventricles and results in the preexcitation (Wolff-Parkinson-White) syndrome. The AV node is richly supplied with fibers from the sympathetic and vagal nerves.

a. Functional Regions.

  (1) The **A-N region** is a transitional area between the atrial muscle fibers and the main part of the AV node. Some P cells exist in this region.

  (2) The **N region** contains the least excitable cells in the heart; these are typical slow-response fibers.

  (a) AV Nodal Delay. The slow (0.02–0.05 m/sec) conduction velocity through the N region delays ventricular depolarization for 100–150 msec after atrial depolarization. This delay provides time for atrial contraction, which enhances ventricular filling especially at rapid heart rates.

  (b) AV Nodal Block. The relative refractory period of cells in the N region extends beyond the end of phase 3 of the AP. A depolarization arriving during the relative refractory period has an abnormally small amplitude and slow conduction. The number of impulses conducted through the AV node per minute, thus, is limited to a maximal rate of about 180 beats during supraventricular tachycardias.

  (c) Retrograde Conduction. Impulses can be conducted from the ventricles to the atria; however, the conduction velocity is slower and AV nodal block occurs at lower rates than with antegrade conduction.

  (3) The **N-H region** is a transitional area between the nodal fibers and the His bundle. The N-H zone contains many P cells and usually serves as the pacemaker during complete AV nodal block.

    b. Autonomic Effects.
        (1) Sympathetic stimulation increases the conduction velocity and shortens the refractory period. These changes enhance conduction through the AV node.
        (2) Parasympathetic (vagal) stimulation slows the conduction velocity and prolongs the refractory period. These effects result from the decreased slow inward Ca^{2+} current caused by ACh.

5. Ventricular Conduction. The bundle of His, located subendocardially on the right side of the interventricular septum, divides into the **right** and **left bundle branches**.
    a. The right bundle branch is located on the right side of the interventricular septum and ramifies into a complex network of Purkinje fibers. A block of the right bundle branch delays activation of the right ventricle and can be detected on an electrocardiogram.
    b. The left bundle branch arises perpendicularly from the bundle of His and perforates the interventricular septum, where it bifurcates into left anterior and left posterior divisions. A block of these fascicles is called a left-anterior or a left-posterior **hemiblock**.
    c. The **Purkinje fibers** ramify throughout the subendocardial layer of both ventricles. These are the largest cardiac cells (70–80 μ in diameter), and they possess the highest conduction velocity in the heart (1–4 m/sec). This rapid conduction of the AP normally ensures nearly synchronous contraction of the ventricles. The Purkinje fibers have pacemaker potential, but the rate of diastolic depolarization is slower than in the AV node which, in turn, is slower than in the SA node.
        (1) Reentry is the basis for many cardiac arrhythmias and extrasystoles. Reentry is defined as reexcitation of an area of the myocardium by a single cardiac impulse.
        (2) The extensive branching of the Purkinje system favors reentry of excitation when partial block exists in some of the terminal fibers.
    d. Ventricular Myocardium. The ventricles are depolarized from endocardium to epicardium. Because the left ventricle has a thicker wall than the right ventricle, the epicardial surface at the base of the left ventricle is the final area of depolarization. The conduction velocity of the AP through the ventricular myocardium is about 0.3–0.4 m/sec.

D. EXCITATION-CONTRACTION COUPLING

1. Structure.
    a. The myofibrils in cardiac muscle are surrounded by an anastomosing network of tubules called the **sarcoplasmic reticulum** (SR). (The SR is much less developed in cardiac muscle than in skeletal muscle.)
    b. The SR contains dilated terminals, called **cisternae**, which are located adjacent to the sarcolemma and the **T tubules**. The T tubules are continuous with the sarcolemma and invaginate into the interior of the cell at the **Z line** of the sarcomere.
    c. The myofibrils are composed of thick and thin filaments called **myosin** and **actin**, respectively. The thin filaments contain two other proteins, **troponin** and **tropomyosin**.
    d. The organization of a sarcomere is depicted schematically in Figure 2-5.

2. Contraction Mechanism.
    a. Effects of Ca^{2+}.
        (1) Between contractions, the sarcoplasmic Ca^{2+} concentration is 10^{-7} mol/L. Each contraction is preceded by an AP that depolarizes the sarcolemma and the T tubules. The AP triggers the release of Ca^{2+} from the SR, and, unlike in skeletal muscle, additional Ca^{2+} diffuses into the cell across the T tubules (slow inward Ca^{2+} current). The AP raises the sarcoplasmic Ca^{2+} concentration to 10^{-6} mol/L.
        (2) At high Ca^{2+} concentrations, four Ca^{2+} bind to each troponin molecule. When troponin is saturated with Ca^{2+}, a conformational change allows cross-bridges to form between actin and myosin. Normally, all of the troponin is not saturated with Ca^{2+} so that factors raising intracellular Ca^{2+} concentration can provide more sites for crossbridge formation (e.g., an increased extracellular Ca^{2+} concentration produces a greater Ca^{2+} influx during the AP and, thus, a more vigorous contraction).
    b. Sympathetic Effects.
        (1) Sympathetic nerve stimulation or the administration of sympathomimetic drugs increases the sarcoplasmic concentration of **cyclic adenosine monophosphate (cAMP)** and results in a more forceful contraction.
        (2) Sympathetic effects increase the slow inward current carried by Ca^{2+}. Norepinephrine also causes a more rapid uptake of Ca^{2+} by the SR, which shortens the duration of the AP and the contractile response.
    c. Digitalis increases the force of cardiac contraction by inhibition of the Na^+-K^+-ATPase exchange pump; this, in turn, increases sarcoplasmic Na^+ concentraion. The increased Na^+ concentration stimulates the Ca^{2+}-Na^+ exchange mechanism, which increases the intracellular Ca^{2+} concentration.

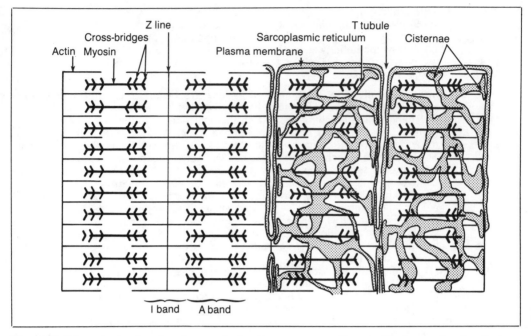

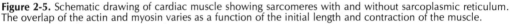

Figure 2-5. Schematic drawing of cardiac muscle showing sarcomeres with and without sarcoplasmic reticulum. The overlap of the actin and myosin varies as a function of the initial length and contraction of the muscle.

d. Length-Tension Relationship (Frank-Starling Effect).

   (1) One of the major determinants of the force of cardiac contraction is the initial **length** or **preload** of the muscle fibers (Fig. 2-6). In the heart, preload is determined by end-diastolic volume.

   (2) An increase in end-diastolic volume elongates the cardiac fibers, exposing more of the myosin heads (cross-bridge sites) to actin.

   (3) Elongation of the sarcomeres also increases the release of Ca^{2+} from the SR. Thus, an increase in preload produces a stronger contraction and a greater work output by the heart.

   (4) The Frank-Starling effect is important in balancing the output of the two sides of the heart over extended periods of time.

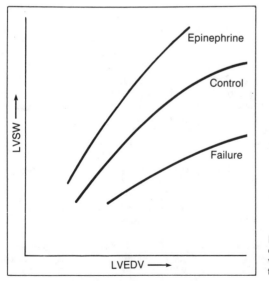

Figure 2-6. Group of Starling curves showing the effects of epinephrine and ventricular failure. *LVEDV* = left ventricular end-diastolic volume; and *LVSW* = left ventricular stroke work.

3. Contractility.

    a. Positive inotropism (increased contractility) is defined as an increase in the force of contraction at a constant preload and contrasts to the Frank-Starling effect, in which an increase in the force of contraction is produced by an increase in preload. Positive inotropic effects are produced by increases in extracellular Ca^{2+} concentration, sympathetic stimulation, glucagon, and digitalis. Figure 2-6 shows a group of ventricular function (Frank-Starling) curves and depicts the ventricular performance as a function of the end-diastolic volume or preload. Positive inotropism causes a shift to the left (i.e., the same cardiac work can be performed at a smaller end-diastolic volume).

    b. Negative inotropism (decreased contractility) is caused by hypoxia, acidosis, infarcts, and myocardial ischemia. Negative inotropism causes a shift to the right in the ventricular function curves (i.e., less work is performed by the heart at any given level of end-diastolic volume). The heart shifts from one curve to another (alters contractility) in response to various physiologic and pathologic events.

III. ELECTROCARDIOGRAPHY. The electrocardiogram (EKG) provides a method of evaluating excitation events, arrhythmias, the presence of ischemia or necrosis, and hypertrophy of the heart. Space does not permit an exhaustive discussion of EKG abnormalities; therefore, this section emphasizes a vectorial analysis of cardiac electrophysiology.

    A. VOLUME CONDUCTION. The body tissues function as a conductor. Electrodes attached to the body surface allow the measurement of voltage changes within the body after adequate amplification. During diastole, cardiac cells are positively charged on the outside, but electrodes on the skin do not detect any voltage changes caused by cardiac activity (Fig. 2-7). Excitation of a portion of the heart reverses membrane potential in this area, and the outside of these cells

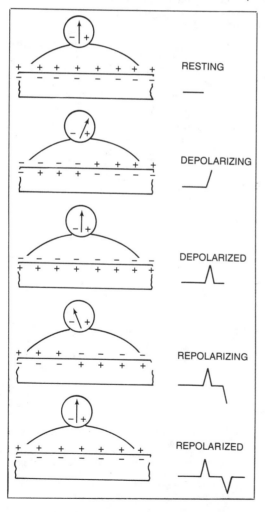

RESTING

DEPOLARIZING

DEPOLARIZED

REPOLARIZING

REPOLARIZED

Figure 2-7. The effects of depolarization and repolarization of isolated, excitable tissue on a galvanometer's deflection, using standardized connections. Note that when repolarization and depolarization are in the same direction, the galvanometer deflections are oppositely directed. In the limb leads of an electrocardiogram, the QRS complex and the T wave normally are in the same direction.

becomes negatively charged with respect to ground. Thus, a potential difference exists between the depolarized cells and the nonexcited cells (see Fig. 2-7). This potential difference can be recorded from surface electrodes and is referred to as the EKG.

1. **Equivalent dipole** is the sum of all of the positive and negative charges within the heart at any instant. The **equivalent** or **cardiac dipole** represents a point source of charge and is a vector because it has both **direction** and **magnitude**. The direction or axis of the cardiac dipole is the line that connects the positive and negative poles. (A vector, by convention, is drawn with its head at the positive pole and its tail at the negative pole.) Depolarization is represented by an arrow pointed in the direction of depolarization; the length of the arrow is determined by the magnitude of the vector.

2. **Surface potential** (i.e., the magnitude of the voltage recorded at the body surface) is a function of the magnitude and direction of the vector as well as the orientation of the vector with the axis of the recording electrodes. By convention, a wave of depolarization approaching a positive electrode produces an upward deflection of the EKG. A wave of depolarization proceeding **parallel** to an electrode axis produces the maximal deflection for that dipole, whereas a depolarization wave **perpendicular** to an electrode axis produces no net deflection of the EKG. Figure 2-7 illustrates these important relationships.

3. **Electrical zero** is the center of Einthoven's triangle (i.e., the intersection of the triaxial and hexaxial reference systems) as well as the electrical center of the heart. The tail of the cardiac vector is located at the electrical zero, with the head pointing in the direction of depolarization.

B. ELECTROCARDIOGRAPHIC LEADS. If electrodes are attached to the left shoulder, right shoulder, and left leg, a triangle is formed (**Einthoven's triangle**). (The right leg serves as a ground conductor.) Lines that bisect each side of the triangle (i.e., at the zero axis of each side) meet at the center of the triangle, at the heart (Fig. 2-8).

1. **Bipolar Leads.** Three leads are formed by measuring the potential differences between any two limb electrodes. These leads are switch-selectable on all standard EKG machines.
   a. Lead I is the potential at the left arm (LA) minus the potential at the right arm (RA) or LA − RA.
   b. Lead II is the potential at the left leg (LL) minus the potential at RA or LL − RA.
   c. Lead III is the potential at LL minus the potential at LA or LL − LA.

2. **Unipolar (V) Leads.** If the three limb leads are connected to a common terminal, the combined potential will be zero. By attaching the common terminal to the negative pole of a galvanometer and a fourth or exploring electrode to the positive pole of a galvanometer, the

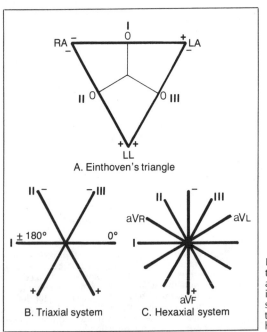

A. Einthoven's triangle

B. Triaxial system

C. Hexaxial system

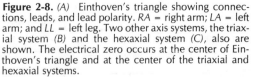

Figure 2-8. *(A)* Einthoven's triangle showing connections, leads, and lead polarity. *RA* = right arm; *LA* = left arm; and *LL* = left leg. Two other axis systems, the triaxial system *(B)* and the hexaxial system *(C),* also are shown. The electrical zero occurs at the center of Einthoven's triangle and at the center of the triaxial and hexaxial systems.

absolute voltages can be recorded from the body surface. This **unipolar electrode** records the precordial or chest leads from standardized sites.

  a. V_1 is in the fourth intercostal space (ICS), just to the right of the sternum.

  b. V_2 is in the fourth ICS, just to the left of the sternum.

  c. V_4 is at the midclavicular line (MCL) in the fifth ICS.

  d. V_3 is halfway between V_2 and V_4.

  e. V_5 is in the anterior axillary line at the same level as V_4.

  f. V_6 is in the midaxillary line at the same level as V_4 and V_5.

3. Augmented Unipolar Leads. Any of the three limb electrodes can be used to record cardiac potentials in comparison to the common terminal. For example, the voltage recorded at RA can be determined by the equation: RA − (RA + LA + LL). The resulting voltage is small, because the potential difference is reduced by the RA potential in the common terminal. Disconnecting the RA lead from the common terminal increases the potential difference by 50 percent and results in the augmented unipolar limb lead, aVR.

  a. **aVR** is the potential difference between RA and (LA + LL).

  b. **aVL** is the potential difference between LA and (RA + LL).

  c. **aVF** is the potential difference between LL and (RA + LA).

  d. Augmented limb lead axes bisect the angles of Einthoven's triangle. The axis of aVR is from $-150°$ to $+30°$; the axis of aVL is from $-30°$ to $+150°$; and the axis of aVF is from $+90°$ to $-90°$.

4. Axial Reference Systems.

  a. A **triaxial reference system** is obtained by moving the sides of Einthoven's triangle without changing their orientation so they intersect at the center of the triangle. Lead I then divides the system into an upper (negative) and a lower (positive) hemisphere (see Fig. 2-8).

  b. A **hexaxial reference system**, with an axis at every 30 degrees, is obtained by superimposing the axes of the augmented unipolar limb leads on the triaxial system (see Fig. 2-8). In this system, lead I and aVF divide the system into quadrants, with the zero isopotential point for all leads at the central intersection.

C. VECTOR LOOPS. The **instantaneous cardiac vector** represents the cardiac dipole generated during the depolarization process. The instantaneous vector originates from the zero isopotential point and inscribes a loop as the tissues are depolarized. Three loops are formed during one cardiac cycle. The **P loop** is caused by atrial depolarization, the **QRS loop** is caused by ventricular depolarization, and the **T loop** is caused by ventricular repolarization (Fig. 2-9). Atrial repolarization is not detectable on surface recordings because of the prolonged time course and the small potentials involved.

  1. The **P loop** normally is small and directed inferiorly toward the left, resulting in a positive deflection (P wave) in the three standard bipolar limb leads. Activation of the SA node or the specialized atrial tracts is not detectable because of the small mass of tissue involved.

  2. The **QRS loop** is inscribed in a counterclockwise direction in the frontal plane, and it is directed inferiorly toward the left and generates the QRS complex on the scalar EKG. The nor-

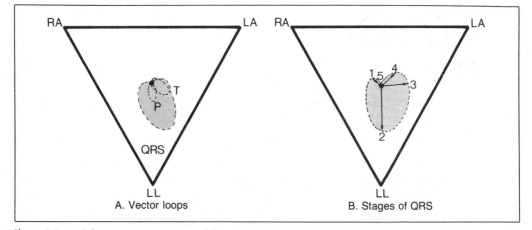

Figure 2-9. *(A)* Schematic representation of the vectorcardiographic loops P, QRS, and T, which can be recorded with appropriate equipment and placement of electrodes. *(B)* A QRS loop that has been arbitrarily divided into five phases.

mal QRS loop is caused by a specific sequence of activation. After the impulse passes through the AV node, it is rapidly conducted by the Purkinje fibers to the endocardial surface of the ventricles. Although it is a continuous process, ventricular depolarization can be divided into five stages (see Fig. 2-9).

  a. **Stage 1.** The initial part of the QRS loop is caused by depolarization of the **interventricular septum**, beginning at the left endocardial surface and spreading superiorly to the right. (Using the clock face to describe the direction in the frontal plane, this initial vector is at about 11 o'clock.) This initial vector produces small negative waves **(Q waves)** in leads I and II and positive waves in the right precordial leads.

  b. **Stage 2.** Next, the depolarization spreads through the remainder of the interventricular septum to the subendocardial layers of both ventricles. This second vector is large, directed inferiorly (at about 6 o'clock), and produces large positive waves in leads II, III, and aVF.

  c. **Stage 3.** During stage 3, the right ventricular wall is completely depolarized; the process continues in the wall of the left ventricle because of its thicker structure. This third vector is directed toward the left (at about 3 o'clock) and produces **R waves** in leads I, II, and aVL as well as the left precordial chest leads.

  d. **Stage 4.** The final area of depolarization is the **epicardial surface** at the base of the left ventricle. This vector is directed toward the left shoulder (at about 1 o'clock) and produces small **S waves** in leads III and aVF.

  e. **Stage 5.** Normally, the vector loop returns to the zero isopotential point at the end of depolarization.

3. **T Loop.** The direction of ventricular repolarization is opposite to that of depolarization, because the inner layers of the myocardium are hypoxic due to the vascular compression occurring during systole. (The hypoxia prolongs the relative refractory period of the endocardial layers and delays the onset of repolarization in these areas.) The first regions to repolarize are the epicardial areas at the apex of the heart. The T wave is the scalar EKG equivalent of the vector T loop. The reversed direction of repolarization compared to depolarization normally produces upright T waves in the bipolar limb leads of the EKG.

D. **FRONTAL, HORIZONTAL, AND SAGITTAL PLANES**

1. **Frontal Plane.** The electrical properties of the heart in the frontal plane can be examined using the six **standard limb leads** (i.e., I, II, III, aVR, aVL, and aVF).

2. **Horizontal Plane.** The **precordial chest leads** (V_1–V_6) provide a visualization of the electrical properties of the heart in the horizontal plane.

3. **Sagittal Plane.** Special electrode positions can be used to examine the electrical properties of the heart in the sagittal plane.

E. **SCALAR ELECTROCARDIOGRAPHY.** Scalar EKGs are tracings of the surface cardiac potentials recorded against time. Unless otherwise indicated, these tracings are made at a standard recording speed (25 mm/sec) and amplification (1 mV = 1 cm deflection). With standard EKG paper, each small horizontal division represents 0.04 second, each large division is 0.2 second, and each small vertical division represents 0.1 mV. The scalar EKG is a reflection of the vector loops on a particular lead. Each lead represents an electrical view of the heart from a different orientation, but all leads reflect the same vector loop in any one individual unless there is a change in the sequence of excitation.

1. **EKG Waves and Intervals** (Fig. 2-10).

  a. The **P wave** normally is positive (upright) in the standard limb leads and inverted in aVR, measures less than 2.5 mm in height, and lasts less than 0.11 second.

    (1) **P Mitrale.** Left atrial hypertrophy, as in mitral stenosis, results in a broad, notched P wave. More time is required to depolarize the hypertrophied left atrium, and so the P wave duration is increased.

    (2) **P Pulmonale.** Right atrial enlargement, which frequently accompanies chronic lung disease, creates tall, peaked P waves. This voltage increase results from the summation of voltage due to depolarization of the enlarged right atrium and the normal left atrium.

  b. The **P-R interval** is measured from the onset of the P wave to the onset of the QRS complex and normally lasts between 0.12 and 0.21 second, depending on heart rate. The P-R interval is a measure of the **AV conduction time**, including the delay through the AV node.

  c. The **QRS complex** normally lasts less than 0.08 second. A prolonged QRS indicates an ectopic ventricular pacemaker or an intraventricular conduction block due to blockage of a bundle branch or fascicle. Capital letters denote waves that are greater than 5 mm in height, and small letters denote waves that are less than 5 mm. For example, a **qRs** com-

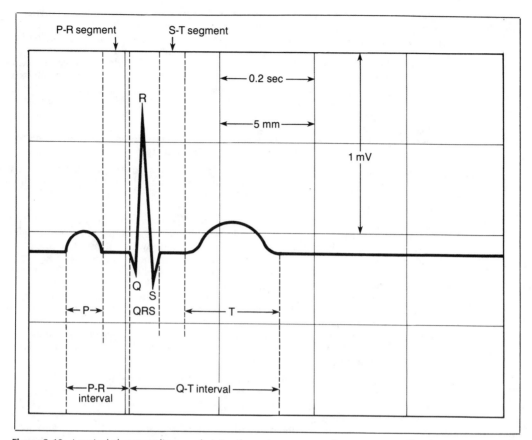

Figure 2-10. A typical electrocardiogram showing the various waves and segments as well as the standard voltages and times.

plex would be characterized by a small initial negative wave, a large positive wave, and a small final negative wave.

   (1) The **Q (q) wave** is an initial negative wave that precedes a positive wave. If the entire QRS complex is negative, it is termed a **QS complex.**

   (2) The **R (r) wave** is the first positive wave; if the deflection becomes negative, the next positive deflection is termed **R′ (r′).** Notching or slurring of the R wave should be indicated.

   (3) The **S (s) wave** is a negative wave following an R wave.

   d. The **T wave** normally is in the same direction as the QRS complex, because repolarization follows a path that is opposite to depolarization. A **vulnerable period** occurs during the downslope of the T wave, when the ventricle is partially repolarized and the myocardial fibers are in a relative refractory state. A ventricular ectopic stimulus at this time will initiate runs of ventricular extrasystoles or ventricular fibrillation due to the likelihood of aberrant conduction and reentry mechanisms within the ventricular myocardium.

   e. The **Q-T interval,** measured from the onset of the QRS complex to the end of the T wave, indicates the duration of electrical systole. The Q-T interval varies with heart rate, and **nomograms** are available to correct the interval for various rates. The corrected Q-T interval should not exceed 0.43 second.

   f. The **P-R segment** extends from the end of the P wave to the onset of the QRS complex and normally is isoelectric.

   g. The **J point** occurs at the termination of the QRS complex, when the entire ventricular myocardium is depolarized. This point normally is isoelectric but is displaced by a current of injury indicating myocardial ischemia, inflammation, or infarction.

   h. The **R-R interval** is the time between successive QRS complexes. Heart rate is equal to 60 divided by the R-R interval in seconds.

   i. The **S-T segment** extends from the J point to the onset of the T wave and is equivalent to phase 2 of the AP. This segment normally is isoelectric but can vary from − 0.5 to + 2.0 mm in the precordial leads. Elevation or depression of the S-T segment indicates myocardial ischemia or strain, and the displacement is due to a current of injury.

   (1) Current of Injury. Because injured muscle has a smaller membrane potential than normal, it is electrically negative with respect to normal myocardium. This difference in potential creates a current during the resting intervals so that the T-P interval (i.e., the time from the end of the T wave to the start of the P wave) recorded from an electrode overlying the injured area is depressed below the true isoelectric line.

   (2) S-T Segment Elevation. When an injured area depolarizes, it is more positive than normal regions and results in the S-T segment elevation in tracings from electrodes overlying the injured area.

   (3) S-T Segment Depression. An electrode on the opposite side of the heart from the injury will reveal a S-T segment depression on this tracing.

2. Derivation of Scalar EKGs. As an example, the QRS complex for the **bipolar limb leads** is constructed below, using the vector loop and Einthoven's triangle. The same approach can be applied to any lead as well as to the P and T loops, if the correct conventions and polarity (see Section III A 1, 2) are used.

   a. Lead I. Figure 2-11 shows the vector loop and the "shadow" it casts on the leads. (The lines bisecting the three sides of the triangle are the zero isopotential lines around which the EKG is inscribed.) The initial vector (stage 1) is directed toward the negative side of lead I, resulting in a small, initial negative (q) wave. The second vector (stage 2), although large, is almost perpendicular to lead I and casts only a small shadow. Vectors 3 and 4 are directed toward the positive pole of lead I and generate the R wave. The terminal vector is directed toward the negative pole of lead I, resulting in a small terminal (s) wave.

   b. Leads II and III. The contour of the QRS complex for leads II and III can be derived in the same manner as for lead I.

   c. Unipolar Leads. QRS complexes for the unipolar leads can be derived by orienting the axes and using the rule that an approaching wave of depolarization generates an upright or positive deflection on the EKG.

3. The **mean electrical axis** (Figure 2-12) or **average vector** is the line that exactly bisects any given vector loop. This axis can be derived by using any two standard limb leads or any two augmented limb leads. An aV and a bipolar limb lead cannot be used together because of the difference in their amplification.

   a. Measurement. Clinically, the mean electrical axis of the QRS complex is determined by algebraically summing the heights of the Q, R, and S waves in each lead, as shown in Figure 2-12. (To be precise, the areas under the three waves should be summed algebraically.)

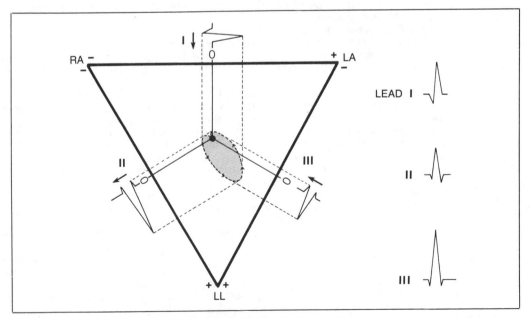

Figure 2-11. Reconstruction of the QRS complex from the depolarization loop for the three bipolar limb leads. *Arrows* indicate the direction of the recording. *RA* = right arm; *LL* = left leg; and *LA* = left arm.

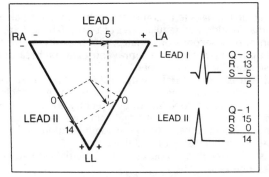

Figure 2-12. Vectorial analysis of mean electrical axis determination. The Q and S waves are summed and added algebraically to the R wave to give the magnitude of the dipole in two leads. The magnitudes are plotted along the respective leads toward the appropriate polarity. Perpendiculars are drawn through the arrow heads, and the intersection marks the head of the mean electrical vector. The vector is drawn from the electrical zero (the center of the triangle) to the intersection of the perpendiculars. The mean electrical axis, in degrees, and the magnitude are given by the direction and length of the mean electrical vector (axis). *RA* = right arm; *LL* = left leg; and *LA* = left arm.

      (1) Plotting the Lead Vectors. The vector from each lead is plotted on an appropriate scale using either Einthoven's triangle or the hexaxial reference system. The tail of the arrow is placed on the zero potential point, and the vector is drawn along the lead, with the head of the arrow pointing to the correct pole. Perpendiculars are drawn to the head of the vectors.

      (2) Drawing the Mean Vector. The head of an arrow is drawn at the intersection of the two perpendiculars, and the tail is at the center of the triangle. The length of this arrow represents the magnitude of the mean electrical axis, and its direction, in degrees, represents the electrical axis in the frontal plane.

   b. Axis Deviation. The mean electrical axis of the QRS complex determines if axis deviation exists.

      (1) Normal Axis. A mean electrical axis of $-15°$ to $+90°$ is normal in an adult individual.

      (2) Right axis deviation (RAD) is defined as a mean electrical axis of $+90°$ to $+180°$. RAD occurs secondary to right ventricular hypertrophy (e.g., pulmonary hypertension, chronic lung disease, and pulmonary valve stenosis) and delayed activation of the right ventricle (e.g., right bundle branch block).

      (3) Left axis deviation (LAD) is defined as a mean electrical axis of $-15°$ to $-90°$. LAD is associated with a horizontal heart due to obesity, left ventricular hypertrophy, or left bundle branch block.

      (4) Indeterminate Axis. A mean electrical axis of $-90°$ to $-180°$ may indicate either extreme left or right axis deviation.

  4. Electrical orientation in the horizontal plane is determined using the precordial chest leads.

   a. The **transition zone** or **null point** is defined as the position in which the QRS complex is biphasic (i.e., when the algebraic sum of positive and negative deflections equals zero). The null point normally occurs between leads V_2 and V_3. A leftward shift in the null point usually is associated with left ventricular hypertrophy; a rightward shift is associated with right ventricular hypertrophy.

   b. The **electrical axis** in the horizontal plane is perpendicular to the null point.

IV. CARDIAC RATE, RHYTHM, AND CONDUCTION DISTURBANCES

A. CARDIAC RATE

  1. Normocardia is defined as a normal heart rate (i.e., 60–100 beats/min).

  2. Tachycardia is defined as a heart rate that exceeds 100 beats/min.

  3. Bradycardia is defined as a heart rate of less than 60 beats/min.

B. CARDIAC RHYTHMS

  1. Sinus rhythm is normal when: the SA node is the pacemaker; each P wave is followed by a normal QRS complex; the P-R and Q-T intervals are normal; and the R-R interval is regular.

   a. Sinus arrhythmia occurs when one criterion, the R-R interval (heart rate), varies with respiration. Normally, heart rate increases during inspiration and slows during expiration due to variations in vagal tone that affect the SA node. Sinus arrhythmia is common in young individuals and in endurance athletes with slow heart rates.

   b. Sinus tachycardia is a normal response to exercise and occurs in the presence of fever and hyperthyroidism and as a reflex response to low arterial pressures.

2. Atrial Rhythms.

a. Atrial tachycardia occurs when an ectopic atrial site becomes the dominant pacemaker.

(1) **Characteristics** of atrial tachycardias include very regular rates of 140–220 beats/min, a very rapid onset, and a duration of only seconds or minutes. Occasionally, an atrial tachycardia may last for hours or days and is termed a **supraventricular tachycardia**, because the actual pacemaker site cannot be determined.

(2) **Cause.** These attacks, termed **paroxysmal atrial tachycardias** (PATs), may be precipitated by overindulgence in caffeine, nicotine, or alcohol or by anxiety attacks. PATs may be caused by discharge of a single ectopic site or by a reentry phenomenon.

(3) **Treatment.** PATs usually are interrupted by increasing vagal tone (via carotid sinus massage or a Valsalva maneuver). Persistent cases may require administration of digoxin or verapamil. Rarely, cardioversion is necessary to end the attack.

b. Premature atrial contractions (PACs) or **atrial extrasystoles** occur if an atrial ectopic site fires and becomes the pacemaker for one beat.

(1) **Significance.** PACs are entirely normal, causing only occasional irregularity in cardiac rhythm.

(2) **Diagnosis.** The condition can be readily diagnosed if it occurs during an EKG, because a premature P wave follows a normal QRS complex and T wave. The premature P wave may have an abnormal configuration; the P-R interval also may change with the altered path of depolarization. The premature impulse discharges the SA node, which then must repolarize and fire after the normal interval. This results in a shift in cardiac rhythm (Fig. 2-13). PACs can be diagnosed at bedside.

c. Atrial flutter occurs when atrial rates are 220–350 beats/min. During atrial flutter, the AV node is unable to transmit all of the atrial impulses. Thus, an AV block develops and the ventricular rate is one-half, one-third, or one-fourth the atrial rate.

(1) **Cause.** Like atrial tachycardia, atrial flutter may be due to a single ectopic focus or to a reentry phenomenon.

(2) **Characteristics.** The ventricular rate may be normal and very regular, although rapid changes in ventricular rate may occur as the AV block changes (e.g., from 4:1 to 3:1). P waves that appear saw-toothed on the EKG are virtually diagnositc of atrial flutter.

(3) **Significance.** Atrial flutter differs from atrial tachycardia only in the atrial rate, and it has a similar prognosis.

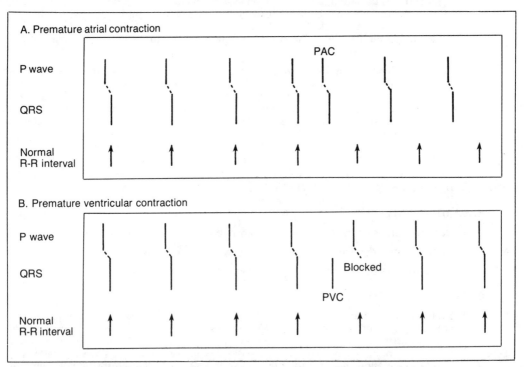

Figure 2-13. The effect of premature contractions on the sequence and timing of atrial and ventricular events. *(A)* A premature atrial contraction *(PAC)* causes a permanent shift in the atrial and ventricular events. *(B)* A premature ventricular contraction *(PVC)* causes no permanent shift in rhythm since the PVC and the subsequent beat typically equal two normal cycle lengths.

d. Atrial fibrillation is a totally irregular, rapid atrial rate, with contraction of only small portions of the atrial musculature at any one time. Few of the atrial impulses reaching the AV node are transmitted to the ventricles, making the ventricular rate completely (irregularly) irregular. The irregularity causes large variations in the aortic pressure so that some of the pulses are not transmitted to the peripheral arteries (pulse deficit).

   (1) Characteristics. The baseline of the EKG shows small, irregular oscillations (F waves) caused by depolarization of small units of the atrial musculature. P waves are not recognizable, and the R-R interval is irregularly irregular. The impulses transmitted through the AV node normally are conducted through the ventricles; thus, the QRS complex and T wave are normal.

   (2) Significance. Atrial fibrillation represents a higher rate of atrial activity than atrial flutter and frequently is associated with an enlargement of the atrium secondary to AV valve disease. Also, cardiac output decreases due to the valve disease and to the loss of an effective atrial contraction. Prolonged atrial fibrillation is associated with thrombosis in the atrial appendages. Atrial thrombi may cause pulmonary emboli (right atrium) or systemic emboli (left atrium).

   (3) Treatment. Atrial fibrillation should be corrected by cardioversion, once atrial thrombosis is ruled out. The prognosis of atrial fibrillation depends on the underlying cause.

3. AV Nodal Rhythms. Pacemaker cells are found only in the A-N and N-H zones of the AV node. The term **junctional arrhythmia** denotes ectopic pacing from the AV node.

   a. Junctional premature beats are characterized by an inverted P wave and a normal QRS complex. They are designated **high** or **low** junctional beats, depending on whether the P wave precedes or follows the onset of the QRS complex. Junctional beats cause only occasional irregularity in cardiac rhythm. If the nodal tissues recover excitability, a retrograde impulse may return, causing a second ventricular depolarization called a **reciprocal beat**.

   b. Transient or permanent junctional arrhythmias arise in normal individuals due to SA node suppression. Permanent junctional rhythms occur secondary to many organic heart diseases.

   c. Junctional tachycardias, which are similar to atrial tachycardias and indistinguishable from them on an EKG, should be included with the supraventricular tachycardias.

4. Ventricular Rhythms.

   a. Premature ventricular contractions (PVCs) occur in any portion of the ventricular myocardium and occasionally in normal individuals. Frequent PVCs accompany all forms of heart disease, especially coronary artery disease, because ischemia increases the irritability of the myocardium.

   (1) Characteristics of a PVC include a prolonged (i.e., longer then 0.1 second), bizarre QRS complex, no preceding P wave, and a T wave that is oppositely directed from the QRS complex. The source of the PVC is determined by vectorial analysis using the initial portion of the QRS complex.

   (2) Compensatory Pause. Retrograde transmission of depolarization to the atria is uncommon, so atrial rate is unaltered. Following a PVC, the next atrial impulse arrives while the AV node is still refractory and, therefore, is not conducted to the ventricles, creating a pause in the ventricular rhythm. This pause is compensatory, so that the R-R interval of the beat preceding the PVC plus the PVC interval is equal to two normal cycle lengths (see Fig. 2-13). The beat following the PVC, stronger than normal due to the added stroke volume, is detected by palpation.

   (3) Interpolated Beats. If the sinus rhythm is slow, a PVC may not alter the normal R-R interval and in this case is termed an interpolated beat.

   b. Ventricular tachycardias result from a rapid, repetitive discharge from a ventricular site, usually due to a reentry phenomenon. Alterations in vagal tone (e.g., via carotid sinus massage and Valsalva maneuver) do not affect ventricular tachycardias, because the ventricles do not receive any vagal innervation.

   (1) Characteristics. The EKG reveals wide, bizarre QRS complexes occurring at a rapid rate. The P waves usually are indistinguishable, although the SA node activity continues independently of the ventricles.

   (2) Significance. Ventricular tachycardias are associated with serious heart disease, digitalis, and quinidine toxicity. They also occur when an intracardiac catheter stimulates the endocardial surface of the ventricles. A slight movement or withdrawal of the catheter may terminate this condition. Cardiac output is reduced during ventricular tachycardias due to decreased filling time and effectiveness of ventricular contraction.

   (3) Treatment. Cardioversion is necessary to terminate an attack of ventricular tachycardia caused by organic heart disease.

c. **Ventricular fibrillation** is due to rapid, irregular, ineffective contractions of small segments of the ventricular myocardium. The peripheral pulse is absent, and cardiac output is zero. The diagnosis of ventricular fibrillation requires an EKG, because this condition is indistinguishable by physical signs from cardiac standstill.

(1) **Characteristics.** The EKG shows undulating waves of varying frequency and amplitude. Ventricular fibrillation often is precipitated by one or more PVCs, one of which falls on the vulnerable interval of the T wave.

(2) **Treatment.** Cardiopulmonary resuscitation must begin immediately until cardioversion (defibrillation) can be performed.

C. CONDUCTION DISTURBANCES

1. **SA nodal block**, or the **sick sinus syndrome**, involves the disappearance of the P wave for many seconds while an ectopic pacemaker, usually a junctional or ventricular site, drives the ventricles. SA nodal block results in bradycardia and commonly occurs in elderly individuals and after a coronary occlusion. Various drugs and implanted pacemakers have been used to normalize the heart rate.

2. **AV Nodal Block.**
   a. **Degree of Block.**
   (1) **First-degree** AV nodal block is characterized by a prolonged P-R interval (i.e., beyond 0.21 second) due to slowed conduction through the AV node. This condition does not affect the pumping ability of the ventricles but does indicate the presence of a conduction disturbance. First-degree block can be caused by: increased vagal tone, quinidine, acute infectious diseases, and many forms of heart disease.
   (2) **Second-degree** AV nodal block is characterized by failure of the AV node to transmit all atrial impulses.
      (a) **Wenckebach block** (Mobitz Type I) is characterized by a progressive lengthening of the P-R interval in successive beats and finally a failure of one impulse to be transmitted.
      (b) **Periodic block** (Mobitz Type II) is characterized by an occasional failure of conduction, which results in such atrial to ventricular rates as 6:5 and 8:7. The P-R interval is constant in this condition.
      (c) **Constant block** is a form of second-degree block that is more severe than the periodic type. The ratio of atrial to ventricular beats is a constant small number (e.g., 2:1 and 3:1).
   (3) **Third-degree** (complete) AV nodal block is complete interruption of AV nodal conduction, which causes the atria and ventricles to beat at independent rates.
   b. **Significance of Block.** The significance of AV nodal block depends on the underlying cause. First-degree block due to increased vagal tone is of no consequence. Second- and third-degree blocks usually stem from organic heart disease. The onset of third-degree AV nodal block may be associated with prolonged ventricular standstill until a ventricular focus begins firing. Cardiac arrest results in cerebral ischemia and syncope, a condition termed **Stokes-Adams syndrome**.

3. **Wolff-Parkinson-White syndrome**, also termed ventricular preexcitation or accelerated AV conduction, is the result of an aberrant conduction pathway between the atria and the ventricles. This pathway has an additional connection or **accessory bundle** that eliminates or reduces the normal delay between atrial and ventricular activation.
   a. **Characteristics.** The EKG reveals a shortened P-R interval (i.e., less than 0.1 second) and a widened QRS complex with a slurred initial upstroke (delta wave). The P-J interval may be normal or shortened, depending on whether the accessory bundle inserts into the ventricular myocardium or the bundle of His, respectively.
   b. **Significance.** The major significance of the accessory bundle is that it predisposes to paroxysmal tachycardias because of a reentry phenomenon. The atrial impulse normally is conducted to the ventricles and returns to the atria via the accessory bundle, thus triggering the reentry mechanism. If localized, the accessory bundle can be surgically removed.

4. **Bundle Branch Block.** Blockage of the right or left bundle branches or the left anterior or posterior fascicles results in an abnormal sequence of ventricular excitation.
   a. **Right bundle branch block** (RBBB) delays the activation of the right ventricle.
   (1) **Characteristics.**
      (a) EKG reveals a prolonged QRS complex (i.e., beyond 0.12 second). The initial QRS vectors are shifted to the left due to unopposed activation of the left ventricle, and the final vectors are directed to the right as excitation spreads to the right ventricle through the myocardium. This sequence of excitation produces the typical rSR' pattern in the right precordial leads, a wide S wave in lead I, and a RAD of the

mean electrical axis. Repolarization of the right ventricle follows the same path as depolarization, causing an inversion of the T wave in the right precordial leads.

    (b) Incomplete RBBB produces similar QRS patterns as RBBB, but with a QRS duration of 0.08–0.10 second.

    (c) Significance. RBBB can occur in a normal individual as a transient or a permanent manifestation. RBBB occurs secondary to chronic pulmonary disease, acutely as a consequence of pulmonary embolism, and in conjunction with some forms of congenital heart disease (e.g., atrial septal defect).

  b. Left bundle branch block (LBBB) occurs in most forms of heart disease and rarely in the absence of organic heart disease. LBBB is commonly associated with coronary artery disease and conditions leading to left ventricular hypertrophy (e.g., hypertension and aortic stenosis). LBBB is best diagnosed using the left precordial leads, which reveal a QRS duration of greater than 0.12 second and initial and final QRS vectors directed to the left. The initial vector is due to septal activation from right to left, and the final vector is due to delayed activation of the left ventricle.

  c. Other forms of intraventricular block include those involving:

    (1) All three fascicles of the bundle branches

    (2) The peripheral Purkinje network

    (3) An area of infarction

D. OTHER CARDIAC DYSFUNCTIONS

  1. Myocardial Ischemia and Myocardial Infarction. Narrowing or occlusion of a coronary artery will reduce or stop blood flow to an area of the myocardium. The most common cause of vessel narrowing is an atherosclerotic plaque, and the most common cause of vessel occlusion is an atherosclerotic coronary thrombosis. Coronary blood flow reduction causes myocardial ischemia or infarction.

  a. Myocardial Ischemia.

    (1) Symptoms. The cardinal sign of coronary artery insufficiency is anginal pain related to either exertion or emotional stress. The pain may be precordial or may radiate to the shoulder, arm, and jaw. The pain is described as a "tight band" or "weight" on the chest, which results in a strangling sensation. Anginal pain typically lasts 1–10 minutes and is relieved by rest.

    (2) Diagnosis. The diagnosis usually is from patient history, as the resting EKG may be within normal limits. EKG monitoring over 24 hours may provide evidence of ischemic changes during anginal attacks. Stress (exercise) tests with EKG monitoring precipitate ischemic episodes for diagnostic purposes.

    (3) EKG Pattern. Myocardial ischemia creates S-T segment and T wave changes.

      (a) The reduced membrane potential in the ischemic area of the myocardium creates a current of injury from the normal regions to the ischemic area. This results in a S-T segment elevation from leads overlying the ischemic area.

      (b) By altering the path of repolarization, ischemia causes T-wave inversion.

    (4) Treatment. The major aim of treatment is to reduce the work of the heart either with vasodilators (e.g., nitroglycerin) or with β-adrenergic blockers that reduce heart rate (e.g., propranolol).

  b. Myocardial infarction occurs when blood flow ceases or is reduced below a critical level. The left ventricle is the common site of myocardial infarction.

    (1) Symptoms of myocardial infarction are similar to those of coronary insufficiency, except the pain is more intense, generally persists longer than 15–20 minutes, may not be related to exertion or exercise, and is not relieved by nitrates.

    (2) Diagnosis. Patient history is critical to diagnosis of myocardial infarction, and the EKG provides supportive evidence. Measurement of serum levels of cardiac enzymes creatine phosphokinase (CPK), glutamic-oxaloacetic transaminase (SGOT), and lactic acid dehydrogenase (LDH) may aid in the diagnosis, but no enzyme is specific for the myocardium.

    (3) EKG Pattern. Following myocardial infarction the EKG undergoes a series of changes, which should be recorded daily for diagnostic purposes. A **transmural** infarct results in wide, deep Q waves from leads overlying the infarcted area, as these leads essentially record an intracavitary potential. These same leads demonstrate an S-T segment elevation and, early after the infarct, demonstrate T wave inversion.

    (4) Treatment. The major aims of treatment are pain relief and support of the circulation if cardiogenic shock develops. Continuous EKG monitoring should be performed in an intensive care unit to detect arrhythmias, especially PVCs, which may trigger bouts of ventricular tachycardia or ventricular fibrillation. Treatment includes cardioversion, to interrupt arrhythmias, and drug therapy (membrane stabilizers), to prevent recurrent arrhythmias.

2. Ventricular hypertrophy occurs if the work of one or both ventricles is increased sufficiently. Ventricular hypertrophy increases the diameter, not the number, of myocardial cells; therefore, the diffusion distance increases for O_2 and other metabolites, a condition that may lead to borderline ischemia.

   a. Diagnostic Criteria for Ventricular Hypertrophy.

   (1) Height of R Wave. There is a direct correlation between the thickness of the ventricular wall and the height of the R wave in the overlying leads. The increased height of the R wave is a reflection of the increased magnitude of the depolarization vector.

   (2) Duration of QRS Complex. The presence of increased muscle mass prolongs the QRS complex to a duration of usually less but occasionally more than 0.12 second.

   (3) S-T Segment and T Wave Alterations. Ventricular hypertrophy causes S-T segment depression and T-wave inversion in epicardial leads. The altered repolarization process may be related to endocardial fibrosis and mild ischemia, commonly present in patients with ventricular hypertrophy. If changes in the S-T segment and T wave are the only manifestations, the condition is termed a **ventricular strain pattern**.

   b. Diagnostic Criteria for Left Ventricular Hypertrophy.

   (1) Axis. The mean electrical axis is shifted superiorly to the left, and the transition zone of the precordial leads is shifted to the left.

   (2) QRS Complex. Left ventricular hypertrophy is present if:

   (a) The R wave in V_5 or V_6 plus the S wave in V_1 is greater than 35 mm

   (b) The R wave in V_5 or V_6 is greater than 25 mm

   (c) The R wave in aVL is greater than 13 mm

   (d) The R wave in aVF is greater than 25 mm

   (3) S-T segment and T wave changes may or may not occur in left ventricular hypertrophy.

   c. Diagnostic Criteria for Right Ventricular Hypertrophy.

   (1) Axis. The mean electrical axis usually is vertical or greater than +110° in the frontal plane.

   (2) QRS Complex. Right precordial leads generally show tall R waves rather than the normal S waves. The QRS complex is prolonged to a duration of 0.08–0.12 second.

   (3) S-T Segment and T Wave. The S-T segment is depressed, and the T wave is inverted in the right precordial leads.

   d. Diagnostic Criteria for Right Ventricular Strain (Acute Cor Pulmonale). The appearance of S-T segment and T wave changes without abnormal R waves in the right precordial leads is diagnostic of acute cor pulmonale. This condition occurs with acute exacerbations of pulmonary disease and is an important sign of pulmonary emboli.

V. CARDIODYNAMICS

A. CARDIAC CYCLE (Fig. 2-14). The cardiac cycle includes both electrical and mechanical events that generate potential and kinetic energy (pressure and flow, respectively) within the system. The electrical events precede and initiate the corresponding mechanical events; therefore, the P wave precedes **atrial contraction** and the QRS complex precedes **ventricular contraction**.

1. Atrial Contraction. The two atria contract almost synchronously due to the specialized tracts that carry excitation between the atria. The increase in atrial pressure resulting from atrial contraction, termed the **a wave**, is reflected back into the venous system; there, it is recorded from the jugular vein with appropriate transducers to aid in cardiac diagnosis. The level of filling of the jugular vein provides an estimate of the right ventricular filling pressure. Blood pressure in the ventricles at the termination of the atrial contraction is termed the **ventricular end-diastolic pressure** (VEDP), which sets the preload for the next ventricular contraction.

   a. Apposition of AV Valves. Blood flow through the AV valves during atrial contraction narrows the opening between the valve leaflets by a Bernoulli effect (see legend to Fig. 2-19).

   b. Alteration of the a Wave.

   (1) Atrial Fibrillation. The a waves disappear in the presence of atrial fibrillation.

   (2) Cannon a waves occur intermittently during ventricular contraction in the presence of complete heart block, whenever the atria contract while the tricuspid valve is closed.

   (3) Giant a Waves. The amplitude of the a waves increases markedly in the presence of tricuspid stenosis due to the impediment imposed on the forward flow of blood into the right ventricle.

2. Ventricular Contraction.

   a. AV Valve Closure. Shortly after the QRS complex begins, the pressure in the ventricular cavities rises due to the onset of ventricular contraction. This reverses the pressure gradient across the AV valves and normally causes complete closure of the valve leaflets.

   (1) The first heart sound (S_1) is a low-frequency sound that occurs just after the onset of ventricular contraction. Final closure of the AV valves from a partially closed position

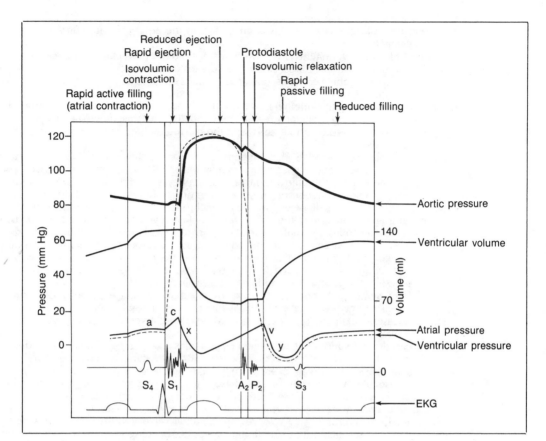

Figure 2-14. The events of the cardiac cycle. S_1, S_2 (A_2 and P_2), S_3, and S_4 represent heart sounds.

produces a low intensity S_1. A louder S_1 occurs when the AV valves are closed from an initially widely separated position. S_1 is louder if the P-R interval is abnormally long or short, because the valve leaflets are widely separated under these conditions.

(2) **Splitting of S_1.** Earlier activation and more vigorous contraction of the left ventricle causes the mitral valve to close as much as 0.04 second before the tricuspid valve, producing an audible splitting of the S_1 components.

b. **Isometric (Isovolumic) Contraction.** Once the AV valves close, the ventricular volume remains constant until the ventricular pressure exceeds that in the arteries; at this point ventricular ejection begins. During isovolumic contraction the pressure gradient across the AV valves causes them to bulge toward the atria. This produces the atrial **c wave**, a mechanical artifact of the valve motion.

c. **Rapid Ejection.** When pressure in the left ventricle exceeds the aortic diastolic pressure (80 mm Hg) and pressure in the right ventricle exceeds pulmonary artery diastolic pressure (8–12 mm Hg), ventricular ejection begins. About two-thirds of the **stroke volume** (i.e., volume ejected per beat) is ejected in the first third of systole, which is termed the **rapid-ejection phase**. This phase ends when the ventricular and arterial systolic pressures peak. Normally, ventricular and arterial systolic pressures are essentially equal because the semilunar valves open widely and represent a low resistance segment. Stenosis of the semilunar valves creates a high resistance, and significant systolic pressure gradients occur between the ventricular and the arterial systems. Semilunar valve stenosis increases ventricular work, reduces the rate of ventricular ejection, and creates an ejection-type murmur.

(1) **Apex Beat.** During the rapid-ejection phase, the ventricles "wring out" their contents. This moves the apex of the heart forward, creating the **apex beat**, which is palpable on the chest wall. Normally, this beat is located in the fifth intercostal space in the midclavicular line. Enlargement of the left ventricle moves the apex beat laterally and inferiorly.

(2) **X Descent of Atrial Pressure.** During the rapid-ejection phase, the base of the heart moves inferiorly, exerting traction on the atrial walls and lowering the internal atrial

pressure. Like the c wave, this is an artifact in the atrial pressure curve created by movement.

d. Reduced Ejection. During the latter two-thirds of systole, the ejection rate declines as do ventricular and arterial pressures. Aortic pressure exceeds ventricular pressure, but blood continues to leave the ventricles powered by leftover kinetic energy from the rapid-ejection phase.

e. Protodiastole. During this brief phase (about 0.015 second), blood flow in the root of the aorta is reversed as ventricular pressure falls rapidly. This backflow slightly increases ventricular volume. Protodiastole ends with the closure of the **semilunar** (i.e., aortic and pulmonary) **valves.**

    (1) The **second heart sound (S_2)** has a high frequency and is caused by the abrupt arrest of blood and the resultant tissue vibrations as the semilunar valves snap closed. This closure causes a notch in the arterial pressure pulse, termed the **incisura**. The incisura and the S_2 indicate the end of systole and the onset of isovolumic relaxation.

    (2) Splitting of S_2. Normally, ejection ends asynchronously in the two ventricles because of beat-to-beat variation in the stroke volume of each ventricle; the aortic component (A_2) of S_2 precedes the pulmonary component (P_2). During inspiration, the right ventricular ejection period is prolonged due to an increased stroke volume secondary to an augmented venous return. This delays P_2, separating A_2 and P_2 by as much as 0.05 second. During expiration, the A_2-P_2 interval is about 0.02 second, which cannot be separated by the human ear and is heard as a single sound.

        (a) Fixed Splitting. A pressure or volume overload on the right ventricle causes a fixed, wide split between A_2 and P_2. This is especially common in atrial septal defect and in cases of pulmonary hypertension when the right ventricle maintains a normal stroke volume.

        (b) Paradoxical splitting occurs when P_2 precedes A_2 and the splitting lessens during inspiration. Such splitting is common in LBBB, aortic stenosis, and cases of systemic hypertension (i.e., whenever left ventricular systole is delayed or prolonged).

f. Isometric (Isovolumic) Relaxation. This phase extends from S_2 to AV valve opening, during which the ventricle is a closed, isolated chamber. Isovolumic relaxation lasts about 0.04 second, ending at the peak of the atrial pressure **v wave** as ventricular pressure falls below atrial pressure and the AV valves open. Mitral valve stenosis (i.e., narrowing by fibrotic cusps) is suspected if an additional sound, the **opening snap**, is heard when this valve opens (Fig. 2-15). The opening snap occurs about 0.04–0.06 second after S_2 and must be differentiated from a split S_2 and a third heart sound.

g. Rapid Passive Filling. During ventricular systole, the venous return continues and the atrial pressure increases from its nadir of the X descent. The rising atrial pressure causes the

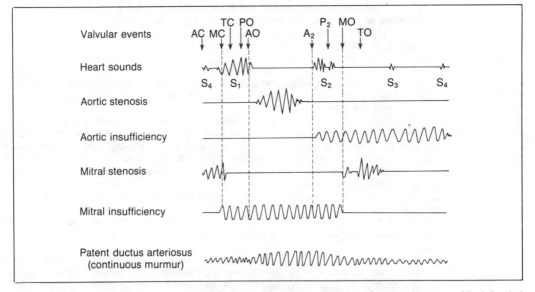

Figure 2-15. Schematic representation of the heart sounds, valvular events, and murmurs generated by left-sided valve lesions. Right-sided valve lesions begin and end in association with right heart valve events (not shown). *AC* = atrial contraction; *MC* = mitral valve closure; *TC* = tricuspid valve closure; *AO* = aortic valve opening; *PO* = pulmonary valve opening; A_2 = aortic valve closure; P_2 = pulmonary valve closure; *MO* = mitral valve opening; and *TO* = tricuspid valve opening.

v wave in the atrial pressure curve, which reaches its peak just as the AV valves open. Once the AV valves open, the atria and ventricles are a common chamber, and the pressure in both cavities falls as ventricular relaxation continues. The decrease in atrial pressure following AV-valve opening is termed the **Y descent**.

h. Reduced Filling and Diastasis. During the latter stages of diastole, the atrial and ventricular pressures rise slowly as blood returns to the heart. Ventricular filling ceases with very slow heart rates, because the ventricles reach their volume limit; this phase is termed **diastasis**. Heart rate is increased primarily by shortening the periods of diastasis and reduced filling. Thus, increases in heart rate to 140–150 beats/min do not significantly decrease ventricular filling. Tachycardia increases cardiac output to rates of 180–200 beats/min. At heart rates above 200 beats/min, cardiac output declines due to encroachment on the rapid-filling phase.

i. Additional Heart Sounds.

   (1) A **third heart sound (S_3)** occurs near the end of the rapid-filling phase and is due to the vibrations in the heart and vessels as the ventricles become distended with blood. S_3 is a low-pitched, low-intensity sound heard best in children and in individuals with thin chest walls.

   (2) A **fourth heart sound (S_4)**, caused by ventricular distension secondary to atrial contraction, is of similar quality and intensity as S_3 but is not heard in normal individuals. S_4 indicates ventricular dilation or ventricular failure.

   (3) **Gallop rhythm** refers to a cadence of three heart sounds (S_1, S_2, S_3 or S_1, S_2, S_4) and usually is heard only with a rapid heart rate. A **summation gallop** refers to the superposition of S_3 and S_4.

   (4) **Ejection "clicks"** often are heard in the presence of systemic or pulmonary hypertension and semilunar valve disease.

j. Aortic and Pulmonary Artery Pressures. The recording of the pulsatile pressure changes in these vessels is termed a **pressure pulse**.

   (1) **Diastole.** During diastole, the arterial pressure declines as blood continues to flow from the arteries to the peripheral vessels. The lowest pressure in the aorta (diastolic pressure) occurs just prior to the onset of ventricular ejection.

   (2) **Systole.** During the period of rapid ventricular ejection, the pressure in the aorta is slightly less than that in the ventricle. The peak of aortic pressure is the arterial systolic pressure and occurs at the end of the rapid-ejection phase.

   (3) **Pulse pressure** is the difference between the systolic and the diastolic pressures and normally is 120 mm Hg – 80 mm Hg or 40 mm Hg.

k. Ventricular volume may be measured angiographically or by radionuclide imaging and computer analysis (see Fig. 2-14). Echocardiography is used to measure ventricular dimensions and ventricular wall and valve motion during the cardiac cycle.

   (1) The **ventricular end-diastolic volume** (VEDV) is the volume of blood in the ventricle just prior to S_1. The normal left VEDV is about 120 ml.

   (2) The **ventricular end-systolic volume** (VESV) is the volume of blood remaining in the ventricle at the end of ejection. The normal left VESV is 40 ml.

   (3) The **stroke volume** is the volume of blood ejected with each beat and equals VEDV minus VESV or 75–80 ml.

B. VALVE LESIONS AND MURMURS (see Fig. 2-15). Abnormalities of the cardiac valves may result from either congenital or acquired heart disease. The pathophysiology of valve lesions is determined by the reduction in blood flow, the additional work of the heart, and the retrograde congestion produced by the particular lesion. Valve abnormalities produce physical signs of dysfunction, discernible during a physical examination, which result from abnormal blood flow and pressure gradients.

1. Stenosis is a narrowing of the valve orifice which impedes the forward flow of blood.

2. Insufficiency of a valve allows the backward flow of blood in response to a pressure gradient (e.g., with aortic insufficiency there is a retrograde blood flow from the aorta to the ventricle during diastole).

3. Turbulence.

   a. Reynold's Number. Blood flow varies with local conditions, as demonstrated by the Reynold's number (Re). This dimensionless quantity is determined by the velocity of flow (V), tube diameter (D), fluid density (d), and fluid viscosity (v) and is expressed as: **Re = VDd/v.**

   b. Normal flow is **laminar** or **streamlined** if Re is less than 2000, which produces a parabolic velocity front across the vessel. During laminar flow, velocity at the vessel wall is zero, whereas maximal flow velocity occurs at the center of the vessel.

c. **Turbulent flow**, indicated when Re exceeds 3000, is favored by high velocity, large diameter, high density, and low viscosity. Turbulence occurs at a critical flow velocity, as shown by the development of vortices in the bloodstream. The vortices cause vibrations in the tissues, which are detected as **murmurs**.

4. **Murmur.** This audible component of turbulent blood flow is an important physical sign of a valve lesion (see Fig. 2-15). The intensity of a murmur is related to the velocity of flow, which is proportional to the pressure gradient across a lesion. Murmurs are classified according to timing, location, radiation, character, and intensity.

a. **Timing.** Murmurs may occur during systole, diastole, continuously throughout the cardiac cycle, or at specific times during the cycle.

(1) **Midsystolic murmurs** (also termed **diamond-shaped** or **ejection murmurs**) begin at the onset of ventricular ejection, reach maximal intensity at midsystole, and stop before S_2. These murmurs are caused by stenosis of the aortic or pulmonary valve.

(2) **Pansystolic** or **holosystolic murmurs** begin with S_1 and end at or just after S_2. These murmurs are caused by AV valve insufficiency or a ventricular septal defect, since this is the only time a pressure gradient exists across these structures.

(3) **Early diastolic murmurs** start at S_2 and diminish in intensity as diastole continues. These murmurs are due to aortic or pulmonic valve insufficiency. As the pressure gradient diminishes during diastole, so does the intensity of these murmurs.

(4) **Mid-diastolic murmurs** begin after the AV valves open and often are preceded by an opening snap. Caused by AV valve stenosis, this murmur occurs during the rapid passive filling phase when blood flow through the stenosed valve is high.

(5) **Presystolic murmurs** also are associated with AV valve stenosis and occur during atrial contraction, another period of increased blood flow. This murmur disappears with the onset of atrial fibrillation, since an effective atrial contraction is lost.

b. **Location and Radiation.** Murmurs are best heard on the chest wall closest to their origin and in the downstream direction of blood flow. For example, the murmur of aortic stenosis is best heard at the right sternal border in the second interspace; this murmur radiates to, and is clearly heard over, the vessels of the neck.

c. The **intensity** of the murmurs generally is graded on a scale of 1 to 6, with 6 being the loudest.

C. **THE HEART AS A PUMP.** The heart consists of two pumps—the right and left ventricles—which are connected by the pulmonary and systemic circulations.

1. The **right ventricle** (RV) is crescent-shaped and ejects blood by shortening the free wall and bulging the ventricular septum during systole. The RV is designed to pump relatively large volumes of blood at relatively low pressures through the pulmonary circuit. Pulmonary hypertension causes the free wall of the RV to thicken and assume a shape more like that of the left ventricle.

2. The **left ventricle** (LV) is thick-walled and ovoid or oblate spheroid. The LV ejects blood by decreasing the transverse diameter and shortening the base-to-apex dimension. The left ventricle pumps effectively against the relatively high pressures of the systemic circulation but responds to hypertensive states by increasing muscle mass and eventually by increasing the left ventricular end-diastolic volume.

3. **Determinants of Ventricular Filling.** The **stroke volume** of the ventricles is dependent on the contractility of the ventricle, the aortic pressure (afterload), and the **ventricular end-diastolic volume** (VEDV). The VEDV is determined by ventricular filling pressure, ventricular compliance, and heart rate.

a. **Ventricular Filling Pressure.** The pressure in the great veins (i.e., central venous pressure) determines the filling of the RV; the filling of the LV is determined by the pressure in the pulmonary veins.

(1) **RV Filling Pressure.** The central venous pressure is monitored clinically in cases of diminished cardiac function (e.g., cardiovascular shock, critical burns, and myocardial infarction).

(2) **LV filling pressure** can be measured by the pulmonary wedge pressure. In this method, a multilumened catheter with a balloon on the end is floated into the pulmonary circulation until it reaches a small pulmonary artery; the balloon then is inflated to completely occlude the vessel. The pressure measured from an opening distal to the balloon is the pressure reflected back from the pulmonary veins and the left atrium.

b. **Ventricular compliance** can be altered by different pathologic conditions, but the effects of these changes on ventricular function are not well understood.

c. **Heart rate** is the primary determinant of the duration of ventricular filling. Slow heart rates are associated with a greater VEDV than rapid heart rates. With increased VEDV comes a more forceful contraction and consequently a greater stroke volume (see Section II D and E). The increased output secondary to changes in VEDV or end-diastolic fiber length is termed the Frank-Starling mechanism or **heterometric autoregulation**.

4. **Force-Velocity Relationship.** Figure 2-16 shows how the initial length, the afterload, and the state of contractility relate in an isolated papillary muscle. An increase in the initial length increases the force of contraction but does not increase the maximal velocity of shortening. Positive inotropic agents (e.g., norepinephrine), however, increase both. In the intact heart, the initial length is analogous to the VEDV; the afterload is equivalent to the aortic pressure.

5. **Cardiac work** can be measured in pressure (P) and volume (V) units that convert to the physical units of force and distance. Figure 2-17 represents one cardiac cycle; the various events of the cycle are indicated in the loop that is formed on the axes. The area enclosed by the loop represents an amount of work. Since both ventricles pump the same average stroke volume and the systolic pressure of the RV is about one-seventh that of the LV, the work of the RV is about one-seventh that of the LV.

   a. **Myocardial O_2 consumption** is determined by the work performed by the heart and is dependent on the myocardial wall tension, contractility, and heart rate.

   (1) **Myocardial wall tension** is a function of the internal pressure and the radius of curvature of the cavity.

   (a) Assuming that the ventricle is a sphere, these parameters relate according to the **law of Laplace**. This law states that:

$$T = Pr/2,$$

   where T = wall tension; P = ventricular pressure; and r = ventricular radius.

   (b) This relationship indicates that a large ventricle requires a higher wall tension (i.e., higher O_2 consumption) to produce a given systolic pressure than a small ventricle. Thus, ventricular dilation in response to hypertension is a mechanical disadvantage since O_2 consumption must rise in order to maintain a normal arterial pressure.

   (c) During ejection, total wall tension decreases in a normal-sized heart but increases in a dilated heart. (An enlarged heart requires a considerably smaller decrease in radius to eject its stroke volume than a normal-sized heart.)

   (2) **Contractility.** Sympathomimetic drugs and digitalis increase the metabolic requirements of the myocardium. Angina frequently is treated with a combination of a nitrite and propranolol. The nitrites provide peripheral vasodilation, which decreases the afterload work of the heart, and propranolol reduces the myocardial contractility and heart rate, which decreases myocardial O_2 consumption.

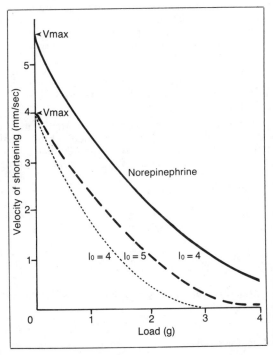

Figure 2-16. The effects of norepinephrine, load, and initial length (*lo*) on the velocity of shortening. Increases in lo increase the velocity at any load but do not increase the maximal velocity (V_{max}). Norepinephrine (*solid line*) does increase the V_{max}, which is considered to be an index of contractility, even though lo is unchanged from control conditions.

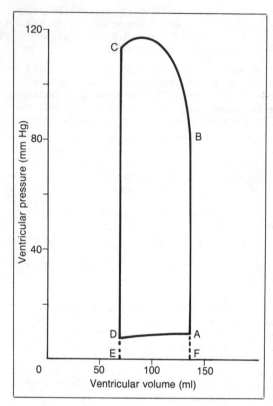

Figure 2-17. Volume/pressure curve of the left ventricle. The area within *ABCD* represents the active work generated by the left ventricle during one contraction. The area *ADEF* is the amount of passive work done on the ventricle by the venous return. *A* = mitral valve closure; *B* = aortic valve opening; *AB* = isovolumic contraction; *C* = aortic valve closure; *BC* = ejection; *D* = mitral valve opening; *CD* = isovolumic relaxation; and *DA* = ventricular filling.

(3) Heart Rate. The largest component of cardiac work is that of generating a static force to raise ventricular pressure to the level of aortic pressure, which is a necessary condition for ventricular ejection. An increase in heart rate translates into more static work performed each minute during the period of isovolumic contraction.

b. Myocardial O_2 Supply. The heart can function anaerobically for brief periods of time only; normal cardiac function depends on an adequate delivery of O_2 and other substrates. The O_2 delivery to any tissue is equal to the arterial O_2 content (i.e., O_2 volume/100 ml blood) times the blood flow. Limitation of O_2 delivery to the heart almost always results from partial or complete coronary artery obstruction secondary to coronary atherosclerosis.

c. Cardiac efficiency can be calculated from the myocardial O_2 consumption. O_2 usage can be determined by measuring coronary blood flow and the A-V difference of O_2. Coronary sinus blood must be used to estimate the venous composition from the left ventricle. The normal heart has an efficiency of 20–25 percent; an increase in the pressure load on the heart reduces efficiency, and a volume load increases efficiency.

6. The **cardiac output** (stroke volume × heart rate) averages about 5 L/min in a normal man and is about 20 percent less in a woman. Over any interval of time, the cardiac output must equal the venous return. The cardiac output is exactly controlled to provide an adequate mean blood pressure for delivery of nutrients and O_2. The cardiac output can be measured using either of the following techniques.

a. The **Fick principle** is an example of the law of conservation of mass, a widely applied principle in physiology. When applied to the cardiac output, the Fick principle states that the amount of O_2 delivered to the tissues must equal the O_2 uptake by the lungs plus the O_2 delivered to the lungs in the pulmonary artery. This is mathematically expressed as:

$$Q = \frac{\dot{V}O_2}{CaO_2 - C\bar{v}O_2},$$

where Q = cardiac output; CaO_2 = arterial O_2 content; $C\bar{v}O_2$ = mixed venous O_2 content; and $\dot{V}O_2$ = O_2 consumption (ml/min).

b. Indicator Dilution.

(1) Purpose.

(a) If a suitable indicator (A = amount injected) is injected into an unknown volume of distribution (V), the V can be estimated from the resultant indicator concentration (c) with the equation: V = A/c.

(b) Similarly, the flow of a fluid (Q) can be measured if the mean concentration of the indicator is determined for the time (t) required for that indicator to pass a given site. This flow is measured with the equation: Q = A/ct.

(2) Technique. In the body, the indicator is injected instantaneously but is dispersed in time due to different transit times between the injection and sampling sites. Usually, recirculation of some indicator occurs before passage of the indicator is complete (Fig. 2-18). After the peak concentration is reached, the indicator concentration follows an exponential time course, so that the curve can be extrapolated to zero concentration and the duration of the indicator passage (t) can be estimated (see Fig. 2-18).

(3) Indicators used for cardiac-output measurement include Evans blue, Cardio-Green, hypertonic or hypotonic saline, ascorbate, and cold saline. Cold saline, used in thermodilution, is the most commonly used indicator in human cardiac-output studies.

(4) Thermodilution uses a multilumened balloon catheter with a thermistor near the distal end. The catheter is inserted into a peripheral vein, and the tip is floated into a branch of the pulmonary artery; cold saline is injected through a proximal opening which is within the RV. The saline mixes with the blood in the RV; the temperature change is measured by the distal thermistor, which is downstream in the pulmonary artery. Thermodilution is a preferred technique because little or no recirculation occurs, cold saline is relatively innocuous, and little temperature change occurs in the surrounding tissues.

VI. HEMODYNAMICS

A. BASIC UNITS

1. Pressure Gradient (dP). Fluid flow depends on a difference in total energy between two points within a system. The energy can be converted from potential energy (i.e., pressure) to kinetic energy (i.e., velocity) and vice versa. Figure 2-19 shows a system of pitot tubes and stand pipes. A pitot tube facing upstream measures potential plus kinetic energy; a pitot tube facing downstream measures potential minus kinetic energy. The level of fluid in the risers provides a measure of the energy monitored. Pressure is a force per unit area (e.g., dynes/cm²) and usually is expressed in terms of the height of a column of fluid that the pressure will support. Common units of pressure are **mm Hg** or **cm H₂O** (1 mm Hg = 1.36 cm H₂O).

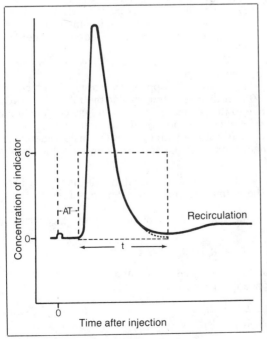

Figure 2-18. Indicator dilution curve, which can be used to measure cardiac output. *AT* = appearance time; *c* = mean dye concentration for one passage of the dye; and *t* = passage time. The *dotted line* is an extrapolation of the indicator curve to zero concentration. The area inside the rectangle equals the area enclosed by the indicator dilution curve. If c = 1.0 mg/L, t = 0.5 min, and a = 2.5 mg, then: Q = a/ct = 2.5/(1) (0.5) = 5 L/min.

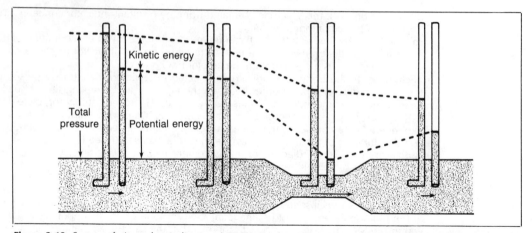

Figure 2-19. System of pitot tubes indicating the variation in total pressure, potential energy, and kinetic energy with change in flow velocity. The *arrows* represent the relative velocities in the segments of the system. The *narrow section*, a high resistance segment, causes a large pressure drop. An increased flow velocity decreases lateral pressure (increases kinetic energy) and is termed a Bernoulli effect.

2. **Flow** (Q) is a volume movement per unit time represented as the average velocity of movement (cm/sec) times the cross-sectional area of the tube (cm²). Flow equals the pressure gradient between two points in a system divided by the resistance to flow. For laminar flow, **Poiseuille's law** states that:

$$Q = \frac{(dP \pi \, r^4)}{(8 \, u \, L)},$$

where dP = pressure gradient; r = radius of the tube; u = viscosity of the fluid; and L = length of the tube.

a. **Effects of Tube Radius.** Any change in tube radius produces large changes in the magnitude of flow. For example, if the radius is reduced to one-half and the pressure gradient is unchanged, the flow will be one-sixteenth of its previous value.

b. **Viscosity** is the fluid's internal friction to flow. Newtonian fluids exhibit a linear relationship between the stress (pressure) and the strain rate (velocity of flow). Blood is a non-Newtonian fluid in that its apparent viscosity varies as a function of the flow rate. Blood with a normal hematocrit has a viscosity about three times the viscosity of water flowing through medium-sized tubes or vessels.

(1) **Anomalous Viscosity.** The viscosity of blood increases markedly as the flow rate slows, presumably due to aggregation of erythrocytes.

(2) The **Fahraeus-Lindqvist effect** is the decrease in apparent viscosity when blood flows through tubes smaller than 200 μ in diameter; the smaller the tube, the lower the viscosity. This effect may be due to the plasma being packed between erythrocytes, which interrupts the lamina and reduces the internal friction of the fluid.

(3) **Hematocrit,** or the percent of blood that is erythrocytes, is the major determinant of blood viscosity. Blood with a hematocrit of 60 has approximately twice the viscosity of blood with a hematocrit of 40.

3. **Resistance (R)** usually is stated as: R = dP/Q, with R measured in units of mm Hg/L/min or **peripheral resistance units** (PRU). The normal systemic circulation has a resistance of approximately 20 PRU.

a. **Series resistances** are additive; thus, the total resistance of the arteries (R_a), capillaries (R_c), and veins (R_v) in an organ is expressed as:

$$R_{total} = R_a + R_c + R_v.$$

b. **Parallel resistances** are added as the reciprocals. In a vascular bed containing three parallel pathways, the total resistance is expressed as:

$$\frac{1}{R_{total}} = \frac{1}{R_1} + \frac{1}{R_2} + \frac{1}{R_3}.$$

c. The resistance of the entire systemic circulation is less than the resistance of one organ. This system of parallel resistances allows different organs to adjust their flow according to local metabolic requirements and provides each organ with essentially the same pressure for perfusion.

B. ARTERIAL PRESSURE is determined by the interaction of four parameters: the elasticity of the arterial system (E), the mean arterial blood volume (V), the pulse pressure (dP), and the systolic uptake of blood by the arteries (dV). These parameters are interrelated by an equation derived from Hook's Law, which states:

$$E = \frac{(dP) (V)}{dV}.$$

1. Parameters.
   a. The **elastic constant** (E) is similar to Young's modulus of elasticity; E increases as an individual ages and the vascular system becomes less compliant (see Fig. 2-1). An increase in E, while stroke volume and arterial resistance remain normal, causes an increase in the pulse pressure. E is less in the pulmonary artery than it is in the aorta.
   b. The **pulse pressure** (dP) is largely determined by the elastic constant and the systolic uptake of blood by the arteries. Increases in stroke volume, with other parameters constant, increase dP.
   c. **Systolic arterial uptake** (dV) is less than the stroke volume, because some blood flows out of the arterial system during systole. Increases in stroke volume increase dV.
   d. The **mean arterial blood volume** (V), when increased, causes an increase in the arterial pressure by distending the vascular system, which then recoils on the contained blood with a greater force.

2. Physiologic Determinants of Arterial Pressure. The body adjusts arterial pressure by altering heart rate, stroke volume, and peripheral resistance. It is important to understand the relationships between these parameters in order to analyze compensatory mechanisms during such stresses as hemorrhage, syncope, and arrhythmias.
   a. **Heart rate** affects arterial pressure by reducing the time for diastolic runoff of blood to the periphery. An increase in heart rate increases the mean arterial blood volume and raises arterial diastolic pressure. If stroke volume remains constant, an increase in heart rate produces an increase in mean arterial pressure and a decrease in pulse pressure.
   b. **Increased peripheral resistance** raises the diastolic arterial volume by reducing the peripheral runoff. If stroke volume is unchanged, an increased resistance will reduce pulse pressure.
   c. **Stroke volume** increments, at a constant heart rate and peripheral resistance, increase mean arterial volume, mean arterial pressure, pulse pressure, and systolic uptake.
   d. The **elastic constant** progressively increases from before birth until the time of death.
      (1) Elevation of the elastic constant produces a greater pulse pressure for any systolic uptake of blood. This increase in the arterial pressure raises the work of the ventricle during ejection.
      (2) Also, an increased elastic constant raises the systolic pressure and lowers the diastolic pressure. The diastolic pressure is above normal in elderly individuals because of atheromata that form in the smaller vessels and increase the peripheral resistance.

C. VENOUS RETURN. The venous system is a low-resistance, low-pressure, highly distensible segment of the vascular system. Veins comprise the body's major blood reservoir; they contain 70–80 percent of the circulating blood volume at any one time. Several factors affect the functioning of the venous system.

1. Gravity. Since veins are highly distensible, small pressure changes can produce large changes in the blood volume in the venous system.
   a. **Peripheral Pooling.** Changing from the supine to the erect position shifts about 500 ml of blood from the pulmonary circulation to the dependent veins of the legs. The hydrostatic pressure generated while in an erect position causes dilation of the veins and peripheral pooling of blood. This peripheral pooling decreases venous return, consequently decreasing cardiac output and arterial blood pressure.
   b. **Hydrostatic Effects.** In the erect position, the arterial pressure in the head is reduced by about 30 mm Hg due to the hydrostatic column of blood from the heart to the brain. Reflex alterations in the smooth muscle tone in the veins as well as increases in heart rate and peripheral vascular resistance compensate for the hydrostatic changes in order to maintain an adequate arterial pressure.

2. **Venous Valves.** The veins of the dependent parts of the body have a system of valves that prevent backflow of venous blood. These valves also support the column of blood to minimize increases in capillary pressure. Insufficiency of the valves allows the hydrostatic column of blood to exert full pressure, which results in edema formation in the dependent parts. Varicose veins are marked dilations of dependent veins for which there is familial and occupational predisposition.

3. **Skeletal-Muscle Pump.** Contraction of skeletal muscle aids venous return by compressing the veins between the contracting skeletal muscle. This effect is greatest when venous valves are competent to prevent backflow during relaxation of the skeletal muscle.

4. **Respiration.** The normal negative intrathoracic pressure enhances venous return by increasing the pressure gradient between the heart and the periphery. Inspiration causes a greater negativity in the intrathoracic (pleural) pressure, which increases venous return by further enhancing the pressure gradient.

5. **Positive end-expiratory pressure** (PEEP) or a **Valsalva maneuver** impedes venous return and produces both direct and reflex changes in the cardiovascular system. PEEP is used to treat respiratory distress syndrome; the positive alveolar pressure stabilizes alveoli and minimizes or eliminates atelectasis.

VII. CARDIOVASCULAR REGULATION.
The cardiovascular system is designed to maintain an adequate perfusion pressure to provide sufficient blood flow to all organs of the body. To accomplish this, the system is designed to control both arterial pressure and blood volume.

A. **CONTROL OF BLOOD VOLUME.** The blood volume is regulated by a complex interaction of neural, renal, and endocrine mechanisms.

1. **Neural Component.** The arterial baroreceptors as well as stretch receptors in the atria (low-pressure baroreceptors) serve as sensors to detect increases in blood pressure or volume. Impulses from these receptors are transmitted to the hypothalamus where the neurohypophysial cells are inhibited from releasing **antidiuretic hormone** (ADH).

2. **Endocrine Component.**
   a. **ADH** is released from the neurohypophysis by increases in the osmolality of the blood perfusing the brain. The increased osmolality is detected by osmoreceptors in the anterior hypothalamus. ADH stimulates increased reabsorption of water from the distal convoluted tubule and the collecting duct, causing water retention and dilution of the solutes.
   b. The **renin-angiotensin-aldosterone system** is the major control mechanism for the electrolytes within the body. A decrease in the mean blood pressure or pulse pressure in the kidney causes the release of the enzyme **renin** from the juxtaglomerular cells of the nephron. Renin in the plasma reacts with a substrate called **angiotensinogen** to form the decapeptide, **angiotension I** (AI). AI is converted to **angiotensin II** (AII), an octapeptide, by a peptidase located on the endothelial surface of the renal and pulmonary vasculature. AII, a potent vasoconstrictor, acts directly on blood vessels to increase resistance and return blood pressure to normal. In addition, AII causes the release of aldosterone from the zona glomerulosa of the adrenals. Aldosterone causes a reabsorption of electrolytes (primarily Na^+) and obligated water in the distal tubule. The increased fluid reexpands circulating blood volume, which tends to increase blood pressure.

B. **CONTROL OF ARTERIAL PRESSURE**

1. **Baroreceptors.**
   a. **Structure and Distribution.** Baroreceptors are highly branched nerve endings located in the carotid sinuses (at the bifurcation of the carotid arteries) and in the aortic arch. The carotid sinus represents the most distensible area of the arterial system, indicating that these receptors respond to stretch of the arterial walls rather than to pressure.
   b. **Function.**
      (1) **Baroreceptor Response.** The baroreceptors generate a receptor potential in response to distension of the walls of the carotid sinuses. The receptor potential results in the generation of an action potential in the vagus and glossopharyngeal nerves. The action potential frequency is linearly related to the receptor potential. The receptor potential has two distinct components.
         (a) The **dynamic receptor response**, the initial receptor potential, is proportional to the rate of change of the arterial pressure.
         (b) The **static receptor response** is proportional to the new steady pressure and continues unchanged as long as the pressure is maintained. These responses provide

information to the cardiovascular control centers in the central nervous system (CNS) regarding the heart rate, the rate of change in blood pressure (related to the ventricular contractility), and the pulse pressure during the cardiac cycle.

(2) **Baroreceptor Reflex.** Afferent neural activity over the sinus and the aortic nerves causes a reflex slowing of the heart and a withdrawal of sympathetic tone to the arterioles, resulting in vasodilation. Massage of the carotid sinus is used clinically to induce a vagally mediated slowing of the heart to interrupt paroxysmal atrial tachycardia. An increased sensitivity of the carotid sinus is seen in some older individuals who experience syncope secondary to the vagally mediated sinus arrest and a prolonged period of ventricular asystole. This condition is termed the **Stokes-Adams syndrome**.

2. CNS Centers.

  a. Medulla. Located within the reticular formation of the brain stem are diffuse collections of cells involved in the integration of cardiovascular function. A pressor area is located in the lateral portion of the reticular formation, whereas stimulation of the medial portion of the reticular formation causes a depressor response. The depressor response is due to the withdrawal of vasoconstrictor influence rather than an active vasodilation. Cardioaccelerator and cardioinhibitory centers also are located in the reticular formation. These latter two centers send impulses to the heart over the cardiac sympathetic and vagal nerves, respectively.

  b. Hypothalamus. The medullary cardiovascular centers receive input from higher levels of the brain. The effect of temperature changes on the hypothalamic centers are relayed to the medulla and cause vasoconstriction (heat conservation) or vasodilation (heat dissipation) in the vessels of the skin. Emotional stresses also influence heart rate and blood pressure, and these effects are relayed from the higher centers to stimulate or inhibit the medullary centers. In some animals, but not in humans, the skeletal muscles receive a sympathetic cholinergic innervation as well as the α-adrenergic innervation that normally controls blood flow. Stimulation of these cholinergic fibers causes a vasodilation, which occurs in anticipation of exercise. The vasodilation opens the **thoroughfare channels** in skeletal muscle; once exercise begins, however, the nutrient channels are dilated by the release of the local vasodilator.

  c. Cortex. Stimulation of various areas of the motor cortex leads to complex responses, including appropriate cardiovascular adjustments. These complex motor patterns involve autonomic responses. More and more frequently, autonomic training is being used in attempts to control heart rate and blood pressure, which emphasizes the control that the higher centers can exert on the autonomic nervous system. The sympathetic cholinergic outflow to the skeletal muscle arises from cerebral areas and follows a separate descending tract than the sympathetic adrenergic outflow.

3. The **autonomic nervous system** is divided into a sympathetic and a parasympathetic component.

  a. Sympathetic preganglionic fibers arise from cell bodies in the intermediolateral gray columns of the spinal cord of the thoracodorsal regions. Axons of these cells exit the CNS through the ventral root and proceed to the paravertebral chain of ganglia, where they synapse with the postganglionic cells of the sympathetic system. All of the preganglionic sympathetic fibers are cholinergic, while sympathetic postganglionic fibers may be either adrenergic or cholinergic. Sympathetic postganglionic cholinergic fibers innervate some exocrine glands and sweat glands and are the vasodilator fibers in skeletal muscles of certain species.

   (1) **Vasomotor Tone.** Most of the sympathetic nerves to blood vessels exhibit some neural activity, inducing about 50 percent of the maximal vasoconstrictor activity in most of the vascular beds. Thus, vasoconstriction or vasodilation is produced by increasing or decreasing the sympathetic neural activity.

   (2) **Denervation hypersensitivity** follows interruption of the sympathetic nerve supply to vascular smooth muscle and certain glands. Normally, receptors are located in the subjacent areas of the nerve terminals. Following denervation, however, receptors develop over the entire membrane, causing the end organs to be extremely sensitive to any circulating catecholamines. There are two types of receptors to sympathetic mediators.

    (a) **Alpha receptors** are stimulated most strongly by epinephrine and norepinephrine; these receptors cause **vasoconstriction**.

    (b) **Beta Receptors.** These exist in two forms, β_1 anb β_2; both are stimulated most by isoproterenol and least by norepinephrine. The β_1 receptors are located in the myocardium (not the coronary arteries) and cause positive inotropic and chronotropic responses. The β_2 receptors are located throughout the body; stimulation of these receptors leads to such reactions as bronchodilation and vasodilation. Pharmacologic agents are available that are specific for stimulating or blocking the β_1 or β_2 receptors.

b. The **parasympathetic nervous system** arises from cranial and sacral outflows of the CNS. The major effect of the parasympathetic system on cardiovascular function is its action, through the vagus nerve, on the SA and AV nodes and its direct reduction of atrial contractility. Marked efferent activity over the vagus nerve can result in extensive slowing of the heart, sinus arrest, and AV nodal block.

VIII. SPECIAL CIRCULATORY BEDS

A. CORONARY CIRCULATION. The heart is supplied by the right and left coronary arteries; the left coronary artery divides after about 1 cm into the anterior descending and circumflex branches. Normally, the coronary arteries function as end arteries; however, the presence of an arterial plaque or occlusion allows anastomoses to become functional. The heart has a very limited capacity for anaerobic metabolism (O_2 debt), so adequate flow is essential to maintain a constant adequate supply of nutrients. The heart metabolizes all substrates in approximate proportion to their vascular concentration but preferentially metabolizes fatty acids when available.

1. Coronary flow rate represents approximately 5 percent of the resting cardiac output or about 60–80 ml of blood/100 g of tissue/min.

2. Myocardial O_2 consumption averages 6–8 ml O_2/100 g of tissue/min; thus, hemoglobin delivers approximately half of its O_2 to the myocardium. In contrast, hemoglobin releases only about one-fourth of the O_2 to the remainder of the body at rest. The venous PO_2 of the heart is about 25–30 mm Hg under resting conditions and is even lower during exercise or stress. During exercise, cardiac O_2 consumption may increase five- to sixfold in well-trained athletes due to increases in coronary flow and to a widening of the A-V O_2 difference.

3. Coronary Organization. The coronary vessels traverse the epicardium of the heart and then subdivide, sending **penetrating branches** through the myocardium. These branches further subdivide into **arcades** that distribute blood to the myocardium. During ventricular systole, myocardial wall tension increases, compressing the vessels and, thus, increasing the resistance to flow. Consequently, the coronary flow pattern is complex due to this extravascular compression (Fig. 2-20). Maximal flow in the left coronary vessels usually occurs during isovolumic relaxation, while arterial pressure is still relatively high and the myocardial wall tension is falling.

a. Left ventricular venous drainage occurs primarily through the coronary sinus, with some small contributions directly through the thebesian veins and coronary luminal connections.

b. Right ventricular venous drainage occurs through the multiple anterior coronary veins. A larger proportion of right ventricular flow occurs through the coronary luminal connections than for the left ventricle.

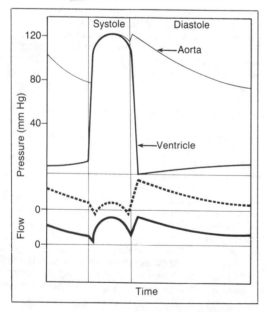

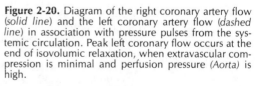

Figure 2-20. Diagram of the right coronary artery flow (*solid line*) and the left coronary artery flow (*dashed line*) in association with pressure pulses from the systemic circulation. Peak left coronary flow occurs at the end of isovolumic relaxation, when extravascular compression is minimal and perfusion pressure (*Aorta*) is high.

4. Control of Coronary Flow.
  a. Adenosine. The myocardium extracts large amounts of O_2 from the coronary blood even at rest; therefore, during exercise it is critical that coronary flow increases to maintain an adequate PO_2 in the myocardium. The major factor causing coronary vasodilation during hypoxic states is adenosine, which is derived from adenosine monophosphate by a reaction catalyzed by the enzyme, 5'-nucleotidase. Release of adenosine into the myocardium causes a strong vasodilator response.
  b. Sympathetic stimulation causes coronary vasodilation primarily due to an increase in the metabolic rate of the myocardium. Such an increase without a compensatory increase in coronary flow causes the breakdown of adenosine triphosphate (ATP) and the release of adenosine. The coronary vessels contain both alpha and beta receptors, but the alpha vasoconstrictor activity is weak. Sympathetic cholinergic vasodilator fibers are absent in the myocardium.

B. CEREBRAL CIRCULATION. Although the brain is surrounded by the rigid skull, vascular diameter is altered in response to changing blood flow needs, because vasodilation on the arterial side is compensated by narrowing of the venous bed. Interruption of blood flow for only 5–10 seconds causes loss of consciousness, and circulatory arrest for only 3–4 minutes causes irreversible brain damage.

  1. Organization. The major vessels supplying the brain in humans are the carotid arteries. The carotid sinus provides a reflex input to maintain blood pressure at an adequate level to maintain flow to the brain.

  2. Cerebral autoregulation effectively controls blood flow over a 80–180 mm Hg range of arterial pressures. Hypercapnea and hypocapnea double and halve, respectively, normal cerebral blood flow. The cerebral vasodilation produced by both hypoxia and hypercapnea appear to be due to the H^+ that arises from lactic and carbonic acids. The reduction in blood flow accompanying hyperventilation is sufficient to produce cerebral hypoxia, which causes dizziness and fainting.

  3. Neural control of cerebral blood flow is limited to the larger vessels and those located within the pia. The cerebral vessels are supplied by both sympathetic and parasympathetic neurons, but their role is unknown.

  4. The **Cushing reflex** is a systemic vasoconstriction in response to an increase in the cerebrospinal fluid (CSF) pressure. This reflex markedly increases systemic pressure in response to increases in the CSF pressure in order to maintain adequate blood flow to the brain. The reflex is initiated by the medullary centers, which are activated by a hypoxic state due to ischemia.

C. CIRCULATION THROUGH SKELETAL MUSCLE. Skeletal muscle comprises 40–50 percent of the total body weight in a normal adult male. Because of this large mass, even resting skeletal muscle uses approximately 20 percent of the total resting O_2 consumption.

  1. Resting skeletal muscle receives a relatively large blood flow for its metabolic rate so that venous O_2 content is about 17–18 ml/dl. Blood flow in resting skeletal muscle varies from 1.5 to 6.0 ml/100 g/min, and the variability is largely due to the type of muscle. Muscles composed mainly of red fibers receive a larger blood flow than muscles composed mainly of white fibers.

  2. Exercising skeletal muscle receives about 80 ml blood/100 g/min and extracts about 80 percent of the O_2 from the arterial blood. In some species, sympathetic cholinergic fibers increase muscle blood flow before the actual onset of exercise. Vasodilation during exercise occurs due to local increases in CO_2, H^+, K^+, adenosine, and osmolality and to a decrease in O_2. Even with marked vasodilation the blood flow is inadequate to maintain aerobic levels in cells during moderate to severe exercise. The **anaerobic threshold** occurs at an O_2 consumption that is approximately 60 percent of the maximal. The anaerobic threshold is the level of exercise that produces a significant rise in the level of lactic acid in the bloodstream.

D. FETAL CIRCULATION is different in many ways from postnatal circulation. The placenta is a major organ in the fetus and receives more than 50 percent of the combined output of both ventricles. At birth, marked changes occur in both the systemic and pulmonary circulations as the placental blood supply is lost.

  1. Pulmonary circulatory changes are primarily due to a marked decrease in pulmonary vascular resistance as the lungs are inflated at birth. After inflation of the lungs, the pulmonary vascular resistance is only about one-tenth of its value during fetal life.
  a. Foramen Ovale. The decrease in pulmonary vascular resistance drops the pressure in the

right atrium below that in the left atrium, a situation that is opposite that during fetal life. The pressure reversal causes the flap-like valve to close over the foramen ovale, which normally fuses to the septum after a few days.

  b. The **ductus arteriosus** conducts blood from the pulmonary artery to the aorta during fetal life. The flow through this vessel reverses after birth as the pulmonary resistance is reduced and the pulmonary artery pressure becomes less than in the aorta. The ductus arteriosus begins to constrict shortly after birth due to the release of prostaglandins. After birth, patency of this vessel can be maintained with administration of aspirin, which inhibits prostaglandin formation. The ductus arteriosus may remain patent for several days, during which time a continuous murmur may be heard due to turbulent blood flow through the narrowed structure (see Fig. 2-15).

  c. **Right ventricular wall** thickness is about equal to that of the left ventricular wall during fetal life. The thickness of the right ventricle decreases over several months, and the electrical axis of the heart swings to the left over a period of several years until it lies within the normal adult quadrant.

2. **Systemic circulatory changes** are caused by the loss of the placental circulation, which makes more blood available to all the other body organs. The umbilical arteries constrict in response to cooling, trauma, circulating catecholamines, and high PO_2. The loss of the placental circulatory bed after birth greatly increases the systemic vascular resistance, which raises the arterial pressure and, consequently, the afterload of the left ventricle. The higher afterload leads to increases in left ventricular and left atrial pressures, which contribute to the closure of the foramen ovale.

IX. RESPONSES TO STRESS

A. **SHOCK** is a complex syndrome characterized by inadequate blood flow to such critical organs as the heart, brain, liver, kidneys, and gastrointestinal tract.

1. **Classification.**
   a. **Primary shock** or **syncope** is due to a vasovagal reflex that produces peripheral vasodilation, cardiac slowing, or both. Syncope usually is triggered by pain, stress, fright, heat, and the sight of blood. The vasovagal reflex causes vasodilation and, thus, a drop in blood pressure. If an individual is sitting, and especially if he or she is standing, cerebral blood flow becomes inadequate and consciousness is lost. Such an individual should be kept recumbent until blood pressure is normal, thus ensuring adequate perfusion of the brain.
   b. **Hypovolemic shock** is caused by the following disorders:
      (1) **External hemorrhage** secondary to trauma
      (2) **Internal hemorrhage** secondary to fractures, hemothorax, and hemoperitoneum
      (3) **Gastrointestinal losses** secondary to diarrhea and vomiting
      (4) **Renal losses** secondary to diabetes mellitus, diabetes insipidus, and excessive use of diuretics
      (5) **Cutaneous losses** secondary to burns, sweating, and exudation
   c. **Cardiac shock** may be due to postinfarction cardiac dysfunction, or it may be secondary to a tachyarrhythmia or a severe bradycardia secondary to complete A-V block.
   d. **Other forms** of shock include:
      (1) **Septic shock** due to septicemias (especially gram-negative bacteria) and to liberated bacterial toxins
      (2) **Neuropathic shock** secondary to anesthesia, drug overdose, ganglion-blocking drugs, spinal shock, and orthostatic hypotension
      (3) **Blood flow obstruction** due to massive pulmonary emboli, tension pneumothorax, positive end-expiratory pressure respirations (PEEP), and cardiac tamponade
      (4) **Shock** caused by endocrine failure (e.g., Addison's disease and myxedema), severe hypoxia, and anaphylaxis

2. **Physical Signs.** Generally, individuals who are in shock initially are agitated and restless. These symptoms later progress to lethargy, confusion, and coma. More specific signs of shock depend on the stage of shock and the severity of the dysfunction.
   a. The **first stage** is characterized by a moderate reduction in cardiac output secondary to fluid loss or a negative inotropic effect on the heart. In general, compensatory reflexes may maintain the blood pressure at normal or slightly below normal levels. These reflex adjustments increase heart rate and sympathetic discharge, as evidenced by decreased blood flow to the skin (pallor) and a cold sweat.
   b. The **second stage** of shock occurs with blood volume losses of 15–25 percent. Intense arteriolar vasoconstriction results, which usually is not adequate to maintain a normal blood pressure since the cardiac output is moderately reduced.

    c. The **third stage** is characterized by small additional losses of blood or cardiac function, which produce a rapid deterioration of the circulation to critical body organs. If this state persists, widespread tissue injury results, which is termed **irreversible shock**.

  3. Treatment. Normal cardiac output should be reestablished by replenishing the circulating blood volume, if this is depleted. Also, normal sinus rhythm should be reestablished and the inotropic state of the ventricles should be increased with digitalis or β-sympathetic agonists.

B. HYPERTENSION generally is defined as a condition characterized by systolic pressure that exceeds 140 mm Hg and diastolic pressure that exceeds 90 mm Hg. Recent evidence indicates that mortality and morbidity rates are increased in patients with diastolic pressures above 85 mm Hg, which may prove to be a new borderline for the diagnosis of hypertension.

  1. Classification.
    a. Essential hypertension accounts for more than 90 percent of all cases of hypertension and is of unknown etiology. Excessive salt intake, resetting of the baroreceptors, increased salt and fluid in the arteriolar walls, and long-standing exposure to stress are some of the theories behind the increased vascular resistance that increases systemic arterial pressures.
    b. Systolic hypertension with a wide pulse pressure is caused by various malformations or conditions that decrease the compliance of the aorta or that increase the cardiac output or stroke volume of the left ventricle. These latter conditions include thyrotoxicosis, patent ductus, A-V fistula, fever, and aortic valve insufficiency.
    c. Systolic and diastolic hypertension has a varied etiology.
      (1) Renal disease, if severe, results in systemic hypertension due to the release of renin and the formation of AII. Destruction of the renal parenchyma secondary to any severe renal disease as well as renovascular stenosis and renal infarction all result in hypertension.
      (2) Endocrine diseases such as Cushing's syndrome, primary hyperaldosteronism, pheochromocytoma, and acromegaly may cause hypertension.
      (3) Miscellaneous causes of hypertension include psychogenesis, coarctation of the aorta (which is important to diagnose because it can be surgically cured), polyarteritis nodosa, and excessive transfusion.
    d. Malignant hypertension occurs when arterial pressure is markedly increased and vascular, CNS, renal, and cardiac deterioration occurs over weeks or months instead of years.

  2. Symptoms of hypertension generally are absent until the disease is far advanced and has caused cardiac, renal, neural, and ocular damage. The disease usually is diagnosed as the result of a blood pressure reading taken during a routine physical examination.
    a. Cardiac symptoms are due to the increased work required for the left ventricle, which initially hypertrophies and eventually, if the hypertension is severe, fails with pulmonary congestion and edema. Hypertension also increases atheromatous plaque formation in vessels, which may precipitate coronary insufficiency (angina) or myocardial infarction.
    b. Neural symptoms include occipital headaches (especially common in the morning), vertigo, tinnitus, dimmed vision, and syncope. Permanent neurologic damage can result from vascular occlusion and hemorrhage secondary to the stress imposed on the vessels by the hypertension.
    c. Renal effects are due to atherosclerotic lesions that affect the afferent and efferent arterioles and the glomerular tufts. These lesions cause a decreased glomerular filtration rate and renal tubular dysfunction, which eventually lead to renal failure.

  3. Treatment of hypertension is critically important to prevent or minimize damage to cerebral, ocular, renal, and cardiac tissues. Treatment of essential hypertension is primarily with diuretics and β-adrenergic blocking drugs. Treatment of secondary hypertension is based on the underlying disease.

C. EXERCISE
  1. General Considerations. The cardiovascular and respiratory response to exercise has gained major interest due to the current concern with physical conditioning. In addition, exercise is used as a stress test to evaluate the respiratory system, coronary circulation, and the cardiac reserve.
    a. Specificity. Endurance type training produces different effects than weight training, and only endurance (aerobic) exercises produce cardiovascular conditioning. Training effects are specific for the particular muscle groups involved.
    b. Characteristics. Muscles conditioned by endurance training exhibit increases in capillary density, myoglobin concentration, glycogen, and mitochondrial enzymes of the citric acid

cycle. These enzymes are involved in the breakdown of long-chain fatty acids, which serve as an energy source during exercise and, therefore, spare muscle glycogen.

2. Cardiac Response to Exercise.

   a. Heart rate increases linearly with work rate up to a maximum that is determined by an individual's age (Fig. 2-21). The maximal heart rate (HR_{max}) can be estimated using the following equation:

$$HR_{max} = 210 - [0.65 \times age\ (yrs)].$$

   Cardiovascular conditioning requires maintaining a heart rate of 60–70 percent of the HR_{max} for 20–30 minutes, 3–4 times a week, for at least 3 months to obtain optimum benefits.

   b. Stroke Volume. Endurance training increases the stroke volume of the heart. Thus, conditioned athletes can maintain any level of cardiac output at a lower heart rate than nonconditioned individuals.

   c. Maximal cardiac output in conditioned athletes is greater than in nonconditioned individuals due to the slower heart rates obtained through conditioning. Conditioning does not have any effect on the maximal heart rate for an individual. The maximal cardiac output is approximately 30–35 L/min in Olympic-class runners, while the maximal cardiac output in nonconditioned adults is approximately 20 L/min.

   d. Anaerobic threshold (AT) occurs at approximately 60 percent of the maximal exercise level. The AT represents the point at which muscle metabolism shifts from an aerobic to an anaerobic state, which liberates lactate into the circulation. Training improves the absolute AT as well as the AT relative to the maximal O_2 consumption.

3. Respiratory Responses to Exercise.

   a. Minute ventilation increases linearly with work rate until the AT is reached. At that point, minute ventilation increases more steeply with work rate due to the additional respiratory drive imposed by the liberated lactic acid (see Fig. 2-22). Endurance training increases the maximal minute ventilation achieved during exercise, but the maximal voluntary ventilation does not improve.

   b. CO_2 output increases linearly with work rate until AT, when the CO_2 output increases more steeply due to the increased respiration. Above the AT, the arterial CO_2 declines as the body stores are depleted since excretion exceeds production.

   c. O_2 consumption increases linearly with work rate and is directly dependent on the work performed. The O_2 consumption does not change with training at any workload. Training does not improve the body's efficiency unless muscle coordination is improved by practice. Conditioning produces a 5–20 percent increase in the maximal O_2 consumption as a result of an increase in the cardiac output and a widening of the A-V O_2 difference at the maximal exercise level.

   d. The **alveolar to arterial (A-a) O_2 gradient** normally remains at 5–10 mm Hg during moderate levels of exercise. Beyond the AT, however, the A-a O_2 gradient widens slightly due to an increase in the alveolar PO_2.

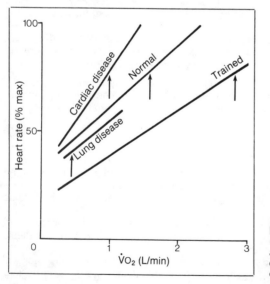

Figure 2-21. The effect of training, heart disease, and lung disease on the heart rate response to endurance (aerobic) exercise. *Vertical arrows* mark the onset of the anaerobic threshold, which occurs at about 60 percent of the maximal exercise level for every individual. $\dot{V}O$ = oxygen utilization.

e. The **respiratory exchange ratio** (R, $\dot{V}CO_2/\dot{V}O_2$) rises to greater than 1 at work levels above the AT. Normally, however, R never exceeds 1.25, even at maximal levels of exercise. The increase in R is due to the conversion of bicarbonate to carbonic acid in the buffering process of lactic acid and to the liberation of the CO_2 at the lungs.

4. Pathologic Responses.
   a. Chronic obstructive lung disease limits exercise mainly due to the onset of severe dyspnea. In these patients, heart rate does not reach the minimal level for training effects, and the AT occurs at very low levels of exercise due to the impairment of O_2 uptake (see Fig. 2-21). Thus, exercise training in these patients has only slight benefits that seem to be due mainly to desensitization of symptoms. In these patients, it is more worthwhile to emphasize training of the respiratory muscles.
   b. Cardiac patients have an early onset of the AT and reach maximal heart rates at much lower levels of exercise than normal individuals (Fig. 2-22).

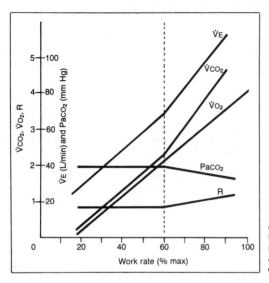

Figure 2-22. The effects of exercise on respiratory parameters of gas exchange. The *dashed line* indicates the onset of the anaerobic threshold. $\dot{V}E$ = minute ventilation; $\dot{V}CO_2$ = carbon dioxide excreted in expired gas per minute; $\dot{V}O_2$ = oxygen utilization per minute; $PaCO_2$ = arterial carbon dioxide tension; and R = the respiratory exchange ratio.

STUDY QUESTIONS

Directions: Each question below contains five suggested answers. Choose the **one best** response to each question.

1. In a recumbent individual, the greatest difference in blood pressure occurs between which two vessels?

(A) The ascending aorta and brachial artery
(B) The saphenous vein and right atrium
(C) The pulmonary artery and left atrium
(D) The arteriolar and venous ends of a capillary
(E) The femoral artery and femoral vein

2. The QRS complex occurs just prior to which of the following events of the cardiac cycle?

(A) The c wave of the atrial pressure curve
(B) The v wave of the atrial pressure curve
(C) The second heart sound
(D) The fourth heart sound
(E) Isovolumic relaxation

3. The principal energy source for the heart of a resting, healthy, fasting individual is

(A) glucose
(B) free fatty acids
(C) lactate and pyruvate
(D) amino acids
(E) polypeptides

4. A complete heart block is recognized on an EKG by which of the following abnormalities?

(A) Prolongation of the P-R interval
(B) Prolongation of the QRS complex
(C) Prolongation of the Q-T interval
(D) Dissociation of the atria and ventricles
(E) Irregularity of the pulse rate

5. Peak ventricular pressure is significantly greater than peak aortic pressure in which of the following conditions?

(A) Hypertension
(B) Aortic insufficiency
(C) Aortic stenosis
(D) Aortic atherosclerosis
(E) Aortic aneurysm

6. With hyperkalemia, the P wave may disappear without any change in heart rate or the configuration of the QRS complex. This may be due to

(A) suppression of the SA node and emergence of an ectopic pacemaker in the AV node
(B) suppression of the SA node and emergence of a ventricular pacemaker
(C) production of a SA nodal block
(D) production of an AV nodal block
(E) inhibition of atrial muscle conduction

7. A prolongation of the AV conduction time is recognized on the EKG by a

(A) long, flat P wave
(B) P-R interval exceeding 0.21 second duration
(C) QRS complex exceeding 0.12 second duration
(D) Q-T interval exceeding 0.18 second duration
(E) T wave exceeding 0.1 second duration

8. Which of the following mechanisms is most important for increasing blood flow to skeletal muscles during exercise?

(A) An increase in aortic pressure
(B) An increase in α-adrenergic impulses
(C) An increase in β-adrenergic impulses
(D) A vasoconstriction in the splanchnic and renal systems
(E) A vasodilation secondary to local metabolites

9. A ventricular rate of less than 60 beats/min in humans may result from all of the following conditions EXCEPT

(A) first-degree AV block
(B) second-degree AV block
(C) third-degree AV block
(D) vagal stimulation
(E) cardiovascular conditioning programs

10. Which of the following factors is most important to blood flow regulation at the local level?

(A) Arteriovenous pressure difference
(B) Metabolic activity of the organ or tissue
(C) Local neurotransmitters
(D) Circulating neurotransmitters
(E) Cardiac output

11. All of the following mechanisms will increase the transcapillary flux of liquids, solids, and gases EXCEPT

(A) an increase in capillary surface area
(B) an increase in the concentration gradient across the capillary wall
(C) an increase in the diffusion distance
(D) an increase in the capillary transit time
(E) a decrease in arteriolar resistance

Directions: Each question below contains four suggested answers of which **one or more** is correct. Choose the answer

A   if **1, 2, and 3** are correct
B   if **1 and 3** are correct
C   if **2 and 4** are correct
D   if **4** is correct
E   if **1, 2, 3, and 4** are correct

12. Premature excitation of the ventricles causes a compensatory pause between ventricular beats, which

(1) is longer than that following premature excitation of the atria
(2) is due to the long refractory period of ventricular structures
(3) allows time for additional ventricular filling
(4) equals twice the length of the normal cardiac cycle

15. A first-degree AV block is revealed on a patient's EKG. Additional features exhibited on this EKG would include

(1) a more rapid atrial than ventricular rate
(2) inverted P waves in the bipolar limb leads
(3) a QRS complex duration that exceeds 0.12 second
(4) a prolonged P-R interval

13. True statements concerning fluid exchange across the microvascular bed include which of the following?

(1) Fluid exchange follows the laws of pressure filtration across a semipermeable membrane
(2) Fluid is filtered outward at the arterial end of the vascular bed
(3) Edema occurs if the plasma protein concentration is reduced drastically
(4) Edema occurs if the arterial blood pressure decreases

16. During moderate exercise, diastolic arterial pressure may not change significantly from that during rest, although cardiac output increases. This is the result of

(1) increased stroke volume
(2) increased heart rate
(3) decreased arteriolar runoff during diastole
(4) decreased total peripheral resistance

14. A patient has right axis deviation (RAD) on an EKG. RAD is consistent with

(1) chronic lung disease
(2) an electrical axis most parallel to lead II
(3) right ventricular hypertrophy
(4) a positive QRS complex in lead I

17. Cutting the afferent nerves from the aortic and carotid baroreceptors will cause which of the following physiologic changes?

(1) A decrease in heart rate
(2) An increase in peripheral resistance
(3) An increase in pulse pressure
(4) An increase in mean arterial pressure

SUMMARY OF DIRECTIONS

A	B	C	D	E
1, 2, 3 only	1, 3 only	2, 4 only	4 only	All are correct

---

18. Cardiovascular disturbances that lead to tissue edema include

(1) venous obstruction
(2) low intravascular protein concentration
(3) lymphatic obstruction
(4) arteriolar vasoconstriction

19. Conditions resulting from increased intrapericardial pressure (cardiac tamponade) include

(1) increased venous pressure
(2) increased arterial pressure
(3) decreased ventricular filling
(4) decreased atrial pressure

Directions: The groups of questions below consist of lettered choices followed by several numbered items. For each numbered item select the **one** lettered choice with which it is **most** closely associated. Each lettered choice may be used once, more than once, or not at all.

Questions 20–22

Match each mechanism that stimulates cardiovascular activity to its associated response.

(A) Starling effect
(B) Increased contractility
(C) Both the Starling effect and increased contractility
(D) Neither the Starling effect nor increased contractility

20. Vagal impulses to the heart

21. An increase in end-diastolic volume

22. Assuming an upright posture

Questions 23-27

Match each sign or symptom of cardiac dysfunction with the appropriate valve lesion or lesions.

(A) Aortic regurgitation
(B) Aortic stenosis
(C) Both aortic regurgitation and aortic stenosis
(D) Neither aortic regurgitation nor aortic stenosis

23. Decreased left atrial pressure

24. EKG that reveals a left axis deviation

25. Greatly increased pulse pressure

26. Diastolic murmur

27. Increased arterial systolic pressure

ANSWERS AND EXPLANATIONS

1. The answer is E. (*I C; Fig. 2-2*) In passing from the femoral artery to the femoral veins, blood must go through the arterioles—the "stopcocks" of the circulation. In a recumbent individual, the pressure in the femoral artery is about 90 mm Hg but only 15–20 mm Hg in the femoral vein, making the pressure gradient 70–75 mm Hg in this location. The pressure loss from the aorta to the distributing arteries is only a few mm Hg. Venous pressure normally is 15–20 mm Hg, and the right atrial pressure is −5 to 0 mm Hg. The pressure in the pulmonary artery averages about 15 mm Hg, and left atrial pressure is 0–12 mm Hg. Arteriolar pressure is about 30–40 mm Hg.

2. The answer is A. (*V A 2 b*) The QRS complex occurs just prior to the onset of ventricular contraction, which is indicated on the atrial pressure curve by the inscription of the c wave. The second heart sound occurs at the end of systole and is followed by the period of isovolumic relaxation. Isovolumic relaxation terminates with the opening of the AV valves, at the peak of the v wave of the atrial pressure curve. The fourth heart sound occurs during atrial systole.

3. The answer is B. (*VIII A*) The heart can metabolize any of the following materials: glucose, free fatty acids, lactate and pyruvate, amino acids, and polypeptides; whichever substance is present in adequate supply is used by the heart. In a resting, healthy, fasting individual, however, the major metabolite used by the heart would be free fatty acids.

4. The answer is D. [*IV C 2 a (3)*] A complete (third-degree) heart block indicates that the atria and ventricles are exhibiting completely independent rates. Prolongation of the P-R interval to more than 0.21 second indicates a first-degree heart block. Prolongation of the QRS complex and the Q-T interval indicates disturbances in depolarization and repolarization, respectively. Irregularity of the pulse rate usually indicates atrial fibrillation or atrial or ventricular premature beats.

5. The answer is C. (*V A 2 c*) The narrowed aortic valve represents a high resistance segment that requires a high pressure gradient to generate flow. The normal aortic valve represents a relatively low resistance segment; the pressure gradient is minimal in this case, so that peak ventricular and aortic pressures are essentially equal. Hypertension, aortic insufficiency, aortic atherosclerosis, and aortic aneurysm all are associated with a normal aortic valve.

6. The answer is E. (*II C 1*) The atrial muscle is more sensitive to depolarization block due to hyperkalemia than the specialized atrial tracts. Both SA and AV nodal blocks are associated with a decreased heart rate. SA-node suppression also is associated with a decrease in heart rate; in addition, cases involving a ventricular pacemaker are characterized by an abnormal QRS complex.

7. The answer is B. (*IV C 2*) The AV conduction time is the time required for a wave of depolarization to travel from the atria to the ventricles. Most of this time is occupied by the transmission through the AV node because conduction is so slow in this tissue. The P-R interval, which includes this period, normally has a duration of 0.12–0.21 second.

8. The answer is E. (*VIII C 2*) During exercise, blood flow to skeletal muscle can increase to 20–30 times blood flow during rest. An increase in the driving force or an increase in the flow resistance in other organs would not be adequate to cause the increased flow. The blood flow increases by a coordinated reflex effect that increases cardiac output and a local effect causing vasodilation in muscles due to an increase in the osmolality and in the concentrations of K^+, H^+, adenosine, and CO_2, and to a reduction in local PO_2.

9. The answer is A. (*IV C 2*) Bradycardia, or a heart rate of less than 60 beats/min, will not result from first-degree AV block, which is simply a prolongation of the P-R interval. The long P-R interval is due to a slowed conduction from the SA node to the ventricular myocardium. Second-degree AV block can result in bradycardia since a fraction of atrial beats are lost. In third-degree AV block, the ventricular pacemaker usually has a rate of less than 60 beats/min. During vagal stimulation and in highly conditioned athletes, the increased vagal tone slows the SA node to bradycardic levels.

10. The answer is B. [*I D 4 b (1)*] Some suggested mediators of local blood flow include low PO_2; high concentrations of CO_2, K^+, and H^+; and increased osmolarity of the interstitial fluids secondary to increased metabolic activity. Circulating and local neurotransmitters do not have the focused activity necessary to control local blood flow. All systemic organs have access to cardiac output, but local changes in vascular resistance allow flow changes.

11. The answer is C. (*I D 4 a*) Recruitment of capillaries causes an increase in exchange area and, therefore, the efflux of substances; an increase in the concentration gradient produces the same effect. A decrease in arteriolar resistance causes an increase in capillary pressure; an increase in the capillary transit time provides additional time for exchange to occur. An increase in diffusion distance reduces the rate of transfer of substances.

12. The answer is A (1, 2, 3). *(IV B 2 b, 4 a; Fig. 2-13)* The compensatory pause after a ventricular premature beat allows ventricular rhythm to continue unchanged after the premature beat. The additional filling time allows the next beat to be more powerful than normal; this is the beat normally felt by the patient. The compensatory pause itself does not equal two cardiac cycles; the premature beat **plus** the succeeding beat equal two cycles.

13. The answer is A (1, 2, 3). *(I D 4 b)* A decrease in arterial blood pressure lowers capillary pressure; this causes a decrease, not an increase, in the outward flux of water. The capillary bed is essentially a semipermeable membrane where water leaves the vascular compartment when the outward filtration forces (primarily the intravascular pressure) exceed the reabsorption forces (mainly the intravascular oncotic pressure). Filtration is most likely to occur at the arterial end of the capillary, where hydrostatic pressure is highest. Reabsorption of fluid is reduced if there is a low protein, especially albumin, concentration.

14. The answer is B (1, 3). *[III E 3 b (1)]* Chronic lung disease, with destruction of the pulmonary vascular bed, is the most common cause of right axis deviation (RAD). This damage to the pulmonary circulation increases the work of the right side of the heart, which leads to right ventricular hypertrophy. The electrical axis in RAD is more parallel to lead III or lead I than it is to lead II, and the QRS complex is negative in lead I.

15. The answer is D (4). *[IV C 2 a (1)]* In first-degree AV block, the P-R interval is extended beyond the normal limit of 0.21 second. The block is caused by prolongation of the conduction time through the AV node. This prolongation is linked to hypoxia and such agents as toxins and drugs. A more rapid atrial than ventricular rate indicates either a second- or third-degree AV block. Inverted P waves indicate an altered direction of atrial depolarization (ectopic pacemaker). Prolongation of the QRS complex indicates an intraventricular block, which is below the level of the AV node.

16. The answer is D (4). *(IX C 2 a–d)* The increased blood flow to skeletal muscle occurs because of a local vasodilation, which increases the rate of runoff of blood from the arterial system. Although stroke volume and heart rate increase during exercise (which would raise arterial diastolic pressure), the increased rate of runoff compensates for these changes, and the diastolic pressure remains near resting levels.

17. The answer is C (2, 4). *(VI B 2; VII B 1 b)* The afferent limb of the moderator reflex normally provides important input to the cardiovascular control centers; cutting these nerves will interrupt this function. Normally, the heart has tonic vagal impulses that slow its rate; therefore, interruption of the reflex will cause an increase in heart rate. Afferent impulses from the baroreceptors also function in the cardioinhibitory center of the medulla. Interruption of these impulses will cause a loss of inhibition and, therefore, an increase in peripheral resistance. From the relationships defined in the elastic modulus, an increase in heart rate and peripheral resistance will reduce the peripheral runoff of blood during each cardiac cycle. This causes the volume of blood in the arterial system to increase, raising mean arterial pressure and reducing the pulse pressure.

18. The answer is A (1, 2, 3). *(I D 4 b; Fig. 2-3)* Edema formation can result from increased capillary pressure (due to venous obstruction), decreased oncotic forces within the vasculature (due to low intravascular protein concentration), and impaired fluid removal from the tissues (due to lymphatic obstruction). Another cause of tissue edema is increased permeability of capillary walls, which is caused by such agents as histamine. Arteriolar vasoconstriction reduces capillary pressure and, therefore, does not lead to edema formation.

19. The answer is B (1, 3). *(V C 3)* Accumulation of fluid in the pericardial cavity compresses the heart and compromises cardiac filling. Thus, stroke volume decreases, which tends to raise the ventricular end-diastolic pressure and the venous pressure. The arterial pressure decreases due to the reduced stroke volume and cardiac output, while atrial pressure rises from the compressive effects of the pericardial fluid.

20–22. The answers are: 20-A, 21-A, 22-B. *(II C 1 b (3), D 2–3; VI C 1)* Vagal impulses cause a slowing of the sinus pacemaker, so that diastolic filling time is prolonged and ventricular volume is increased. This increase in the preload alters the length of myocardial fibers and invokes the Starling effect to increase stroke volume. The vagi do not innervate human ventricular muscle, so there is no effect of vagal discharge on the ventricular myocardium. An increase in end-diastolic volume activates the Starling effect in the heart. Increased contractility results from an increased force of contraction at a constant or reduced end-diastolic volume. Assuming an upright posture causes pooling of blood in the dependent portions of the body, so that central blood volume and ventricular volume are reduced. In order to maintain normal cardiac output, the sympathetic tone to the heart must increase; thus, ventricular contractility increases.

23–27. The answers are: 23-D, 24-C, 25-A, 26-A, 27-A. (*IV D 2 b; V B 4 a*) Both aortic stenosis and aortic regurgitation increase left ventricular end-diastolic pressure and volume, thereby increasing the left atrial pressure. Both lesions also enlarge the left ventricle, causing a prolongation of left ventricular depolarization time. This appears on the EKG as a left axis deviation, a sign that is consistent with left ventricular hypertrophy. Aortic stenosis impedes ejection, causing a small, late pulse called "pulsus parvus et tardus." Aortic regurgitation creates a large stroke volume, which results in a characteristically large pulse pressure. Aortic stenosis causes a systolic murmur; the murmur caused by aortic insufficiency occurs during diastole. The arterial systolic pressure is increased in aortic insufficiency due to the large stroke volume. With aortic stenosis, however, the rate of ventricular ejection is slower than normal; thus, arterial systolic uptake is below normal due to the increased time for runoff during systole.

3

Respiratory Physiology

I. INTRODUCTION. The primary function of the respiratory system is the exchange of oxygen (O_2) and carbon dioxide (CO_2) between the body and the environment. Secondary roles of the respiratory system include: aiding in acid-base balance, serving in the defense reactions of the body, acting as a circulatory filter, and functioning in hormonal regulation of several endocrine systems.

A. THE CONDUCTING ZONE. The upper airways, from the nose and mouth to the terminal airways, do not function in gas exchange but serve as conduits for the passage of air.

    1. Organization.

        a. The conducting system begins as a relatively narrow passageway with its smallest **cross-sectional area** at the level of the larynx. The system undergoes an irregular dichotomous branching for approximately 16 generations. (The branching continues for 7 additional generations, but these are included in the respiratory zone of the lungs.) Each subsequent generation of airways increases the total cross-sectional area of the system (Fig. 3-1).

        b. Because of the tremendous increase in the cross-sectional area of the terminal airways, the linear velocity of gas molecules is extremely low in this area. This low velocity of flow is important, because it produces laminar gas flow and a low flow resistance. The flow resistance of airways less than 2 mm in diameter represents only about 10 percent of the total airway resistance (see Section V B 2).

    2. Functions.

        a. Conditioning of Air. Inspired air normally is adjusted to body temperature and saturated with water vapor by the time it reaches the trachea during nasal breathing. Mouth breathing or a tracheostomy allows cool, dry air to reach the lower airways and can result in drying of secretions and impaction of mucous plugs.

        b. Removal of Foreign Material. Particulate matter in inspired air is removed by several mechanisms.

            (1) Filtration. As inspired gases pass through the nose, the nasal vibrissae filter out particles that are larger than 50 μ in diameter.

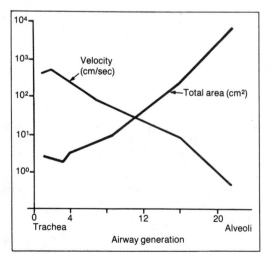

Figure 3-1. The reciprocal relationship between cross-sectional area and velocity of flow in the airways. Turbulence is most likely to occur in the central airways, where flow velocity is high. The low velocity in the terminal airways favors laminar flow and requires a small pressure difference (low resistance).

(2) Impaction. Most particles that are 2–50 μ in diameter are removed from inspired gas by impaction on the walls of the upper airways. The momentum of the particles is sufficient to prevent them from traversing these curved and branching airways so they stick to the mucous lining.

(3) Sedimentation. Particles that are 0.01–2 μ in diameter are largely deposited in the small airways and alveoli because of the effects of gravity.

(4) Diffusion. Particles less than 0.01 μ in diameter may reach the walls of the terminal airways and alveoli by diffusion. Deposition of irritating particles in these areas leads to the development of inflammation and the activation of macrophages. A large proportion of particles in this size range never settle on the respiratory surfaces and are expired.

(5) Solution. Some vapors, such as sulfur dioxide, are extremely soluble and go into solution in the moist linings of the upper airways. Soluble substances may dissolve in the tissues and reach the blood, where they can alter permeability and produce inflammation as they back-diffuse into the alveoli.

c. Reaction to Foreign Material.

(1) Irritant receptors in large bronchi are stimulated by foreign material, leading to **reflex bronchoconstriction, cough,** and increased **secretions of mucus**. Toxic material also can cause paralysis of cilia, leading to stasis of secretions and predisposing to infection.

(2) Deposition of foreign material in small airways produces **mucous gland hypertrophy** and **hyperplasia** as well as **edema** and **inflammatory cell infiltration** in the airways. These changes narrow the airway lumen, which increases airway resistance and predisposes to premature closure of the small airways during expiration. Premature airway closure also may be produced by an increased secretion of mucus, which can increase the surface tension of the mucous membranes.

(3) Foreign particles and vapors that reach the alveolar surface lead to the presence of an increased number of **alveolar macrophages**. These macrophages have an impaired ability to phagocytize material but release proteolytic enzymes into lung tissues. An increase in the proteolytic enzyme concentration or an impaired ability to inactivate these enzymes (α_1-antitrypsin deficiency) can cause tissue destruction, which may progress to emphysema.

B. THE RESPIRATORY ZONE

1. Organization. The respiratory portion of the lung includes the **respiratory bronchioles, alveolar ducts,** and **alveolar sacs** or **alveoli.** The functional unit of the lung is the **acinus,** which includes all of the structures distal to a terminal bronchiole. This distinction is functionally important because the conducting airways receive their blood supply from the bronchial arteries (systemic blood), whereas the respiratory portion is supplied by branches of the pulmonary artery.

2. Functions—Gas Exchange. Gas exchange across the respiratory membrane occurs by the process of diffusion. The structures are perfectly suited for this process; that is, an extremely large surface area (50–70 m²) is separated from the blood in the pulmonary capillaries by an extremely thin membrane averaging about 0.2 μ in thickness.

a. Diffusion.

(1) Definition. Diffusion, the process by which gas exchange occurs between alveolar gas and pulmonary capillary blood, is a random movement of molecules that causes a net transfer of material from areas of high concentration to areas of low concentration. Each species of gas molecule moves along its own concentration gradient independent of other gases.

(2) The Fick equation defines the factors governing diffusion and can be expressed as:

$$\dot{V}x = \frac{DxA (Pax - Pcx)}{t},$$

where $\dot{V}x$ = the volume of x transferred per minute; Dx = the diffusion properties of the membrane for x; A = the area of the exchange surface; t = thickness of the membrane; Pax = alveolar tension of x; and Pcx = capillary tension of x.

b. Diffusion Capacity of the Lung. In the lung, the ratio of the surface area and the diffusion distance is unknown but represents the critical factor in the transfer of gases. The effects of a change in this ratio (A/t) can be evaluated by measuring the diffusion capacity of the lung (DL), also termed the **transfer factor**, which is defined as:

$$DL = \frac{\dot{V}x}{Pax - Pcx} = \frac{DxA}{t}.$$

diffusion capacity of lung

(1) **Carbon Monoxide (CO).** CO is the only gas that is used to measure DL. Its extreme affinity for hemoglobin (200 times that of O_2) means that the CO tension in pulmonary capillary blood essentially remains zero during the measurement of DL for CO (DLCO), which is expressed as:

$$DLCO = \dot{V}CO/PACO,$$

where PACO = the CO tension in alveolar gas. Normally, DLCO measures 25–30 ml CO/min/mm Hg. Several factors alter DLCO.

(a) Gas transfer in the lung is altered by the fact that **the diffusion barrier is not homogeneous**.

$$1/DLCO = 1/DMCO + 1/(VC \times \Theta),$$

where DMCO = the diffusion capacity of the alveolar-capillary membrane alone; VC = the pulmonary capillary blood volume; and Θ = the reaction rate of CO with hemoglobin.

(b) Normally, DLCO is **increased during exercise** to approximately 75 ml CO/min/mm Hg due to an increase in the effective area of gas exchange secondary to recruitment of pulmonary capillaries and an increase in the rate of blood flow through the lungs.

(c) DLCO is **reduced in emphysema** due to the destruction of alveolar walls and to the associated loss of the capillary bed.

(d) DLCO is **reduced by restrictive lung disease** due to the destruction of the pulmonary parenchyma, which reduces the surface area available for exchange, and because of an increase in the thickness of the interstitium, which increases the diffusion distance.

(e) DLCO commonly is **reduced by severe ventilation: perfusion imbalance**, because an uneven matching of ventilation and perfusion reduces the membrane area available for gas exchange.

(2) O_2 diffuses slightly faster than CO because of differences in solubility and molecular weight. DL of O_2 (DLO$_2$) = 1.24 DLCO.

(3) CO_2, although heavier than O_2, diffuses approximately 20 times faster than O_2 because of greater solubility in tissue fluids. Consequently, CO_2 transfer may be completely normal while O_2 diffusion is markedly reduced in lung disease.

c. **Perfusion Limitation.** Gas transfer is perfusion limited if the gas tension in the blood reaches an equilibrium with the alveolar gas tension early in the pulmonary capillary (Fig. 3-2). During perfusion-limited conditions, further gas uptake can be achieved only by replacing the equilibrated blood with fresh blood (i.e., by increasing the perfusion rate).

d. **Diffusion Limitation.** Gas transfer that is diffusion limited does not exhibit an equilibrium in gas tension between the pulmonary capillaries and the alveolar gas (see Fig. 3-2). Nitrous oxide (N_2O) uptake is an example of perfusion-limited uptake, while CO is diffusion limited because of its high affinity for hemoglobin. O_2 normally is perfusion limited, although when the diffusion capacity is reduced, it becomes diffusion limited (see Fig. 3-2).

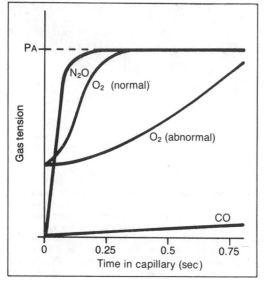

Figure 3-2. Alteration in gas tensions during transit through the pulmonary capillary; total transit time is 0.75 second. Nitrous oxide (N_2O) is perfusion limited because equilibration occurs between the gas and blood before the end of the capillary is reached. Oxygen (O_2) normally is perfusion limited but may be diffusion limited due to loss of exchange surface area secondary to lung disease. Carbon monoxide (CO) is diffusion limited due to its high affinity for hemoglobin. PA = the alveolar gas tension of each gas. (Adapted from West JB: *Respiratory Physiology*. Baltimore, Williams and Wilkins, 1979, p 24.)

II. GAS TRANSPORT

A. GAS TENSIONS. In a mixture of gases, the partial pressure of any gas is equal to the product of its mole fraction and the total dry gas pressure (Dalton's law). Each gas exerts a pressure or tension independent of the other gases.

B. GAS CONTENT OF FLUIDS. Gas molecules are in constant motion and exert pressure when they strike the walls of a container. In the alveoli, some gas molecules go into solution in the liquid layer lining the alveoli. The volume (content) of gases that is dissolved in liquid is dependent on the liquid itself, the temperature, the solubility of the gas in the liquid (solubility coefficient), and the partial pressure of the gas (Henry's law).

1. The **solubility coefficient** is unique for each gas and each solvent. The solubility of O_2 in plasma at body temperature is 0.003 ml O_2/dl plasma/mm Hg PCO_2, whereas the solubility of CO_2 in plasma is 0.072 ml CO_2/dl plasma/mm Hg PCO_2. CO_2 is 24 times more soluble than O_2 in plasma.

2. The **presence of hemoglobin** markedly increases the O_2 content of blood. Each g of hemoglobin can combine with 1.39 ml O_2 at high PO_2. Normally there are 12–15 g of hemoglobin/dl of blood; the arterial blood normally contains approximately 20 ml O_2/dl blood. The presence of hemoglobin increases the carriage of O_2 to approximately 60 times that in plasma.

C. O_2 TRANSPORT

1. **O_2-Dissociation Curve.** The combination of O_2 with hemoglobin in erythrocytes is dependent on the PO_2 and results in a sigmoid function (Fig. 3-3).
   a. The **O_2 capacity** is equal to: g hemoglobin × 1.39 ml O_2/g hemoglobin in 1 dl of blood. The O_2 capacity varies with and is determined by the amount of hemoglobin in blood.
   b. The **O_2 content** is the volume of O_2 actually present in blood (combined with hemoglobin plus physically dissolved) and usually is expressed in units of volumes percent (i.e., ml O_2/dl blood). The O_2 content depends on the amount of hemoglobin present, the PO_2, and the O_2 affinity of hemoglobin.
   c. The **hemoglobin saturation** is the percent of hemoglobin that is combined with O_2.
   d. The **P_{50}** is defined as the PO_2 that produces a hemoglobin saturation of 50 percent. The P_{50} is an inverse function of the hemoglobin affinity for O_2. The normal P_{50} for arterial blood is 27 mm Hg. An increased P_{50} indicates a reduced affinity of hemoglobin for O_2; thus hemoglobin releases O_2 at a higher PO_2, which tends to raise tissue PO_2. A decrease in the P_{50} is caused by an increased affinity of hemoglobin for O_2, which depresses the tissue PO_2 but aids in loading O_2 onto hemoglobin in the lungs.

2. **O_2 Association with Hemoglobin.** As blood enters the pulmonary capillary, its PO_2 is approximately 40 mm Hg. O_2 diffuses from the alveolus through the alveolocapillary membrane, the plasma, and then into the erythrocyte. Within the erythrocyte, O_2 combines loosely and reversibly with one of the coordination valences of heme iron. Hemoglobin is regarded as an allosteric enzyme that reacts with O_2; the affinity of these molecules for each other is affected by other ligands and temperature (Fig. 3-4).
   a. **Bohr Effect.** Both CO_2 and hydrogen ions (H^+) can react with various units of the hemoglobin molecule, and their presence decreases the affinity of hemoglobin for O_2 (see

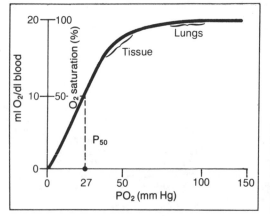

Figure 3-3. Oxygen (O_2)-dissociation curve, showing the volume, percent saturation, and gas tension of O_2 (PO_2); also illustrated is the PO_2 at which there is 50 percent saturation of hemoglobin with O_2 (P_{50}). In the lungs, changes in PO_2 cause only small changes in O_2 content as a result of the flat portion of the dissociation curve. In the tissues, small changes in PO_2 cause a large change in the release of O_2 from hemoglobin.

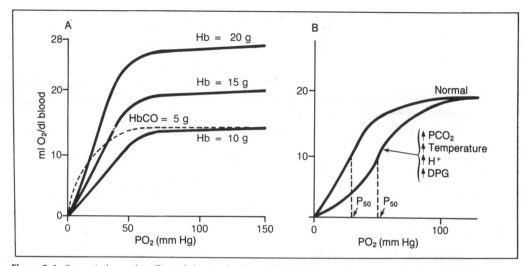

Figure 3-4. *Curve A* shows the effect of altering the hemoglobin (*Hb*) concentration (expressed in g/dl of blood) on oxygen (O_2) content. The *broken line* illustrates a situation in which one-third of the normal Hb is bound to carbon monoxide (*CO*); not only is the O_2 capacity reduced, but the curve is shifted to the left, indicating a decreased P_{50}. *Curve B* shows the effect of various ligands on the O_2-dissociation curve. $PO_2 = O_2$ tension. (Adapted from West JB: *Respiratory Physiology*. Baltimore, Williams and Wilkins, 1979, p 72.)

Fig. 3-4). The increase in the P_{50} caused by the reaction with CO_2 is known as the Bohr effect.

b. Organic Phosphates. Adenosine triphosphate (ATP) and **2,3-diphosphoglycerate** (DPG) can bind to the hemoglobin molecule on a mole-for-mole basis. Their presence stabilizes the deoxy form of hemoglobin, resulting in an increased P_{50}. DPG is produced in erythrocytes by glycolysis, and its rate of production increases during hypoxic and alkalotic conditions.

c. Carboxyhemoglobin. CO has an affinity for hemoglobin that is more than 200 times that of O_2. A CO mole fraction of approximately 0.1 percent in inspired gas eventually results in 50 percent of available hemoglobin combining with CO. CO combines with the most labile hemoglobin molecules, resulting in a decrease in P_{50} for the remaining functional hemoglobin (see Fig. 3-4).

d. Methemoglobin is formed when heme iron is oxidized to the ferric form. Methemoglobin is incapable of carrying O_2 and has a bluish color, which may impart a cyanotic color to tissues. Methemoglobin is continually being formed but is converted back to functional hemoglobin by reducing compounds formed within erythrocytes by glycolysis. Many drugs and compounds in foods can produce methemoglobin, including nitrites, primaquine, and phenacetin.

e. Abnormal Hemoglobins. More than 100 different genetic variants of hemoglobin have been described. Many of these variants have an altered affinity for O_2, which can affect the delivery of O_2 to the tissues.

f. Effect of Temperature. An increase in temperature, as occurs in actively metabolizing tissues, causes an increase in P_{50}. This shift of the O_2-dissociation curve facilitates O_2 release at the tissues (see Fig. 3-4).

D. CO_2 TRANSPORT. CO_2 diffuses from the tissues where it is formed into the capillary blood, which raises the blood PCO_2 from the normal arterial value of 40 mm Hg to approximately 46 mm Hg.

1. Forms of CO_2 Transport.

a. Approximately 5 percent of CO_2 is carried physically dissolved in the plasma.

b. Approximately 5 percent of CO_2 reacts with terminal amino groups of hemoglobin and other proteins to form NH_2-COOH, which is called a **carbamino compound**.

c. Approximately 90 percent of CO_2 enters erythrocytes where, in the presence of carbonic anhydrase, it rapidly reacts with water to form carbonic acid. The carbonic acid largely dissociates, and the bicarbonate ion (HCO_3^-) formed diffuses out of the cells back into the plasma. Thus, **most of CO_2 is transported as HCO_3^-** in the plasma. The H^+ formed from the dissociation of carbonic acid is largely buffered by deoxyhemoglobin within erythrocytes. Deoxyhemoglobin is a weaker (less dissociated) acid than oxyhemoglobin; as O_2 is released at the tissues, deoxyhemoglobin combines with the H^+, minimizing the change of pH of the blood.

2. Chloride Shift. As HCO_3^- is formed within erythrocytes at the tissues, it establishes a concentration gradient so that HCO_3^- diffuses into the plasma. The erythrocyte membrane is relatively impermeable to cations; in order to maintain electrical neutrality, chloride diffuses from the plasma into the erythrocyte in exchange for HCO_3^-. The whole process is reversed when CO_2 is liberated at the lungs.

3. The Haldane effect is the increased ability of deoxyhemoglobin to combine with CO_2. As O_2 is released at the tissues, hemoglobin combines more readily with CO_2 than does fully oxygenated hemoglobin; thus, the CO_2 content of the capillary and venous blood increases but with minimal change in the PCO_2 and pH.

III. RESPIRATORY STATICS: PRESSURE-VOLUME (P-V) DIAGRAMS.

The respiratory system normally functions as a negative pressure pump. Characterization of the respiratory pump requires description of the relationship between the distending pressures and the volume of the respiratory system.

A. TRANSMURAL PRESSURES. Each of the components of the respiratory system (e.g., the lungs, the chest wall, or the entire system) can be examined by plotting the transmural pressure of each versus the volume of air that is present. The transmural pressure is the difference in pressure between the inside and the outside of the structure. The absolute or total pressure is not relevant; it is the pressure difference that is important.

1. Transrespiratory pressure (P_{RS}) equals the pressure inside (alveolar pressure, P_A) minus the pressure outside (body surface pressure, P_{BS}) of the respiratory system. Thus, $P_{RS} = P_A - P_{BS}$ (Fig. 3-5).

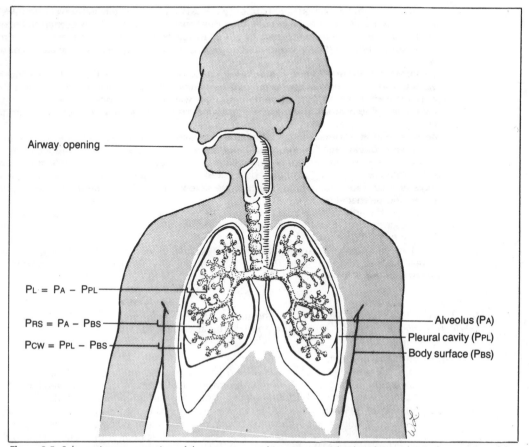

Airway opening

$P_L = P_A - P_{PL}$

$P_{RS} = P_A - P_{BS}$

$P_{CW} = P_{PL} - P_{BS}$

Alveolus (P_A)

Pleural cavity (P_{PL})

Body surface (P_{BS})

Figure 3-5. Schematic representation of the structures and pressures involved in the mechanical properties of the respiratory system. P_{RS} = transrespiratory pressure; P_L = transpulmonary pressure; and P_{CW} = transthoracic pressure.

2. **Transpulmonary pressure (PL)** equals the pressure inside the lungs (PA) minus the pressure outside the lungs (pleural pressure, PPL). Thus, PL = PA − PPL (see Fig. 3-5). The PPL in humans is approximated by measuring the esophageal pressure through a balloon-tipped catheter.

3. **Transthoracic pressure (PCW)** represents the pressure difference across the chest wall; thus, PCW = PPL − PBS (see Fig. 3-5). Because the pressure at the body surface normally is atmospheric or zero, PCW always equals − PPL. It can be shown that PRS = PL + PCW.

B. RELAXATION P-V CURVE. Figure 3-6 shows the P-V relationships of a normal respiratory system from which a number of important measurements can be derived.

1. **Total Lung Capacity (TLC).** At high distending pressures, it is apparent that a volume limit is being approached, because the slope of the PRS line approaches zero. This upper volume limit represents the TLC and is determined by the distensibility of the system and the strength of the inspiratory muscles.

2. **Residual Volume (RV).** The lower volume limit of the respiratory system is determined by the strength of the expiratory muscles and the mobility of the chest cage. With advancing age, the RV also is determined by airway closure, which limits the reduction in lung volume.

3. **Compliance.** The slope of the P-V curve at any point is the compliance or **distensibility** of the structure. The normal compliance of the respiratory system at midlung volume is 0.1 L/cm H_2O. If only one lung is functional, however, then the compliance is 0.05 L/cm H_2O. **Specific compliance** is a measure of the absolute distensibility of a structure, and it corrects for differences in volume. The specific compliance equals compliance divided by a reference volume, usually taken as functional residual capacity (see B 4, below). While the compliance of an elephant lung is much greater than a mouse lung, the specific lung compliance of all mammals is essentially equal.

4. **Functional residual capacity (FRC)** represents the volume of the respiratory system when PRS equals zero. The FRC is the volume attained by the system when the airways are open and the respiratory muscles are relaxed; the FRC is the normal end-expiratory volume of the respiratory system.

C. SUBDIVISIONS OF THE LUNG. For convenience, the lung is subdivided into four volumes and four capacities. Figure 3-7 shows the relationship between the P-V curve and lung volumes.

1. **Lung Volumes.**
   a. **RV** is the volume of gas remaining in the lungs after a maximal expiration.
   b. **Expiratory reserve volume (ERV)** is the maximal volume of gas that can be expired after a normal expiration.
   c. **Tidal volume (VT)** is the volume of gas either inspired or expired with each breath.
   d. **Inspiratory reserve volume (IRV)** is the maximal volume of gas that can be inspired after a normal tidal inspiration.

2. **Lung Capacities.** A lung capacity is a combination of two or more lung volumes.
   a. **TLC** is the volume of gas in the respiratory system after a maximal inspiration; thus, TLC = RV + ERV + VT + IRV.
   b. **Vital capacity (VC)** is the maximal volume of gas that can be expired after a maximal inspiration; thus, VC = ERV + VT + IRV.

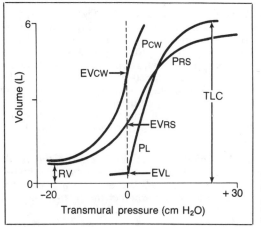

Figure 3-6. Pressure-volume curves of the respiratory system (*PRS*), the lung (*PL*), and the chest wall (*PCW*). *EV* = equilibrium volume or the volume of gas present when the transmural pressure is zero; *RV* = residual volume; and *TLC* = total lung capacity.

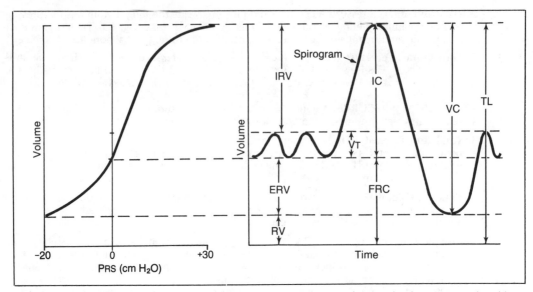

Figure 3-7. *Left:* A pressure-volume curve of the respiratory system (*P*RS), showing the determinants of total lung capacity (*TLC*), functional residual capacity (*FRC*), and residual volume (*RV*). *Right:* A spirogram showing the subdivision into lung volumes and capacities. *IRV* = inspiratory reserve volume; *ERV* = expiratory reserve volume; *IC* = inspiratory capacity; *V*T = tidal volume; and *VC* = vital capacity.

  c. FRC is the volume of gas in the respiratory system after a normal expiration; thus, FRC = RV + ERV.

  d. Inspiratory capacity (IC) is the maximal volume of gas that can be inspired after a normal expiration; thus, IC = VT + IRV.

D. PLEURAL CAVITY

 1. Dynamics. The relaxation P-V curve for the respiratory system is the result of two opposing forces: the expansile tendency of the chest wall and the retractile tendency of the lungs. These opposing forces are exerted across the pleural cavity. At normal lung volumes, these forces create a negative (i.e., below atmospheric) pressure in the interpleural fluid. If the lungs are inflated passively (e.g., during intermittent positive pressure breathing), the pleural pressure becomes positive (i.e., is increased above atmospheric pressure).

  a. The Chest Wall. The volume of the respiratory system when the pleural pressure is zero is the equilibrium volume of the chest wall. This volume in a young adult is 70–80 percent of the TLC (see Fig. 3-6).

  b. The Lung.

   (1) The retractile force of the lung increases with increasing lung volume.

   (2) The **equilibrium volume** or **minimal volume of the lung** is less than the RV in a young adult (see Fig. 3-6). The minimal volume of the lung increases with age and is markedly increased in patients with emphysema.

 2. Fluid Volume. The volume of the interpleural fluid, normally only a few milliliters, is determined by the balance between fluid filtration from the parietal pleural capillaries and the absorptive forces present in the visceral pleural capillaries and the pleural lymphatics. Increased capillary pressures or lymphatic obstruction may lead to transudation of fluid into the pleural cavity. This condition is termed **hydrothorax** or **pleural effusion** and occurs in association with left ventricular failure, mitral valve disease, and lymphatic carcinomatosis of the lungs.

  a. Pyothorax. A closed infection or the rupture of an abscess into the pleural cavity can result in the presence of pus (exudate), termed pyothorax or **empyema**.

  b. Hemothorax. Bleeding into the pleural cavity can result from trauma or rupture of any intrathoracic organ. This blood may clot, and organization of the clot into a fibrous layer can result in restriction of lung volume.

 3. Pneumothorax. Air in the pleural cavity, termed pneumothorax, can be the result of: penetrating chest injuries; blunt trauma resulting in rupture of the esophagus; rupture of a major airway; or rupture of a subpleural bleb or abscess. The effect of a pneumothorax depends on the volume of air in the pleural cavity and whether the pleural cavity communicates with either the lung or the ambient air through the wound.

     a. Open Pneumothorax. A sucking chest wound is one that communicates with the pleural cavity and prevents ventilation of the ipsilateral lung by uncoupling the lung from the chest wall. Pleural pressure remains at or near zero, and the lung collapses to its minimal volume.

     b. Tension Pneumothorax. If the tissues, usually the visceral pleura, act as a valve and allow air to enter the pleural cavity but prevent its egress, the pleural pressure gradually rises and becomes positive. The positive pressure causes unilateral overexpansion of the thorax, compression of the lung, and a shift of the mediastinum to the contralateral side. The mediastinal shift can produce torsion of the great veins, which occludes venous return, and the rapid onset of shock. Decompression of a tension pneumothorax with a large-gauge needle is a lifesaving emergency procedure.

E. ELASTIC FORCES IN THE LUNG

  1. Tissue Forces. The lung contains large amounts of **elastin**, **collagen**, and **actomyosin** fibrils, but the role of these substances in generating retractile force is not understood. Although these substances probably provide support to the airways, parenchyma, and vasculature, simply stretching these fibers requires little force. Filling the lungs with liquid results in a severalfold increase in compliance compared to filling the lungs with air (Fig. 3-8). The conclusion is that the surface forces represent a more powerful determinant of the lung's retractile forces than do the tissue components.

  2. Surface Forces.

     a. The surface forces are a direct function of the lung surface area and the surface tension at the alveolar-air interface. The work required to enlarge the surface area of the lungs is the integral of the pressure change for any increment in lung volume. Alveolar surface area increases in a complex manner with increases in lung volume.

     b. Surface Tension. The actual surface tension at the alveolar-air interface is not known. However, investigators have revealed experimentally that the alveolar surface tension is much lower at a small lung volume than at a high lung volume. This decreased surface tension is due to the presence of **pulmonary surfactant**.

  3. Pulmonary Surfactant (PS).

     a. Components. PS is a complex mixture of phospholipids, cholesterol, and a specific protein; however, the major component of PS is the phospholipid, **dipalmitoyl phosphatidylcholine**.

     b. Functions.

        (1) Reduction of Surface Tension. By replacing water molecules with surface-active lipid-protein complexes, PS reduces surface tension at the alveolar-air interface. This reduction increases compliance and decreases the work of respiration.

        (2) Increase of Alveolar Radius. PS smooths out irregularities in the alveolar cell surface, which increases the mean radius of the alveoli. The increase in alveolar radius also increases lung compliance through a Laplace effect.

        (3) Reduction of Capillary Filtration. The presence of normal PS reduces the transpulmonary pressure required to maintain lung volume. The decrease in transpulmonary pressure minimizes the negative pulmonary interstitial pressure, which decreases the filtration forces across the pulmonary capillary.

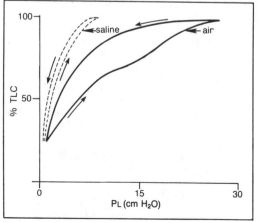

Figure 3-8. Pressure-volume loops of the lung when filled with either saline or air, showing the direction of volume change (*arrows*), the percent of total lung capacity (% *TLC*), and the transpulmonary pressure (P_L). Note the marked increase in the retractive force during air filling.

(4) Stabilization of Alveoli.
  (a) PS Stabilization. In accordance with Laplace's law $(P = 2T/r)$, variations in the radius of individual air spaces (r) should lead to instability of alveoli unless there is a proportional change in the surface tension (T), because alveolar pressure (P) is equal throughout the lung. If surface tension were similar in large and small air spaces that were interconnected, the small air spaces would collapse, and the large air spaces would become overinflated. How the surface tension is adjusted in the individual alveoli of the lung is unknown.
  (b) Tissue interdependence also has been proposed to be a mechanism for stabilization of alveoli. Adjacent alveoli share a common wall; therefore, in order for one alveolus to collapse, the neighboring alveolus must expand. Because of these tissue connections, the alveoli, and also larger lung segments, tend to stabilize each other, and atelectasis is prevented.
(5) Atelectasis (i.e., the absence of air in the lungs) can involve individual alveoli (microatelectasis), entire lobes, or a total lung. The following are causes of atelectasis.
  (a) Compression. Collapse of air spaces may be caused by extrinsic pressure (e.g., a pleural effusion), which compresses lung tissue and reduces lung volume.
  (b) Absorption.
    (i) If an airway is occluded or if the ventilation to a lung segment is critically reduced, the gas absorption into the pulmonary capillaries may exceed the ingress of gas into the area. Consequently, lung volume decreases and the region eventually becomes gas free. This is due to the total gas tension in the pulmonary arterial blood being less than 1 atmosphere and to the presence of a gradient to promote gas absorption from alveoli.
    (ii) Supplemental O_2 (especially 100 percent O_2) markedly increases the tendency for absorption atelectasis.
    (iii) Absorption atelectasis may be minimized, especially in diseased lungs, by the development of interlobular channels (canals of Lambert) or interalveolar channels (pores of Kohn). These channels provide alternate paths for ventilation of air spaces **(collateral ventilation)**.
  (c) Abnormal PS. Conditions such as infantile or adult respiratory distress syndrome, pulmonary emboli, shock, and O_2 toxicity result in an abnormally high surface tension of pulmonary lavage fluid, which reflects the properties of PS. These conditions all result in the formation of atelectatic areas in the lungs.

IV. VENTILATION AND ALVEOLAR GAS TENSIONS

A. MINUTE VENTILATION (V̇E) is the volume of gas expired per minute. The inspired volume and expired volume usually are not equal, because the volume of O_2 absorbed exceeds the volume of CO_2 expired by about 50 ml/min. V̇E equals the tidal volume (VT) times the respiratory frequency. VT normally is 500 ml, and the normal respiration rate is 12–15 breaths/min; thus, V̇E normally is 6–7.5 L/min.

B. ALVEOLAR VENTILATION (V̇A) equals V̇E minus the ventilation of the dead space. In other words, V̇A represents the part of V̇E that reaches the respiratory zone of the lungs.

C. DEAD SPACE VENTILATION (V̇D). The total dead space represents the volume of the conducting system of airways **(anatomic dead space)** plus the volume of air spaces that are ventilated but not perfused **(alveolar dead space)**. Normally, the alveolar dead space is zero, so that the total dead space and the anatomic dead space are equal. The normal anatomic dead space is about 150 ml.

  1. Measurement by Fowler's Method. If an individual inspires 100 percent O_2, at the end of inspiration all of the dead space is filled with O_2. During the subsequent expiration, the nitrogen (N_2) tension in the expired gas is initially zero, then rises, and eventually reaches a plateau as all of the supplemental O_2 is washed from the dead space. The volume expired at the midpoint of the rising phase of N_2 tension represents the volume of the dead space [see single-breath N_2 test, Section VI B 3 b (1)–(2)].

  2. Measurement by Bohr's Method. This technique requires measuring PCO_2 in the mixed expired gases and in end-tidal samples, which theoretically represent pure alveolar gas. From Dalton's law and the above relationships:

$$\dot{V}A = \dot{V}E \ (FECO_2)/FACO_2 \text{ or } \dot{V}D = \dot{V}E \ (1 - FECO_2/FACO_2),$$

  where $FECO_2$ and $FACO_2$ are mole fractions of expired and alveolar CO_2, respectively.

D. ALVEOLAR AND ARTERIAL PCO$_2$. An important relationship is revealed by rearranging the above formulas to yield the following equation:

$$P_{ACO_2} = (\dot{V}CO_2) \, K \, /\dot{V}_A,$$

where K = a constant that includes the barometric pressure and conversion factors to account for changes in temperature and humidity. Since there is no significant difference between alveolar PCO$_2$ (P$_{ACO_2}$) and arterial PCO$_2$ (P$_{aCO_2}$), this equation also can be written:

$$P_{aCO_2} = (\dot{V}CO_2) \, K \, /\dot{V}_A.$$

This simple formula shows that the primary determinants of P$_{ACO_2}$ or P$_{aCO_2}$ are CO$_2$ production ($\dot{V}CO_2$) and alveolar ventilation ($\dot{V}_A$). If PCO$_2$ is below normal, then $\dot{V}_A$ is excessive for the $\dot{V}CO_2$ rate. If PCO$_2$ is above normal, then $\dot{V}_A$ is inadequate. When ventilation: perfusion abnormalities impair CO$_2$ excretion, $\dot{V}_A$ usually increases so that P$_{aCO_2}$ returns to normal, although at the expense of increased $\dot{V}_E$ and increased work of breathing.

E. ALVEOLAR PO$_2$

1. **Alveolar Gas Equation.** The average alveolar PO$_2$ (P$_{AO_2}$) can be calculated for specific conditions by means of the equation:

$$P_{AO_2} = (P_B - P_{H_2O}) \, (FIO_2) - (P_{ACO_2}) \, (k),$$

where P$_B$ = barometric pressure (760 mm Hg at sea level); P$_{H_2O}$ = water vapor tension (47 mm Hg); FIO$_2$ = mole fraction of inspired O$_2$ (0.21 with air); P$_{ACO_2}$ = alveolar PCO$_2$ (40 mm Hg, normally); and k = the correction factor to account for the difference in inspired and expired volumes [1.2, normally, and exactly equal to FIO$_2$ + (1 − FIO$_2$)/R, where R = $\dot{V}CO_2/\dot{V}O_2$]. Thus, when breathing air at sea level:

$$P_{AO_2} = (760 - 47) \, (0.21) - (40) \, (1.2) = 103 \text{ mm Hg.}$$

2. **Alveolar-Arterial PO$_2$ Difference (A-a O$_2$ Gradient).**
   a. The **A-a O$_2$ gradient** is the best single indicator of the gas exchange properties of the respiratory system. The mean P$_{AO_2}$ is calculated using the alveolar gas equation; P$_{aO_2}$ is obtained from arterial blood gas analysis.
   b. **O$_2$ and CO$_2$ Reciprocity.** Consideration of the alveolar gas equation shows that O$_2$ and CO$_2$ must change reciprocally when breathing any one gas mixture. The only way for O$_2$ to increase in the alveoli without changing the inspired gas composition is for PCO$_2$ to decrease. According to the alveolar gas equation, PO$_2$ also can be increased by increasing the O$_2$ fraction in the inspired air or by increasing the barometric pressure.

V. PULMONARY DYNAMICS

A. EVENTS DURING A RESPIRATORY CYCLE (Fig. 3-9)

1. **Inspiration** is an active process brought about by contraction of the inspiratory muscles, primarily the **diaphragm**. Contraction of the diaphragm increases the dimensions of the chest cavity in all directions; the dome of the diaphragm descends, and the muscular insertions simultaneously elevate the lower ribs which causes a forward movement of the sternum, due to a "bucket-handle" effect.

2. **Expiration.** Eupneic expiration is essentially passive and occurs when the inspiratory muscles relax, allowing the elastic forces of the lungs to return lung volume to the FRC.

3. **Lung Volume.** The V$_T$ can be used to divide the respiratory cycle into inspiratory and expiratory phases. At the peak of inspiration, lung volume is constant for an instant, which means the gas flow and the pressure gradient between the alveoli and the mouth (transairway pressure) also are zero at this time.

4. **Alveolar Pressure.** During inspiration, lung volume increases faster than gas flows through the airways; thus, alveolar pressure becomes negative (i.e., below atmospheric pressure). The negative alveolar pressure represents the gradient that causes inspiratory flow, and flow continues until the pressure gradient disappears at the end of inspiration. During expiration, as the inspiratory muscles relax, the elastic forces from the lungs compress the alveolar gas, increasing alveolar pressure and creating a pressure gradient that expels gas from the lungs. The alveolar pressure is a function of flow rate and airway resistance.

5. **Pleural Pressure.**
   a. Pleural pressure changes as a function of both resistance and elastance forces in the respiratory system. The largest component of the resistance forces is airway resistance; tissue resistance represents only 10–20 percent of the total. The elastance forces of the

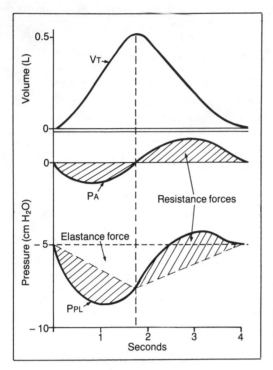

Figure 3-9. Events of a respiratory cycle. V_T = tidal volume; P_A = alveolar pressure; and P_{PL} = pleural pressure. Resistance forces on the P_A curve result from airway resistance only; resistance forces on the P_{PL} curve include both airway and tissue resistance. At the peak of inspiration, flow is zero, P_A is zero, and the change in P_{PL} is due to elastance force only.

lungs can be estimated at the end of inspiration, because resistance forces normally are zero at this instant.

    b. Dynamic lung compliance ($\Delta V/\Delta P$) can be measured during normal respiration by the changes in V_T and pleural pressure during inspiration. In normal individuals, dynamic compliance equals static compliance at all respiratory frequencies, but in patients with increased airway resistance, dynamic compliance decreases at rapid respiratory rates. This phenomenon is referred to as **frequency-dependent compliance** (see Section VI B 2 a). The presence of frequency-dependent compliance indicates that there is increased airway resistance rather than changes in the elastic properties of the lungs.

B. WORK OF BREATHING. Work is performed by the respiratory muscles whenever lung volume changes; **work** = $\Delta P \times \Delta V$. The magnitude of respiratory work for any given V_T depends on **elastance forces** and **resistance forces**.

    1. Elastance Forces. Elastance is the reciprocal of compliance; thus, patients with restrictive lung disease (low compliance) must overcome large elastance forces when their lungs are stretched during inspiration. The greater the V_T, the greater is the elastic work of breathing. Patients with restrictive lung disease minimize the work of breathing by reducing V_T and breathing at a higher frequency to maintain a normal alveolar ventilation.

    2. Resistance Forces. The largest component of total respiratory resistance is **airway resistance**; additional resistance arises because of the need to produce tissue movement in the lungs and the thoracic wall, termed **tissue resistance**. Airway resistance equals the transairway pressure gradient divided by the flow rate; the reciprocal of resistance is termed **airway conductance**. The following factors control airway resistance.

    a. Flow rate ($\dot{V}$) and airway radius (r) are the major determinants of airway resistance. Transairway pressure (P_{AW}), $\dot{V}$, and r relate according to the equation:

$$P_{AW} = K_1\dot{V}/r^4 + K_2\,\dot{V}^2/r^5,$$

where K_1 = a constant related to the laminar flow component; and K_2 = a constant related to the turbulent flow component.

    (1) A high linear velocity or **flow rate** requires excessively high pressure due to the $\dot{V}^2$ term in the equation. High velocity occurs in the large, central airways (trachea and main bronchi), where the cross-sectional area is relatively small (see Fig. 3-1). The high velocity and large diameter of the central airways favor turbulent flow conditions, which require higher pressure gradients to produce a given flow rate than do laminar flow conditions.

(2) The **radius** of an individual airway is an extremely important determinant of airway resistance because of the fourth and fifth power functions (r^4, r^5) in the equation. Changes in airway radius are the major causes of alterations in airway resistance. As the airways divide, each daughter branch is only slightly smaller than the parent generation. Thus, the total cross-sectional area of the airways increases exponentially as gas moves toward the periphery of the lung. The velocity of gas molecules becomes extremely low in the region of the respiratory bronchioles, which leads to the development of laminar flow conditions. Less than 10 percent of the total airway resistance is located in airways less than 2 mm in diameter. Thus, large increases in resistance must occur in these airways before any change can be detected in most pulmonary function tests.

b. Lung Volume. Increases in lung volume cause a reduction in airway resistance by dilation of the airways, a result of increased radial traction exerted on airway walls by the alveolar septa. The increased lung volume associated with obstructive lung disease tends to correct the increased airway resistance that is synonymous with these diseases.

c. The phase of respiration is an important determinant of airway resistance, because airway diameter is altered during the breathing cycle. During inspiration, the airways are dilated, because the alveolar pressure is more negative than the pressures within the airways. During expiration, the positive alveolar pressures compress the airways, reduce airway diameter, and increase airway resistance.

d. Autonomic Effects.

(1) Vagal Effects. Efferent vagal impulses to the lungs produce bronchoconstriction by contraction of the bronchial smooth muscle. Vagal impulses also produce an increased secretory activity of the bronchial mucous and serous glands. Some patients with asthma have a significant cholinergic component to their disease and obtain relief from atropine (a cholinergic-blocking drug) administered either systemically or by means of a nebulizer into the inspired air.

(2) Sympathetic Effects. The airways are heavily supplied by beta$_2$ receptors but contain very few alpha receptors; thus, the response to sympathetic agonists is bronchodilation. The administration of β-adrenergic blocking drugs to a patient with asthma may trigger an attack of status asthmaticus.

e. Decreased elastic recoil, as seen in elderly individuals and patients with emphysema, reduces the radial traction exerted on the walls of the airways and causes a reduction in airway radius. The airways are narrowed and also are more collapsible (i.e., they close at higher lung volumes).

f. Local PCO$_2$. A decrease in the PCO$_2$ of expired gas causes a local bronchoconstriction as the result of a direct, local effect on bronchial smooth muscle. The decreased PCO$_2$ is caused by regional overventilation; the resultant bronchoconstriction tends to return regional ventilation to normal by increasing regional airway resistance. This local mechanism is important in adjusting the regional balance between ventilation and perfusion in the lungs.

g. Pathologic Effects. Histamine, which is released from mast cells, and other mediators (e.g., slow-reacting substance of anaphylaxis, eosinophil chemotactic factor, and prostaglandins) produce contraction of the bronchial smooth muscle. In addition, release of these substances is associated with an increase in vascular permeability, which leads to edema formation, inflammation, and an outpouring of mucus. All these factors can narrow the airway lumen and produce an increase in airway resistance.

C. EFFORT-INDEPENDENT FLOW

1. At volumes less than 80 percent TLC, expiratory flow is largely independent of effort. This relationship is shown in Figure 3-10, where expiratory flow is plotted as a function of pleural pressure (which under these conditions is dependent mainly on effort) at various lung volumes.

2. The mechanism that controls expiratory flow during forced expiration is dynamic compression of the airways, shown in Figure 3-11. A number of factors are involved in determining airway resistance during forced expiration, and all of these interact to control the flow rate.

a. Elastic Recoil. The diameter of intrapulmonary airways is a direct function of the elastic recoil forces that provide radial traction on the airway walls. The elastic recoil of the lung is increased by increases in lung volume and by reductions in the compliance of the lung (restrictive lung disease). Lung recoil is reduced by emphysema and the aging process. A reduction in lung recoil force produces airway narrowing, early airway closure, and a greater airway collapsibility, all of which reduce flow rates by increasing airway resistance.

b. Lung volume is the most important determinant of expiratory air flow, because changes in

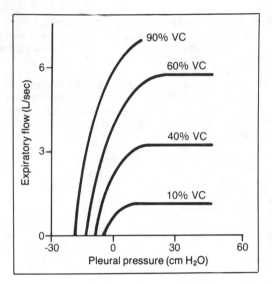

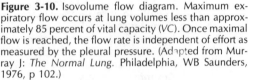

Figure 3-10. Isovolume flow diagram. Maximum expiratory flow occurs at lung volumes less than approximately 85 percent of vital capacity (VC). Once maximal flow is reached, the flow rate is independent of effort as measured by the pleural pressure. (Adapted from Murray J: *The Normal Lung.* Philadelphia, WB Saunders, 1976, p 102.)

lung volume alter a number of other factors that control airway resistance [see Section VI A 1 c (1)–(2)].

    c. Increased airway resistance, caused by conditions such as chronic bronchitis, emphysema, and asthma, increases the pressure drop that occurs in the small peripheral airways. This increased pressure drop produces a greater compressive effect across airway walls, leading to airway narrowing and a further reduction in flow rate at any given lung volume. Patients with obstructive lung disease can reduce respiratory work by breathing slowly, to reduce gas velocity and the pressure gradient, and with a larger VT to maintain a normal minute ventilation.

VI. VENTILATION: PERFUSION RATIO ($\dot{V}A/\dot{Q}C$).

Abnormalities of $\dot{V}A/\dot{Q}C$ are the most common cause of **hypoxia** in clinical medicine and can be caused by a wide range of cardiopulmonary disorders. At the alveolar level, the $\dot{V}A/\dot{Q}C$ is important in determining gas exchange. Normally, alveolar ventilation is 4.5–5.0 L/min, and cardiac output is 5.0 L/min, so that the overall $\dot{V}A/\dot{Q}C$ is 0.9–1.0. Alveoli with ratios significantly below 0.9 are relatively overperfused. The excess blood flow is not fully oxygenated and is considered to be shunted blood **(physiologic shunt)**. Alveoli with ratios greater than 1.0 are relatively overventilated, but this excess ventilation cannot significantly

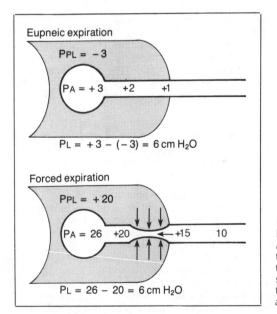

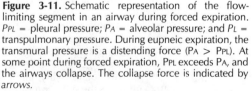

Figure 3-11. Schematic representation of the flow-limiting segment in an airway during forced expiration. P_{PL} = pleural pressure; P_A = alveolar pressure; and P_L = transpulmonary pressure. During eupneic expiration, the transmural pressure is a distending force ($P_A > P_{PL}$). At some point during forced expiration, P_{PL} exceeds P_A, and the airways collapse. The collapse force is indicated by *arrows.*

increase the O_2 content of the blood because of the plateau of the O_2-hemoglobin dissociation curve. The excess ventilation is wasted and considered to be added dead space (**alveolar dead space**).

A. PULMONARY CIRCULATION

1. **Pulmonary Vascular Resistance.** The entire output of the right ventricle normally is distributed to the pulmonary vasculature at a mean pulmonary arterial pressure of 12–15 mm Hg. Because the left atrial pressure is normally about 5 mm Hg, the pressure gradient across the pulmonary circulation is 7–10 mm Hg, and because the same blood flow traverses the systemic and pulmonary circulations, the vascular resistance of the lung is about 1/10 that of the systemic circuit.

   a. **Vascular Structure.** At birth, the pulmonary vessels contain a thick layer of smooth muscle that normally regresses by the time an infant is 6 months old. Residence at high altitude, chronic hypoxia due to cardiac or pulmonary disease, and pulmonary hypertension cause smooth muscle hypertrophy and a return to the fetal-like pulmonary vascular structure.

   b. **Hypoxic Vasoconstriction.** Alveolar hypoxia causes localized pulmonary vasoconstriction that is potentiated by hypercapnia and acidosis. The mechanism of this phenomenon is unknown, but the effects produced are the opposite of the effects produced by the same levels of O_2, CO_2, and H^+ on the systemic circulation. Regional hypoxic vasoconstriction is an important mechanism in balancing ventilation and perfusion, because the local vasoconstriction causes blood to be shifted toward regions of higher PO_2. If the hypoxia is generalized, as in hypoventilation or as a result of living in an area of high altitude, the hypoxic vasoconstriction increases the total pulmonary vascular resistance, causing pulmonary hypertension. Pulmonary hypertension increases the work of the right side of the heart, which causes ventricular hypertrophy, electrocardiographic right-axis deviation, and eventually may cause right ventricular failure.

   c. **Lung Volume.** An alteration in lung volume is the most significant factor affecting pulmonary vascular resistance. Pulmonary vascular resistance is minimal at approximately FRC and increases at both higher and lower lung volumes. The effect of lung volume on pulmonary vascular resistance is complex, because several categories of vessels are affected differently by volume changes.

   (1) **Extra-alveolar vessels**, the distributing vessels of the lung, are dilated by increases in lung volume and constricted by reduced lung volume because of the radial traction exerted on their walls by parenchymal septa.

   (2) **Alveolar vessels** are compressed between the enlarging alveoli during inspiration. The vascular resistance of the alveolar vessels is increased by increases in lung volume. Due to these mechanisms, the increased right ventricular stroke volume associated with inspiration is stored in the extra-alveolar vessels, and the venous return to the left atrium does not normally increase until the expiratory phase. The alternate fluctuation in the venous return to the two sides of the heart is the major factor that alters the normal splitting of the second heart sound.

   d. **Recruitment and Dilation.** The pulmonary vasculature is very distensible; increases in the transmural pressure produce vascular dilation which reduces the resistance to flow. In addition, increased flow (e.g., during exercise) is associated with an opening of previously closed portions of the vascular bed (recruitment) which further lowers pulmonary vascular resistance.

2. **Flow Distribution.** Pulmonary blood flow varies in different parts of the lung because of the low pressure head that is present, the distensibility of the vessels, and the hydrostatic effects of gravity. The normal distribution of pulmonary flow is shown in Figure 3-12.

   a. **Normal pattern.** The hydrostatic pressure of a column of fluid is determined by the equation: $\mathbf{h} \times \mathbf{d} \times \mathbf{g}$, where h = the height of the fluid column above a reference plane; d = the density of the fluid; and g = the gravitational constant. The reference plane usually is taken at the level of the atria, which is approximately the middle of the lung. Assuming the normal lung is 40 cm from base to apex, the mean pulmonary arterial pressure (Pa) must be 20 cm H_2O to support a column of blood from the reference plane to the apex of the lung. The normal Pa is approximately 30/10 cm H_2O (22/7 mm Hg); the normal pulmonary venous pressure (Pv) is 8–10 cm H_2O. (Normally, there is no pressure gradient between the pulmonary artery and the left atrium at end-diastole, and flow is very pulsatile through the pulmonary vasculature.) From these relationships it is apparent that Pa is insufficient to perfuse the lung apex during the latter portion of diastole.

   (1) **Zone 1** is defined as the **apical area** of the lung, where alveolar pressure (Pa) exceeds Pa. Under these conditions, the pulmonary capillaries are collapsed, and blood flow is zero. Zone 1 is increased by factors that reduce Pa and is decreased by any factor that raises Pa or reduces the vertical height of the fluid column (e.g., lying down).

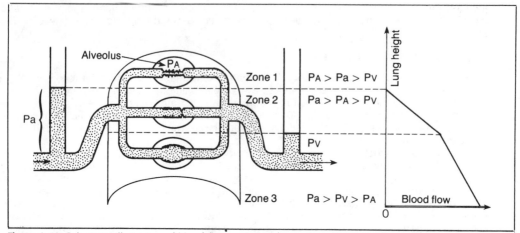

Figure 3-12. Schematic illustration of blood flow variation down an upright lung. *PA* = alveolar pressure; *Pa* = arterial pressure; and *Pv* = venous pressure. (Adapted from West JB, et al: Distribution of blood flow in the isolated lung: relation to vascular and alveolar pressure. *J Appl Physiol* 19:723, 1964.)

    (2) Zone 2 is defined as the **midregion** of the lung where (Pv) is less than PA. Here, the pressure gradient producing flow is Pa–PA rather than Pa–Pv. For each cm of vertical distance down the lung there is a 1 cm H_2O increment in the driving pressure; therefore, flow increases linearly in zone 2.

    (3) Zone 3 is defined as the **base region** of the lung; here, Pa exceeds Pv, which in turn exceeds PA. Blood flow increases linearly down zone 3, but the increased flow in this zone is a result of distension and recruitment of vessels. The distension and recruitment result from the high hydrostatic pressures in both the arteries and the veins, which lower the regional vascular resistance.

    (4) Zone 4, when present, is a region at the **lung base**, where blood flow is reduced below that of the areas just above. Zone 4 occurs in the presence of an increased Pv as seen in patients with mitral stenosis or left ventricular failure. The reduced blood flow associated with zone 4 is thought to be caused by perivascular cuffing from edema fluid, which reduces the vascular lumen and increases the resistance to flow.

  b. Abnormal Pattern. Increases or decreases in flow are caused by variations in vascular resistance produced by pulmonary disease. Variations in flow can be a result of infections, hypoventilation, emboli, tumors, and blebs. The pulmonary vasculature can be evaluated clinically by several methods.

    (1) Flow-directed catheters (e.g., a Swan-Ganz catheter) can be inserted into the pulmonary artery to measure pulmonary artery and wedge pressures. **Pulmonary wedge pressure** is a reflected pressure from the left atrium and is used to evaluate the function of the left side of the heart.

    (2) Pulmonary angiograms are obtained by injecting radiographic contrast media into the right side of the heart while taking rapid-sequenced radiographs. Pulmonary angiograms allow evaluation of vascular distortion, destruction, or obstruction from tumors, emboli, or parenchymal lesions.

    (3) Perfusion scans involve intravenous injection of radioactively labeled macroaggregated albumin followed by gamma imaging and are used to evaluate the distribution of pulmonary blood flow. The labeled albumin accumulates in the pulmonary capillaries in proportion to the local blood flow. Only a fraction of the capillaries are occluded, and the albumin is rapidly cleared.

B. VENTILATION. The distribution of the VT is determined by the pleural pressure gradient, the time constant of the lung, and airway closure.

  1. Pleural Pressure Gradient. The pleural pressure varies from the apex to the base of the lung within the intact thorax when exposed to gravity. While the cause of this variation is incompletely understood, the lung is considered to be a low-density fluid (because of its air content) contained within the pleural cavity. The density of the normal lung is about 0.25 g/ml, which produces a pressure difference of 7.5 cm H_2O between the apex and the base of a 30-cm lung in a vertical position.

    a. Regional Lung Volume. The pleural pressure at the apex and the base of the lung in a vertical position is − 10 and − 2.5 cm H_2O, respectively. Since the alveolar pressure is zero

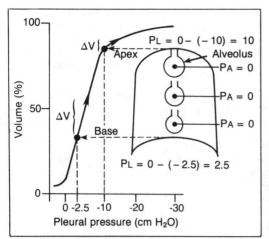

Figure 3-13. The transpulmonary pressure (P_L) varies from the apex to the base of the upright lung due to the pleural pressure gradient. This gradient places the basal alveoli on a more compliant segment of the pressure-volume curve than the apical alveoli. During inspiration, pleural pressure becomes more negative, and the lung expands. The change in volume (ΔV) is indicated by *arrows*. Apical alveoli increase in volume to a lesser extent than basal alveoli, resulting in an uneven distribution of the tidal volume. P_A = alveolar pressure.

throughout the lung under static conditions, the transpulmonary pressure is 10 cm H_2O at the apex and 2.5 cm H_2O at the base. These two transpulmonary pressures place the apex and base on very different positions of the P-V curve (Fig. 3-13). The apical alveoli are distended to 85–90 percent of their maximal volume; the basal alveoli contain only 35–40 percent of their maximal volume.

    b. V_T Distribution. Because of the position on the P-V curve, the compliance of the basal alveoli is greater than that of the apical alveoli (see Fig. 3-13). In a lung in the vertical position, the basal alveoli accept a greater portion of the V_T than do the apical alveoli. A linear increase is found in the regional ventilation from the apex to the base of a lung in a vertical position and from the dorsal to the ventral aspects of a lung in a prone position, because of the effects of gravity. The distribution of ventilation is in the same direction as the gradient for blood flow, but the blood flow gradient is steeper than the ventilation gradient. The difference in these gradients results in the apex being overventilated (hypoperfused) and the base being overperfused (hypoventilated); therefore, gas concentrations in the apex and base of the lung are different.

2. Time Constant of the Lung. The functional unit of the lung, the **acinus**, is composed of a respiratory bronchiole and all structures distal to it. Each acinus can be visualized as a balloon at the end of a tube, the balloon providing compliance (C) and the tube providing resistance (R) characteristics. The time constant is the product of R and C and is measured in seconds. The RC sets the time course of acinar filling and emptying. Variations in the RC of different units result in alterations in the distribution of the V_T. The normal lung behaves as if all of the lung units have essentially equal RCs.

    a. An **increased RC** is primarily a function of an increase in airway resistance. An increased R results in a slower rate of both filling and emptying of the acini (Fig. 3-14C). The degree of filling in lung units with a prolonged RC is primarily determined by the duration of inspiration. This effect results in the phenomenon termed **frequency-dependent compliance**, because the apparent compliance of the lung is decreased when respiratory frequency in-

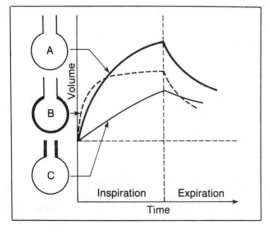

Figure 3-14. Schematic models of acini showing (A) normal filling as well as (B) the effects of decreased compliance and (C) the effects of increased airway resistance on the time course of filling. With airway resistance, acini filling is slow and greatly dependent on the duration of inspiration. In normal cases as well as those involving decreased compliance, acini reach plateaus of volume at normal respiratory frequencies. (Adapted from West JB: *Ventilation/Blood Flow and Gas Exchange*. Oxford, Blackwell Scientific, 1970, p 52.)

creases (shorter duration of inspiration). The presence of frequency-dependent compliance indicates the presence of airway disease rather than parenchymal disease.

    b. A **decreased RC** is primarily the result of reduced lung compliance secondary to conditions such as destruction of lung parenchyma, fibrosis, and pulmonary edema. Decreased compliance causes these areas to fill rapidly, but to a lesser extent than normal (see Fig. 3-14*B*).

  3. Airway Closure. Patency of small airways is determined by **transpulmonary pressure**, **airway diameter**, and **airway collapsibility**. The portion of the vital capacity (VC) remaining when airway closure begins is the **closing volume**; closing capacity equals closing volume plus residual volume (RV).

    a. Detection of airway closure is by the **single-breath N_2 test**.

      (1) In this test, an individual breathes out to RV, takes a VC inspiration of 100 percent O_2, then expires slowly and steadily to RV. The N_2 tension in the expired air is recorded during the second expiration (Fig. 3-15).

      (2) The recorded N_2 tension reveals four phases of expiration.

        (a) Phase 1 is characterized by purely dead space gas filled with 100 percent O_2 and, therefore, containing no N_2.

        (b) Phase 2 is characterized by gas from the dead space and some alveoli.

        (c) Phase 3 or the **alveolar plateau** is characterized by purely alveolar gas. The slope of phase 3 is directly proportional to the unevenness of the distribution of inspired gas.

        (d) The onset of **phase 4**, termed the **terminal rise**, identifies the onset of airway closure. Phase 4 is characterized by higher N_2 tension, which is due to gas from the apex of the lung becoming less diluted with O_2 as the basal airways close off. The apex of the lung has a higher N_2 tension than the base because it expands less, so that the N_2 is less diluted after inspiration of O_2.

    b. Early Closure.

      (1) Transpulmonary pressure is lowest and airways are narrowest at the base of the lung. Airway closure begins in the dependent portion of the lung and progresses upward as lung volume decreases during expiration. The airways close at higher lung volumes during expiration in patients with a loss of lung recoil **(decreased transmural pressure)**, such as is produced by aging and emphysema, than in young, healthy individuals.

      (2) A **reduction in airway diameter** caused by contraction of smooth muscle, edema, and intraluminal accumulation of mucus causes premature (early) airway closure.

C. RANGE OF V̇A/Q̇C IN THE LUNG. Both alveolar ventilation and blood flow are distributed unevenly to the lung. Compared to the apex, the base of the lung in a vertical position receives a greater portion of both VT and blood flow. However, the blood flow gradient is steeper than the ventilation gradient, which causes overperfusion of the base of the lung (i.e., low V̇A/Q̇C) and overventilation of the apex of the lung [i.e., high V̇A/Q̇C] (Fig. 3-16).

  1. Effect of Low V̇A/Q̇C.

    a. Zero Ventilation. When ventilation is zero but blood flow continues, PAO_2 and $PACO_2$ rapidly equilibrate with pulmonary arterial blood and gas exchange ceases. (Eventually the

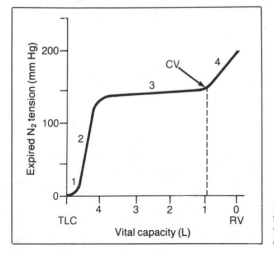

Figure 3-15. Record of expired nitrogen (N_2) tension following a vital capacity inspiration of 100 percent oxygen. *CV* = closing volume; *TLC* = total lung capacity; and *RV* = residual volume.

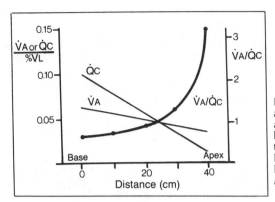

Figure 3-16. Diagram showing the distribution of alveolar ventilation ($\dot{V}_A$), pulmonary blood flow ($\dot{Q}_C$), and the ventilation: perfusion ratio ($\dot{V}_A/\dot{Q}_C$) in a normal lung. Because the blood flow gradient is steeper than the ventilation gradient, $\dot{V}_A/\dot{Q}_C$ is low at the base of the lung but high at the lung apex. $\%V_L$ = the percent of the lung volume. (Adapted from West JB: *Ventilation/Blood Flow and Gas Exchange.* Oxford, Blackwell Scientific, 1970, p 33.)

alveoli become atelectatic because gas absorption continues as long as the total gas tension in mixed venous blood is below atmospheric pressure.) As a result, blood leaving a nonventilated lung has the same composition as mixed venous blood and, thus, cannot be differentiated from blood that bypasses the lungs completely. This blood flow is termed an **anatomic** or **true shunt**.

  b. **Below Normal $\dot{V}_A/\dot{Q}_C$.** The lung acts as an inert gas exchanger, with complete gas equilibration occurring between individual alveoli and their respective capillaries.
  (1) **Effect on $PaCO_2$.** A low $\dot{V}_A/\dot{Q}_C$ means that alveolar ventilation is inadequate for blood flow and less CO_2 than normal is removed from the alveoli. Consequently, $PACO_2$ rises toward (but never exceeds) the mixed venous PCO_2. Because gas tensions equilibrate between alveoli and capillaries, the end-capillary PCO_2 also rises. This sequence also raises $PaCO_2$; however, medullary chemoreceptors are sensitive to increases in PCO_2 and respond by increasing the minute ventilation. The increased ventilation raises the $\dot{V}_A/\dot{Q}_C$ in other areas of the lung, which lowers the end-capillary PCO_2. As a result, $PaCO_2$ usually remains normal in the presence of regional $\dot{V}_A/\dot{Q}_C$ reductions.
  (2) **Effect on PaO_2.**
    (a) A low $\dot{V}_A/\dot{Q}_C$ raises $PACO_2$, which decreases PAO_2 (see alveolar gas equation, IV D). The decreased PAO_2 reduces the O_2 content and, thus, the PO_2 of the end-capillary blood. PaO_2 is reduced in areas of low $\dot{V}_A/\dot{Q}_C$. This mechanism is the most common cause of **arterial hypoxia**.
    (b) The excess blood flow that lowers $\dot{V}_A/\dot{Q}_C$ is termed a **physiologic shunt** and is differentiated from an anatomic shunt by the administration of 100 percent O_2 (see VII C 1 d). The fraction of the cardiac output representing the physiologic shunt equals:

$$\frac{Q_S}{Q_T} = \frac{CiO_2 - CaO_2}{CiO_2 - C\bar{v}O_2},$$

where Q_S = the shunted blood flow; Q_T = the total cardiac output; CiO_2 = the ideal O_2 content equivalent to the existing PAO_2; CaO_2 = the actual arterial O_2 content; and $C\bar{v}O_2$ = the mixed venous O_2 content.

2. **Effect of High $\dot{V}_A/\dot{Q}_C$.** When ventilation exceeds blood flow, $PACO_2$ decreases and PAO_2 increases. The excess ventilation is wasted effort, because the O_2 content of pulmonary end-capillary blood is not raised significantly due to the plateau of the O_2-hemoglobin dissociation curve (see Fig. 3-3).
  a. **$\dot{V}_A/\dot{Q}_C$ approaches infinity** when alveolar ventilation continues and pulmonary capillary blood flow ceases. Under these conditions, local gas exchange ceases, and the alveolar gas composition is equal to that of inspired gas. Ventilation to the affected area of the lung is termed **alveolar dead space ventilation**; the sum of alveolar dead space and anatomic dead space is termed **physiologic** or **total dead space**.
  b. A **$\dot{V}_A/\dot{Q}_C$ that is greater than normal but less than infinity** also is characterized by an increase in alveolar dead space, because the excess ventilation is not completely effective in gas exchange. The fraction of the V_T that is dead space equals:

$$(PaCO_2 - PECO_2)/PaCO_2,$$

where $PECO_2$ = mixed expired CO_2 tension.

 D. CAUSES OF $\dot{V}_A/\dot{Q}_C$ IMBALANCE. The range of $\dot{V}_A/\dot{Q}_C$ in the normal lung is about 0.6–3.0. Any disease that alters either ventilation or perfusion can produce an imbalance; in the diseased lung, the $\dot{V}_A/\dot{Q}_C$ can range from zero to infinity.

1. **Pulmonary emboli** arising from systemic venous thrombi can occlude portions of the pulmonary circulation and produce alveolar dead space.

2. **Pneumonia, pulmonary abscess, and pulmonary edema** impair regional ventilation, produce an increase in the physiologic or anatomic shunt, and decrease PaO_2 and hemoglobin saturation.

3. **Chronic obstructive pulmonary disease** increases the time constant of acini, increases the resistive work of breathing, and causes complex changes in both ventilation and perfusion.

4. **Restrictive lung disease** decreases lung compliance, which increases the elastic work of breathing; regional ventilation is altered, and the diffusion properties of the respiratory membrane are reduced.

VII. HYPOXIA is defined as an inadequate O_2 supply to the body tissues. The O_2 supply can be restricted anywhere along the respiratory pathway from the atmosphere to the mitochondria. Hypoxia occurs distal to the restriction, whereas PO_2 will be normal proximally.

A. SYMPTOMS OF HYPOXIA depend on the rate of onset and the severity of the malfunction.

1. **Fulminant hypoxia** occurs during exposure to very low PO_2 (e.g., altitudes in excess of 30,000 feet or atmospheres in which the O_2 is displaced by foreign gases or consumed by combustion). Unconsciousness may occur in as little as 15 seconds (the circulation time from the lungs to the brain). When PO_2 is extremely low, it produces a condition called anoxia, and clinical brain death is possible within 3–4 minutes.

2. **Acute Hypoxia.** Symptoms of this condition generally occur within minutes of exposure to PO_2 equivalent to that at an altitude of 18,000–25,000 feet. The symptoms of acute hypoxia are similar to those of alcohol intoxication (i.e, euphoria, incoordination, slurred speech, and slowed reflexes and mental processes). Coma may occur within minutes to hours, and death may follow if compensatory mechanisms are not adequate.

3. **Chronic Hypoxia.** Exposure to moderately reduced PO_2 produces symptoms resembling severe fatigue. Most cases of clinical hypoxia fall into this category. Affected patients may experience dyspnea and mild-to-severe exercise limitation. Respiratory arrhythmias may occur in these patients, especially during sleep.

B. SIGNS OF HYPOXIA

1. **Cyanosis** is the bluish color of tissues caused by the presence of **deoxyhemoglobin** (reduced hemoglobin) in a concentration greater than 5 g/dl of capillary blood. **Methemoglobin** also can cause cyanosis because of its slate-gray color. Cyanosis is most readily seen in the lips, mucous membranes, nail beds, and ear lobes. There is no absolute PO_2 level that produces cyanosis, because the level of deoxyhemoglobin also depends on the amount of blood flow, the amount of hemoglobin, and the P_{50}. The ability to recognize cyanosis depends on the observer, the affected individual's skin pigmentation, and the lighting. Some types of hypoxia never produce cyanosis.

2. **Tachycardia and Tachypnea.** Hypoxia is a potent stimulus to both the cardiovascular and respiratory control centers through its effect on the peripheral chemoreceptors. Hypoxia should always be considered in the presence of an unexplained tachycardia. A normal PaO_2 does not rule out all causes of hypoxia.

C. CLASSIFICATION OF HYPOXIA

1. **Arterial (hypoxic) hypoxia** (Fig. 3-17A) results from inadequate oxygenation of arterial blood. It is caused either by breathing gas mixtures with a low PO_2 (e.g., high altitudes or reduced mole fraction of inspired O_2) or by one or more of four pathophysiologic mechanisms.

a. **Hypoventilation** reduces PAO_2 and PaO_2 because of the hypercapnia it produces (see Section IV D). Hypercapnia is the pathognomonic sign of hypoventilation; a patient with a PCO_2 greater than 45 mm Hg is, by definition, hypoventilating.

(1) **Causes.**

(a) **Depression of the respiratory center** by drugs such as barbiturates, heroin, and anesthetics can decrease alveolar ventilation. **Chronic hypercapnia** itself leads to respiratory depression, because CO_2 in high concentrations actually is an anesthetic.

(b) **Increased work of breathing**, secondary to chronic lung disease, can induce respiratory muscle fatigue and reduce respiratory muscle power. Affected patients reduce the work of respiration by lowering the minute ventilation, which then results in hypoxemia and hypercapnia.

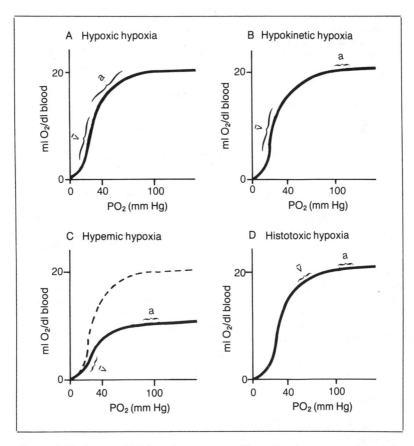

Figure 3-17. Oxygen (O_2)-dissociation curves illustrating four types of hypoxia. PO_2 = O_2 tension; a = arterial blood; and $\bar{v}$ = mixed venous blood. In *curve C*, the *dashed line* illustrates a normal curve (15 g of hemoglobin), and the *solid line* shows the effect of 7.5 g of hemoglobin. Note that the arterial PO_2 is below normal only in hypoxic hypoxia (*curve A*).

    (c) Impairment of the respiratory center by head trauma, infections (e.g., encephalitis and abscess), and neoplasms that involve the brain stem can reduce the respiratory drive.

    (d) Interruption of the Neural Path. High cervical dislocation or trauma, poliomyelitis, Guillain-Barré syndrome, and diphtheria can interfere with neural transmission to the respiratory muscles.

    (e) Neuromuscular or skeletal disease affecting the **thorax,** such as myasthenia gravis, flail chest, kyphoscoliosis, and muscular dystrophy, can impair ventilation.

  (2) Treatment. Adequate therapy for the underlying cause of hypoventilation is paramount, and treatment of the hypoxia and hypercapnia is secondary. Supplemental O_2 must be administered cautiously to patients with chronic hypercapnia because they have lost their CO_2 drive for respiration. The respiratory drive in these patients is primarily due to the peripheral chemoreceptors' response to the hypoxic stimulus; administration of 100 percent O_2 abolishes the hypoxic drive for respiration, and a further decrease in ventilation ensues **(oxygen paradox).** Only small increases in the inspired O_2 fraction are necessary to correct the hypoxia of hypoventilation. CO_2 retention can be corrected only by increasing the effective alveolar ventilation, usually by means of a mechanical respirator.

b. Diffusion Impairment. Arterial hypoxia can occur secondary to a loss of alveolar surface area or to an increase in the diffusion distance. Because CO_2 diffuses 20 times more rapidly than O_2 in the lungs, the PCO_2 can remain normal or even be slightly reduced as a result of increased ventilation secondary to the hypoxic stimulus. The diffusing capacity of the lung is significantly reduced below the normal level of 25–28 ml/min/mm Hg. Administration of supplemental O_2 to these patients can increase the diffusion gradient to maintain a normal PaO_2.

c. **Anatomic Shunt.** A right-to-left shunt through septal defects in the heart, pulmonary arterial-venous anastomoses, and blood flow through atelectatic lungs all produce arterial hypoxia by diluting arterialized (oxygenated) blood with mixed venous blood. Administration of 100 percent O_2 to a patient with an anatomic shunt does not raise the PaO_2 to the normal value (greater than 600 mm Hg), but it can significantly increase the O_2 content of arterial blood. The O_2 content increases as a result of the O_2 that is physically dissolved in the plasma of the nonshunted blood. When exposed to 100 percent O_2, the hemoglobin of the nonshunted blood becomes maximally saturated, and 2 ml O_2 is dissolved in each dl of blood.

d. **Physiologic Shunt** (see VI C 1 b). The administration of 100 percent O_2 can be used to distinguish between anatomic and physiologic shunts. In the presence of a low but finite $\dot{V}A/\dot{Q}C$ (i.e., a physiologic shunt), the administration of 100 percent O_2 eventually flushes all of the N_2 from the alveoli and results in a PAO_2 of approximately 670 mm Hg—the same value that occurs in normal individuals. (In the presence of an anatomic shunt, PaO_2 is reduced.) The administration of 100 percent O_2 to patients with $\dot{V}A/\dot{Q}C$ abnormalities causes regions of the lung with very low ventilation rates to convert into areas of **true shunt (zero ventilation)** due to atelectasis. Atelectasis occurs because the stabilizing effect of N_2 is absent, and gas absorption into the blood exceeds the gas ingress into these areas. Collateral ventilation (through pores of Kohn and larger channels) from contiguous areas tends to minimize the atelectasis.

2. **Hypemic (anemic) hypoxia** (see Fig. 3-17*B*) is caused by an inadequate amount of functional **hemoglobin** in the blood. PaO_2 remains normal in this condition, even though the arterial O_2 content is reduced. The reduced O_2 content causes the PO_2 in the systemic capillaries, tissues, and venous blood to reach hypoxic levels after O_2 is released at the tissues. The tissue PO_2 can be maintained by increasing blood flow, because O_2 delivery is a function of both O_2 content and cardiac output.

a. **Insufficient Hemoglobin.** Inadequate body stores of iron, vitamin B_{12}, and folic acid, as well as certain hemoglobinopathies and hemolytic disorders, can reduce the concentration of circulating hemoglobin. These conditions reduce the O_2 capacity of the blood and, thus, the O_2 content of arterial blood.

b. **Carboxyhemoglobin.** CO has an affinity for hemoglobin that is more than 200 times the affinity of O_2. Consequently, CO displaces O_2 from hemoglobin and inactivates the hemoglobin as an O_2-carrying molecule. In addition, CO combines with the most labile hemoglobin molecules, so that the affinity of the remaining functional hemoglobin is increased, and the P_{50} is reduced. The reduced P_{50} causes the tissue PO_2 to fall further than expected (see Fig. 3-4). The signs and symptoms of severe CO poisoning include air hunger (dyspnea), which is caused by stimulation of the peripheral chemoreceptors, and the presence of a characteristic cherry-red color.

c. **Methemoglobin.** Oxidation of heme iron to the ferric state causes the formation of nonfunctional methemoglobin. A low level of methemoglobin normally is formed continually in erythrocytes, appears as Heinz bodies, and then is converted back to hemoglobin. Reduction of iron to the ferrous form depends on the formation of reducing agents during anaerobic metabolism in erythrocytes. Methemoglobinemia can be produced by the administration of oxidizing agents; it occurs in patients with glucose-6-phosphate dehydrogenase deficiency after the ingestion of fava beans, aspirin, and other oxidizing compounds. **Cyanosis** is a prominent sign of methemoglobinemia because of the slate-gray color of methemoglobin.

3. **Hypokinetic (stagnant) hypoxia** (see Fig. 3-17*C*) is caused by reduced blood flow to a localized area (arterial or venous occlusive disease). In general, PaO_2 is normal, and tissue and venous PO_2 are at hypoxic levels because of the reduced blood flow.

a. **Decreased Cardiac Output.** A generalized hypokinetic hypoxia occurs in patients with congestive heart failure, various cardiac valvular lesions, circulatory shock, syncope resulting from bradycardia, and during severe tachycardias. In all of these conditions the cardiac output is reduced, either acutely or chronically, and the O_2 supply to the body is inadequate. Symptoms and treatment vary depending on the initiating cause.

b. **Vascular Obstruction.** Both venous and arterial obstruction result in inadequate regional blood flow. In general, venous obstruction, such as a venous thrombosis, results in an engorged, edematous, cyanotic region, whereas arterial insufficiency in an extremity produces a pale, cool limb without palpable pulses. Atheromatous plaques are the most common cause of arterial insufficiency.

4. **Histotoxic hypoxia** (see Fig. 3-17*D*) is a result of **inactivation of metabolic enzymes** by a chemical, such as cyanide. In this condition, PaO_2 is normal, and, because the tissues are not using a normal amount of O_2, PvO_2 is above normal. Tachypnea is a prominent symptom because peripheral chemoreceptors are stimulated by cyanide.

D. RESPONSE TO CHRONIC HYPOXIA. Physiologic adjustments occur in the body during hypoxic conditions in an attempt to return the O_2 supply to the tissues to normal. The total O_2 delivery to the tissues is equal to the arterial O_2 content times the cardiac output.

1. **Accommodation** refers to the initial adjustments by the body to hypoxia.
   a. **Hyperventilation.** Hypoxia stimulates the peripheral chemoreceptors, and a reflex increase in the rate and depth of respiration occurs. Increased ventilation reduces the CO_2 stores in the body and causes an alkalosis, which limits the increase in respiration. The alkalosis is slowly corrected by renal excretion of bicarbonate; thus, respiration rate and depth continue to increase during the first week of hypoxic exposure as the alkalosis is corrected.
   b. **Tachycardia.** Hypoxia reflexly increases heart rate and cardiac output. In humans, the cardiac output returns to normal after about 2 weeks at a high altitude.
   c. **Increased DPG Concentration.** Both hypoxia and alkalosis cause an increase in the concentration of free DPG within the erythrocytes. The increased DPG concentration causes an increase in the P_{50} of the O_2-dissociation curve, which raises tissue PO_2 toward the normal level.

2. **Acclimatization** refers to changes in the body in response to long-term exposure to hypoxia.
   a. **Polycythemia.** A decrease in tissue PO_2 causes the elaboration of **renal erythropoietic factor**, which acts on a plasma globulin to form **erythropoietin**. Erythropoietin stimulates the production of erythrocytes by the bone marrow, which eventually increases both the number of circulating erythrocytes and the hematocrit.
   b. **Pulmonary hypertension** is caused by chronic hypoxia because of **hypoxic vasoconstriction**. The increased pulmonary arterial pressure produces a more even distribution of pulmonary blood flow, reducing the range of $\dot{V}A/\dot{Q}C$. The pulmonary hypertension can induce right-axis deviation and incomplete right bundle branch block and can lead to right ventricular failure.
   c. **Increased enzyme activity** and increased cellular density of **mitochondria** result from hypoxic exposure. These changes increase cells' ability to function at low PO_2.
   d. **Increased capillary density** occurs, especially in muscle tissue, under hypoxic conditions. The increased number of capillaries decreases the diffusion distance and the diffusion gradient between capillaries and cells.
   e. **Life-long exposure** to high altitude or hypoxia (cyanotic heart disease) causes additional bodily alterations that may improve respiratory performance.
      (1) **Decreased Hypoxic Drive.** Individuals exposed to hypoxia from birth have a lower respiratory drive than normal individuals. This reduced drive causes a higher PCO_2 and a lower PO_2 than is found in newcomers to high altitude, but it diminishes the work of respiration.
      (2) **Total Lung Capacity.** High-altitude natives have a significantly larger lung volume and diffusing capacity than their sea-level counterparts.

3. **Acute Mountain Sickness.** Fatigue, nausea, headache, dyspnea, and palpitations frequently affect people for several days after arriving at altitudes in excess of 9000–10,000 feet. Exercise during the first several days at this altitude may produce severe pulmonary edema as a result of high blood flow coupled with hypoxic vasoconstriction, which markedly increases pulmonary arterial pressure. Pulmonary vasoconstriction is thought to be uneven throughout the lung, and edema is thought to occur in areas where the vessels are less constricted. In areas of less constriction, the arterial pressure is transmitted to the pulmonary capillaries, which increases the transudation of fluid into the interstitium and alveoli.

4. **Chronic Mountain Sickness.** Long-term residence at high altitudes occasionally leads to malaise, fatigue, and exercise intolerance—a condition termed chronic mountain sickness.

VIII. RESPIRATORY CONTROL

A. NEURAL MECHANISMS. Motor neurons control the respiratory muscles via separate voluntary and involuntary pathways. The voluntary pathway follows the corticospinal tract and subserves functions such as talking, voluntary breath-holding, and playing musical instruments. The involuntary pathway follows the bulbospinal tract, and its function is required for respiration during sleep and anesthesia and for the automatic responses during most of the waking day (Fig. 3-18).

1. **Medullary Centers.** Respiration is dependent on the rhythmic generation of impulses within the brain stem and transmission of the impulses to the respiratory muscles. This inherent rhythmicity arises within the medulla oblongata, but its exact source and mechanism are unknown. The rhythmicity is influenced by impulses arising in the pons, hypothalamus,

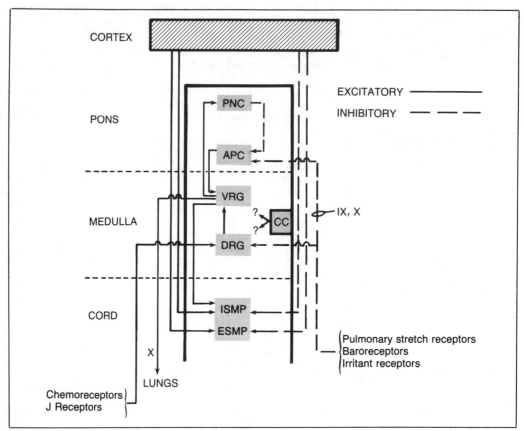

Figure 3-18. Schematic representation of major respiratory control processes. *PNC* = pneumotaxic center; *APC* = apneustic center; *VRG* = ventral respiratory group; *DRG* = dorsal respiratory group; *CC* = central chemoreceptors; *ISMP* and *ESMP* = inspiratory and expiratory spinal motor neuron pools; and *IX* and *X* = the glossopharyngeal and vagus nerves.

reticular activating system, and cortex, as well as by afferent input from the vagus, glossopharyngeal, and other nerves.

  a. Dorsal Respiratory Group (DRG).
    (1) The DRG is a group of cells located in the ventrolateral nuclei of the **tractus solitarius**.
    (2) The DRG is the initial site for processing afferent input from the vagus and glossopharyngeal nerves. The DRG may be the primary rhythm generator through a reciprocal inhibition with other cell groups. The motor output from the DRG goes primarily to contralateral phrenic neurons and to the ventral respiratory group.

  b. Ventral Respiratory Group (VRG).
    (1) The VRG is associated with the **nuclei ambiguus** and **retroambigualis** and lies ventrolaterally and rostrally to the DRG.
    (2) The VRG cells are active during both inspiration and expiration. The VRG receives impulses from the DRG, and its cells are the upper motor neurons for the accessory muscles of respiration and the vagal motor neurons.

2. Pontine Centers.
  a. Pneumotaxic Center (PNC).
    (1) The PNC is located within the **rostral pons** in the region of the **nucleus parabrachialis**.
    (2) The PNC adjusts the respiratory system's sensitivity to various stimuli and limits the duration of inspiration. Midpontine transsection or ablation of the PNC coupled with bilateral vagotomy results in inspiratory breath-holding (apneusis). Thus, the PNC has an overall inhibitory effect on inspiration.

  b. Apneustic Center (APC).
    (1) The APC, located in the **caudal pons**, has not been associated with any specific cell group.
    (2) The APC apparently is the site of origin of the long reticulospinal neurons that impinge on and facilitate the respiratory spinal motor neurons.

3. Central Chemoreceptors.

   a. H$^+$. Areas on or just beneath the ventral surface of the medulla oblongata influence respiration by responding to changes in the local H$^+$ concentration. When increased, H$^+$ stimulates respiration through an effect on the central chemoreceptors; when decreased, H$^+$ depresses respiration. Athough H$^+$ is the stimulus that activates the central chemoreceptors, H$^+$ does not readily cross the blood-brain barrier.

   b. CO$_2$ readily crosses the blood-brain barrier and affects respiration primarily by acidifying the cerebrospinal and interstitial fluids surrounding the central chemoreceptors. Approximately 85 percent of the resting respiratory drive results from the CO$_2$ stimulus to the central chemoreceptors.

   c. Cerebrospinal Fluid (CSF) Buffering. Normally, CSF has only small amounts of protein (20 mg/dl compared to 6000 mg/dl for plasma); thus, the bicarbonate-carbonic acid (HCO$_3^-$/H$_2$CO$_3$) system provides important buffers. Although H$^+$ and HCO$_3^-$ cross the blood-brain barrier only slowly, passage of these ions between blood and CSF can normalize the pH of CSF over the course of several days. Loss of HCO$_3^-$ from CSF decreases the buffer capacity, so that H$^+$ formation from CO$_2$ becomes an even stronger respiratory stimulus. Conversely, the increased HCO$_3^-$ concentration that occurs in chronic respiratory acidosis can reduce the responsiveness of the central chemoreceptors by increasing the buffer capacity.

B. PERIPHERAL CHEMORECEPTORS, located in the **carotid** and **aortic bodies**, respond to changes in the O$_2$, CO$_2$, and H$^+$ concentrations of the arterial blood.

   1. O$_2$. The blood flow to the carotid bodies is 2000 ml/min/100 g tissue, which results in a very small arteriovenous O$_2$ difference (0.15 ml O$_2$/dl blood), even though the metabolic rate of this tissue is one of the highest in the body. Because of these relationships, the carotid bodies essentially monitor the PO$_2$ of the plasma, and changes in hemoglobin levels do not alter the response of these organs. The peripheral chemoreceptors are the only receptors that monitor O$_2$ levels in the body. Afferent impulses from the carotid bodies begin when PO$_2$ falls below 500 mm Hg, and the rate of impulse formation rises rapidly when PO$_2$ falls below 100 mm Hg.

   2. H$^+$ stimulates the peripheral chemoreceptors and results in an increased minute ventilation. During a metabolic acidosis, the ventilation is increased by the increased drive from the peripheral chemoreceptors, which are stimulated by the H$^+$ from fixed acids. Increased alveolar ventilation produces a reduction in PaCO$_2$ which tends to compensate for the reduced pH. In addition, the decreased PaCO$_2$ tends to lessen the respiratory drive, so that the respiratory rate is not as high as it would be if PCO$_2$ remained normal.

   3. CO$_2$ also stimulates the peripheral chemoreceptors, but its major action in respiration is on the central chemoreceptors.

C. REFLEXES

   1. Stretch receptors located in the walls of **small airways** are stimulated by increases in lung volume. Afferent vagal impulses from these receptors inhibit medullary and pontine centers and function to **terminate inspiration**. This reaction is termed the **Hering-Breuer reflex**. Vagotomy causes the respiratory pattern to become slower and deeper, but minute ventilation may remain normal.

   2. Irritant receptors, located between **epithelial cells** in the **large airways**, are stimulated by smoke, noxious vapors, and particles in inspired gas. Cough, bronchoconstriction, and breath-holding occur reflexly after stimulation of these receptors.

   3. J-receptors, so-named because they are **juxtacapillary** receptors, are stimulated by distension and distortion of the pulmonary microvasculature. When these receptors are stimulated by emboli or vascular congestion (pulmonary edema), they cause reflex tachypnea.

   4. Chest wall receptors, located in the chest wall muscles and their tendons, transmit afferent information on chest wall position and respiratory effort to the central nervous system. Input from these stretch receptors may give rise to the sensation of dyspnea when the respiratory effort is excessive for the volume change produced.

D. EXERCISE represents the second strongest respiratory stimulus, second only to voluntary hyperventilation. The mechanism of the respiratory stimulus during exercise is not fully understood.

   1. Blood Gas Composition. PaO$_2$ and PaCO$_2$ are virtually unchanged during light and moderate exercise. During heavy exercise, the arterial pH decreases due to an increased lactic acid concentration, which causes a further increase in minute ventilation. During this intensity of exercise, PaCO$_2$ drops below the normal value of 40 mm Hg.

2. Neural Component. Afferent neural impulses from proprioceptors and joint receptors reflexly increase respiration. Respiration can be increased by input from these receptors even after passive limb movements.

3. Voluntary and Training Effects. Experienced athletes increase their respiration just prior to the start of an event. Repetitive training improves the ventilatory response, so that trained athletes can adjust their minute ventilation more rapidly to the appropriate level.

4. Anaerobic Threshold. There is a specific level of exercise (which is determined by an individual's physical fitness) at which the **lactate concentration** in the arterial blood rises rapidly. The added acid load causes a decrease in the arterial pH, which in turn causes a further stimulation of respiration. The level of exercise at which the increase in lactate concentration occurs is termed the anaerobic threshold.

5. Temperature. The minute ventilation remains elevated for some time after exercise. It returns to normal at about the same rate as body temperature.

E. DYSFUNCTION OF RESPIRATORY CONTROL (Fig. 3-19)

1. Cheyne-Stokes respiration is characterized by waxing and waning tidal volumes separated by periods of apnea. This respiratory pattern occurs during conditions of delayed feedback of sensory output between peripheral chemoreceptors and brain stem centers (e.g., in congestive heart failure); it also is common following central depression, such as that seen following a cerebrovascular accident or during sleep for the first several days at high altitude.

2. Biot's breathing, another type of periodic breathing, is characterized by one or more strong respiratory efforts followed by a prolonged period of apnea. This type of breathing is associated with many types of disease or injury to the brain.

3. Ondine's curse occurs when the involuntary respiratory tracts are interrupted or destroyed. In this condition, involuntary respiration is not possible, and apnea occurs unless a conscious effort is made to breathe. Treatment of this condition is by means of a mechanical respirator.

4. Sleep-apnea syndromes recently have been recognized as severe disorders of respiratory control in which marked hypoxia and hypercapnia occur during sleep. This condition has been implicated as the cause of sudden infant death syndrome (crib death).
  a. Obstructed Apnea. Some individuals experience loss of upper airway muscle tone during REM sleep, resulting in partial or complete occlusion of the hypopharynx. Partial occlusion is signified by snoring. During complete airway occlusion there are respiratory movements, but no air moves into or out of the airway. The individual usually awakens, muscle tone is reestablished, and the airway reopens. These individuals usually complain of chronic fatigue and sleepiness due to the disturbed sleep patterns.
  b. Nonobstructed Apnea. Affected individuals experience prolonged periods of apnea during sleep without any respiratory muscle contractions. Examination of family members of these patients indicates that many have a reduced chemoreceptor sensitivity to O$_2$ and CO$_2$. This syndrome appears to represent a familial trait of chemoreceptor hyposensitivity.

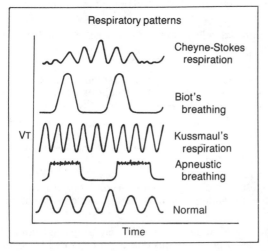

Figure 3-19. Diagram illustrating various breathing patterns. V$_T$ = tidal volume.

F. RESPIRATORY FAILURE

1. **Definition.** Respiratory failure occurs in the presence of **hypoxia** (PaO_2 less than 60 mm Hg), **hypercapnia** ($PaCO_2$ greater than 50 mm Hg), or both.

2. **Causes.** Respiratory failure can be due to malfunction of any part of the **external respiratory mechanism** (i.e., the central controller, the chemoreceptors, the muscles, the motor or sensory nerves, the thoracic wall, the airways, the lung parenchyma, and the pulmonary vasculature).

3. **Examples.**
   a. **Respiratory Depression.** Administration of central nervous system depressants causes a decreased sensitivity of the controller mechanism, which results in hypoventilation.
      (1) **Characteristic Blood Gases.** Hypoventilation results in acute retention of CO_2 and hypoxia. The acute increase in PCO_2 produces an acidosis which, by definition, is a respiratory acidosis. Until renal compensation occurs, the arterial pH remains acidotic and the bicarbonate concentration remains within the normal range.
      (2) **Treatment.** Adequate treatment of respiratory acidosis secondary to hypoventilation consists of returning the minute ventilation to normal by means of a mechanical respirator. Administration of O_2 is not sufficient treatment and may further depress respiration by eliminating or diminishing the hypoxic drive from the peripheral chemoreceptors.
   b. **The Pickwickian syndrome** is the combined presence of massive obesity, hypoventilation, and somnolence. The massive obesity causes a form of restrictive lung disease and provides a high work load on the inspiratory muscles. Most, if not all, affected patients have reduced chemoreceptor sensitivity. The somnolence may be related to the hypercapnia, because CO_2 in high concentrations is an anesthetic.
      (1) **Characteristic Blood Gases.** Affected individuals experience **hypercapnia** and consequently **hypoxia** when breathing air. Because the condition is chronic, renal compensation usually is adequate, the arterial pH is normal, and the serum bicarbonate concentration is increased above normal.
      (2) **Treatment** of acute symptoms is based on correcting the hypercapnia and hypoxia to relieve the pulmonary hypertension which invariably is present due to asphyxia. Treatment of the chronic condition itself must be based on diet control and weight loss to correct the pulmonary mechanics. Individuals who attain a normal weight are able to maintain normal blood gas composition.
   c. **Chronic obstructive pulmonary disease** is a broad category of lung diseases in which the primary derangement is an increase in resistance of airways. In some of these diseases, such as emphysema, the lung parenchyma also may be involved in the disease process.
      (1) **Characteristic Blood Gases.** The usual derangement in these conditions is a severe $\dot{V}A/\dot{Q}C$ **imbalance.** The arterial blood has a decreased PO_2, whereas PCO_2 is maintained at the normal level by an increased minute ventilation (increased alveolar dead space). Progression of the disease or an acute exacerbation may precipitate an excessive level of respiratory work, so that minute ventilation cannot be maintained. Under these conditions, the hypoxia worsens as the PCO_2 rises secondary to the reduced minute volume.
      (2) **Treatment** of severe chronic obstructive pulmonary disease represents a formidable challenge to the clinician. A major aim of treatment is to maximize airway function by removing the initiating agent (e.g., cigarette smoke, noxious vapors, and dust) and eliminating infections and inflammation. Acute episodes of bronchial spasm are treated with bronchodilators (beta sympathomimetics).

IX. PULMONARY FUNCTION TESTS.

Hundreds of different tests have been proposed to evaluate different aspects of pulmonary function; only a few of the more commonly used tests are described here. These tests can be subdivided into several different categories.

A. STATIC TESTS

1. **Lung Volumes.**
   a. Measurement of functional residual capacity (FRC), total lung capacity (TLC), and vital capacity (VC) and comparison to predicted values for a patient's height, age, and sex are an important part of a pulmonary evaluation.
   b. Any division of the lung volume that includes residual volume (RV) cannot be measured by simple spirometry. Rather, RV and therefore FRC must be measured by a total body plethysmograph or by a dilution measurement.

(1) A suitable indicator gas (e.g., helium), which can be measured readily, is placed in a spirometer. A patient then breathes into the system until equilibration between the lungs and the spirometer occurs.

(2) FRC is obtained from the following relationship:

$$(V_s) \, (F_i) = (V_s + FRC) \, (F_f) \text{ or } FRC = (V_s) \, (F_i)/F_f - V_s,$$

where F_i and F_f = the initial and final fractions of the indicator gas, respectively; and V_s = the volume in the spirometer.

2. Pulmonary compliance can be measured with an esophageal catheter, which provides an estimate of the change in pleural pressure and volume that occurs during tidal breathing. The normal pulmonary compliance is 0.2 L/cm H_2O. Pulmonary compliance is reduced by restrictive lung diseases that affect the lung parenchyma and is increased by emphysema and aging.

3. Maximal expiratory pressures can be measured with a mercury or aneroid manometer. An individual with normal expiratory muscles can generate pressure of at least 100 mm Hg at TLC. Inability to produce normal expiratory force indicates dysfunction of the neuromuscular or chest wall components of the respiratory system.

B. DYNAMIC TESTS

1. Maximal voluntary ventilation (MVV), formerly called **maximal breathing capacity (MBC)**, is extremely dependent on an individual's cooperation and has been superseded by other tests. The MVV represents the maximal ventilation that can be achieved during a 12-second interval and is expressed as L/min. The normal value for a young adult male is about 180 L/min, which far exceeds the ventilation achieved by any other respiratory stimulus.

2. Forced expiratory volume (FEV) is the test most commonly used to evaluate the mechanical properties of the lung. The FEV is simply a maximally rapid expiratory VC maneuver. Measurement of various segments of the forced vital capacity (FVC) give somewhat different insights into different processes.

a. One-second timed VC (i.e., forced expiratory volume in 1 second or FEV_1) is the volume expired in the first second of a FVC (Fig. 3-20). The FEV_1 normally is greater than 80 percent of the VC; thus, $FEV_1/FVC \times 100 > 80$ percent.

(1) Patients with **restrictive lung disease** (Fig. 3-21) have a reduced VC and TLC due to decreased lung compliance. These patients can readily expire greater than 80 percent of their VC, because increased radial traction from their restricted lungs tends to dilate their airways. Expressing the results as FEV_1/predicted VC gives a value below 0.8 because of the reduction in their actual VC.

(2) Patients with **chronic obstructive disease** (see Fig. 3-21) generally have a large TLC but a reduced VC, because the RV is increased. Obstructive lung disease is synonymous with an increase in airway resistance which necessarily reduces the expiratory flow rates. The low flow rates occur in spite of the fact that these individuals generate high pressures which are dissipated by the high resistances.

b. Maximal midexpiratory flow (MMEF) represents the average flow rate for the middle 50 percent of the VC (see Fig. 3-20). Normal values are 5–6 L/sec. The MMEF has been proposed to be more sensitive than the FEV_1 to small changes in airway resistance.

c. Peak expiratory flow (PEF), the maximal flow achieved during a FVC, is dependent on voluntary effort. The PEF is limited by the central, large airways which may be entirely normal in obstructive lung disease. (This disease generally affects the peripheral airways

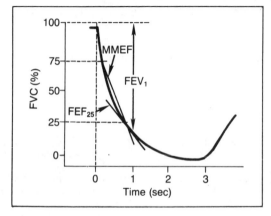

Figure 3-20. Forced vital capacity (*FVC*) maneuver showing three measurements commonly used to evaluate lung functions. *FEV₁* = forced expiratory volume in 1 second; *MMEF* = maximal midexpiratory flow, which measures the flow rate (L/sec) for the middle 50 percent of the FVC; and *FEF₂₅* = forced expiratory flow at 25 percent of the VC, which is obtained by measuring the slope of the FVC curve at 25 percent of the VC.

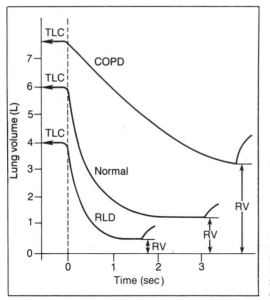

Figure 3-21. Forced vital capacity maneuvers from a normal individual and from patients with chronic obstructive pulmonary disease (*COPD*) and restrictive lung disease (*RLD*). The ordinate represents the absolute lung volume; *TLC* = total lung capacity; and *RV* = residual volume.

which are less than 2 mm in diameter.) One advantage of using PEF is that it can be measured by portable instruments.

    d. Forced expiratory flow at 25 percent VC (FEF_{25}) is considered the most sensitive indicator of small airways disease, because the flow-limiting segment moves peripherally as lung volume is reduced (see Fig. 3-20). At 25 percent of the VC, the expiratory flow is effort independent, and the flow is limited by the peripheral airways.

3. Dynamic compliance is measured using V_T and the change in pleural (esophageal) pressure during a respiratory cycle.

    a. Normal. In normal lungs at the end of inspiration, the dynamic compliance equals static compliance, because alveolar pressure and air flow are zero at this point. This equilibration results from the low and essentially equal time constants in all areas of the normal lung.

    b. Chronic Obstructive Pulmonary Disease.

        (1) The increased airway resistance in this condition is not uniform throughout the lungs. Consequently, there is a prolongation and variation in the time constants in different lung segments. Under these conditions, alveolar pressure and air flow do not equilibrate at the end of inspiration and expiration, which results in a decrease in V_T for any given change in pressure (reduced compliance) when compared to static compliance.

        (2) Frequency-dependent compliance means that dynamic compliance decreases as respiratory frequency increases. It indicates that there is obstructive airway disease rather than an alteration in the distensibility of the lungs.

C. TESTS OF GAS EXCHANGE

1. Diffusion capacity of the lung (see Section I B 2 b) is reduced in the presence of severe $\dot{V}_A/\dot{Q}_C$ imbalance, emphysema, and restrictive disease. In all of these conditions there is a reduction in the effective area of gas exchange.

2. Alveolar-arterial PO_2 gradient (A-a O_2 gradient) is the best single test of gas exchange properties of the lung (see Section IV E 2). The A-a O_2 gradient can be sequentially determined to follow the progress of lung disease; an increasing gradient indicates a deterioration of lung function. Administration of positive end-expiratory pressure (PEEP) using a respirator reduces the A-a O_2 gradient by stabilizing alveoli and improving $\dot{V}_A/\dot{Q}_C$.

3. $PaCO_2$. The average P_ACO_2 and average $PaCO_2$ are essentially equal, so that $PaCO_2$ provides an insight into the effective alveolar ventilation. Hypercapnia indicates inadequate alveolar ventilation, which may be due to:

    a. A reduction in overall minute ventilation secondary to respiratory center depression or increased work of breathing

    b. A $\dot{V}_A/\dot{Q}_C$ imbalance with a decrease in the effective ventilated lung volume (increase in alveolar dead space)

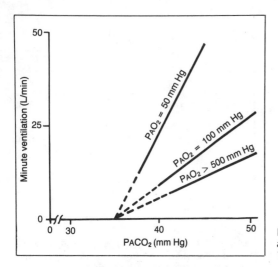

Figure 3-22. Carbon dioxide response curves obtained at alveolar oxygen tension (P_AO_2) values of 50 mm Hg, 100 mm Hg, and > 500 mm Hg.

D. TESTS OF RESPIRATORY CONTROL

1. **Chemical Factors.** Respiratory response curves can be obtained after altering inspired O_2 or CO_2 or arterial pH, keeping the other two factors constant. Figure 3-22 shows a typical response of minute ventilation to this type of test. The slope of the response curves represents the sensitivity of the control system, while the P_ACO_2 intercept represents the threshold for the respiratory drive. A decreased responsiveness to chemical stimuli has been seen in patients with primary hypoventilation and in some of their family members. A decreased respiratory response also occurs in patients with an increased work of breathing.

2. **The P_{100}** is the negative pressure generated in the large airways and is measured in a mouthpiece within the first 100 msec of initiating inspiration against a completely occluded airway. The pressure so measured correlates with the motor output of the respiratory centers. The time limit of 100 msec prevents a reflex compensation for the airway occlusion. Obviously, a patient must be unaware when the occlusion is produced until after the initiation of the inspiration.

3. **Electromyographic (EMG) activity** provides a measure of the muscle activity generated by the neural outflow from the respiratory centers. EMG activity is reduced in patients suffering from neural or muscular disease and also in those with a reduced central respiratory drive.

STUDY QUESTIONS

Directions: Each question below contains five suggested answers. Choose the **one best** response to each question.

1. A patient with restrictive lung disease exhibits a respiratory pattern characterized by which of the following features?

Respiratory Frequency	Tidal Volume
(A) high	small
(B) low	large
(C) normal	normal
(D) high	large
(E) normal	large

2. Blood flow to the lungs is best described as being

(A) finely regulated by the parasympathetic nervous system
(B) finely regulated by the sympathetic nervous system
(C) higher at the base than at the apex of the upright lung
(D) directed mainly to areas that are hypoxic
(E) constant throughout the duration of the respiratory cycle

3. A healthy, 30-year-old man with a P_ACO_2 of 40 mm Hg doubles his minute ventilation for 2 minutes. After 2 minutes, what is the P_ACO_2 level?

(A) > 40 mm Hg
(B) 40 mm Hg
(C) 20–40 mm Hg
(D) 0–20 mm Hg
(E) 0 mm Hg

4. Which of the following mechanisms produces the greatest minute ventilation?

(A) A decreased PO_2 plus an increased PCO_2
(B) Hypotension
(C) Hyperthermia
(D) Voluntary effort
(E) Exercise

5. In the presence of alveolar dead space, the regional airway resistance is increased due to which of the following mechanisms?

(A) A reflex triggered by the central chemoreceptors
(B) A reflex triggered by the peripheral chemoreceptors
(C) The Hering-Breuer reflex
(D) The local effect of low PCO_2 on the airways
(E) The local effect of low PO_2 on the airways

6. Which of the following ranges of hemoglobin saturation from venous to arterial blood represents the normal case?

(A) 40–97 percent
(B) 26–75 percent
(C) 75–97 percent
(D) 40–75 percent
(E) 60–90 percent

7. Which of the following statements is true concerning the vital capacity (VC) of a patient with restrictive lung disease?

(A) VC is below normal due to increased airway resistance
(B) VC is below normal due to reduced lung compliance
(C) VC is above normal due to the destruction of alveolar septa
(D) VC is above normal due to a greater radial traction exerted on airway walls
(E) VC is normal because lung volumes are not altered in this condition

Directions: Each question below contains four suggested answers of which **one or more** is correct. Choose the answer

    A if **1, 2, and 3** are correct
    B if **1 and 3** are correct
    C if **2 and 4** are correct
    D if **4** is correct
    E if **1, 2, 3, and 4** are correct

8. Functions of pulmonary surfactant (i.e., the layer lining the alveoli) include a reduction of

(1) the filtration forces from pulmonary capillaries
(2) the surface tension in the lungs
(3) transpulmonary pressure
(4) alveolar radius

9. The pleural pressure gradient from the base to the apex of the lung causes the

(1) basal alveoli to be less inflated than apical alveoli
(2) apical alveoli to receive the largest portion of the tidal volume
(3) basal airways to close off first during expiration
(4) pulmonary blood flow to be unevenly distributed

10. Conditions that reduce 1-second timed vital capacity (FEV_1) include

(1) restrictive lung disease
(2) obstructive lung disease
(3) increased airway resistance
(4) weakness of the expiratory muscles

11. Conditions that produce a decrease in static lung compliance include a reduction of the

(1) activity of the pulmonary surfactant
(2) diameter of the airspace
(3) amount of functional lung tissue
(4) elastic forces in the lung due to emphysema

Directions: The groups of questions below consist of lettered choices followed by several numbered items. For each numbered item select the **one** lettered choice with which it is **most** closely associated. Each lettered choice may be used once, more than once, or not at all.

Questions 12–16

For each of the following expressions of gas tension, choose the pressure value that represents the norm.

(A) 150 mm Hg
(B) 103 mm Hg
(C) 90–98 mm Hg
(D) 60 mm Hg
(E) 40 mm Hg

12. Mixed venous PO_2

13. Inspired PO_2 at sea level

14. Mean P_{AO_2} at sea level

15. $PaCO_2$

16. PaO_2

Questions 17–21

For each of the following descriptions of respiratory adjustments, choose the condition with which it is associated.

(A) Exercise
(B) High altitude
(C) Emphysema
(D) Restrictive lung disease
(E) None of the above

17. Increased airway resistance

18. Decreased compliance of the lung

19. Increased pulmonary vascular resistance

20. Increased diffusing capacity of the lung

21. Decreased work of respiration

Questions 22–26

Match each of the following descriptions of gas volume to the appropriate expression of lung capacity.

(A) Total lung capacity
(B) Inspiratory capacity
(C) Functional residual capacity
(D) Vital capacity
(E) Closing capacity

22. Volume of gas in the lungs at the onset of phase 4 of the single-breath N_2 test

23. Volume of gas in the lungs at the end of a normal expiration

24. Maximal volume of gas inspired after a normal expiration

25. Maximal volume of gas expired after a maximal inspiration

26. Volume of gas in the lungs when the retractive force of the lungs equals the expansile force of the chest wall

ANSWERS AND EXPLANATIONS

1. The answer is A. (*V B 1*) Restrictive lung disease increases the work component needed to overcome the elastance forces. The respiratory pattern that is characteristic of this disorder minimizes the work of respiration. Thus, to minimize the elastic work of breathing, the tidal volume is reduced, and the respiratory frequency is increased to maintain a normal alveolar ventilation.

2. The answer is C. (*VI A 1, 2*) Blood flow to the lungs is strongly affected by gravity due to the normally low pulmonary artery pressure. Sympathetic influence on the pulmonary circulation is minimal and is overpowered by gravitational forces and alterations in transmural vascular pressures. Hypoxia produces vasoconstriction, not vasodilation, in the pulmonary circulation. Pulmonary blood flow increases during inspiration, due to the more negative intrathoracic pressure, and decreases during expiration. Injection of acetylcholine into the pulmonary circulation can cause some pulmonary vasodilation, but the vagus nerve normally does not exert any control over pulmonary blood flow.

3. The answer is C. (*IV B, D*) The normal P_ACO_2 (40 mm Hg) is reduced by hyperventilation in proportion to the increase of alveolar ventilation. Because the new alveolar ventilation is unknown and because 2 minutes is not enough time to achieve an equilibrium, the P_ACO_2 at the end of 2 minutes must be between 20 and 40 mm Hg.

4. The answer is D. (*VIII D; IX B; Fig. 3-19*) Maximal respiratory minute volume is achieved by voluntary effort. The second most powerful respiratory stimulus is exercise, followed by the stimulatory effects of hypoxia plus hypercapnia. The combination of hypoxia and hypercapnia can increase the minute ventilation to about 80–90 L/min; however, voluntary ventilation can achieve rates in excess of 200 L/min for short periods in a normal man.

5. The answer is D. (*V B 2 f; VI C 2*) The presence of alveolar dead space indicates that there is inadequate perfusion for the ventilation; thus, the PCO_2 is reduced. A low PCO_2 causes a direct effect on airway smooth muscle, which produces bronchoconstriction. The lung does not contain adequate nerves to pinpoint local changes; therefore, it relies primarily on local mechanisms rather than on reflexes, especially those triggered by receptors located external to the respiratory system itself.

6. The answer is C. (*II C 1; Fig. 3-3*) Mixed venous blood normally has a PO_2 of 40 mm Hg, which is equivalent to a hemoglobin saturation of 75 percent, as shown on the hemoglobin dissociation curve. The saturation of hemoglobin depends on the PO_2 and the P_{50}, which determines the position of the dissociation curve along the abscissa. Arterial blood has a PO_2 of 90–98 mm Hg, which corresponds to a saturation of 97 percent.

7. The answer is B. (*V C 2 a; IX A*) The vital capacity (VC) is reduced in restrictive disease of the lungs. The VC is low because of the increased elastance forces, which decrease the compliance of the lungs. The airway resistance in restrictive disease is less than normal for equivalent lung volumes, because the increased elastance forces tend to dilate the airways.

8. The answer is A (1, 2, 3). (*III E 3 b*) Pulmonary surfactant (PS) is a complex protein-lipid mixture, which has a primary role in decreasing the surface tension at the alveolar-air interface. In addition, PS increases the mean radius of the alveoli; both of these effects increase lung compliance. Normal PS also reduces the transpulmonary pressure that maintains lung volume. This decrease in turn decreases the filtration forces across the pulmonary capillary.

9. The answer is B (1, 3). (*VI B 1, 3*) The basal alveoli are located on the steep portion of their P-V curve (i.e., they have a high compliance) and change volume the most during a tidal volume. It is the force of gravity, not the pleural pressure gradient, that causes the pulmonary blood flow to be unevenly distributed.

10. The answer is E (all). (*IX B 2*) The 1-second timed vital capacity (FEV_1) is lower than normal in restrictive lung disease because the entire lung volume and vital capacity are reduced. An increased airway resistance, which is synonymous with obstructive lung disease, causes a lower rate of air flow, which decreases the volume expired in 1 second. Weakness of the expiratory muscles decreases FEV_1 because flow rate is reduced due to the lowered expiratory pressures that are achieved.

11. The answer is A (1, 2, 3). (*III B 3, E 1–3*) When pulmonary surfactant is reduced, surface tension is increased, which means that a greater transpulmonary pressure is needed to expand the lung. A reduced airspace diameter increases transpulmonary pressure and reduces lung compliance due to the effects of the law of Laplace. Lung compliance depends on the amount of functional lung tissue; thus, the compliance of an infant lung is much less than that of an adult lung, although both lungs have the same specific compliance. Lung compliance in emphysema is increased due to loss of some tissue elastic forces.

12–16. The answers are: 12-E, 13-A, 14-B, 15-E, 16-C. (*II C 2, D; IV E 1–2*) The mixed venous PO_2 is determined by the O_2 consumption, the cardiac output, and the O_2 capacity of the blood, which is a function of the hemoglobin concentration. The mixed venous PO_2 is decreased by a decreased cardiac output (O_2 consumption constant) and by an increased metabolic rate (cardiac output constant). The inspired PO_2 equals the ambient pressure minus 47 mm Hg times the fraction of inspired O_2 (0.21 breathing air). The mean PAO_2 is determined by the alveolar gas equation. PCO_2 is consistently maintained at the normal level by the respiratory control mechanism, which is primarily the effect of the central chemoreceptors. The difference between alveolar and arterial PO_2 (A-a PO_2 gradient) is due to diffusion defects, $\dot{V}/\dot{Q}$ imbalance, and anatomic shunts. The A-a PO_2 gradient serves as the best measure of the gas exchange mechanisms of the respiratory system.

17–21. The answers are: 17-C, 18-D, 19-B, 20-A, 21-E. (*I B 2 b; V B 1, 2 e; VII D 2 b*) Emphysema is one type of obstructive lung disease; by definition, therefore, emphysema is associated with an increase in airway resistance. Restrictive lung disease is associated with a reduced distensibility or compliance of the lung. The hypoxia experienced at high altitude causes pulmonary vasoconstriction, so that pulmonary arterial pressure rises due to increased pulmonary resistance. The diffusing capacity of the lung increases during exercise because there is a recruitment of pulmonary capillaries. This is due to the increased blood flow that increases the surface area available for exchange.

22–26. The answers are: 22-E, 23-C, 24-B, 25-D, 26-C. (*III B–C, D 1; VI B 3*) The closing capacity is determined by the single-breath N_2 test and is the lung volume at the onset of phase 4. Closing capacity increases (i.e., occurs earlier in expiration) with age, cigarette smoking, and airway disease. Since expiration normally is passive, the lung volume at a normal end-expiration is an equilibrium volume or functional residual capacity determined by the elastic forces of the lungs and the chest wall. From a normal end-expiratory position, maximal inspiration draws in the tidal volume and the inspiratory reserve volume, which together comprise the inspiratory capacity. The vital capacity requires that the individual starts at total lung capacity (which requires a maximal inspiration) and then makes maximal expiratory effort to reach residual volume.

I. INTRODUCTION

A. KIDNEY FUNCTION. The kidneys maintain the constancy of the body's internal environment by regulating the volume and composition of the extracellular fluids. To accomplish this, the kidneys balance precisely the intake, production, excretion, and consumption of many organic and inorganic compounds. This balancing requires that the kidneys perform the following tasks.

1. **Excretion of Inorganic Compounds.** The renal excretion of sodium ion (Na^+), potassium ion (K^+), calcium ion (Ca^{2+}), magnesium ion (Mg^{2+}), hydrogen ion (H^+), and bicarbonate ion (HCO_3^-) exactly balances the intake and excretion of these substances through other routes (e.g., the gastrointestinal tract and the skin).

2. **Excretion of Organic Waste Products.** Normally, the kidneys excrete such waste products as urea and creatinine in amounts that equal their rate of production.

3. **Blood pressure regulation** by the kidneys involves the formation and release of renin—a major component of the renin-angiotensin-aldosterone mechanism. In addition, the renal control of fluid volume is essential to the regulation of blood pressure.

4. **Erythrocyte volume regulation** occurs through the formation and release of renal erythropoietic factor.

5. **Vitamin D Activation.** Dietary vitamin D must undergo two hydroxylations in order to be useful to the body. The first step is performed by the liver. The final hydroxylation, by the kidneys, converts this vitamin to its most biologically active form.

B. KIDNEY STRUCTURE. The kidneys are located retroperitoneally in the upper dorsal region of the abdominal cavity. Each human kidney is composed of approximately 1 million nephrons.

1. **Nephron.** This basic functional unit of the kidney is composed of a glomerulus, with its associated afferent and efferent arterioles, and a renal tubule.
   a. **Types of Nephrons.**
      (1) **Cortical nephrons** comprise about 85 percent of the nephrons in the kidney and have glomeruli located in the renal **cortex**. These nephrons have short **loops of Henle**, which descend only as far as the outer layer of the renal **medulla**.
      (2) **Juxtamedullary nephrons** are located at the junction of the cortex and the medulla of the kidney. Juxtamedullary nephrons have long loops of Henle, which penetrate deep into the medulla and sometimes reach the tip of the **renal papillae**. These nephrons are important in the **countercurrent system**, by which the kidneys concentrate urine.
   b. **Glomerulus.** The glomerulus consists of a tuft of 20–40 capillary loops protruding into **Bowman's capsule**, which is the beginning of the renal tubule. The capillary endothelium is fenestrated with an incomplete basement membrane, and together these structures provide a minimal resistance for filtration of plasma while providing a sieving function for retention of plasma proteins.
   c. **Renal Tubule.** The renal tubule begins as Bowman's capsule, which is an expanded, invaginated bulb surrounding the glomerulus. The epithelium of Bowman's capsule is an attenuated layer that is about 400 Å in thickness. The renal tubule consists of the **proximal convoluted tubule**, the **loop of Henle**, the **distal convoluted tubule**, and the **collecting duct** that carries the final urine to the renal pelvis and the ureter.

2. **Renal Blood Vessels.**
   a. **Renal Arteries.** Each kidney receives a renal artery, which is a major branch from the aorta. The kidneys receive approximately 25 percent of the total resting cardiac output or about 1.25 L blood/min. The sympathetic tone to renal vessels is minimal at rest but increases during exercise to shunt renal blood flow to exercising skeletal muscles.

b. Afferent and Efferent Arterioles. Each renal artery subdivides into progressively smaller branches, and the smallest branches give off a series of afferent arterioles. Each afferent arteriole forms a tuft of capillaries, which protrudes into Bowman's capsule. These capillaries come together and form a second arteriole, the efferent arteriole, which divides shortly after to form the peritubular capillaries that surround the various portions of the renal tubule.

c. Peritubular capillaries differ in organization depending on their association with different nephrons.

    (1) The efferent arterioles of cortical nephrons divide into peritubular capillaries that connect with other nephrons, forming a rich meshwork of microvessels. This meshwork functions to remove water and solutes that have diffused from the renal tubules.

    (2) Efferent arterioles of juxtamedullary nephrons also form peritubular capillaries, a special portion of which are the **vasa recta**. These small vessels descend with the long loops of Henle into the renal medulla and return to the area of the glomerulus. The vasa recta form capillary beds at different levels along the loop of Henle.

d. Renal veins are formed from the confluence of the peritubular capillaries and exit the kidney at the **hilus**.

II. BODY FLUID COMPARTMENTS

A. WATER CONTENT AND DISTRIBUTION (Fig. 4-1)

    1. Total body water (TBW) constitutes 55–60 percent of the body weight in young men and 45–50 percent of the body weight in young women. (The lower percentage in women largely is due to the relatively greater amount of adipose tissue in women than in men.)

      a. Distribution of TBW is as follows.

        (1) Muscle (50 percent)

        (2) Skin (20 percent)

        (3) Other organs (20 percent)

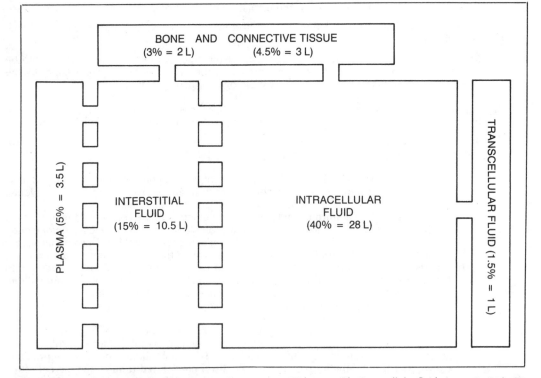

Figure 4-1. Distribution of the total body water in an average 70-kg man. The intracellular fluid compartment is not subdivided, but the extracellular fluid compartment consists of four subdivisions: plasma, interstitial fluid, transcellular fluid, and the water of bone and connective tissue. Relative volumes are expressed as percents of the body weight, assuming a fat content of about 10 percent. Fat has relatively little water associated with it. (Reprinted with permission from Bauman JW Jr, Chinard FP: Body Fluids. In *Renal Function: Physiological and Medical Aspects.* St. Louis, CV Mosby, 1975, p 19.)

(4) Blood (10 percent)
b. Although the percentage of TBW declines with advancing age and with obesity, this percentage for any individual (regardless of sex) represents a constant 70 percent of that individual's **lean body mass** (LBM or fat-free mass). Based on this constant relationship, the amount of body fat can be determined as:

$$\text{body fat (\%)} = 100 - \frac{\text{percentage of TBW}}{0.7},$$

and the LBM can be estimated as:

$$\text{LBM (kg)} = \frac{\text{TBW (L)}}{0.7}.$$

c. About one-third of the TBW is in the extracellular fluid (ECF) compartment, and the remaining two-thirds is in the intracellular fluid (ICF) compartment. (The TBW distribution in a young, 70-kg man is summarized in Table 4-1.)

2. The ECF compartment includes several subcompartments.
a. Plasma volume is about 3.5 L in a young, 70-kg man and represents about 25 percent of the ECF.
(1) Blood volume constitutes about 8 percent of the total body weight and can be obtained from plasma volume and the hematocrit as:

$$\text{blood volume (L)} = \text{plasma volume (L)} \times \frac{100}{(100 - \text{hematocrit})}.$$

(2) Blood volume occupies approximately 80 ml/kg of body weight.
b. Interstitial fluid volume is about 10.5 L in a young, 70-kg man and accounts for approximately 75 percent of the ECF. Interstitial fluid is the fluid between cells and includes lymph. Lymph constitutes 2–3 percent of the body weight. Cells are bathed essentially in the same environment, the ECF, which is composed of the interstitial fluid and the plasma and constitutes the internal environment or **milieu intérieur** of the body.
c. Transcellular fluid volume is about 1 L in most humans and occupies about 15 ml/kg of body weight. This ECF subcompartment includes digestive secretions, sweat, cerebrospinal fluid (CSF), pleural fluid, synovial fluid, intraocular fluid, pericardial fluid, bile, and luminal fluids of the gut, thyroid, and cochlea.
(1) Gastrointestinal luminal fluid constitutes about half of the transcellular fluid and occupies about 7.4 ml/kg of body weight.
(2) CSF occupies about 2.8 ml/kg of body weight.
(3) Biliary fluid volume is about 2.1 ml/kg of body weight.

3. The volume of the ICF compartment varies but usually constitutes 30–40 percent of the body weight.

B. VOLUME MEASUREMENT IN THE MAJOR FLUID COMPARTMENTS

1. Indicator Dilution Principle. The volume of water in each fluid compartment must be measured indirectly by the indicator dilution principle. This principle is based on the relationship of the amount of a substance introduced (**Q**), the volume in which that substance is distributed (**V**), and the final concentration attained (**C**).
a. The equation for this relationship is:

$$C = \frac{Q}{V} \quad \text{or} \quad V = \frac{Q}{C}, \tag{1}$$

where V = the volume (in ml or L) in which the quantity, Q, (in g, kg, or mEq) is distributed to yield the concentration, C, (in g/ml or L or in mEq/ml or L).

Example: If 25 mg of glucose are added to an unknown volume of distilled water and the final concentration of glucose after mixing is 0.05 mg/ml, then the volume of solvent is:

$$V = \frac{25 \text{ mg}}{0.05 \text{ mg/ml}} = 500 \text{ ml}.$$

b. Volume measurement by the dilution principle requires that the introduced substance be distributed evenly in the body fluid compartment being measured.
(1) Equation (1) must be altered if:
(a) The solute leaves the compartment by excretion in the urine or by transfer to another compartment where it exists in a different concentration

Table 4-1. Distribution of Body Water in a Young, 70-kg Man

Compartment	Volume (L)	Percent Body Weight	Percent Lean Body Mass	Percent Body Water
Total body water	42*	60†	70	100
Extracellular fluid	14	20	24	33
Plasma	3.5	5	6	8
Interstitial fluid	10.5	15	18	25
Intracellular fluid	28	40	48	67

* Total body water is 35 L in a young, 70-kg woman.
† Body water accounts for 50 percent of the total body weight of a young, 70-kg woman.

    (b) The solute is metabolized
    (c) The solute is vaporized from the skin and respiratory tract
  (2) The amount of substance lost from the fluid compartment, then, is subtracted from the quantity administered, as:

$$V = \frac{Q \text{ administered} - Q \text{ removed}}{C}.$$

Example: A 60-kg woman is infused with 1 millicurie (mCi) of tritium oxide (^3H_2O). After 2 hours, 0.4 percent of the administered dose is lost in the urine and by vaporization from the skin and respiratory tract. The radioactivity of a plasma sample is measured by liquid scintillation spectrometry and indicates a concentration of 0.03 mCi/L of plasma water. Since the concentration of ^3H_2O in the major compartments should be the same as in plasma, the TBW can be calculated as:

$$V = \frac{Q \text{ infused} - Q \text{ excreted}}{C}$$

$$= \frac{1 \text{ mCi} - (1 \text{ mCi} \times 0.004)}{0.03 \text{ mCi/L}}$$

$$= \frac{0.996}{0.03}$$

$$= 33.2 \text{ L}.$$

  c. ^3H_2O is an unstable isotope and the substance of choice for measuring TBW. It is a weak beta emitter with a **biologic half-life** of 10 days but a **physical half-life of 12 years**. Other substances used to measure TBW include:
    (1) Antipyrine and *N*-acetyl-4-amino antipyrine (NAAP), which rarely are used
    (2) Deuterium oxide (^2H_2O), which is a stable isotope
    (3) Urea and thiourea

2. The ECF Volume.
  a. Substances used to measure the ECF volume are of two types:
    (1) Saccharides such as inulin, sucrose, raffinose, and mannitol
    (2) Ions such as thiosulfate, thiocyanate, and the radionuclides of sulfate (SO_4^{2-}), chloride (Cl^-), bromide (Br^-), and Na^+
  b. The interstitial fluid volume cannot be measured directly, because no substance is distributed exclusively within this compartment. Therefore, the interstitial fluid volume is determined as the difference between the ECF volume and plasma volume.
  c. Plasma volume is measured using either of two dilution methods.
    (1) The first method employs substances that neither leave the vascular system nor penetrate the erythrocytes. Such substances include:
      (a) Evans blue dye (T-1824)
      (b) Radioiodinated human serum albumin (RISA)*
      (c) Radioiodinated gamma globulin and fibrinogen, which generally do not leak out of the bloodstream

---

*RISA slowly leaks out of the circulation into the interstitial fluid; therefore, the plasma volume is slightly overestimated when RISA is used.

(2) The second method is based on the fact that the radioisotopes of phosphorus (^{32}P), iron (^{55,59}Fe), and chromium (^{51}Cr) penetrate and bind to erythrocytes. The tagged cells are injected intravenously and their volume of distribution is measured. Plasma volume (PV), then, is calculated from the measured erythrocyte volume (BV) and hematocrit (Hct) as:

$$PV \text{ (in L)} = BV \text{ (in L)} \times \frac{(100 - \text{Hct})}{100}.$$

3. The ICF volume cannot be measured directly by dilution, because no substance is distributed only in this compartment. The ICF volume is obtained by subtracting the ECF volume from the TBW.*

C. DISTURBANCES OF VOLUME AND CONCENTRATION (OSMOLALITY) OF BODY FLUIDS (Fig. 4-2 and Table 4-2)

1. General Considerations.
a. The general clinical terms for volume abnormalities are **dehydration** and **overhydration**.
b. The adjectives **isosmotic**, **hyperosmotic**, and **hyposmotic** are used to describe the changes in volume (i.e., dehydration and overhydration) and refer to the ECF in its new steady state.
c. The distribution of body water in the ECF and ICF compartments is determined by **osmotic forces**. Water flows from a region of low osmolality to a region of higher osmolality.

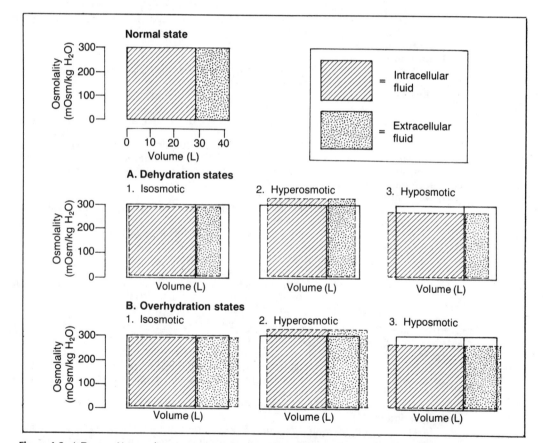

Figure 4-2. A Darrow-Yannet diagram representing the volume (*abscissa*) and osmolality (*ordinate*) of the body fluid compartments in a 70-kg man. The combined plasma and interstitial fluid spaces are represented by the extracellular fluid compartment. This diagram is useful to simplify the clinical analysis of fluid balance. The area of each fluid compartment rectangle represents the total milliosmoles of solutes in that compartment. In all diagrams, the normal state is indicated by *solid lines* and the shifts from normality are indicated by *dashed lines*. (Reprinted with permission from Valtin H: *Renal Function: Mechanism Preserving Fluid and Solute Balance in Health*, 2nd ed. Boston, Little, Brown, 1983, p 272.)

---

*There is a good correlation between the ICF volume and total exchangeable K$^+$. ICF volume also is related to muscle mass, which decreases with age.

Table 4-2. Steady State Changes in Volume and Osmolal Concentration of Body Fluids

Type of Change	Volume (L)		Osmolality (mOsm/kg H$_2$O)	
	ICF	ECF	ICF	ECF
Contraction (dehydration)				
Isosmotic	0	↓	0	0
Hyperosmotic	↓	↓	↑	↑
Hyposmotic	↑	↓	↓	↓
Expansion (overhydration)				
Isosmotic	0	↑	0	0
Hyperosmotic	↓	↑	↑	↑
Hyposmotic	↑	↑	↓	↓

Note.—The changes in volume and in osmolality refer to the **ECF compartment** in the new steady state.

   2. Dehydration (Volume Contraction) States.
      a. Isosmotic Dehydration (see Fig. 4-2 *A1*).
         (1) Description.
            (a) Initially, fluid is lost from the plasma and then is repleted from the interstitial space. No major change occurs in the osmolality of the ECF; therefore, no fluid shifts into or out of the ICF compartment.
            (b) Finally, the volume of the ECF is reduced with no change in osmolality.
         (2) Causes of isosmotic dehydration include hemorrhage, plasma exudation through burned skin, and gastrointestinal fluid loss (e.g., vomiting and diarrhea).
      b. Hyperosmotic Dehydration (see Fig. 4-2 *A2*).
         (1) Description.
            (a) Initially, fluid is lost from the plasma, which becomes hyperosmotic, causing a fluid shift from the interstitial fluid to the plasma.
            (b) The rise in interstitial fluid osmolality causes fluid to shift from the ICF to the ECF compartment.
            (c) Finally, the ECF and ICF volumes both are decreased, and the osmolality of both major fluid compartments is increased.
         (2) Causes of hyperosmotic dehydration include water deficits due to decreased intake, diabetes insipidus (neurogenic or nephrogenic), alcoholism, administration of lithium salts, fever, and excessive evaporation from the skin and breath.
      c. Hyposmotic Dehydration (see Fig. 4-2 *A3*).
         (1) Description.
            (a) Initially, loss of sodium chloride (NaCl) causes a loss of water. This is followed by water retention but a continued loss of NaCl.
            (b) A net loss of NaCl in excess of water loss results in a decreased osmolality of the ECF and a subsequent shift of fluid from the ECF to the ICF compartment.
            (c) Finally, the ECF volume is decreased, the ICF volume is increased, and the osmolality of both major fluid compartments is decreased.
         (2) Causes of hyposmotic dehydration include loss of NaCl due to heavy loss of hypotonic sweat and renal loss of NaCl due to adrenal insufficiency (e.g., primary hypoadrenocorticalism or Addison's disease).

   3. Overhydration (Volume Expansion) States.
      a. Isosmotic Overhydration (see Fig. 4-2 *B1*).
         (1) Description. Isosmotic overhydration is characterized by an overall expansion of the ECF volume with no change in the osmolality of the ICF and ECF compartments.
         (2) Causes of isosmotic overhydration are edema and oral or parenteral administration of a large volume of isotonic NaCl.
      b. Hyperosmotic Overhydration (see Fig. 4-2 *B2*).
         (1) Description.
            (a) Initially, there is water retention followed by NaCl retention in excess of water.
            (b) The rise in plasma osmolality causes a shift of water from the ICF to the ECF compartment.
            (c) Finally, the ECF volume is expanded, the ICF volume is contracted, and the osmolality of both major fluid compartments is increased.
         (2) Cause. Oral or parenteral intake of large amounts of hypertonic fluid causes hyperosmotic overhydration.
      c. Hyposmotic Overhydration (see Fig. 4-2 *B3*).
         (1) Description.
            (a) Initially, water enters the plasma, causing a decline in the plasma osmolality, a shift of water into the interstitial space, and a decrease in the interstitial fluid osmolality.

        (b) The decrease in interstitial fluid osmolality causes water to shift from the ECF to the ICF compartment.

        (c) Finally, the ECF and ICF volumes increase and the osmolality of both major fluid compartments decreases.

    (2) Causes of hyposmotic overhydration include ingestion of a large volume of water and renal retention of water due to the syndrome of inappropriate antidiuretic hormone (ADH) secretion (SIADH).

D. IONIC COMPOSITION OF BODY FLUIDS (Fig. 4-3 and Table 4-3)

  1. General Considerations.

    a. Ions constitute about 95 percent of the solutes in the body fluids.

    b. The sum of the concentrations (in mEq/L) of the cations equals the sum of the concentrations (in mEq/L) of the anions in each compartment, making the fluid in each compartment electrically neutral.

    c. Physicians rely on the changes in electrolyte concentrations in the ECF compartment, particularly in the plasma, in the treatment of patients.

  2. Cations.

    a. The monovalent elements Na^+ and K^+ are the predominant cations of the ECF and ICF compartments, respectively. Essentially all of the body K^+ is in the exchangeable pool, whereas only 65–70 percent of the body Na^+ is exchangeable. Only the exchangeable solutes are osmotically active.

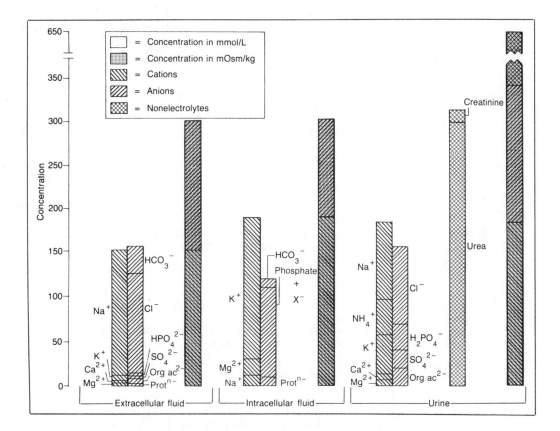

Figure 4-3. Millimolar and osmolal concentrations of body fluids. Total electrolyte concentrations are expressed in mmol/L to emphasize the qualitative heterogeneity of the body fluids. In the extracellular fluid (ECF), the major portion of cations and anions is univalent, and osmotic inactivation of electrolytes by protein is minimal. Therefore, the cation and anion concentrations are almost equal and exert similar osmotic effects. In contrast, in the intracellular fluid (ICF), most anions are bivalent or multivalent, and protein binding of some electrolytes makes them osmotically inactive. Therefore, the millimolar cation concentration is greater than the anion concentration, and cations exert a proportionately greater osmotic effect. However, the osmotic activity of the ECF and the ICF are equal. Although more ions exist in the cells than in the ECF, there is no difference in osmolality because many of the cellular ions are bound to protein and are osmotically inactive. X^- = undetermined anions. (Note that urine contains more divalent anions than divalent cations, making the column for anions lower than that for cations.)

Table 4-3A. Electrolyte Concentration of Body Fluids

Ion	Plasma (mg/L)	Plasma (mEq/L)	Plasma (mmol/L)	Interstitial Fluid (mEq/L)	ICF (mEq/L)	Urine (mEq/L)
Cations						
Na^+	3266	142	142	145	10	50–130
K^+	156	4	4	4	145	20–170
Ca^{2+}	100	5	2.5	3	. . .	5–12
Mg^{2+}	24	2	1	2	40	2–18
NH_4^+	. . .	. . .	. . .	. . .	. . .	30–50
Total cations	3546	153	149.5	154	195	. . .
Anions						
Cl^-	3692	104	104	117	5	50–130
HCO_3^-	1525	25	25	27	10	. . .
Phosphate*	106	2	1.1	2	100	20–40
Protein	65,000	15	1.1	1	60	. . .
SO_4^{2-}†	16	1	0.5	1	20	30–45
Organic acids	175	6	3	6	. . .	20–50‡
Total anions	70,514	153	134.7	154	195	. . .
Total electrolytes	. . .	306	284.2	308	390	. . .

Note.—At the pH of body fluids, the proteins have multiple charges per molecule (average valence of − 15). Hence, the ICF has more total charges than does the ECF. The total concentration of cations in each compartment must equal the total concentration of anions in each compartment when expressed in mEq/L. (After Frisell WR: Intra- and Extracellular Electrolytes. In *Acid-Base Chemistry in Medicine.* New York, Macmillan, 1968, p 5.)
* HPO_4^{2-} (ECF, ICF); $H_2PO_4^-$ (urine).
† As free sulfur.
‡ Assuming average valence of − 2.

Table 4-3B. Nonelectrolyte Concentration of Body Fluids

Solute	Plasma (mg/L)	Plasma (mmol/L)	Urine (mg/L)	Urine (mmol/L)
Urea	150	2.5	12,000–24,000	200–400
Creatinine	11	0.08	678–2260	6–20
Osmolarity	. . .	. . .	. . .	500–800

   b. The divalent cations Mg^{2+} and Ca^{2+} exist in body fluids in relatively low concentrations. Almost all of the body Ca^{2+} (in bone) and most of the body Mg^{2+} (in bone and cells) is nonexchangeable.
   (1) After K^+, Mg^{2+} is the main cation of the ICF.
   (2) After Na^+, Ca^{2+} is the main cation of the ECF.

   3. Anions. The chief anions of the body fluids are Cl^-, HCO_3^-, phosphates, organic ions, and polyvalent proteins.
   a. The main intracellular anions are phosphates, proteins, and organic ions.
   b. Cl^- and HCO_3^- are the predominant extracellular anions.

   4. Summary. Na^+ and Cl^- are the principal electrolytes in the ECF, K^+ and phosphates are the principal electrolytes in the ICF, and K^+ and Cl^- are the principal electrolytes in the erythrocytes.

III. OVERVIEW OF RENAL TUBULAR FUNCTION. The constancy of the body's internal environment is maintained, in large part, by the continuous functioning of its roughly 2 million nephrons. As blood passes through the kidneys, the nephrons **clear** the plasma of unwanted substances (e.g., urea) while simultaneously retaining other, essential substances (i.e., water). Unwanted substances are removed by glomerular filtration and renal tubular secretion and are passed into the urine. Substances that the body needs are retained by renal tubular secretion and are returned to the blood by reabsorptive processes.

   A. GLOMERULAR FILTRATION is the initial step in urine formation. The plasma that traverses the glomerular capillaries is filtered by the highly permeable **glomerular membrane**, and the resultant fluid, the **glomerular filtrate,** is passed into Bowman's capsule. **Glomerular filtration rate**

(GFR) refers to the volume of glomerular filtrate formed each minute by all of the nephrons in both kidneys.

B. RENAL TUBULAR SECRETION AND REABSORPTION (Table 4-4) refer to the direction of transport and not to the differences in the underlying mechanisms of transport.

1. Secretion refers to the active (or passive) transport of solutes from the peritubular capillaries into the tubular lumen.

2. Reabsorption denotes the active transport of solutes and the passive movement of water from the tubular lumen into the peritubular capillaries.

IV. RENAL CLEARANCE

A. DEFINITIONS

1. Clearance is a measure of the volume of **plasma** completely freed of a given substance per minute by the kidneys.
   a. It does not designate a filtered mass, and it is not a real volume but a virtual volume of plasma from which the substance is removed per minute.
   b. It is an empiric measure of the ability of the kidneys to remove a substance from the blood plasma; that is, clearance is the efficiency with which the plasma is cleared of a given substance.

2. Renal clearance of a given substance is the ratio of the renal excretion rate of the substance to its concentration in the blood plasma. This ratio is expressed as:

$$C_x = \frac{U_x \cdot \dot{V}}{P_x}, \tag{2}$$

where C_x = clearance of the substance (in ml/min); U_x and P_x = concentration (in mg/ml) of the substance in urine and plasma, respectively; $\dot{V}$ = urine minute volume or volume of urine output per minute (in ml/min); and $U_x \cdot \dot{V}$ = amount of substance excreted per time (in mg/min). [Equation (2) is an expression of the indicator dilution principle, which is discussed in Section II B 1.]

Table 4-4A. Tubular Function: Proximal Tubules

Reabsorption				Secretion*	
Active		**Passive**	**Nonreabsorption**	**Active**	**Passive**
Na^+	Urate	Cl^-	Inulin	$H^{+\,\dagger}$	NH_3
K^+	Glucose	$HCO_3^{-\,\ddagger}$	Creatinine	Urate	
Ca^{2+}	Amino acids	HPO_4^{2-}	Sucrose	PAH	
Mg^{2+}	Protein	Water	Mannitol	Penicillin	
HPO_4^{2-}	Acetoacetate	Urea		Sulfonamides	
SO_4^{2-}	β-Hydroxybutyrate			Creatinine	
NO_3^-	Vitamins				

* Most secretion involves active transport into the proximal tubule.
† Coupled with Na^+ secretion.
‡ HCO_3^- reabsorption occurs by H^+ secretion, and, in this sense, HCO_3^- is active. However, CO_2, formed by the association of H^+ and HCO_3^- in the tubular lumen, diffuses into the tubular cell passively and forms HCO_3^- within the cell, which is passively reabsorbed. HCO_3^- is largely reabsorbed into the tubular cell as CO_2.

Table 4-4B. Tubular Function: Distal Tubules

Reabsorption			Secretion	
Active	**Passive**	**Nonreabsorption**	**Active**	**Passive**
Na^+	Cl^-	Urea*	$K^{+\,\dagger}$	$K^{+\,\ddagger}$
Ca^{2+}	HCO_3^-		H^+	NH_3
Mg^{2+}	Water			
Urate				

* Urea is passively reabsorbed from the thick segment of the ascending limb of the loop of Henle and from the lower part of the collecting tubule.
† Coupled with Na^+ reabsorption.
‡ K^+ secretion in the distal tubule occurs primarily by passive electrical coupling of Na^+-K^+ on the luminal side and secondarily by active Na^+-K^+ exchange on the peritubular side of the cell.

a. Equation (2) can be used to measure the clearance of any substance that is present in the plasma and excreted by the kidneys.

b. Plasma concentrations rather than whole blood concentrations are used in calculations of renal clearance, because only the plasma is cleared by filtration. An analysis of this relationship demonstrates that the unit for clearance is ml/min, as:

$$C_x = \frac{U_x \cdot \dot{V}}{P_x} = \frac{(mg/ml) \ (ml/min)}{(mg/ml)} = ml/min.$$

c. Because substances can be cleared by filtration, tubular secretion, or a combination of these, the magnitude of the tubular clearance depends on the plasma concentration and the tubular transport capacity (Tm) for reabsorption or secretion. (The phenomenon of Tm is discussed in more detail in Section VI A 1.)

3. Basically, clearance is a comparison of the amount of substance removed from plasma per unit time with the amount of substance excreted in the urine per unit time, which is demonstrated by a simple rearrangement of equation (2); that is:

$$C_x \cdot P_x = U_x \cdot \dot{V}. \tag{3}$$

B. INULIN CLEARANCE

1. As a Measure of GFR.

a. Inulin clearance (C_{in}) is a measure of GFR because the volume of plasma completely cleared of inulin per unit time equals the volume of plasma filtered per unit time (i.e., C_{in} = GFR). The following characteristics of inulin account for this quality.

(1) Inulin is only freely filtered. Because no inulin is secreted or reabsorbed, all excreted inulin must come from the plasma.

(2) Inulin is biologically inert and nontoxic.

(3) Inulin is not bound to plasma proteins, and it is not metabolized to another substance.

b. Although inulin is used most commonly, other substances can measure GFR. The most frequently used agents include mannitol, sorbitol, sucrose (intravenous), iothalamate, radioactive cobalt-labeled vitamin B_{12}, ^{51}Cr-labeled edetic acid (EDTA), and radioiodine-labeled Hypaque. (**Endogenous creatinine** clearance is used clinically as an **estimate** of GFR, as discussed below in Section IV D 2 a, because a small amount of creatinine is **secreted** in humans.)

2. As an Indicator of Plasma Clearance Mechanisms. A comparison of the clearance of a given substance (C_x) with the clearance of inulin (C_{in}) provides information about the renal mechanisms used to remove the substance from plasma.

a. When C_x **equals** C_{in}, excretion is by filtration alone. In terms of equation (3), this means that the mass of the substance excreted in the urine per unit time equals the mass of the substance filtered during the same time, or:

$$U_x \cdot \dot{V} = C_x \cdot P_x.^*$$

b. When C_x **is less than** C_{in}, excretion is by filtration and reabsorption. In this case, the mass of the substance excreted in the urine is less than the mass of the substance filtered during that time, or:

$$U_x \cdot \dot{V} < C_x \cdot P_x.$$

c. When C_x **is greater than** C_{in}, excretion is by filtration and secretion. In this case, the mass of the substance excreted in the urine is greater than the mass of the substance filtered during the same time, or:

$$U_x \cdot \dot{V} > C_x \cdot P_x.$$

C. APPLICATION OF THE CLEARANCE CONCEPT TO TUBULAR FUNCTION (Fig. 4-4)

1. Clearance Ratios.

a. Definition. Clearance ratio is the ratio of the clearance of any substance (C_x) to the clearance of inulin (C_{in}).[†]

(1) The clearance ratio has the formula:

$$\frac{C_x}{C_{in}} = \frac{U_x \cdot \dot{V}}{P_x} \div \frac{U_{in} \cdot \dot{V}}{P_{in}}.$$

---

*The mass filtered per unit time (also called the **amount filtered** or **filtered load**) is equal to GFR $\cdot$ P_x (or $C_{in} \cdot P_x$) and is expressed in mg/min (see Section VI A 3).

[†]Creatinine clearance also can be used in the clearance ratio instead of inulin clearance (see Section IV D).

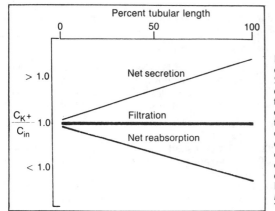

Figure 4-4. Idealized ratios of K^+ clearance (C_{K+}) to the clearance of inulin (C_{in}), as a function of tubular length. Since inulin is neither secreted nor reabsorbed, an increase in the C_{K+}/C_{in} ratio represents tubular reabsorption of K^+. In evaluating solute transport, the TF/P concentration ratio of a solute divided by the TF/P concentration ratio of inulin has the same significance as the clearance ratio. In fact, the TF/P clearance ratio is the clearance ratio for the site of sampling. When the TF/P concentration ratio of a solute is divided by the TF/P concentration ratio of inulin, it is called the **tubular clearance ratio**. (Reprinted with permission from Rose BD: Evaluation of Micropuncture Data. In *Clinical Physiology of Acid-Base and Electrolyte Disorders*. New York, McGraw-Hill, 1977, p 67.)

(2) Clearance ratio is a **double ratio** (i.e., a ratio of two ratios), and it can be determined without measuring urine flow ($\dot{V}$). Because $\dot{V}$ cancels out, the clearance ratio is reduced to:

$$\frac{C_x}{C_{in}} = \frac{U_x}{P_x} \div \frac{U_{in}}{P_{in}}. \qquad (4)$$

b. Tubular Clearance Ratio.
    (1) The local tubular clearance of a solute compared to that of inulin at a specific site along the nephron is termed the tubular clearance ratio. This relationship is expressed by equation (4):

$$\frac{C_x}{C_{in}} = \frac{TF_x}{P_x} \div \frac{TF_{in}}{P_{in}},$$

    where TF = the tubular fluid concentration at the local site of micropuncture.
    (2) The tubular clearance ratio is used for two purposes.
        (a) It is used to measure the fraction of a solute remaining in the lumen at a specific tubular segment along the nephron.
        (b) It is used to determine the nature of tubular solute transport (i.e., reabsorption or secretion). Changes in C_x/C_{in} ratios represent net solute reabsorption from, or solute secretion into, the lumen because inulin is neither reabsorbed nor secreted. Thus, to examine solute transport, the effect of water transport is **eliminated** by comparing the TF/P solute concentration to the TF/P concentration of a substance that remains in the tubule (i.e., inulin).
c. Interpretation of Clearance Ratios.
    (1) Case 1: When C_x/C_{in} = 1. A ratio of 1 indicates that the amount of the substance excreted per unit time equals the amount of the substance filtered in the same time. This is simply a restatement of equation (3):

$$(U_x \bullet \dot{V}) = (C_{in} \bullet P_x).$$

        (a) A clearance ratio of 1 indicates that, **on a net basis**, the substance is neither reabsorbed nor secreted and, therefore, is only filtered.
        (b) Substances with clearance ratios close to 1 include mannitol, sorbitol, thiosulfate, ferricyanide, iothalamate, vitamin B_{12}, and sucrose (intravenous).
    (2) Case 2: When C_x/C_{in} < 1. A clearance ratio of less than 1 demonstrates that the amount of the substance excreted per unit time is less than the amount filtered per unit time, or:

$$(U_x \bullet \dot{V}) < (C_{in} \bullet P_x).$$

        (a) A clearance ratio of less than 1 indicates that, **on a net basis**, the substance undergoes reabsorption.
        (b) Substances with clearance ratios of less than 1 include glucose, xylose, and fructose.
        (c) Example. The C_{Na+}/C_{in} ratio at the end of the proximal tubule is about 0.3 in the rat. This means that 30 percent of the filtered Na^+ remains in the lumen at that site, and 70 percent of the filtered Na^+ was reabsorbed along the proximal tubule. Since the TF/P Na^+ concentration ratio (TF/P [Na^+]) equals 1, Na^+ has been reabsorbed with water in a concentration similar to that in plasma (isosmotic reabsorption).

(3) Case 3: When $C_x/C_{in} > 1$. A clearance ratio that is greater than 1 indicates the net excretion of the substance over time is greater than the quantity filtered over time, or:

$$(U_x \cdot \dot{V}) > (C_{in} \cdot P_x).$$

(a) A clearance ratio of more than 1 represents **net secretion** of the substance into the lumen; therefore, the substance is cleared by filtration and secretion.

(b) Substances with clearance ratios greater than 1 include para-aminohippuric acid (PAH), phenol red, iodopyracet or diodone (Diodrast), certain penicillins, and creatinine (in humans).

2. Tubular Fluid Concentration Ratio.

a. Definition. A comparison of the tubular fluid (TF) solute concentration to the plasma (P) solute concentration is called the **TF/P concentration ratio**. The TF/P concentration ratio is not compared to that of inulin and is, therefore, a **single** ratio.

b. Uses of the TF/P Concentration Ratio.

(1) This ratio measures the tubular solute concentration along the nephron, which is a function of solute transport (i.e., reabsorption or secretion) as well as water reabsorption. The following example illustrates this function.

(a) The TF/P concentration ratio for K^+ (TF/P [K^+]) increases in the collecting duct, which indicates either water reabsorption or K^+ secretion.

(b) However, the C_{K^+}/C_{in} (i.e., clearance ratio) remains constant, indicating that K^+ is neither secreted nor reabsorbed.

(c) Therefore, the elevated TF/P [K^+] represents water reabsorption in the collecting duct.

(2) TF/P solute concentration ratios also can represent **total solute concentration** rather than the concentration of a single solute. When measuring the total solute concentration, the TF/P or urine-to-plasma (U/P) ratio is expressed in terms of osmolality (osmolarity) and can be interpreted as follows.

(a) Case 1: When $TF_{osm}/P_{osm} = 1$ or $U_{osm}/P_{osm} = 1$, the tubular fluid or urine is **isosmotic** with respect to plasma.

(b) Case 2: When TF_{osm}/P_{osm} or $U_{osm}/P_{osm} < 1$, the tubular fluid or urine is **hyposmotic** with respect to plasma.

(c) Case 3: When TF_{osm}/P_{osm} or $U_{osm}/P_{osm} > 1$, the tubular fluid or urine is **hyperosmotic** with respect to plasma.

(d) Example. States of hydropenia (dehydration) lead to ADH secretion. This, in turn, leads to increased water reabsorption and an elevation of U_{osm}, which has an upper limit in humans of about 1400 mOsm/kg. With a P_{osm} of about 300 mOsm/kg, this is equivalent to a maximal U_{osm}/P_{osm} of approximately 4.7.

D. CREATININE CLEARANCE AND CREATININE CONCENTRATION

1. Normal Levels and the Effect of Aging.

a. Creatinine clearance has a normal range of 80 to 110 ml/min per 1.73 m² body surface area and declines with age in healthy individuals. Because creatinine clearance is an index of GFR (see Section IV D 2 a), normally GFR also declines with age.

b. Plasma creatinine concentration remains remarkably constant through life, averaging from 0.8 to 1.0 mg/dl in the absence of disease.

c. Because creatinine clearance normally declines with age and plasma creatinine concentration does not, the decline in clearance is attributed to a parallel decrease in the excretion rate of creatinine. (The decline in creatinine clearance is not due to a decreased tubular secretion of creatinine, because the clearance ratio of creatinine to inulin is a quite constant 1.2.)

d. The decline in creatinine excretion with age is due to a primary decline in renal function and a secondary decline in muscle mass.

(1) The decline in GFR is due to declines in renal plasma flow, cardiac output, and renal tissue mass. (Decreased GFR is the most clinically significant renal functional deficit occurring with age.)

(2) The decline in creatinine excretion with a decline in muscle mass over time accounts for the relatively constant plasma creatinine concentration among healthy individuals.

2. Estimating GFR.

a. Creatinine Clearance.

(1) As discussed earlier, inulin clearance can be used to **measure** GFR. In clinical practice, however, it is more common to determine the 24-hour endogenous creatinine clearance as an **estimate** of GFR. (This is true even though the clearance ratio of creatinine to inulin is approximately 1.2, indicating that creatinine is cleared by filtration and secretion.)

    (2) Creatinine clearance determinations do not require administration of exogenous creatinine, as creatinine is a product of muscle metabolism.

  b. Creatinine Concentration. The tendency to rely on serum creatinine **concentration** rather than creatinine **clearance** as an estimate of renal function is a serious error, because plasma creatinine concentration overestimates renal function (i.e., GFR).

    (1) Using serum creatinine concentration as a basis for drug dosages can lead to drug overdose, especially in elderly individuals. Overdose is a particularly serious risk with drugs that are cleared primarily by renal mechanisms, such as digoxin and aminoglycoside antibiotics. The dosages for these drugs frequently are based on serum creatinine concentration.

    (2) To avoid this error, patient age should be considered together with measurements of creatinine clearance and blood levels of the drug. Drug dosages should be adjusted to the 30–40 percent decrease in GFR that normally occurs in individuals between the ages of 30 and 80 years.

V. GLOMERULAR FILTRATION AND GFR

A. HYDROSTATICS OF GLOMERULAR FILTRATION

  1. Fluid movement is proportional to the membrane permeability and to the balance between hydrostatic and osmotic (oncotic) forces across the glomerular membrane. When hydrostatic exceeds oncotic pressure, **filtration** occurs. Conversely, when oncotic exceeds hydrostatic pressure, **reabsorption** occurs.

    a. Effective filtration pressure (EFP) refers to the **net** driving forces for water and solute transport across the glomerular membrane. EFP is the function of two variables: the **hydrostatic pressure gradient** driving fluid **out** of the glomerular capillary and into Bowman's capsule and the **colloid osmotic pressure gradient** bringing fluid **into** the glomerular capillary. Using these variables, EFP is expressed as:

$$EFP = (P_{cap} - P_{Bow}) - COP_{cap,}{}^*$$

    where $(P_{cap} - P_{Bow})$ = the outward forces **promoting** filtration; and COP_{cap} = the inward forces **opposing** filtration.

    b. The **filtration coefficient** (K_f) of the glomerular membrane is a function of the capillary surface area and the membrane permeability. K_f is expressed in ml/min/mm Hg.

    c. Using EFP and K_f, then, the GFR is expressed as:

$$GFR = K_f \times EFP.$$

    (1) K_f normally equals about 12.5 ml/min/mm Hg.

    (2) EFP normally equals about 10 mm Hg, which can be determined from the normal values for P_{cap} (45 mm Hg), P_{Bow} (10 mm Hg), and COP_{cap} (25 mm Hg) using the equation for EFP (see above).

    (3) The normal GFR for both kidneys is about 125 ml/min per 1.73 m² of body surface area.

  2. Although the principle of opposing forces described above accounts for the magnitude and direction of flow, the main mechanism that alters intracapillary hydrostatic pressure probably is not the resistance along the length of the capillary but the contractile state of the **precapillary sphincters**.

B. GLOMERULAR FILTRATION AND SYSTEMIC FILTRATION: A COMPARISON.
The amount of filtration performed by the renal glomerular capillaries is better appreciated when this filtration is compared to that of the extrarenal (systemic) capillaries.

  1. Capillary Exchange Area and K_f.

    a. The total **glomerular** capillary exchange area is estimated to be 0.5–1.5 m²/100 g of renal tissue.

    b. In the adult human, if the entire systemic capillary bed were patent at a given time, it would afford a total capillary exchange area of 1000 m². In a man at rest, approximately 25–35 percent of the systemic capillaries are open at any time, and the **effective pulmonary** capillary surface area is about 60 m².

    c. The K_f of the glomerulus is 50–100 times greater than that of a muscle capillary.

  2. GFR and Systemic Filtration Rate.

    a. About 20 L of fluid are filtered daily from the systemic capillaries. Of these 20 L/day, 16–18 L/day are reabsorbed in the venular ends of the capillaries, and the remaining 2–4 L/day

---

*The colloid osmotic pressure in Bowman's capsule is not expressed in this equation, though it is understood, because the glomerular filtrate normally contains little protein.

represent lymph flow. (**Diffusional exchange** across the entire systemic capillary bed, however, is about 80,000 L/day!)

    b. Approximately 180 L of fluid are filtered daily from the glomerular capillaries. Therefore, the transtubular flow of fluid out of the glomerular capillaries (GFR = 180 L/day) far exceeds the filtration from systemic capillaries (20 L/day).

C. DETERMINANTS OF GFR. The major determinant of GFR is the hydrostatic pressure within the glomerulus. In addition, the **renal blood flow** (RBF) through the glomeruli has a great effect on GFR; when the rate of RBF increases, so does GFR. The following factors determine the RBF.

    1. Autoregulation of Arterial Pressure (Fig. 4-5).

        a. Over a wide range of renal arterial pressures (i.e., from 90 to 190 mm Hg), the GFR and renal plasma flow (RPF) remain quite constant. This intrinsic phenomenon observed in the renal capillaries also occurs in the capillaries of muscles and is termed **autoregulation**. Intrarenal autoregulation is virtually absent, however, at mean arterial blood pressures below 70 mm Hg.

            (1) Autoregulation has been observed to persist after renal denervation, in the isolated perfused kidney, in the transplanted kidney, after adrenal demedullation, and even in the absence of erythrocytes.

            (2) Autoregulation of RBF is necessary for the autoregulation of GFR.

        b. This phenomenon probably is mediated by changes in preglomerular (afferent) arteriolar resistance. Thus, from the equation relating flow (F) to a pressure gradient (ΔP) and resistance (R):

$$F = \frac{\Delta P}{R},$$

where ΔP = the difference between renal arterial and renal venous pressures; and R = renal vascular resistance. Clearly, to have a constant blood flow with a concomitant increase in renal perfusion pressure (ΔP), there must be a commensurate increase in renal vascular resistance.

    2. Sympathetic Stimulation (Table 4-5).

        a. In addition to the autoregulatory response to hypotension (i.e., vasodilation), the reflex increase in sympathetic tone results in both afferent and efferent arteriolar vasoconstriction and an increase in renal vascular resistance together with a decrease in GFR.

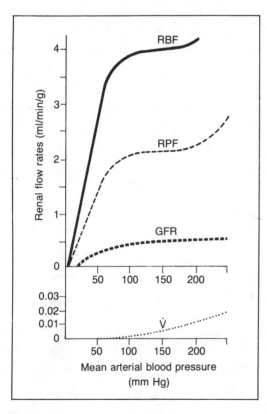

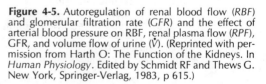

Figure 4-5. Autoregulation of renal blood flow (*RBF*) and glomerular filtration rate (*GFR*) and the effect of arterial blood pressure on RBF, renal plasma flow (*RPF*), GFR, and volume flow of urine (V̇). (Reprinted with permission from Harth O: The Function of the Kidneys. In *Human Physiology*. Edited by Schmidt RF and Thews G. New York, Springer-Verlag, 1983, p 615.)

Table 4-5. Effect of Changes in Renal Vascular Resistance with a Constant Renal Perfusion on GFR, RPF, and FF

Renal Arteriolar Vascular Resistance		GFR (ml/min)	RPF (ml/min)	FF (ml/min)
Afferent	Efferent			
↑	. . .	↓	↓	no change
↓	. . .	↑	↑	no change
. . .	↑	↑	↓	↑
. . .	↓	↓	↑	↓

Note.—GRF = glomerular filtration rate; RPF = renal plasma flow; and FF = filtration fraction. Note that GFR and RPF exhibit parallel shifts with changes in afferent arteriolar resistance but exhibit divergent shifts with changes in efferent arteriolar resistance. Increases in vascular resistance always lead to a decline in RPF, and decreases in arteriolar resistance always lead to an increase in RPF. Upward arrows denote the effect of vasoconstriction, and downward arrows denote the effect of vasodilation.

    b. Conversely, the autoregulatory response to increases in renal arterial blood pressure (i.e., vasoconstriction) causes the reflex decrease in sympathetic tone, which results in renal arteriolar vasodilation and a decrease in renal vascular resistance together with an increase in GFR.

  3. Arteriolar Resistance.
    a. The **afferent arteriole** is the larger diameter arteriole and the major site of autoregulatory resistance. When resistance is altered in the afferent arterioles, GFR and RPF change in the same direction. Therefore, increases and decreases in afferent arteriolar resistance **do not affect** the **filtration fraction** or **FF** (i.e., the ratio of GFR to RPF).*
      (1) Constriction of the afferent arteriole decreases both RPF and GFR.
        (a) An increase in vascular resistance proximal to the glomeruli leads to a decrease in RPF.
        (b) A decrease in hydrostatic pressure within the glomerular capillaries leads to a decrease in GFR.
      (2) Dilation of the afferent arteriole increases both RPF and GFR.
        (a) A decrease in vascular resistance proximal to the glomeruli leads to an increase in RPF.
        (b) An increase in glomerular capillary perfusion pressure leads to an increase in GFR.
    b. When resistance is altered in the **efferent arterioles**, GFR and RPF change in opposite directions. Therefore, increases and decreases in efferent arteriolar resistance **do affect** the **FF.**
      (1) Constriction of the efferent arteriole decreases RPF and increases GFR.
        (a) An increase in vascular resistance distal to the glomeruli leads to a decrease in RPF.
        (b) An increase in the hydrostatic pressure within the glomerular capillaries leads to an increase in GFR.
        (c) Following an increase in efferent arteriolar resistance, the FF increases.
      (2) Dilation of the efferent arteriole increases RPF and decreases GFR.
        (a) A decrease in vascular resistance distal to the glomeruli leads to an increase in RPF.
        (b) A decrease in the hydrostatic pressure within the glomerular capillaries leads to a decrease in GFR.
        (c) Following a decrease in efferent arteriolar resistance, the FF decreases.

VI. RENAL TUBULAR TRANSPORT: REABSORPTION AND SECRETION

A. DEFINITIONS

  1. Renal tubular transport maximum (Tm) refers to the maximal amount of a given solute that can be transported (reabsorbed or secreted) per minute by the renal tubules.
    a. The highest attainable rate of **reabsorption** is called the **maximum tubular reabsorptive capacity** and is designated Tm or Tr. Substances that are reabsorbed by the kidneys and that have a Tm include phosphate ion (HPO_4^{2-}), SO_4^{2-}, glucose (and other monosaccharides), many amino acids, uric acid, and (most likely) albumin.
    b. The highest attainable rate of **secretion** is called the **maximum tubular secretory capacity** and is designated Tm or Ts. Substances that are secreted by the kidneys and that have a Tm include penicillin, certain diuretics, salicylate, PAH, and thiamine (vitamin B_1).

---

*FF is a measure of the fraction of the entering plasma volume (RPF) that is removed through the glomeruli as filtrate. FF refers to the bulk volume of fluid (not solutes) and normally measures about 20 percent.

    c. Some solutes have no definite upper limit for unidirectional transport and, hence, have no Tm.

        (1) The reabsorption of Na^+ along the nephron has no Tm.

        (2) The secretion of K^+ by the distal tubules has no Tm.

2. Threshold concentration refers to the plasma concentration at which a solute begins to appear in the urine and is characteristic of that substance.

3. Filtered Load and Excretion Rate.

    a. Reabsorption and secretion are not directly measured variables but are derived from the measurements of the amount of solute filtered (filtered load) and the amount of solute excreted (excretion rate).

        (1) The **filtered load** is the amount of a substance entering the tubule by filtration per unit time and is mathematically equal to the product of GFR (or C_{in}) and the plasma concentration of the substance (P_x), as:

$$\text{filtered load} = \text{GFR} \cdot P_x; \text{(ml/min) (mg/ml)} = \text{mg/min}$$
$$= C_{in} \cdot P_x; \text{(ml/min) (mg/ml)} = \text{mg/min.}$$

        (2) The **excretion rate** is the amount of a substance that appears in the urine per unit time and is mathematically equal to the product of urine flow rate ($\dot{V}$) and the urine concentration of the substance (U_x), as:

$$\text{excretion rate} = U_x \cdot \dot{V}; \text{(mg/ml) (ml/min)} = \text{mg/min.}$$

    b. If the excretion rate exceeds the filtered load [or if the clearance of the substance (C_x) is greater than C_{in}], net tubular **secretion** of that substance had occurred. Thus, secretion is expressed as:

$$U_x \cdot \dot{V} > C_{in} \cdot P_x \text{ or } C_x > C_{in}.$$

    c. If the filtered load exceeds the excretion rate (or if C_x is less than C_{in}), net **reabsorption** of that substance has occurred. Thus, reabsorption is expressed as:

$$C_{in} \cdot P_x > U_x \cdot \dot{V} \text{ or } C_x < C_{in}.$$

    d. Calculation of Tm (Tr or Ts). Tm is the difference between the filtered load ($C_{in} \cdot P_x$) and the excretion rate ($U_x \cdot \dot{V}$). Tr and Ts, then, are expressed as:

        (1) $Tr = C_{in} \cdot P_x - U_x \cdot \dot{V}$ (in mg/min)

        (2) $Ts = U_x \cdot \dot{V} - C_{in} \cdot P_x$ (in mg/min)

B. GRAPHIC REPRESENTATION OF RENAL TRANSPORT PROCESSES: RENAL TITRATION CURVES (Fig. 4-6)

1. Construction of Renal Titration Curves.

    a. A titration curve is constructed by plotting the following pairs of variables.*

        (1) The filtered load ($C_{in} \cdot P_x$) against the plasma concentration (P_x)

        (2) The excretion rate ($U_x \cdot \dot{V}$) against P_x

        (3) The difference between the filtered load and the excretion rate (i.e., Tr or Ts; see Section VI A 3 d) against P_x

    b. The plotted renal titration curve, therefore, can be used to determine the plasma concentration (P_x) at which the renal tubular membrane carriers are fully saturated (Tm).

2. Glucose Transport: Reabsorption.

    a. Characteristics of Glucose Transport.

        (1) Glucose is an essential nutrient that is actively reabsorbed into the proximal tubule by a Tm-limited process. The Tm for glucose (Tm_G) is about 340 mg/min (or about 2 mmol/min) per 1.73 m^2 body surface area.

        (2) Because the Tm_G is nearly constant and depends on the number of functional nephrons, it is used clinically to estimate the number of functional nephrons, or the tubular reabsorptive capacity (Tr).

    b. Glucose Titration Curve. Figure 4-6*A* and *C* illustrates that glucose transport and excretion processes are functions of the plasma glucose concentration (P_G).

        (1) Increasing the P_G results in a progressive linear increase in the filtered load ($C_{in} \cdot P_G$).

        (2) At a low P_G, the reabsorption of glucose is complete; hence, the clearance of glucose (C_G) is zero.

        (3) When P_G in humans reaches the renal threshold concentration of 180–200 mg/dl (10–11 mmol/L), glucose appears in the urine (glycouria). (Note that the threshold con-

---

*The dependent variables (i.e., filtered load, excretion rate, and Tr or Ts) are plotted on the ordinate as a function of the independent variable (i.e., P_x).

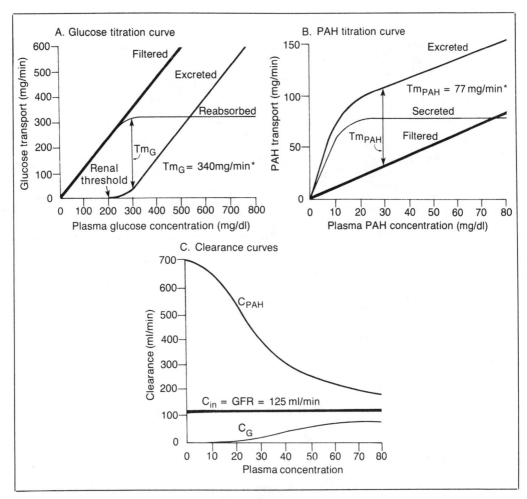

Figure 4-6. Relationships of renal titration curves and clearance curves to plasma concentration. The splay (curved portions) observed in the glucose reabsorption and excretion curves (*graph A*) and the splay in the para-aminohippuric acid (PAH) secretion and excretion curves (*graph B*) are due to the kinetics of tubular transport and the heterogeneity of the nephron population in terms of number, length, and functional variations in transport. *Graph C* demonstrates that the clearance of glucose (C_G) increases and approaches the clearance of inulin (C_{in}) asymptotically as the plasma glucose concentration increases beyond the tubular transport maximum for glucose (Tm_G). Conversely, *graph C* shows that the clearance of PAH (C_{PAH}) declines and approaches the C_{in} asymptotically as the plasma PAH concentration increases above the tubular transport maximum for PAH (Tm_{PAH}). In *graph C*, the plasma PAH concentration is in mg/dl and the plasma glucose concentration is plotted as one-tenth of the actual concentration (i.e., mg/dl × 10^{-1}). * = per 1.73 m^2 body surface area. (Adapted from Selkurt EE: Renal Function. In *Physiology*, 5th ed. Edited by Selkurt EE. Boston, Little, Brown, 1984, pp 424 and 429.)

centration is not synonymous with the P_G that completely saturates the transport mechanism, which is about 300 mg/dl.)

(4) As the Tm_G is approached, the urinary excretion rate increases linearly with increasing P_G.

(5) When the Tm_G is reached, the quantity of glucose reabsorbed per minute remains constant and is independent of P_G.

(6) When the Tm_G is exceeded, the C_G becomes increasingly a function of glomerular filtration; therefore, the C_G approaches C_{in}, and the amount reabsorbed becomes a smaller fraction of the total amount excreted.

3. PAH Transport: Secretion.
a. Characteristics of PAH Transport.
(1) PAH is a weak organic acid that is actively secreted into the proximal tubule by a Tm-limited process. PAH is a foreign substance that is not stored or metabolized and is ex-

creted virtually unchanged in the urine. By filtration and secretion, PAH is almost entirely cleared from the plasma in a single pass through the kidney, if the Tm for PAH (Tm_{PAH}) has not been reached. (The Tm_{PAH} in humans is about 80 mg/min per 1.73 m^2 body surface area.

(2) Because about 10 percent of the plasma PAH is bound to plasma proteins, PAH is not entirely freely filterable and the concentration of PAH in the plasma (P_{PAH}) is greater than that in the glomerular filtrate. However, PAH does not need to be dissolved in plasma to be transported (secreted) by the carrier into the tubular fluid, and protein binding does not diminish the effectiveness of tubular excretion.

(3) Since the Tm_{PAH} is nearly constant, it is used clinically to estimate tubular secretory capacity (Ts).

b. PAH Titration Curve. Figure 4-6*B* and *C* illustrates that the filtration and secretion of PAH are functions of the P_{PAH}. The amount of PAH excreted per minute ($U_{PAH} \bullet \dot{V}$) exceeds the amount filtered ($C_{in} \bullet P_{PAH}$) at any P_{PAH}.

(1) When the P_{PAH} is low, virtually all of the PAH that is not filtered is secreted, and PAH is almost completely cleared from the plasma by the combined processes of glomerular filtration and tubular secretion.

(2) When P_{PAH} is above 20 mg/dl, the transepithelial secretory mechanism becomes saturated, and the Tm_{PAH} is reached.

(3) When the Tm_{PAH} is reached, the quantity of PAH secreted per minute remains constant and is independent of P_{PAH}.

(4) When the Tm_{PAH} is exceeded, the C_{PAH} becomes progressively more a function of glomerular filtration; hence, the C_{PAH} approaches C_{in} (see Fig. 4-6*B*), and the constant amount of PAH secreted becomes a smaller fraction of the total amount excreted.

c. PAH Clearance and RPF.

(1) Fick Principle. According to the Fick principle, blood flow (BF) through an organ can be determined with the equation:

$$BF = \frac{R}{A_x - V_x} , \qquad (5)$$

where BF = the blood flow (in ml/min); R = the rate of removal (or addition) of a substance (in mg/min) from (or to) the blood as it flows through an organ; A_x = the concentration of that substance (in mg/ml) in the blood entering the organ; and V_x = the concentration of that substance (in mg/ml) in the blood leaving the organ.

(2) Since C_{PAH} can be used to measure RPF, equation (5) can be modified to:

$$RPF = \frac{U_{PAH} \bullet \dot{V}}{A_{PAH} - V_{PAH}} , \qquad (6)$$

where RPF = renal plasma flow (in ml/min); $U_{PAH} \bullet \dot{V}$ = the rate of PAH excretion (in mg/min); and A_{PAH} and V_{PAH} = concentrations of PAH (in mg/ml) in the renal artery and the renal vein, respectively.

(a) At **low concentrations** of PAH in arterial plasma, the renal clearance is nearly complete; therefore, as a first approximation, the PAH concentration in renal **venous** plasma may be taken to be zero.

(i) About 10–15 percent of the total RPF perfuses nonexcretory (nontubular) portions of the kidney, such as the renal capsule, perirenal fat, the renal medulla, and the renal pelvis; therefore, this plasma cannot be completely cleared of PAH by filtration and secretion.

(ii) Because about 10 percent of the PAH remains in the renal venous plasma, the RPF calculated from C_{PAH} **underestimates** the actual flow by about 10 percent. Accordingly, the C_{PAH} actually measures the **effective renal plasma flow** (ERPF), and:

$$ERPF = \frac{U_{PAH} \bullet \dot{V}}{A_{PAH}} = C_{PAH}. \qquad (7)$$

(b) Since PAH is not metabolized or excreted by any organ other than the kidney, a sample from any peripheral **vein** can be used to measure arterial plasma PAH concentration, and the equation is written as:

$$ERPF = \frac{U_{PAH} \bullet \dot{V}}{P_{PAH}} = C_{PAH} .$$

(3) Effective renal **blood** flow (ER*B*F) is calculated from the relationship between plasma volume (PV), hematocrit (Hct), and blood volume (BV), as:*

$$BV = \frac{PV \times 100}{(100 - Hct)},$$

and, therefore:

$$ERBF = \frac{ERPF \times 100}{(100 - Hct)}.$$

(4) True renal plasma flow (TR*P*F) can be determined if the extraction ratio (E) is known, where:

$$E = \frac{A_{PAH} - V_{PAH}}{A_{PAH}}, \text{ and}$$

$$TRPF = \frac{ERPF}{E}; TRBF = \frac{TRPF \times 100}{(100 - Hct)}.$$

C. RENAL TRANSPORT OF COMMON SOLUTES AND WATER (see Table 4-4)

1. General Considerations.

  a. The proximal tubules reabsorb 70–85 percent of the filtered Na^+, Cl^-, HCO_3^-, and water and virtually all of the filtered K^+, HPO_4^{2-}, and amino acids. In addition, glucose is reabsorbed almost completely by the proximal tubules and begins to appear in the urine when the renal threshold is exceeded.[†]

  b. Reabsorption of water is passive, and reabsorption of solutes can be passive or active; solute reabsorption generates an osmotic gradient, which causes the passive reabsorption of water (osmosis).

2. Na^+ Reabsorption.

  a. Proximal Tubules. Transtubular reabsorption of Na^+ across the proximal tubules occurs in two steps.

  (1) Step 1: Through the Luminal Membrane. Na^+ is transported through the luminal (adluminal) membrane **passively** down an electrochemical gradient by facilitated diffusion. This first step involves two processes.

  (a) First, Na^+ undergoes **passive cotransport** with the **active cotransport** (secondary active transport) of glucose and amino acids. Thus, the energy for the uphill transport of glucose from lumen to cell is derived from the simultaneous downhill movement (influx) of Na^+ along a concentration gradient, which is maintained by the primary active Na^+ pump.

  (b) Next, electrically coupled **cation exchange** (i.e., Na^+ **reabsorption**) occurs, coupled with H^+ **secretion** across the luminal membrane. (The transport of Na^+ in the proximal segment is not associated with K^+ secretion.)

  (i) To maintain electroneutrality, Na^+ is reabsorbed in exchange for the secretion of H^+.

  (ii) Because tubular H^+ is buffered by HCO_3^-, the influx of Na^+ into the cell must be electrically balanced by an anion.[‡] This is accomplished by passive inward diffusion of Cl^- down its concentration gradient or by regeneration of HCO_3^- from cellular CO_2 and water in the presence of carbonic anhydrase.

  (2) Step 2: Through the Basolateral Membrane.

  (a) Na^+ is **actively** transported through the basolateral (abluminal) membrane (primary active transport) via the Na^+-K^+ pump. Na^+ reabsorption is equivalent to Cl^- reabsorption + (H^+ + K^+) secretion. Approximately 50 percent of the proximal reabsorption of Na^+ is based on the active transport process.

  (b) The remaining percentage of Na^+ reabsorption across the basolateral membrane of the proximal tubule occurs **passively** by solvent drag.

  (i) The efflux of Na^+ into the intercellular spaces creates an electrical gradient for the efflux of Cl^- and HCO_3^-, which, in turn, generates an osmotic gradient for water transport into the interspaces. About 20 percent of the Na^+ movement is loosely coupled to the active secretion of H^+, resulting in the reabsorption of HCO_3^-.

---

*In humans, the C_{PAH} is 600–700 ml/min and, when corrected to ER*B*F (assuming a hematocrit of 45 percent), is approximately 1100–1200 ml/min, which is about 20 percent of the resting cardiac output.
†i.e., approximately 200 mg glucose/dl arterial plasma, or 11 mmol/L.
‡The transcellular transport of Na^+ is passive.

(ii) The osmotic movement of water generates a small increment in hydrostatic pressure, and some bulk flow of water containing Na^+, Cl^-, and HCO_3^- proceeds into the peritubular capillaries.

(c) As a result of these active and passive transport processes, Na^+ reabsorption together with anions along the proximal tubule allows for the isosmotic reabsorption of fluid.

b. Distal and Collecting Tubules.

(1) At the luminal membrane of the distal and collecting tubules, Na^+ is reabsorbed by two electrically linked cation-exchange processes.*

(a) Na^+-H^+ Exchange. This process involves Na^+ reabsorption and H^+ secretion.

(b) Na^+-K^+ Exchange. This process involves Na^+ reabsorption and K^+ secretion. The quantity of Na^+ reabsorbed distally is much greater than the amount of K^+ secreted.

(c) These cation-exchange processes are competitive (i.e., K^+ competes with H^+ for Na^+) and, therefore, are not one-to-one. Both cation-exchange processes, however, are enhanced by aldosterone (see Section IX B).

(2) Na^+ also is transported by the distal and collecting tubules without exchange for H^+ or K^+ but by reabsorption largely in association with Cl^- reabsorption.

(a) When the availability of either K^+ or H^+ is reduced, the exchange of the more available ion for Na^+ is increased.

Example 1: In hypokalemic alkalosis induced by chronic vomiting or hyperaldosteronemia, the exchange of H^+ for Na^+ is greater than the exchange of K^+ for Na^+, the urine becomes more **acid**, and the plasma becomes more **alkaline**.

Example 2: In hypoaldosteronemia, the reduced secretion of K^+ and H^+ (due to the decreased rate of distal Na^+ reabsorption) causes hyperkalemia and acidosis.

(b) The Na^+-H^+ exchange mechanism is not mandatory for H^+ secretion.

c. Nephron. Na^+ reabsorption along the nephron occurs in the following proportions:

(1) Proximal tubule (75 percent)

(2) Ascending limb of the loop of Henle (22 percent)[†]

(3) Distal tubule (4–5 percent)

(4) Collecting tubule (2–3 percent)

3. K^+ Transport.

a. General Considerations.

(1) K^+ is the only plasma electrolyte that is both reabsorbed and secreted into the renal tubules. Virtually all of the K^+ is actively reabsorbed by the proximal tubule, whereas K^+ secretion is a passive process and is largely a function of the distal tubule.

(2) The net secretion of K^+ by the nephron can cause the excretion of more K^+ than was filtered, as in hyperaldosteronemia and in metabolic alkalosis. (Elevated K^+ excretion also can occur, in part, by decreased K^+ reabsorption, as observed in the osmotic diuresis seen with uncontrolled diabetes mellitus.)

(3) The distal tubule has the capacity for both net secretion and net reabsorption of K^+, depending on the plasma [K^+] and the acid-base state of the body.

(a) Metabolic acidosis usually is associated with cellular K^+ efflux, a decreased K^+ secretion, and hyperkalemia.

(b) Metabolic alkalosis usually is associated with cellular K^+ influx, an increased K^+ secretion, and hypokalemia.

b. K^+ Reabsorption. K^+ is completely filterable at the glomerulus, and tubular reabsorption occurs by active transport in the proximal tubule, ascending limb of the loop of Henle, distal tubule, and collecting duct.

c. K^+ secretion involves an active and a passive process.

(1) The critical step is the active transport of K^+ from the interstitial fluid across the basolateral membrane into the distal tubular cell. This active transport step is associated with active Na^+ extrusion via the Na^+-K^+-ATPase system.

(2) The elevated intracellular [K^+] achieved by the basolateral pump favors the net K^+ diffusion at the luminal membrane down a concentration gradient into the tubular lumen. The transcellular transport of K^+ is passive.

(3) Aldosterone influences K^+ secretion in the following ways.

---

*Only a small amount of the filtered Na^+ that is reabsorbed in the distal and collecting tubules is by cation exchange with H^+ and K^+.

[†]Na^+ is **passively** reabsorbed in the thin and thick segments of the ascending limb of the loop of Henle.

        (a) It increases the activity of the basolateral Na^+-K^+-ATPase pump, which increases K^+ entry into the distal cells and the $[K^+]$ gradient from cell to lumen.

        (b) Aldosterone increases the permeability of the luminal membrane to K^+, which also increases K^+ diffusion (secretion).

4. H^+ Secretion and HCO_3^- Reabsorption (see Fig. 5-2). H^+ secretion is the process by which the filtered HCO_3^- is reabsorbed and the tubular fluid becomes acidified. The process of H^+ secretion and HCO_3^- reabsorption occurs throughout the nephron, except in the descending limb of the loop of Henle.

  a. H^+ secretion into the tubular lumen occurs by active transport and is coupled to Na^+ reabsorption; therefore, for each H^+ secreted, one Na^+ and one HCO_3^- are reabsorbed. The renal epithelium secretes about 4300 mEq (mmol) of H^+ daily.

    (1) Approximately 85 percent of the total H^+ secretion occurs in the proximal tubules.

    (2) Approximately 10 percent of the total H^+ secretion occurs in the distal tubules, and about 5 percent occurs in the collecting ducts.

  b. HCO_3^- Reabsorption.

    (1) Most of the H^+ secreted into the tubular fluid react with HCO_3^- to form carbonic acid (H_2CO_3), which is dehydrated (via carbonic anhydrase) to form CO_2 and water. The CO_2 diffuses back into the proximal tubular cells, where it is rehydrated to H_2CO_3, which dissociates into HCO_3^- and H^+.

    (2) The buffering of secreted H^+ by filtered HCO_3^- is not a mechanism for H^+ **excretion**. The CO_2 formed in the lumen from secreted H^+ returns to the tubular cell to form another H^+, and **no net H^+ secretion occurs.***

  c. Nonbicarbonate Buffering of H^+. The H^+ secreted in excess of those required for HCO_3^- reabsorption are buffered in the tubular fluid either by nonbicarbonate buffer anions (chiefly, HPO_4^{2-}) or by NH_3 to form NH_4^+.

    (1) About 20 mEq (mmol) of H^+ per day are buffered by HPO_4^{2-} (and organic ions) and excreted as **titratable acid**. The nonbicarbonate buffering of H^+ occurs mainly in the distal and collecting tubules.

    (2) About 40 mEq (mmol) of H^+ per day are buffered by NH_3 and excreted as NH_4^+.

    (3) The total amount of H^+ excreted daily by an individual with a normal diet equals the sum of titratable acid and NH_4^+ excreted, or about 60 mEq (mmol) of H^+ per day. Thus, only a minute concentration of free H^+ normally exists in the final urine despite the 4300 mEq (mmol) of H^+ secreted daily!

5. Cl^- Transport.

  a. General Considerations.

    (1) Cl^- reabsorption occurs at a rate of 20,000 mEq/day and is inversely related to $[HCO_3^-]$ reabsorption; that is, when HCO_3^- reabsorption increases, Cl^- reabsorption decreases and vice versa. Therefore, the plasma $[Cl^-]$ varies inversely with the HCO_3^- reabsorption rate.

    (2) Cl^- is the chief anion to accompany Na^+ through the renal tubular epithelium. Since Na^+ reabsorption is under control of aldosterone, the plasma $[Cl^-]$ is secondarily influenced by aldosterone.

    (3) Cl^- is excreted with NH_4^+ to eliminate H^+ in exchange for Na^+ reabsorption.

  b. Passive Transport. When Na^+ is actively reabsorbed across the luminal membrane of the proximal tubule, Cl^- is passively reabsorbed along an electrochemical gradient.

    (1) Passive proximal diffusion (reabsorption) of Cl^- is aided by the increased $[Cl^-]$ within the luminal fluid of the late proximal tubule.

    (2) This increased $[Cl^-]$ is due to the reabsorption of $NaHCO_3$ and, therefore, water into the early proximal tubule.

  c. Active Transport. Cl^- is actively reabsorbed in the thick segment of the ascending limb of the loop of Henle.

    (1) Cl^- reabsorption induces the passive reabsorption of Na^+ in the thick segment.

    (2) The reabsorption of NaCl in the thick and thin segments together with the water impermeability of the ascending limb lead to the dilution of the tubular fluid.

6. Water Reabsorption. Water movement across membranes is determined by hydrostatic and osmotic pressure gradients. Water is reabsorbed passively by diffusing along an osmotic gradient, which primarily is established by the reabsorption of Na^+ and Cl^-.

  a. Proximal Tubules.

    (1) In humans, about 75–80 percent of the reabsorption of filtered water occurs in the proximal tubule.

---

*The reabsorption of HCO_3^- across the luminal membrane does not occur in the conventional way (i.e., via an active HCO_3^- pump); instead, HCO_3^- is transported across the luminal membrane mainly as CO_2.

(2) The proximal reabsorption of water is invariant and involves no change in osmolality.

b. Loop of Henle.

(1) Unlike the proximal tubule, the loop of Henle reabsorbs considerably more solute than water. Only about 5 percent of the reabsorption of filtered water occurs here.

(2) The tubular fluid entering the descending limb of the loop of Henle always is **isosmotic**, regardless of the hydration state.

c. Distal and Collecting Tubules.

(1) The distal and collecting tubular reabsorption of water occurs only in the presence of ADH and in the following amounts:*

(a) Late distal tubules (15 percent), by ADH effect

(b) Collecting tubules (4 percent), by ADH effect

(2) The tubular fluid entering the distal tubule always is **hyposmotic** to plasma, regardless of the hydration state. In the late distal tubule and collecting ducts, the osmolality in the tubular fluid changes according to the water permeability of the tubule.

VII. RENAL CONCENTRATION AND DILUTION OF THE URINE

A. FUNCTIONAL CONSIDERATIONS

1. **Purpose.** The kidney can alter the composition of the urine in response to the body's daily needs, thereby maintaining the osmolality of body fluids. When it is necessary to conserve body water, the kidney excretes urine with a high solute concentration. When it is necessary to rid the body of excess water, the kidney excretes urine with a dilute solute concentration.

2. **Role of ADH.** The principal regulator of urine composition is the hormone ADH. In the absence of ADH, the kidney excretes a large volume of dilute urine; when ADH is present in high concentration, the kidney excretes a small volume of concentrated urine.

3. **Components of the Concentrating and Diluting System.** The formation of urine that is dilute (hyposmotic to plasma) or concentrated (hyperosmotic to plasma) is achieved by the countercurrent system of the nephron (Fig. 4-7).† This system includes:

   a. Descending limb of the loop of Henle (DLH)

   b. Thin and thick segments of the ascending limb of the loop of Henle (ALH)

   c. Medullary interstitium

   d. Distal convoluted tubule

   e. Collecting duct

   f. Vasa recta, which are the **vascular elements** of the juxtamedullary nephrons

4. **Mechanisms of Dilution and Concentration.** The kidney forms a dilute urine in the absence of ADH. In the presence of ADH, the kidney forms a concentrated urine via the functioning of the **countercurrent multipliers** (i.e., the loop of Henle and collecting duct) and the **countercurrent exchangers** (i.e., the vasa recta). Regardless of ADH, however, the fluid osmolality in the loop of Henle, vasa recta, and medullary interstitium always increases progressively from the corticomedullary junction to the papillary tip (see Fig. 4-7D).

   a. The fundamental processes involved in the excretion of a dilute or concentrated urine include:

   (1) Variable permeability of the nephron to the passive back-diffusion (reabsorption) of water along an osmotic gradient and of urea along its concentration gradient

   (2) Passive reabsorption of NaCl by the thin segment of the ALH

   (3) Active reabsorption of Cl^- and passive reabsorption of Na^+ by the thick segment of the ALH

   b. The formation of a hyperosmotic urine involves the following steps.

   (1) The medullary interstitium becomes hyperosmotic by the reabsorption of NaCl and urea.

   (2) The urine entering the medullary collecting ducts equilibrates osmotically with the hyperosmotic interstitium, resulting in the excretion of a small volume of concentrated urine in the presence of ADH.

   c. The final osmolality of the urine is determined by the permeability of the collecting ducts to water.

   d. The vascular loop (vasa recta) prevents the dissipation of the osmotic (hypertonic) "layering" in the interstitium (see Section VII A 4).

---

*In the presence of a maximal ADH effect, 99 percent of the water is reabsorbed; in the absence of ADH, 88 percent of the water is reabsorbed.

†Only the juxtamedullary nephrons, with their long loops of Henle, contribute to the medullary hyperosmolality.

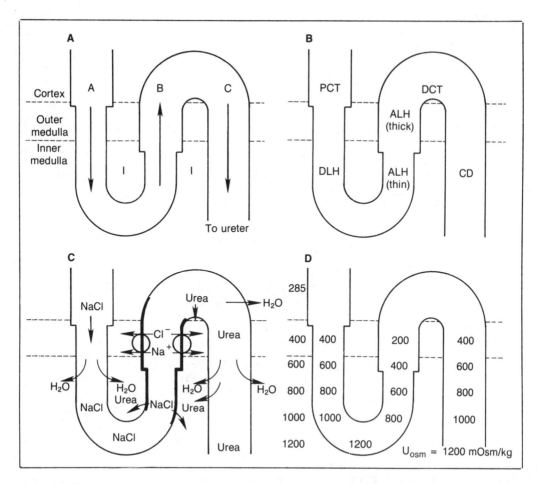

Figure 4-7. The S-shaped model of the countercurrent multiplier system of the juxtamedullary nephron. *Model A* represents the directions of tubular flow (countercurrent); *A* = concentrating segment; *B* = diluting segment; *C* = collecting duct; and *I* = interstitium. *Model B* represents the major components of the nephron; *PCT* = proximal convoluted tubule; *DLH* = descending limb of the loop of Henle; *ALH* = ascending limb of the loop of Henle (thick and thin segments); *DCT* = distal convoluted tubule; and *CD* = collecting duct. *Model C* represents solute and solvent transfer; the *heavy line* indicates water impermeability. *Model D* represents the vertical (longitudinal) and horizontal (transverse) osmotic gradients in the presence of antidiuretic hormone. (After Jamison R, Maffly RH: *N Engl J Med* 295: 1059–1067 1976.)

B. COUNTERCURRENT MULTIPLIERS (Fig. 4-8; see Fig. 4-7)

   1. General Considerations.

   a. The countercurrent multiplier system is analogous to a three-limb (S-shaped) model consisting of the two limbs of the loop of Henle, which are connected by a hairpin turn, and the collecting duct. The fluid flow through the three limbs is **countercurrent** (i.e., in alternating opposite directions, see Fig. 4-7A).

   b. The countercurrent multiplier process has two important consequences (see Figs. 4-7D and 4-8).

   (1) At equilibrium in the presence of ADH, a maximal vertical gradient of 900 mOsm/kg is established between the corticomedullary junction (300 mOsm/kg) and the renal papillae (1200 mOsm/kg). This explains the origin of the term "countercurrent multiplier."

   (2) At any given level in the renal medulla, the osmolality is almost the same in all fluids except that in the ALH. The fluid in the ALH is less concentrated than that in either the DLH or the interstitium and becomes hyposmotic to plasma (see Fig. 4-7D).

   2. Loop of Henle.

   a. The DLH is the concentrating segment of the nephron (see Fig. 4-7C and D and Fig. 4-8). The following characteristics of the DLH account for this.

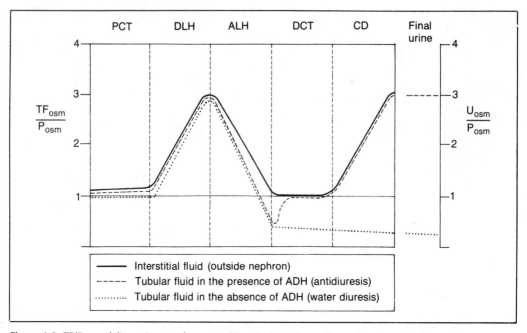

Figure 4-8. TF/P osmolality ratios as a function of length along the nephron. A ratio of 1 indicates isosmolality, a ratio of more than 1 indicates hyperosmolality, and a ratio of less than 1 indicates hyposmolality. Solute-free water reabsorption occurs in the early distal tubule, in the descending limb of the loop of Henle, and in the collecting duct. *PCT* = proximal convoluted tubule; *DLH* = descending limb of the loop of Henle; *ALH* = ascending limb of the loop of Henle; *DCT* = distal convoluted tubule; *CD* = collecting duct; and *ADH* = antidiuretic hormone. (Adapted from Koch A: Kidney Function and Body Fluids. In *Physiology and Biophysics, 20th ed: Circulation, Respiration and Fluid Balance.* Edited by Ruch TC and Patton HD. Philadelphia, WB Saunders, 1974, p 445.)

    (1) The DLH is highly permeable to water. Solute-free water leaves the DLH, causing the fluid in the DLH to become concentrated to a degree that is consistently higher than that of the fluid in the ALH.
    (2) The DLH has a low permeability to NaCl and urea. The predominant solute is NaCl, with relatively small amounts of urea existing in the DLH.
    (3) The fluid in the DLH nearly attains the osmolality of the adjacent medullary interstitium. [The interstitial osmolality is maintained by solvent-free solute (NaCl) that is transported out of the ALH.]
  b. The ALH. The thick and thin segments of the ALH constitute the diluting segments of the nephron. Both segments are impermeable to water and both are permeable to NaCl; in addition, the thin segment of the ALH is permeable to urea.
    (1) As fluid passes up the thin segment of the ALH, NaCl diffuses down its concentration gradient into the interstitial fluid. This contributes to the increased interstitial osmolality and renders the tubular fluid hyposmotic to the peritubular interstitium.
    (2) The thick segment of the ALH actively transports Cl^- out of the lumen, with Na^+ following the electrical gradient established by Cl^- transport. The active Cl^- transport from the ALH to the interstitium is the most important single effect in the countercurrent system.
    (3) The active and passive transport of NaCl from the ALH to the interstitium forms a horizontal osmotic gradient of up to 200 mOsm/kg between the tubular fluid of the ALH and the combined fluid of the interstitium and the DLH (Fig. 4-9; see Fig. 4-7C and D). This leaves behind a smaller volume of hypotonic fluid rich in urea, which flows into the distal convoluted tubule and the cortical and medullary collecting ducts.
    (4) The fluid in the ALH becomes diluted by the net loss of NaCl in excess of the net gain of urea by diffusion. This efflux of NaCl causes the tubular fluid within the ALH to become hyposmotic and that in the interstitial fluid to become hyperosmotic.

3. Simultaneous Events in the Interstitium.
  a. Fluid in the interstitium contains hyperosmotic concentrations of NaCl and urea.
    (1) As the collecting ducts join in the inner medulla, they become increasingly permeable to urea (especially in the presence of ADH), allowing urea to flow passively along its concentration gradient into the interstitium (see Figs. 4-7C and 4-8).

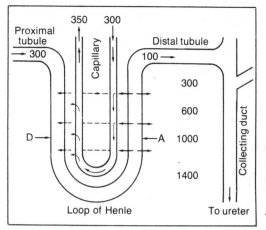

Figure 4-9. Countercurrent system for production of urine. *Numerals* denote osmolality; *D* = descending limb of the loop of Henle; *A* = ascending limb of the loop of Henle; and the *capillary* denotes the vasa recta countercurrent exchange system. (Reprinted with permission from Wilatts SM: Normal Water Balance and Body Fluid Compartments. In *Lecture Notes on Fluid and Electrolyte Balance.* Boston, Blackwell Scientific, 1982, p 19.)

(2) The increase in medullary osmolality causes water to move out of the adjacent tubules, the terminal collecting ducts, and the DLH.
   b. Water is reabsorbed in the last segment of the distal convoluted tubule and in the collecting duct in the cortex and outer medulla.
   c. In the inner medulla both water and urea are reabsorbed from the collecting duct (see Fig. 4-7C).
   (1) Some urea reenters the ALH but at a slower rate than the efflux of NaCl.
   (2) This **medullary recycling of urea**, in addition to solute trapping by countercurrent exchange, causes urea to accumulate in large amounts in the medullary interstitium, where it osmotically abstracts water from the DLH and thereby concentrates NaCl in the DLH fluid.

4. The Distal Convoluted Tubule (see Figs. 4-7 and 4-9).
   a. General Considerations. Before reaching the collecting duct, which is the next major component of countercurrent multiplication, the tubular fluid must pass through the distal convoluted tubule, where the role of ADH again is apparent.
   b. Role of ADH. Whether in the presence or the absence of ADH, fluid entering the distal convoluted tubule always is hypotonic to plasma in cortical as well as juxtamedullary nephrons (see Fig. 4-8).
   (1) During diuresis (i.e., in the absence of ADH), this fluid remains hypotonic to plasma as it traverses the distal convoluted tubule and collecting duct; in fact, the solute concentration of the final urine may be lower than that of the fluid entering the distal convoluted tubule (see Fig. 4-9). Thus, in the absence of ADH, Na$^+$ reabsorption is not retarded.
   (2) During antidiuresis (i.e., in the presence of ADH), the hypotonic fluid entering the distal convoluted tubule becomes isosmotic and the fluid entering the collecting duct is isosmotic (see Fig. 4-8). This is because ADH increases the water permeability of the cells of the late distal convoluted tubule and collecting duct.
   (3) The osmolality of the glomerular filtrate, of the cortical interstitial fluid, and of the fluid leaving the distal convoluted tubule (in the presence of ADH) is isosmotic (300 mOsm/kg).

5. The Collecting Duct. The cortical and upper medullary collecting ducts are relatively impermeable to water, urea, and NaCl.
   a. The impermeability to water occurs in the absence of ADH.
   b. The relative impermeability to NaCl permits the high interstitial concentration of NaCl to act as an effective osmotic gradient between the tubular fluid and the interstitium. (Some Na$^+$ and Cl$^-$ are reabsorbed from the distal convoluted tubule and collecting duct in the presence or absence of ADH.)
   c. When the kidney forms a concentrated urine, the collecting duct receives an isosmotic fluid from the distal convoluted tubule, the collecting duct is made permeable to water, and the urine equilibrates with the hyperosmotic medullary interstitium, resulting in the excretion of a low volume, hypertonic urine.
   d. If the collecting duct is impermeable to water (ADH absent), the dilute tubular fluid entering the collecting duct from the distal convoluted tubule remains hypotonic and is excreted as a higher volume, hypotonic urine.

C. COUNTERCURRENT EXCHANGERS (Fig. 4-10; see Fig. 4-9). The vasa recta function as countercurrent diffusion exchangers, making the concentrating mechanisms far more efficient.

 1. **Functional Considerations of the Vasa Recta** (see Fig. 4-10).
    a. The vasa recta are derived from the efferent arterioles of the juxtamedullary glomeruli and function to maintain the hyperosmolality of the medullary interstitium, The vasa recta are in juxtaposition with the loops of Henle.
    b. These vessels are permeable to solute and water and reach osmotic equilibrium with the medullary interstitium.
    c. The vasa recta return the NaCl and water reabsorbed in the loops of Henle and collecting ducts to the systemic circulation.
    d. The concentrations of Na$^+$ and urea in the medullary interstitium are kept high by the slow blood flow in the vasa recta.

 2. **Countercurrent Exchange Effects of the Vasa Recta** (see Fig. 4-10).
    a. Na$^+$, Cl$^-$, and urea are passively reabsorbed from the ALH, diffuse across the interstitial fluid into the descending limb of the vasa recta, and are returned to the interstitium by the ascending limb of the vasa recta.
       (1) These solutes recirculate in the vasa recta capillary loops and increase the medullary osmolality.
       (2) Na$^+$, Cl$^-$, and urea also recirculate in the loop of Henle.
    b. Water diffuses from the descending limb of the vasa recta across the interstitial fluid and in-to the ascending limb of the vasa recta.* As a result the cortex receives blood that is only slightly hypertonic to plasma (see Fig. 4-10).

D. ROLE OF UREA. The main function of urea in the countercurrent system is to exert an osmotic effect on the DLH, promoting the abstraction of water and raising the intraluminal concentration of NaCl.

 1. In the medulla, ADH enhances both water and urea permeabilities of the collecting duct.
    a. Urea diffuses into the thin segment of the ALH via secretion but is more concentrated in the fluid entering the distal convoluted tubule (concentrated in a smaller volume) than in the filtrate entering the proximal convoluted tubule.
    b. In the presence of ADH, additional water is reabsorbed from the collecting duct, which further increases the urea concentration in the tubular fluid (urine).
    c. The **thin segment** of the ALH is less permeable to urea than to NaCl, and the **thick segment** of the ALH is impermeable to urea.

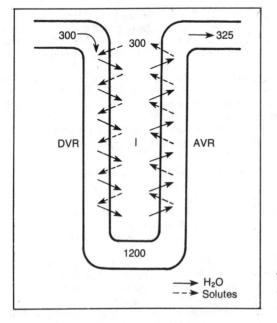

Figure 4-10. Vasa recta as countercurrent exchangers. *AVR* = ascending limb of the vasa recta; *DVR* = descending limb of the vasa recta; *I* = interstitium. (Adapted from Pitts RF: *Physiology of the Kidney and Body Fluids,* 3rd ed. Chicago, Year Book, 1974, p 131.)

---

*Water is continually removed by the ascending limb of the vasa recta.

2. Urea recirculates in the loops of Henle and the vasa recta. The vasa recta are permeable to urea so that urea diffusing out of the papillary collecting duct is trapped in the medullary interstitium.

  a. Of the 1200 mOsm/kg of solute present in the renal papillary loop during **antidiuresis**, half is composed of NaCl and half of urea.

  b. During water **diuresis**, about 10 percent of the total medullary solute concentration is urea.

  c. The maximum urine osmolality (U_{osm}) cannot exceed that in the interstitium, and the ability to conserve water by excreting a concentrated urine is reduced when the papillary concentration is reduced.

E. MEASUREMENT OF RENAL WATER EXCRETION AND CONSERVATION (Fig. 4-11).

  1. Renal Water Excretion. The quantitative measure of the kidney's ability to excrete water is termed **free-water clearance**.

  a. General Considerations.

   (1) Free-water clearance is not a true clearance, because no osmotically free water exists in plasma.

   (2) Free-water clearance denotes the volume of **pure** (i.e., solute-free) **water** that must be removed from, or added to, the flow of urine (in ml/min) to make it isosmotic with plasma.

  b. Measurement.

   (1) Free-water clearance (C_{H_2O}) is calculated using the following equation:

$$\dot{V} = C_{osm} + C_{H_2O}$$
$$C_{H_2O} = \dot{V} - C_{osm},$$

where:

$$C_{osm} = \frac{U_{osm}\ \dot{V}}{P_{osm}}.$$

Substituting for C_{osm}, the equation becomes:

$$C_{H_2O} = \dot{V} - \frac{U_{osm}\ \dot{V}}{P_{osm}},$$

and by factoring out $\dot{V}$, the final equation is:

$$C_{H_2O} = \dot{V}\left(1 - \frac{U_{osm}}{P_{osm}}\right),$$

(8)

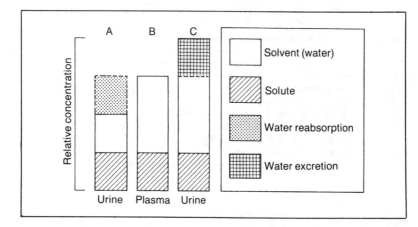

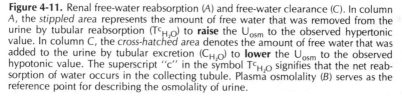

Figure 4-11. Renal free-water reabsorption (A) and free-water clearance (C). In column A, the *stippled area* represents the amount of free water that was removed from the urine by tubular reabsorption ($T^c_{H_2O}$) to **raise** the U_{osm} to the observed hypertonic value. In column C, the *cross-hatched area* denotes the amount of free water that was added to the urine by tubular excretion (C_{H_2O}) to **lower** the U_{osm} to the observed hypotonic value. The superscript "c" in the symbol $T^c_{H_2O}$ signifies that the net reabsorption of water occurs in the collecting tubule. Plasma osmolality (B) serves as the reference point for describing the osmolality of urine.

where $\dot{V}$ = urine volume per unit time (in ml/min); U_{osm} = urine osmolality; P_{osm} = plasma osmolality; and C_{osm} = osmolal clearance, which is the volume of plasma (in ml) completely cleared of osmotically active solutes that appear in the urine each minute.

(2) From equation (8), during **water diuresis** (i.e., in a hydrated state and in the absence of ADH), the U_{osm} is less than the P_{osm} (hyposmotic urine), and the U_{osm}/P_{osm} ratio is less than 1. Therefore, C_{H_2O} is **positive**, indicating that water is being eliminated by the excretion of a large volume of dilute urine. The maximum C_{H_2O} in humans is 15–20 L/day (10–15 ml/min).

(3) From equation (8), during **antidiuresis** (i.e., in a dehydrated state and in the presence of ADH), the U_{osm} is greater than the P_{osm} (hyperosmotic urine), and the U_{osm}/P_{osm} ratio is greater than 1. Therefore, C_{H_2O} is **negative**, indicating that water is being conserved by the excretion of a small volume of concentrated urine. [To avoid the use of the term "negative free-water clearance," the symbol, $T^c_{H_2O}$, is used to denote **free-water reabsorption** (see below).]

2. **Renal Water Conservation.** A quantitative measure of the ability of the kidney to reabsorb water is termed **free-water reabsorption**.

    a. **General Considerations.** Free-water reabsorption denotes the volume of free water reabsorbed per unit time, or, the amount of free water that must be removed from the urine by tubular reabsorption to make it hyperosmotic with plasma.

    b. **Measurement.**

        (1) Free-water reabsorption ($T^c_{H_2O}$) is calculated using the following equation:

$$T^c_{H_2O} = C_{osm} - \dot{V},$$

where:

$$C_{osm} = \frac{U_{osm}\,\dot{V}}{P_{osm}}.$$

Substituting for C_{osm}, the equation becomes:

$$T^c_{H_2O} = \frac{U_{osm}}{P_{osm}} - \dot{V},$$

and by factoring out $\dot{V}$, the final equation is:

$$T^c_{H_2O} = \left(\frac{U_{osm}}{P_{osm}} - 1\right)\dot{V}. \tag{9}$$

        (2) From equation (9), during **antidiuresis** (i.e., in a dehydrated state and in the presence of ADH), the U_{osm} exceeds P_{osm}, and the U_{osm}/P_{osm} ratio is greater than 1. Therefore, $T^c_{H_2O}$ is **positive**, indicating that water is being conserved by the elimination of a small volume of concentrated urine.

3. **Antidiuresis and diuresis** (Table 4-6) can be examined in terms of the volume of urine excreted per unit time ($\dot{V}$). Recall that $\dot{V}$ is the algebraic sum of osmolar clearance (C_{osm}) and either free-water clearance (C_{H_2O}) **or** free-water reabsorption ($T^c_{H_2O}$).

    a. **Antidiuresis** (i.e., a decrease in $\dot{V}$) can result from:

        (1) Water deprivation, which leads to an increase in plasma osmolality

        (2) Reduced circulating blood volume

    b. **Diuresis** (i.e., an increase in $\dot{V}$) can result from:

        (1) Reduced osmotic reabsorption of water, leading to increased solute-free water clearance (C_{H_2O}) and **water diuresis**

        (2) Reduced solute reabsorption (primarily Na^+ with associated anions), leading to increased osmolar clearance (C_{osm}) and **osmotic diuresis**

Table 4-6. Renal Water Excretion and Reabsorption

Renal State	$\dot{V}$ (ml/min)	C_{osm} (ml/min)	C_{H_2O} (ml/min)	$T^c_{H_2O}$ (ml/min)	U_{osm} (mOsm/kg)
Antidiuresis	↓	↑	. . .	↑	↑
Water diuresis	↑	↓	↑	. . .	↓
Osmotic diuresis	↑	↑	↑	. . .	↓

Note.—$\dot{V}$ = urine excretion; C_{osm} = osmolar clearance; C_{H_2O} = free-water clearance; $T^c_{H_2O}$ = free-water reabsorption; and U_{osm} = urine osmolality.

VIII. ADH

A. SYNTHESIS

1. ADH (vasopressin) is a hypothalamic hormone synthesized mainly by the cell bodies of the neurosecretory neurons of the supraopticohypophysial tract that terminate in the pars nervosa of the posterior lobe of the pituitary gland.
   a. Neurosecretory neurons are peptidergic neurons that conduct action potentials like all neurons and, unlike ordinary neurons, synthesize and secrete peptide hormones. The hypothalamic nuclei that represent the main source of ADH are the supraoptic nuclei, which are called **magnocellular neurosecretory neurons**.
   b. ADH is stored (but not synthesized) in the pars nervosa. In the absence of a pars nervosa, the newly synthesized hormone still can be released into the circulation from the ventral diencephalon (hypothalamus).
   c. The biologically active form of ADH in humans and most other mammals is **arginine vasopressin**.*

2. ADH is classified as a polypeptide. Specifically, it is an octapeptide or a nonapeptide with a molecular weight of approximately 1000.
   a. If the two cysteine residues are considered as a single cystine residue, ADH is an octapeptide.
   b. ADH is a nonapeptide if the two cysteine residues are numbered individually.

3. The neurophysins are the physiologic **carrier proteins** for the intraneuronal transport of ADH and are released into the circulation with the neurosecretory products (ADH and oxytocin) without being bound to the hormone.

4. The biologic half-life of ADH is 16–20 minutes.

B. CONTROL OF ADH SECRETION: STIMULI AND INHIBITORS

1. **Osmotic Stimuli.**
   a. Under usual conditions, a 1–2 percent increase in plasma osmolality (i.e., hyperosmolality) is the prime determinant of ADH secretion, and the most common physiologic factor altering the osmolality of the blood is water depletion or water excess. (In clinical medicine, volume deficits are much more prevalent than volume excesses.)
      (1) The stimulation of the osmoreceptors located in the anterior hypothalamus cause the reflex secretion of ADH.
      (2) However, not all solutes (e.g., urea) stimulate the osmoreceptors, despite increasing plasma osmolality. Only those solutes to which cells are relatively impermeable make up the effective osmotic pressure, in response to which ADH is secreted.
         (a) Na^+ and mannitol, which cross the blood-brain barrier relatively slowly, are potent stimulators of ADH release. Since the plasma $[Na^+]$ accounts for 95 percent of the effective osmotic pressure, the osmoreceptors normally function as plasma Na^+ receptors.
         (b) Hyperglycemia is a less potent stimulus for ADH production and secretion than hypernatremia for the same level of osmolality.
   b. The control of ADH secretion by osmolality can be overridden by volume disturbances. For example, marked hyponatremia will be tolerated in order to maintain circulating blood volume.

2. **Osmotic Inhibitors.** Expansion of the intracellular volume of the osmoreceptors secondary to hyposmolality of the ECF (water ingestion) inhibits ADH secretion.

3. **Nonosmotic Stimuli.**
   a. Decreased **effective** circulating blood volume (i.e., **hypovolemia**) is a more potent stimulus to ADH release than is hyperosmolality. A 10–25 percent decrease in blood volume will evoke ADH release. A 10 percent decrease in blood volume is sufficient to cause the release of enough ADH to participate in the immediate regulation of blood pressure. Contraction of blood volume without an alteration in the tonicity of body fluids may cause ADH release.
      (1) A lower osmotic threshold is required to cause ADH secretion in the volume-depleted animal.

---

*The term vasopressin denotes an excitatory action on the blood vessels, causing vasoconstriction of the arterioles and an increase in systemic blood pressure. This effect is observed only when relatively large quantities of vasopressin are released from the posterior lobe (e.g., during hemorrhage) or when pharmacologic amounts are injected. Therefore, the vasopressor effect usually is not considered to be a physiologic effect.

(2) A decrease in the stretch (tension) of the volume receptors (low-pressure baroreceptors) located in the left atrium, vena cavae, great pulmonary veins, carotid sinus, and aortic arch causes an increase in ADH secretion.

(3) Any reduction in intrathoracic blood volume (e.g., due to blood loss, quiet standing, upright body position, and positive-pressure breathing) leads to ADH secretion. Upon standing, the left atrial pressure falls markedly, leading to an increase in ADH release and antidiuresis.

b. Other nonosmotic stimuli for ADH release include pain and such drugs as nicotine, morphine, barbiturates, acetylcholine, chlorpropamide, and β-adrenergic agonists.

c. Excess ADH secretion commonly is associated with pneumonia, tuberculosis, stroke, meningitis, subdural hematoma, and a variety of pulmonary and central nervous system disorders.

d. The postsurgical geriatric hyponatremic patient usually exhibits oversecretion of ADH, which causes water retention, and decreased aldosterone secretion with depression, confusion, lethargy, and weakness.

e. Angiotensin II also stimulates ADH release.

4. Nonosmotic Inhibitors.

a. ADH release is inhibited by increased arterial pressure secondary to vascular or ECF volume expansion.

(1) Thus, ADH release is inhibited by increased tension in the left atrial wall, great veins, or great pulmonary veins secondary to increased intrathoracic blood volume due to hypervolemia, a reclining position, negative-pressure breathing, and water immersion up to the neck.

(2) In the recumbent position, the increase in central blood volume leads to an increase in left atrial pressure and inhibition of ADH release. During sleep, the production of a concentrated urine is by and large due to a reduction of blood pressure, which offsets the effect of reduced ADH secretion due to a change in body position.

b. ADH release also is inhibited by drugs including anticholinergic agents (e.g., atropine), ethanol, phenytoin, lithium, and caffeine.

c. CO_2 inhalation inhibits ADH release.

C. ROLE IN REGULATION OF RENAL WATER EXCRETION

1. The sites of action of ADH are the distal tubules and collecting ducts, where ADH increases the permeability of these structures to water. ADH, through its effect on increased water reabsorption, leads to the production of a urine that has a decreased volume and an increased osmolality. ADH also decreases medullary blood flow.

2. In addition, ADH stimulates the release of adrenocorticotropic hormone (ACTH) from the anterior lobe of the pituitary gland. ACTH plays a relatively small role in controlling aldosterone secretion (see Section IX D 1).

D. ADH-RELATED DISTURBANCES

1. Syndrome of Inappropriate ADH Secretion (SIADH).

a. Pathology. The basic disturbance in this syndrome is an excessive or inappropriate secretion of ADH from the posterior lobe or from an ectopic (nonhypothalamic) source, such as a malignant tumor (e.g., bronchogenic carcinoma). This syndrome is not caused by excessive water intake.

b. Clinical Characteristics. Excessive secretion of ADH has the following effects when water is ingested.

(1) Water retention occurs with an expansion of the blood and ECF volume.

(2) Aldosterone secretion is suppressed, which, in turn, causes increased urinary excretion of Na^+ (hypernatriuria and hyponatremia).

(3) Serum osmolality is decreased (hyposmolality) due to increased water retention and urinary loss of Na^+. (It is the increased ADH secretion despite the presence of hyposmolality that is inappropriate.)

(4) Urine osmolality is increased due to decreased urinary excretion of water and continued excretion of Na^+. The U_{osm} becomes higher than the P_{osm}, and the urinary $[Na^+]$ exceeds 20 mEq/L. (If water intake is restricted, water retention does not occur and the ADH excess has no effect on the plasma $[Na^+]$. Edema does not occur because the increase in volume causes a suppression of aldosterone secretion and an increased Na^+ excretion.)

(5) The retained water in SIADH first enters the plasma, causing its osmolality to decline. This causes a shift of water into the interstitial space (edema), and as the osmolality of that space decreases, there is a further shift of water into the ICF. SIADH leads to water intoxication (overhydration or a dilution syndrome).

    c. Treatment for SIADH is with demeclocycline (Declomycin), which blocks the effect of ADH on the kidney. This tetracycline as well as lithium carbonate, amphotericin B, and methoxyflurane anesthesia are among the etiologic agents in nephrogenic diabetes insipidus.

  2. Diabetes Insipidus (DI).
    a. Pathology. DI is characterized by a complete or partial failure either of ADH secretion (central DI or **neurogenic DI**) or of a renal response to ADH **(nephrogenic DI).**
    b. Clinical Characteristics. Regardless of the cause, DI is characterized by a decrease in renal water reabsorption by the **collecting ducts**. This results in a diuresis of dilute urine up to a volume of 3–20 L/day, a condition referred to as **polyuria**. Because of the stimulation of thirst and increased water intake (polydipsia), however, most DI patients maintain water balance with a near normal plasma $[Na^+]$. In nephrogenic DI, the urine output is directly related to the volume of water delivered to the collecting ducts.
    c. Diagnosis of DI is confirmed in a polyuric patient by the demonstration of insignificant antidiuresis or by the production of hypertonic urine following water restriction, hypertonic saline infusions, and nicotine administration (defined as rapidly smoking three or four cigarettes).
      (1) A response to injected vasopressin documents the DI as neurogenic.
      (2) Lack of a response to vasopressin is indicative of nephrogenic DI.
    d. Treatment.
      (1) Therapy for neurogenic DI is accomplished by the administration of ADH as pitressin tannate or nasal lysine vasopressin. Nonhormonal therapy for neurogenic DI includes oral hypoglycemic agents (e.g., chlorpropamide), thiazide diuretics with sodium restriction, carbamazepine, and clofibrate. A major side effect of chlorpropamide therapy is hypoglycemia, a common problem with hypopituitarism.
      (2) Thiazide diuretics also are quite effective in treating nephrogenic DI, which does not respond to treatment with pitressin or chlorpropamide.
        (a) Thiazides are effective in treating DI because they inhibit Na^+ reabsorption in the diluting segment (ALH).
        (b) Therefore, U_{osm} does not fall below 300 mOsm/kg, and urine volume can be reduced by 50 percent. Urine osmolality can be increased sixfold from a minimal of 50 mOsm/kg.

IX. ALDOSTERONE has the primary function of promoting Na^+ retention. In this way, aldosterone plays a major role in regulating water and electrolyte balance. Together with ADH, aldosterone influences urine composition and volume.

  A. SYNTHESIS, SECRETION, AND INACTIVATION

    1. Synthesis.
      a. Aldosterone is a C-21 (21 carbon atoms) corticosteroid that is synthesized in the outermost area of the adrenal cortex, the **zona glomerulosa**.
      b. Aldosterone represents less than 0.5 percent of the corticosteroids, and like all of the corticosteroids, it is stored in very small quantities. However, aldosterone is the major **mineralocorticoid** in humans.
      c. The circulatory half-life of aldosterone is about 30 minutes in humans during normal activity.
      d. Cholesterol (esterified) is the precursor for steroidogenesis and is stored in the cytoplasmic lipid droplets in the adrenocortical cells. (Free plasma cholesterol appears to be the preferred source of cholesterol for corticosteroid synthesis.)

    2. Secretion. Most of the secreted aldosterone is bound to **albumin**,* with a lesser amount bound to **corticosteroid-binding globulin** (CBG or transcortin). (CBG preferentially binds **cortisol**, which is a **glucocorticoid**.)

    3. Inactivation. Aldosterone is metabolized mainly in the liver, where greater than 90 percent of this corticosteroid is inactivated during a single passage.
      a. Most aldosterone inactivation is by saturation (reduction) of the double bond in the A-ring.
      b. The major metabolite is **tetrahydroaldosterone**, most of which is conjugated with glucuronic acid at the carbon-3 position of the A-ring. The resultant **glucuronides** are more polar and, therefore, more water-soluble, making them readily excreted by the kidney.

  B. PHYSIOLOGIC EFFECTS

    1. Conservation of Na^+.
      a. Aldosterone stimulates Na^+ reabsorption in the distal and collecting tubules.

---

*Protein-bound hormones are biologically inactive.

(1) By reducing Na^+ excretion (and Na^+ clearance), aldosterone causes the formation of a hyperosmotic urine.

(2) Of the amount of filtered Na^+ (25,000 mEq/day), only 1–2 percent (250–500 mEq or 6–12 mg/day) are actively reabsorbed via an aldosterone-dependent mechanism in the distal nephron.*

b. Aldosterone also promotes Na^+ reabsorption in the epithelial cells of the sweat glands, salivary glands, and the gastrointestinal mucosa.

c. By restricting the renal excretion of Na^+, which is the main determinant of plasma osmolality, aldosterone regulates the ECF volume. Therefore, aldosterone regulates the total body Na^+ **content**, while ADH regulates the plasma Na^+ **concentration**.

2. Secretion and Excretion of K^+.

a. Aldosterone promotes K^+ secretion as a secondary effect of its action on Na^+ reabsorption.

(1) In the distal tubule, the linked Na^+ reabsorption–K^+ secretion is referred to as the distal **Na^+-K^+ exchange process**.

(2) Although K^+ appears to be secreted in exchange for Na^+, the distal secretion of K^+ is only indirectly related to Na^+ reabsorption. Distal Na^+ reabsorption is linked to the secretion of both K^+ and H^+.†

(3) Aldosterone increases the $[K^+]$ in sweat glands and saliva.

b. Stimulation of K^+ excretion greatly depends on dietary Na^+, as indicated by a lack of K^+ excretion after aldosterone administration in animals with Na^+-deficient diets.

c. More than 75 percent of the K^+ excreted in the urine is attributed to distal K^+ secretion.

(1) Excess aldosterone secretion causes a decline in the **urinary** Na^+/K^+ concentration ratio (from a normal value of about 2) because it decreases Na^+ excretion and increases K^+ excretion.

(2) Elevated aldosterone release increases the **plasma** Na^+/K^+ concentration ratio (from a normal value of 30) due to the increased excretion of K^+. Aldosterone can cause an **isotonic** expansion of the ECF volume with no change in the plasma $[Na^+]$ or a **hypertonic** expansion of the ECF volume with a rise in the plasma $[Na^+]$. In any case, the total amount of body Na^+ is increased.

3. Water Excretion and ECF Volume Regulation.

a. The effect of aldosterone on water excretion is not important physiologically, as aldosterone treatment fails to correct the impaired water excretion by patients with hypofunctional adrenal glands (due to Addison's disease or adrenal insufficiency) and by patients who have undergone adrenalectomy.

b. Aldosterone has no direct effect on GFR, RPF, and renin production; however, by stimulating Na^+ reabsorption aldosterone causes water retention, and the resultant expansion of the ECF volume then leads to an increase in GFR and RPF and a decrease in renin production.

c. A high circulating aldosterone level is a common finding in edema and is due primarily to the increased aldosterone secretion induced by the depletion of the effective circulating blood volume.

(1) It is unlikely that the plasma $[Na^+]$ is a major regulator of aldosterone secretion because the plasma $[Na^+]$ during Na^+ depletion is normal in humans.

(a) Hyponatremia often is a consequence of ADH secretion and is accompanied by an increase in the ECF volume, which tends to suppress rather than stimulate aldosterone secretion. When ADH and water are given to Na^+-depleted individuals, aldosterone secretion falls despite a fall in plasma $[Na^+]$.

(b) By and large, hyponatremia in the clinical setting is due to **excess body Na^+** caused by a decreased effective blood volume. The low blood volume leads to an increase in ADH secretion and aldosterone secretion which, in turn, lead to edema. **Over 90 percent of the cases of hyponatremia should be treated by restriction of NaCl and water.**

(2) Therefore, it is the volume of the ECF rather than the plasma $[Na^+]$ that influences aldosterone secretion in most circumstances.

C. ALDOSTERONE AND ACID-BASE BALANCE. Aldosterone affects acid-base balance through its control of K^+ secretion. As stated above, aldosterone promotes increased distal tubular **secretion** and, therefore, **excretion** of K^+. Aldosterone also promotes the excretion of H^+ and NH_4^+.

1. Hyperaldosteronemia (i.e., increased aldosterone secretion) is one cause of K^+ depletion, a condition termed **hypokalemia**.

---

*Filtered load refers to the amount of Na^+ in the glomerular filtrate and **not** in the distal nephron.

†Na^+ transport in the **proximal** tubule is not associated with K^+ or H^+ exchange processes.

a. Hypokalemia is characterized by an **increase** in intracellular $[H^+]$, which favors distal tubular secretion of H^+ over K^+ and results in a state of **metabolic alkalosis**.

b. Conversely, if metabolic alkalosis is the primary event, there is a **decrease** in intracellular $[H^+]$. This results in an increase in distal tubular intracellular $[K^+]$, which favors increased urinary loss of K^+ over H^+ and results in a state of **hypokalemic metabolic alkalosis** (as occurs in hyperaldosteronemia).

2. **Hypoaldosteronemia** (i.e., decreased aldosterone secretion) is one cause of excess K^+, a condition termed **hyperkalemia**.

a. Hyperkalemia is characterized by an increase in intracellular $[K^+]$, which favors distal tubular secretion of K^+ over H^+ and results in a state of **metabolic acidosis**.

b. On the other hand, when metabolic acidosis is the primary event there is an increase in intracellular $[H^+]$. This results in an increase in distal tubular intracellular $[H^+]$, which favors increased secretion of H^+ over K^+ and results in a state of **hyperkalemic metabolic acidosis** (as occurs in hypoaldosteronemia).

3. **Hypokalemia** causes the following conditions.

a. **Renal concentrating ability** is impaired by hypokalemia, leading to the formation of hyposmotic urine and polyuria that is resistant to ADH.

b. **Reduced carbohydrate tolerance** also results from hypokalemia and leads to a decline in insulin secretion.* As a rule, fasting hyperglycemia is not present with reduced insulin secretion.

D. **CONTROL OF ALDOSTERONE SECRETION.** At least three well-defined mechanisms control aldosterone secretion: adrenocorticotropic hormone (ACTH or corticotropin), plasma $[K^+]$, and the renin-angiotensin system.

1. **Extrarenal Control Mechanisms.** The following mechanisms cause the release of aldosterone by direct action on the adrenal cortex.

a. **Hypothalamic-Hypophysial-Adrenocortical Axis.** Under normal conditions, ACTH is not a major factor in the control of aldosterone synthesis or secretion. However, the pituitary gland plays an important role in the maintenance of the growth and biosynthetic capacity of the zona glomerulosa.

(1) **ACTH** is a 39 amino acid polypeptide that supports steroidogenesis in the zona glomerulosa. ACTH enhances aldosterone production by stimulating the early biosynthetic pathway (i.e., the 20,22-desmolase enzyme complex that catalyzes the conversion of cholesterol to pregnenolone). ACTH also plays a minor role in mediating the diurnal rhythmic secretion of all of the corticosteroids.

(2) Corticotropin releasing hormone (CRH) is a hypothalamic **hypophysiotropic** polypeptide made up of 41 amino acid residues. CRH is secreted into the hypophysial portal system and causes release of ACTH from the pituitary gland. Since CRH secretion is regulated by higher brain centers (e.g., the limbic system), these centers play a role in Na^+ balance.

(a) Hypophysectomized patients and those with pituitary insufficiency exhibit normal aldosterone secretion on a moderate salt intake; however, these individuals demonstrate a suboptimal aldosterone response to Na^+ restriction.

(b) Normal individuals injected chronically with ACTH show an acute rise in aldosterone secretion followed by a return to control level or below in 3–4 days, despite ACTH administration over a period of 7–8 days.

(c) When **aldosterone** is administered for several days to normal individuals, the kidney "escapes" from the Na^+-retaining effect but not from the K^+-excreting effect. The escape phenomenon prevents the appearance of edema in individuals treated with aldosterone for prolonged periods and in patients with primary aldosteronism.

b. **Hyperkalemia.** A 10 percent **increase** in plasma $[K^+]$ can stimulate the synthesis and release of aldosterone by a **direct** action on the zona glomerulosa. In the anephric human, K^+ appears to be the major regulator of aldosterone even though aldosterone levels are low.

(1) This release probably occurs by the depolarization of the glomerulosa cell membrane by the elevated plasma $[K^+]$.

(2) Stimulation of aldosterone secretion by K^+ loading will be limited by the simultaneous reduction in renin release.

(3) K^+ stimulates an early step in the biosynthetic pathway for aldosterone synthesis.

---

*The reduction in carbohydrate tolerance due to hypokalemia occurs in only about half of patients with elevated plasma aldosterone.

(4) K$^+$ loading increases the width of the zona glomerulosa layer in experimental animals; prolonged aldosterone administration results in atrophy of this layer due to depressed renin secretion.

c. Hyponatremia. A 10 percent **decrease** in plasma [Na$^+$] also appears to stimulate the synthesis and release of aldosterone directly at the level of the zona glomerulosa. However, this effect usually is overridden by changes in the effective circulating volume. Thus, aldosterone secretion is **increased** in the hyponatremic patient who is volume-depleted but is **reduced** in the hyponatremic patient who is volume-repleted.

2. Intrarenal Control Mechanism. Aldosterone secretion also is regulated by the renin-angiotensin system, the major component of which is the **juxtaglomerular apparatus (JGA)**. The renin-angiotensin-aldosterone system is regulated by the **sympathetic nervous system**.

  a. Anatomy of the JGA (Fig. 4-12). The JGA is a combination of specialized tubular and vascular cells located at the vascular pole where the afferent and efferent arterioles enter and leave the glomerulus. The JGA is composed of three cell types.

  (1) Juxtaglomerular (JG) cells are specialized **myoepithelial** (modified vascular smooth muscle) cells located in the **media** of the afferent arteriole, which synthesize, store, and release a proteolytic enzyme called **renin**. Renin is stored in the granules of the JG cells.

   (a) The JG cells are **baroreceptors** (tension receptors) and respond to changes in the transmural pressure gradient between the afferent arteriole and the interstitium. They are innervated by sympathetic nerve fibers.

   (b) These vascular "volume" receptors monitor renal perfusion pressure.

  (2) Macula densa cells are specialized renal tubular epithelial cells located at the transition between the thick segment of the ALH and the distal convoluted tubule.

   (a) These cells are in direct contact with the mesangial cells (see below), in close contact with the JG cells, and contiguous with both the afferent and efferent arterioles as the tubule passes between the arterioles supplying its glomerulus of origin.

   (b) The macula densa cells are characterized by prominent nuclei in those cells on the side of the tubule that is in contact with the mesangial and vascular elements of the JGA.

   (c) The macula densa cells function as **chemoreceptors** and are stimulated by a **decreased** Na$^+$ (NaCl) load. This inverse relationship between Na$^+$ load and renin release provides a reasonable explanation for the clinical problems involving a decreased filtered load of Na$^+$ and Cl$^-$ in association with increased renin release.

  (3) Mesangial cells also are referred to as the **polkissen** (asymmetrical cap) and are the interstitial cells of the JGA. Mesangial cells are in contact with both the JG cells and the macula densa cells. A **decreased** intraluminal Na$^+$ load, Cl$^-$ load, or both in the region of the macula densa stimulates the JG cells.

  b. Role of the Sympathetic Nervous System. The sympathetic nervous system plays an important role in the control of renin release via the **renal nerves**.

  (1) The JG cells of the afferent arterioles are innervated directly by the sympathetic postganglionic fibers (unmyelinated). In the absence of renal nerves, the renal response to Na$^+$ depletion is attenuated.

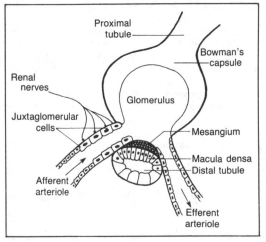

Figure 4-12. The anatomic components of the juxtaglomerular apparatus. (Reprinted with permission from Yates FE, et al: The Adrenal Cortex. In *Medical Physiology*, 14th ed. Edited by Mountcastle VB. St. Louis, CV Mosby, 1980, p 1590.)

(2) Circulating catecholamines (i.e., epinephrine and norepinephrine) and stimulation of the renal nerves produce vasoconstriction of the afferent arterioles, which causes renin release by a decrease in perfusion pressure.

    (a) This renin response caused by catecholamines and renal nerve stimulation is mediated via the β-adrenergic receptor and can be elicited by the synthetic sympathomimetic amine, **isoproterenol**, which is a β-agonist.

    (b) Renal denervation and β-adrenergic receptor blockade by the β-blocker, **propranolol**, inhibits the release of renin.

(3) Renal innervation is not a requisite for renin release because the denervated kidney can adapt to a variable salt intake.

(4) In humans, the assumption of an upright posture and exercise increase renal sympathetic activity, which produces renal arteriolar vasoconstriction and an increase in renin release. Thus, the sympathetic nervous system, by modulating the secretion of renin, has an **indirect** effect on aldosterone secretion.

c. Role of Renin: Stimuli for Release.

(1) Renin has a circulatory half-life of 40–120 minutes in humans. The common denominator for renin release by the intrarenal mechanism is a decrease in the effective circulating blood volume, which is induced by:

    (a) Acute hypovolemia associated with hemorrhage, diuretic administration, or salt depletion

    (b) Acute hypotension associated with ganglionic blockade or a change in posture (postural hypotension)

    (c) Chronic disorders associated with edema (i.e., cirrhosis with ascites, congestive heart failure, and nephrotic syndrome)

(2) Renin release is increased by K^+ depletion, epinephrine, norepinephrine, isoproterenol, and standing.

d. Role of Renin: Inhibition of Renin Secretion.

(1) Renin release is inhibited by angiotensin II, angiotensin III, ADH, hypernatremia, and hyperkalemia.

(2) **K^+ loading** leads to the inhibition of renin release and to the direct stimulation of the glomerulosa cells to secrete aldosterone. In contrast to its effect on normal individuals, aldosterone administration to patients with heart failure, cirrhosis with ascites, and nephrosis causes Na^+ retention **without** K^+ excretion because of greater proximal reabsorption of Na^+ with less Na^+ available for the distal exchange with K^+.

e. Angiotensin Synthesis.

(1) Renin is secreted into the bloodstream, where it combines with the renin substrate, angiotensinogen, which is an α-2 globulin synthesized in the liver.

    (a) Renin is not saturated with its substrate in normal plasma; the same amount of renin generates more angiotensin I if the substrate concentration is increased above normal.

    (b) Oral contraceptives increase plasma angiotensinogen concentration and decrease plasma renin concentration.

(2) The only physiologic effect of renin is to convert angiotensinogen to the biologically inactive decapeptide, angiotensin I.

(3) Angiotensin I is converted primarily in the lung by pulmonary endothelial cells to the physiologically active octapeptide, angiotensin II (Fig. 4-13).

    (a) The enzyme that forms angiotensin II is a peptidase (dipeptidyl carboxypeptidase) called angiotensin-converting enzyme. It is found chiefly in pulmonary tissue and to a lesser degree in renal tissue and blood plasma.

    (b) The converting enzyme is identical to kininase II, which converts the nonapeptide vasodilator, bradykinin (kallidin 9), to inactive peptides, thereby diminishing the circulating levels of a vasodepressor substance and enhancing the vasoconstrictive action of angiotensin II.

(4) Angiotensin II, with a circulatory half-life of 1–3 minutes, has several important physiologic actions.

    (a) It functions as the tropic hormone for the zona glomerulosa and stimulates the secretion (and synthesis) of aldosterone. Angiotensin II is the **aldosterone-stimulating hormone**.

    (b) It is a potent local vasoconstrictor (vasopressor) of the renal arterioles at low plasma concentrations; at higher concentrations, angiotensin II exerts a general vasopressor effect on the smooth muscle cells of arterioles throughout the cardiovascular system, leading to an elevation of systemic mean arterial blood pressure.

    (c) Angiotensin II stimulates the secretion of ADH and ACTH.

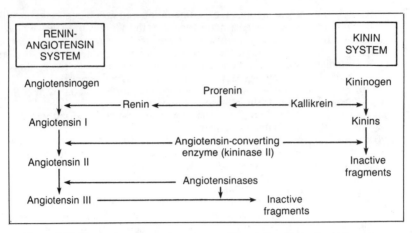

Figure 4-13. The renin-angiotensin system and its relationship to the kinin system.

  (d) Angiotensin II stimulates thirst, which leads to drinking behavior. It also stimulates the release of epinephrine and norepinephrine from the adrenal medulla.
 (5) Angiotensin II is inactivated by two angiotensinases (peptidases), and it can be converted to angiotensin III, which is a heptapeptide and a very potent stimulator of aldosterone secretion. Angiotensin III is not an effective vasoconstrictor in contrast to angiotensin II.
 (6) Renin, converting enzyme, angiotensinogen, and angiotensin II have been found in brain tissue.
 f. Plasma Renin Activity and Plasma Renin Concentration. Plasma renin activity is defined as the rate of angiotensin I formation when plasma renin acts on **endogenous** substrate. **Plasma renin concentration** is measured when **exogenous** substrate is added to plasma to saturate the enzyme and increase the velocity of angiotensin I formation to a maximal rate.
 (1) Oral contraceptive administration is known to increase plasma renin activity and aldosterone secretion via a marked increase in renin substrate concentration, whether the woman is normotensive or hypertensive prior to the therapy.
 (2) As a result of the increased angiotensin II formation, plasma renin concentration is suppressed. Thus, the increase in plasma renin activity in this situation occurs without an increase in renin concentration.
 (3) Oral contraceptives are a cause of hypertension in women through this mechanism.

E. ALDOSTERONISM refers to a condition of excessive aldosterone secretion.

 1. Primary Hyperaldosteronism (Conn's Syndrome).
  a. Etiology. Primary hyperaldosteronism results from a tumor of the zona glomerulosa.
  b. Characteristics of primary hyperaldosteronism include:
   (1) Elevated plasma (and urinary) aldosterone
   (2) Hypertension due to Na^+ and water retention
   (3) Hypokalemic alkalosis with a K^+ excretion rate of greater than 40 mEq/day*
   (4) Decreased levels of angiotensin and renin
   (5) Decreased hematocrit due to the expansion of the plasma volume
   (6) Polyuria and dilute urine due to secondary nephrogenic DI
   (7) Absence of peripheral edema
   (8) Decreased plasma colloidal osmotic (oncotic) pressure

 2. Secondary Hyperaldosteronism.
  a. Etiology. Secondary hyperaldosteronism is caused by the following extra-adrenal factors.
   (1) Diuretic therapy, which is the most common cause
   (2) Extravascular loss of Na^+ and water, which is associated with edema (due to such underlying factors as nephrosis, cirrhosis, and congestive heart failure), an increase in the total ECF volume, and a loss of effective blood volume
   (3) Hyperreninism caused by a tumor of the JG cells
   (4) Renovascular disease (e.g., renal artery stenosis)
  b. Characteristics. Secondary hyperaldosteronism is characterized by edema and Na^+ retention. Urinary K^+ excretion is not increased because there is a reduced flow of fluid into and through the distal segments of the nephron. This low fluid flow reduces K^+ secretion

---

*Hypokalemia is not always concomitant with hypermineralocorticoidism (e.g., in 11β-hydroxylase deficiency).

and offsets the stimulating effect of aldosterone. Also, with decreased Na^+ and water delivery to the distal tubule, the quantity of K^+ (and H^+) secreted in the urine is limited. Additional characteristics of secondary hyperaldosteronism include:

(1) Increased plasma (and urinary) aldosterone
(2) Hypertension with edema (due to Na^+ retention and water accumulation in the interstitial fluid compartment) and a decrease in plasma volume
(3) Hypokalemic alkalosis
(4) Increased angiotensin and plasma renin activity*
(5) Peripheral edema

3. **Licorice ingestion** in excessive amounts can mimic primary aldosteronism, because licorice contains the salt-retaining substance, glycyrrhizinic acid. Patients with this condition present with:
   a. Hypertension and hypokalemic alkalosis
   b. Suppressed plasma renin levels
   c. Reduced aldosterone secretion due to chronic volume expansion

F. ALDOSTERONE ANTAGONISM: SPIRONOLACTONE

1. Spironolactone is a steroidal aldosterone antagonist, which blocks the Na^+-retaining and K^+-excreting effects that aldosterone exerts on the distal renal tubule. As a result, spironolactone leads to an increase in urinary Na^+ excretion and a decrease in K^+ excretion. This antagonist is efficacious only in the presence of aldosterone or another mineralocorticoid; spironolactone is without effect in adrenalectomized individuals.

2. Because the drug enhances Na^+ diuresis, spironolactone is effective in potentiating the action of many antihypertensive drugs whose dosage should be reduced in its presence.

3. Spironolactone is useful in the differential diagnosis of primary and secondary hyperaldosteronism.
   a. If both plasma $[K^+]$ and blood pressure are returned to normal with spironolactone, primary hyperaldosteronism is suspected.
   b. If spironolactone causes the plasma $[K^+]$ to return to normal without the antihypertensive effect, secondary hyperaldosteronism is suspected.

X. MICTURITION in the adult is a voluntary action involving both reflex and voluntary neural activities.

A. INNERVATION OF THE BLADDER. Neural control of bladder function is dependent on both parasympathetic and sympathetic innervation.

1. **Parasympathetic nerves** from the sacral outflow innervate the detrusor muscle and the posterior urethra and provide the neural stimuli to initiate micturition.

2. **Sympathetic fibers** from the lower dorsal and upper lumbar segments supply the blood vessels of the bladder wall and the muscles of the bladder neck. These nerves close the bladder neck to prevent reflux of semen into the bladder at the time of orgasm.

3. In addition, the **internal pudendal nerve** supplies sensory and motor fibers to the external sphincter, urethra, and perineal muscles.

B. THE MICTURITION REFLEX

1. Stress relaxation occurs in the bladder muscle to keep the internal pressure at relatively low and constant pressure until the bladder reaches its normal capacity of 400–500 ml.

2. The desire to micturate is initiated by stretch receptors in the bladder wall, which transmit afferent impulses over the pelvic nerves.

3. Micturition occurs following voluntary relaxation of the perineal muscles, parasympathetic stimulation of the detrusor muscle, contraction of the abdominal wall muscles and the diaphragm, and, finally, relaxation of the bladder neck. The evacuation of the bladder is aided by the rise in the intra-abdominal pressure as well as the increase in bladder pressure, which may reach 150 cm H_2O.

C. ABNORMALITIES OF MICTURITION

1. **Polyuria** is defined as an increased frequency of urination.

---

*The elevated renin level is the characteristic that differentiates secondary aldosteronism from the primary form.

    a. Polyuria of **diabetes** occurs secondary to the osmotic diuresis that characterizes this disease.

    b. **Cystitis**, causing inflammation of the bladder epithelium, results in frequency and **dysuria** (i.e., painful urination).

    c. **External compression** of the bladder secondary to such conditions as pregnancy, pelvic tumors, and cysts can increase urinary frequency by reducing bladder capacity.

2. Incontinence refers to involuntary urination.

    a. **Congenital** incontinence is caused by a variety of developmental defects involving the bladder or the neural mechanisms that control micturition.

    b. **Acquired** incontinence occurs secondary to spinal cord lesions resulting from trauma, tumors, tabes, or primary neural diseases.

    c. **Paradoxic** incontinence is the frequent, involuntary voiding due to overflow of a distended bladder. This condition occurs secondary to mechanical obstruction or neural malfunction.

3. Enuresis (bedwetting) refers to incontinence that occurs at night. Enuresis has a varied etiology, but it is important to exclude organic disease as a cause. Only 5 percent of children exhibit enuresis by the age of 9, and almost all of these are cured by the onset of puberty.

STUDY QUESTIONS

Directions: Each question below contains five suggested answers. Choose the **one best** response to each question.

1. A normal individual on a diet high in K^+ exhibits increased K^+ excretion. The major cause of this increased renal excretion of K^+ is

(A) increased secretion of K^+ by the distal tubule and collecting duct
(B) decreased reabsorption of K^+ by the proximal tubule
(C) decreased reabsorption of K^+ by the loop of Henle
(D) decreased aldosterone secretion
(E) increased GFR

2. The filtered HCO_3^- within the proximal tubular lumen is reabsorbed mainly in the form of

(A) H_2CO_3
(B) H^+
(C) CO_2
(D) HCO_3^-
(E) OH^-

3. Which of the following pressure changes leads to an increased GFR?

(A) Increased arterial plasma colloid osmotic pressure
(B) Increased glomerular capillary hydrostatic pressure
(C) Increased turgor pressure in the renal interstitium
(D) Increased hydrostatic pressure in Bowman's capsule
(E) Decreased net filtration pressure

4. After being perfused with plasma containing inulin in a concentration of 10 mg/ml, an isolated skeletal muscle is found to contain a total of 25 mg of inulin. The total ECF volume of this muscle is

(A) 0.4 ml
(B) 2.5 ml
(C) 4.0 ml
(D) 25 ml
(E) 250 ml

5. Which of the following substances has the lowest clearance in the human kidney?

(A) Creatinine
(B) PAH
(C) K^+
(D) Cl^-
(E) Glucose

6. All of the following factors lead to a decrease in the filtration fraction EXCEPT

(A) increased ureteral pressure
(B) increased efferent arteriolar resistance
(C) increased plasma protein concentration
(D) decreased glomerular capillary pressure
(E) decreased filtration area

7. Resistance to blood flow through the kidney can be determined by

(A) measuring the hydrostatic pressure difference between the renal artery and renal vein
(B) measuring the RBF
(C) dividing the RPF by the hematocrit
(D) dividing the renal arteriovenous pressure difference by the RBF
(E) measuring the renal clearance of PAH

8. A patient with an aldosterone-secreting tumor is likely to exhibit all of the following signs or symptoms EXCEPT

(A) hypertension
(B) alkalosis
(C) edema
(D) hypokalemia
(E) low plasma renin activity

9. Antidiuretic hormone exerts its action on the kidney by increasing

(A) free-water clearance
(B) renal medullary blood flow
(C) free-water reabsorption
(D) proximal tubular reabsorption of Na^+
(E) proximal tubular reabsorption of water

10. The release of antidiuretic hormone is caused by all of the following factors EXCEPT

(A) dehydration
(B) hypovolemia
(C) caffeine
(D) nicotine
(E) positive-pressure breathing

11. A drug that inhibits the active transport of Cl^- exerts its action primarily on the

(A) proximal convoluted tubule
(B) concentrating segment of the nephron
(C) collecting duct
(D) thick segment of the ascending limb of the loop of Henle
(E) vasa recta

12. Identify the renal tubular segment that functions to reabsorb a small fraction of the filtered NaCl and water and to secrete H^+, NH_4^+, and K^+.

(A) Proximal convoluted tubule
(B) Descending limb of the loop of Henle
(C) Ascending limb of the loop of Henle
(D) Distal convoluted tubule
(E) Collecting duct

13. The hydrostatic pressure in the efferent arteriole is best described as being

(A) less than that in the afferent arteriole
(B) higher than the glomerular capillary pressure
(C) less than the peritubular capillary pressure
(D) equal to the glomerular capsular pressure
(E) higher than the oncotic pressure in the glomerular capillary

14. Which of the following findings best supports the diagnosis of primary aldosteronism as opposed to secondary aldosteronism?

(A) Decreased plasma renin activity
(B) Hypokalemia
(C) Hyperaldosteronemia
(D) Increased plasma aldosterone level
(E) Hypertension

Questions 15–17

The renal function of a 70-kg man is tested by measuring the plasma and urine concentrations of three administered solutes, and the data obtained from these measurements are shown in the table below.

| | Concentration | | | |
	PAH (pg/ml)	Inulin (mg/ml)	Creatinine (mg/dl)	Hematocrit (%)
Plasma	0.2	0.1	0.8	40
Urine	56	6.0	80	. . .

15. Based on the data above and assuming a urine flow rate of 1.5 ml/min, the effective RPF for this man is

(A) higher than the true RPF
(B) higher than the effective RBF
(C) 420 ml/min
(D) 520 ml/min
(E) 700 ml/min

16. Using the data shown above and given a urine flow rate of 1.5 ml/min, the GFR for this man is determined to be

(A) greater than inulin clearance
(B) greater than PAH clearance
(C) equal to the volume of blood from which inulin is completely cleared
(D) 60 ml/min
(E) 90 ml/min

17. If this man has a urine flow rate of 1.5 ml/min, his 24-hour endogenous creatinine excretion is

(A) 21.6 mg/kg body weight
(B) 15.12 mg/kg body weight
(C) 17.28 mg/kg body weight
(D) 24.7 mg/kg body weight
(E) the same as his 24-hour inulin excretion

Directions: Each question below contains four suggested answers of which **one or more** is correct. Choose the answer

A   if **1, 2, and 3** are correct
B   if **1 and 3** are correct
C   if **2 and 4** are correct
D   if **4** is correct
E   if **1, 2, 3, and 4** are correct

18. H^+ is excreted by several renal mechanisms including combination with

(1) HCO_3^- via the enzyme carbonic anhydrase
(2) NH_3 to form NH_4^+
(3) Cl^- to form HCl
(4) divalent HPO_4^{2-} to form monovalent $H_2PO_4^-$

19. The clinical findings of a serum $[Na^+]$ of 155 mEq/L with a urine osmolality of 50 mOsm/kg of water can be explained by such conditions as

(1) lack of antidiuretic hormone
(2) volume expansion with isotonic saline
(3) nephrogenic diabetes insipidus
(4) excessive ingestion of water

20. Substances needed to measure the interstitial fluid volume include

(1) tritium oxide
(2) inulin
(3) antipyrine
(4) radioiodinated human serum albumin

21. A patient is treated with a drug that causes the formation of a large volume of urine with an osmolarity of 100 mOsm/L. The actions of this drug include

(1) inhibition of renin secretion
(2) inhibition of antidiuretic hormone secretion
(3) increased water permeability of the distal tubule and collecting duct
(4) decreased active Cl^- reabsorption by the thick segment of the ascending limb of the loop of Henle

22. A decrease in the urine to plasma (U/P) ratio of inulin concentration with a constant GFR indicates an increase in

(1) free-water clearance
(2) inulin clearance
(3) urine flow
(4) plasma inulin concentration

23. True statements regarding creatinine measurements include which of the following?

(1) A 24-hour endogenous creatinine excretion is useful in assuring a complete 24-hour urine collection
(2) Creatinine clearance overestimates the GFR
(3) The creatinine-to-inulin clearance ratio is greater than 1
(4) Creatinine clearance decreases with age

24. A decrease in the effective circulating blood volume leads to an increase in the

(1) fractional reabsorption of Na^+ by the renal tubules
(2) secretion of antidiuretic hormone
(3) secretion of aldosterone
(4) rate of vagal afferent discharge from the atrial and venous volume receptors

Directions: The groups of questions below consist of lettered choices followed by several numbered items. For each numbered item select the **one** lettered choice with which it is **most** closely associated. Each lettered choice may be used once, more than once, or not at all.

Questions 25–28

The table below shows blood data obtained from a normal individual and from five patients designated by the letters A–E. For each disorder that follows, select the patient whose blood data best characterize that condition.

	$[Na^+]$ (mEq/L)	$[K^+]$ (mEq/L)	$[Cl^-]$ (mEq/L)	$[HCO_3^-]$ (mEq/L)	Creatinine (mg/dl)	BUN (mg/dl)
Normal	142	4.0	103	24	0.8–1.6	10–20
(A)	146	2.9	104	32	1.2	10
(B)	155	4.4	115	25	1.6	18
(C)	120	3.0	88	23	0.8	8
(D)	128	5.6	88	19	2.0	28
(E)	125	2.9	80	10	3.2	50

25. Diabetes insipidus

26. Syndrome of inappropriate antidiuretic hormone secretion

27. Hypoaldosteronemia

28. Primary hyperaldosteronemia

Questions 29–32

The diagrams below represent various states of abnormal hydration. In each diagram, the normal state (*solid lines*) is superimposed on the abnormal state (*dashed lines*) to illustrate the shifts in the volumes (width of *rectangles*) and total osmolalities (height of *rectangles*) of the ECF and ICF compartments.

Each of the numbered phrases that follow describes a condition that leads to abnormal hydration. Match each phrase to the lettered diagram that represents the appropriate state of abnormal hydration.

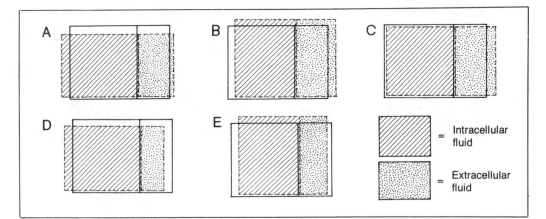

29. Ingestion of a large quantity of water

30. Infusion of isotonic saline

31. Infusion of hypertonic saline

32. Adrenal insufficiency

ANSWERS AND EXPLANATIONS

1. The answer is A. (*VI C 3; IX B 2, C; Fig. 4-4; Table 4-4*) The active reabsorption of K^+ by the proximal tubule and the ascending limb of the loop of Henle represents the reabsorption of 90 percent of the filtered K^+ and occurs at virtually the same rate regardless of changes in body K^+. Since the distal and collecting tubules can both reabsorb and secrete K^+, K^+ excretion is regulated mainly by alterations in the rate of distal K^+ secretion. The major determinants of distal K^+ secretion and urinary K^+ excretion are dietary K^+ intake and aldosterone secretion. An increase in either plasma $[K^+]$ or aldosterone can promote the active transport of K^+ from the peritubular capillary, across the basolateral (abluminal) border, and into the renal tubular cell. The distal secretion of K^+ appears to be passive and varies directly with the cellular $[K^+]$; therefore, K^+ loading favors K^+ secretion and excretion. An increase in plasma $[K^+]$ also stimulates the secretion of aldosterone, which has a kaliuretic effect. K^+ is both reabsorbed from and secreted into the renal tubules. Foods high in K^+ include milk, orange juice, bananas, chocolate, meat, and vegetables.

2. The answer is C. (*VI C 4 b*) The bulk of HCO_3^- reabsorption occurs in the proximal tubule. The reabsorption of HCO_3^- and the formation of titratable acid and NH_4^+ occur by the active secretion of H^+ into the tubular lumen. The key element in the reabsorption of HCO_3^- is the exchange of secreted H^+ for reabsorbed Na^+.

　　The secreted H^+ combines with the filtered HCO_3^- to form H_2CO_3, and the H_2CO_3 dehydrates into CO_2 and water. Finally, the CO_2 and water both are reabsorbed, thus making HCO_3^- reabsorption an indirect process. The CO_2 diffuses into the tubular cell and regenerates HCO_3^- by the action of carbonic anhydrase. It is important to recall that the secreted H^+, which combines with filtered HCO_3^-, is not excreted.

3. The answer is B. (*V A, C*) Of the pressure changes listed, only the increase in glomerular capillary pressure would cause an increase in GFR. All of the other pressure changes listed would lead to a decrease in filtration.

4. The answer is B. (*II B 1*) The volume of a fluid compartment can be measured indirectly by the dilution principle, which expresses the relationship of the concentration of a substance in solution (C), the volume of distribution of that substance (V), and the quantity or amount of substance present (Q), as: V = Q/C. Substituting the given values for Q (25 mg) and C (10 mg/ml), the ECF volume is determined as:

$$V = \frac{25 \text{ mg}}{10 \text{ mg/ml}} = 2.5 \text{ ml.}$$

5. The answer is E. (*VI B 2 a–b, C 1 a*) Under normal conditions, all the filtered glucose is reabsorbed in the proximal tubule and returned to the systemic circulation via the peritubular capillaries. Therefore, at normal plasma glucose concentrations (i.e., below 180 mg/dl or 10 mmol/L), the clearance of glucose is zero.

6. The answer is B. (*V C 3 b; Table 4-5*) When efferent arteriolar resistance increases, the GFR increases and the RPF decreases (i.e., GFR and RPF change in opposite directions). This leads to an increase in the GFR/RPF ratio, which is called the filtration fraction. All of the other factors listed also affect the filtration fraction, but these factors decrease the GFR and, therefore, decrease the filtration fraction.

7. The answer is D. (*V C 1 b*) The simplest equation that relates flow (F) to a pressure gradient (ΔP) and resistance (R) is Ohm's law, which states: $R = \Delta P/F$. Ohm's law can be rewritten for fluid flow as:

$$F \text{ (ml/min)} = \frac{\Delta P \text{ (mm Hg)}}{R \text{ (mm Hg/ml/min)}} .$$

The only choice that includes all three of the variables (renal arterial pressure, renal venous pressure, and renal blood flow) is D. Choices A, B, and C contain only one or two of the three variables required for the calculation of resistance in the equation. The clearance equation does not have a resistance term and, therefore, cannot be used to calculate that variable.

8. The answer is C. [*IX D 1 a (2) (c), F 1 b*] The major locus of action of aldosterone is the distal tubule, where Na^+ reabsorption and K^+ secretion are linked and occur simultaneously. Aldosterone enhances secretion of K^+ and H^+ into the tubular fluid, and these effects are heightened with increased aldosterone secretion, which leads to hypokalemia and alkalosis, respectively. The impairment of the renal concentrating ability leads to polyuria, which is insensitive to antidiuretic hormone. The increased tubular reabsorption of Na^+ leads to the increased retention of water and the isotonic expansion of the ECF compartment, which dilutes the plasma proteins. The aldosterone-stimulated renal reabsorption of Na^+ and water leads to hypertension and low plasma renin activity. When aldosterone is administered for several days or is chronically elevated, the kidney escapes from the Na^+-retaining effect but not from the K^+-excreting effect of the hormone. Under these circumstances, the proximal reabsorption of Na^+ is inhibited, the GFR is increased, and the combination of these events produces an increased excretion of NaCl and water in the urine. The Na^+ escape is believed to be responsible for the absence of edema in patients with primary aldosteronism. Na^+ escape does not occur in edematous patients with secondary hyperaldosteronism (e.g., hyperaldosteronism due to heart failure, cirrhosis with ascites, or nephrosis). These conditions are associated with an expanded total ECF with a reduction in the effective circulating volume. The kidney in these situations continues to retain Na^+.

9. The answer is C. (*VIII C; Figs. 4-8 and 4-11*) Antidiuretic hormone (ADH) has no direct effect on Na^+ reabsorption. It promotes urinary concentration by increasing the water permeability of both the late distal tubule and the collecting duct. Medullary hypertonicity is produced by NaCl reabsorption without water in the ascending limb of the loop of Henle (ADH-independent) and by the medullary accumulation of urea (ADH-dependent). This allows the tubular fluid to reach osmotic equilibrium with the hypertonic interstitium. The reabsorbed water returns to the systemic circulation via the capillaries of the vasa recta.

Water diffuses from the descending limb of the loop of Henle and the vasa recta across the interstitium into the ascending limb of the vasa recta. ADH causes the conservation of solute-free water (i.e., it decreases free-water clearance).

10. The answer is C. (*VIII B 1, 3*) In addition to increased extracellular osmolality (hypertonicity), stimuli for antidiuretic hormone (ADH) release include hypovolemia, pain, nicotine, and cholinergic agents. The normal control of ADH secretion by osmolality can be overridden by volume disturbances. Thus, volume depletion is a more potent stimulus for ADH release than is hyperosmolality.

11. The answer is D. (*VI C 5 c; VII B 2 b; Table 4-4*) The thick segment of the ascending limb of the loop of Henle actively transports Cl^- out of the tubular lumen, which causes the passive transport of Na^+ out of the lumen. Both the thin and thick segments of the ascending limb transport NaCl without water into the interstitium. In contrast, the descending limb, the concentrating limb of the nephron, is relatively impermeable to NaCl but is permeable to water.

Drugs used clinically to block NaCl reabsorption from the thick segment of the ascending limb of the loop of Henle are called loop diuretics and include ethacrynic acid, furosemide, and organomercurials. These diuretics tend to cause production of large volumes of relatively isotonic urine. Drugs used to block NaCl reabsorption act mainly on the loop, leading to an increase in the osmolality of the tubular fluid and an osmotic diuresis.

12. The answer is D. (*VI C 1-6; Table 4-4*) Na^+ and Cl^- reabsorption is discontinuous along the nephron but does occur beyond the proximal convoluted tubule and descending limb of the loop of Henle in the distal and collecting tubules. Water permeability varies along the distal convoluted tubule; water permeability of the early distal tubule is extremely low in contrast to that of the late distal tubule. The major determinant of high water permeability in the late distal tubule and the collecting duct is antidiuretic hormone.

The loops of Henle reabsorb approximately 25 percent of the filtered Na^+ and Cl^- and 15 percent of the filtered water. The ascending limb of the loop of Henle is impermeable to water.

Both the distal tubule and collecting duct have the ability to secrete as well as reabsorb K^+. About 65 percent of the total filtered K^+ is reabsorbed by the proximal convoluted tubule, and an additional 20–30 percent is reabsorbed by the ascending limb of the loop of Henle. The proximal convoluted tubule does not secrete K^+.

NH_3 and H^+ are secreted by the proximal and distal convoluted tubules and the collecting duct. NH_3 secretion represents the main renal adaptive response to an acid load. H^+ cannot be excreted as free H^+ but must be buffered with urinary buffers (to form titratable acid) and with NH_3 (to form NH_4^+). The process of H^+ secretion (and HCO_3^- reabsorption) occurs throughout the nephron with the exception of the descending limb of the loop of Henle. NH_3 production and secretion are confined to the distal convoluted tubule and collecting duct.

13. The answer is A. (*V C*) The afferent arteriole is the only source of blood to the glomerular capillaries, which consist of 20–40 parallel capillary loops. Thus, the kidney represents the only organ in which true capillaries are both supplied and drained by arterioles. Pressure gradient is one of the principal determinants of blood flow. In order for blood to flow from an afferent to an efferent arteriole, the pressure within the efferent arteriole must be less than that within the afferent arteriole. The pressure drop across the glomerular capillary bed (i.e., from afferent to efferent arteriole) is very small (about 1 mm Hg).

In contrast to the plasma osmotic pressure in systemic capillaries, which stays relatively constant, that in glomerular capillaries rises along the length of the capillary due to the transfer of fluid by ultrafiltration. However, the hydrostatic pressure at the end of the glomerular capillary (beginning of the efferent arteriole) does not exceed the oncotic pressure.

14. The answer is A. (*IX E 1 b, 2 b*) The diagnosis of primary aldosteronism begins with the finding of hypertension together with an unprovoked hypokalemia. However, hypokalemia as a diagnostic criterion for primary aldosteronism is not a necessity. The diagnosis is established by the demonstration of elevated aldosterone secretion (or excretion) together with a subnormal plasma renin activity.

Secondary aldosteronism is not a specific entity but rather a component of the pathophysiology of a multitude of clinical disorders. It is the demonstration of an elevated aldosterone secretion rate in combination with a high plasma renin activity that establishes the diagnosis of secondary aldosteronism. Both types of aldosteronism are associated with hypokalemia; therefore, plasma $[K^+]$ does not provide for a differential diagnosis.

15–17. The answers are 15-C, 16-E, 17-D. [*IV B 1, D 2 a; VI B 3 c (1), (2)*] Effective RPF (ERPF) is equal to the clearance of PAH (C_{PAH}). Using the given data, ERPF is calculated as:

$$\text{ERPF} = C_{PAH} = \frac{U_{PAH} \bullet \dot{V}}{P_{PAH}}$$

$$= \frac{56.0 \text{ pg/ml} \bullet 1.5 \text{ ml/min}}{0.2 \text{ pg/ml}}$$

$$= 420 \text{ ml/min.}$$

The C_{PAH} measures the ERPF only when the plasma concentration of PAH is fairly low. If PAH was cleared completely from all of the plasma perfusing the kidneys, then C_{PAH} would measure total RPF. Since 10–15 percent of the total RPF perfuses nonsecreting renal tissue, the ERPF is approximately 85–90 percent of the total RPF.

GFR is equal to the clearance of inulin (C_{in}). Using the given data, GFR is calculated as:

$$\text{GFR} = C_{in} = \frac{U_{in} \bullet \dot{V}}{P_{in}}$$

$$= \frac{6.0 \text{ mg/ml} \bullet 1.5 \text{ ml/min}}{0.1 \text{ mg/ml}}$$

$$= 90 \text{ ml/min.}$$

The C_{in} is a measure of GFR because the volume of plasma cleared of inulin is equal to the volume of plasma filtered. Thus, the amount (mass) of inulin excreted in the urine ($U_{in} \bullet \dot{V}$) is equal to the mass filtered ($C_{in} \bullet P_{in}$) during the same period of time.

Creatinine excretion is calculated as:

$$\text{excretion} = U_{cr} \bullet \dot{V}$$

$$= 80 \text{ mg/dl} \bullet 1.5 \text{ ml/min}$$

$$= 800 \text{ mg/L} \bullet 2.16 \text{ L/day}$$

$$= 1728 \text{ mg/day}$$

$$= \frac{1728 \text{ mg}}{70 \text{ kg}}$$

$$= 24.7 \text{ mg/kg.}$$

This indicates that this patient's 24-hour urine collection was complete because it is within the normal range. Creatinine is formed from muscle creatine, and its plasma concentration does not change significantly during a 24-hour period. Therefore, a single determination of plasma creatinine concentration is necessary.

Since it is an endogenous substance, creatinine frequently is used to estimate GFR even though a small amount is secreted, mainly by the proximal tubules. Creatinine clearance ranges from 80 to 110 ml/min per 1.73 m² body surface area. Measurement of 24-hour creatinine excretion is useful to ensure that the urine collection was complete (i.e., 24-hour).

18. The answer is C (2, 4). (*VI C 4 a–c*) Essentially all of the excreted H^+ originates from tubular secretion. The H^+ secreted by the tubules can take one of two pathways: it can combine with filtered HCO_3^-, a process that accomplishes the proximal reabsorption of 90 percent or more of the filtered

HCO_3^- as a result of Na^+-for-H^+ exchange, or it can combine with either filtered nonbicarbonate (mainly HPO_4^{2-}) to form titratable acid or with NH_3 to form NH_4^+. H^+ is not excreted with HCO_3^-, and it does not combine with Cl^- to form HCl.

19. The answer is B (1, 3). (*VIII D 2 a–c*) Diabetes insipidus is characterized by the complete or partial failure of either antidiuretic hormone (ADH) secretion (neurogenic diabetes insipidus) or the renal response to ADH (nephrogenic diabetes insipidus). In both types of diabetes insipidus, water reabsorption is reduced and urine output is increased. At a plasma [Na^+] of 155 mEq/L, endogenous ADH secretion should be maximal. However, the low urine osmolality suggests continuous water diuresis. The diagnosis of neurogenic diabetes insipidus can be confirmed by the administration of vasopressin, which should lower the urine volume, increase the urine osmolality, and correct the hypernatremia. Patients with nephrogenic diabetes insipidus are unresponsive to vasopressin administration; however, a reduction in urine volume is achieved by the induction of hypovolemia with thiazide diuretics.

20. The answer is C (2, 4). (*II B 1, 2*) The interstitial fluid volume is not measured directly because no known substance is distributed exclusively within that fluid compartment. The interstitial fluid volume is determined as the difference between the ECF volume and the plasma volume. The volume of each of these compartments is measured by the dilution technique using inulin (to measure ECF volume) and using radioactive serum albumin (to measure plasma volume). The difference between the ECF volume and plasma volume is an estimate of the interstitial fluid volume.

21. The answer is C (2, 4). (*VI C 5 c; VII B 2 b*) The blockade of active Cl^- reabsorption by the thick segment of the ascending limb of the loop of Henle would lead to an osmotic diuresis (i.e., the formation of a large volume of hypotonic urine) due to NaCl excretion. When the secretion of antidiuretic hormone (ADH) is inhibited, solute-free water clearance increases. Therefore, the inhibition of ADH secretion also would lead to the generation of a large volume of dilute urine.

22. The answer is B (1, 3). [*IV B 1 a, C 2 b (b)*] The excretion of a concentrated urine or a dilute urine depends on an individual's degree of hydration. The ratio of the osmolal concentration of solutes in the urine to that in the plasma is expressed as the U_{osm}/P_{osm} ratio. When the body has excess water, the urine osmolality falls below the plasma osmolality, thereby lowering the U_{osm}/P_{osm} ratio.

For a substance like inulin, which is filtered but is not secreted or reabsorbed by the tubule, a filtrate-urine mass balance must apply to the movement of filtrate through the tubule. Thus, the rate of inulin influx into the filtrate (GFR $\times P_{in}$) must equal the rate of inulin efflux into the final urine ($U_{in} \times \dot{V}$) or:

$$GFR \times P_{in} = U_{in} \times \dot{V}.$$

If GFR is constant and U_{osm} decreases (i.e., the U_{osm}/P_{osm} ratio decreases), then the volume flow of urine must have increased because the product of $U_{in} \times \dot{V}$ is constant. (The clearance of inulin is constant and does not vary with plasma inulin concentration.)

23. The answer is E (all). (*IV D 1, 2*) Since creatinine is filtered and secreted, the clearance of creatinine is slightly higher than that of inulin. Although creatinine clearance declines with age, the creatinine-to-inulin clearance ratio is unchanged through life (in the absence of disease). The age-related decrease in creatinine clearance is accompanied by a parallel reduction in daily creatinine excretion, which reflects a decrease in muscle mass. The plasma creatinine concentration is remarkably constant with age (in the absence of disease); therefore, the decrease in creatinine clearance with age represents a reduction in GFR and a decline in renal tubular secretion of creatinine. Renal function should be estimated by creatinine clearance and not by creatinine concentration because creatinine clearance is slightly higher than GFR and, therefore, overestimates renal function. In most circumstances, the BUN and plasma creatinine concentrations vary inversely with the GFR, increasing as the GFR falls. Because creatinine clearance is relatively constant, the measurement of 24-hour creatinine excretion is useful in assuring a complete urine collection. The normal 24-hour creatinine clearance for men is 20–25 mg/kg of body weight and for women is 15–20 mg/kg of body weight.

24. The answer is A (1, 2, 3). (*VIII B 3 a*) Hypovolemia (i.e., decreased effective circulating blood volume) is a potent stimulus to antidiuretic hormone (ADH) secretion, which varies inversely with left atrial pressure and aldosterone secretion. ADH secretion is caused by a decrease in afferent impulses from the volume receptors (i.e., the low-pressure baroreceptors located in the left atrium, vena cavae, and great pulmonary veins). In response to volume depletion, GFR decreases and the fractional proximal reabsorption of Na^+ and water increases as the kidney corrects the volume deficit by retaining Na^+ and water through the activation of the renin-angiotensin-aldosterone system. The right atrial stretch-sensitive receptors play a minor role in the release of renin, while the afferent arteriolar stretch-sensitive receptors assume a major role in renin secretion. Effective volume depletion reduces free-water clearance by stimulating ADH release and by reducing the delivery of water to the diluting segment of the nephron (i.e., the ascending limb of the loop of Henle), which limits the formation of free water. By this mechanism, relatively less filtered water than solute is excreted, which is equivalent to adding pure water to the body.

25–28. The answers are: 25-B, 26-C, 27-D, 28-A. (*VIII D 1, 2; IX C 1–3, F 1*) The blood data for **patient (B)** are characteristic of diabetes insipidus, in which a reduced osmotic reabsorption of water leads to increased free-water clearance. This in turn leads to water diuresis (polyuria) and polydipsia. Generally, high plasma urea and electrolyte concentrations demonstrate the effects of dehydration. In this patient, high $[Na^+]$ (hypernatremia) is caused by dehydration and is responsible for the hyperosmolal state. The BUN, which is at the upper range of normality, is not responsible for the hyperosmolal state, because the urea nitrogen also is high inside the cells.

The data for **patient (C)** are consistent with the syndrome of inappropriate antidiuretic hormone (ADH) secretion (SIADH). SIADH is not due to the usual stimuli of hyperosmolality and hypovolemia. This endocrinopathy is characterized by increased water reabsorption (decreased free-water clearance), resulting in the dilution of the body fluids, which is demonstrated in this patient by low $[Na^+]$ (hyponatremia) and hypo-osmolality; edema is not present in SIADH. In contrast to diabetes insipidus, the urine osmolality is high in SIADH due to the elevated urinary $[Na^+]$. **Patients with SIADH have a normal ECF volume** and normal renal and adrenal function. If water intake is restricted, water retention does not occur and the ADH excess has no effect on the plasma $[Na^+]$.

The data for **patient (D)** characterize hypoaldosteronemia. Adult patients with hypoaldosteronemia usually present with high $[K^+]$ (hyperkalemia) and cardiac complications (e.g., heart block). Clinical signs of this condition may not appear in adult patients unless they are challenged by salt deprivation. Salt restriction does not enhance aldosterone secretion in these patients; therefore, the patients exhibit salt wasting and volume depletion (hypovolemia). Chronic heparin administration can cause hypoaldosteronemia by blocking aldosterone synthesis. Note that hypoaldosteronemia is a cause of metabolic acidosis (due to low plasma $[HCO_3^-]$).

The data for **patient (A)** are characteristic of primary hyperaldosteronemia (Conn's syndrome), the dominant feature of which is decreased $[K^+]$ (hypokalemia). Additional clinical symptoms include hypertension and alkalosis together with polydipsia and polyuria. Peripheral edema is rare in primary hyperaldosteronemia due to the Na^+ escape phenomenon.

29–32. The answers are: 29-A, 30-C, 31-B, 32-D. (*II C 2 c, 3 a–c; Fig. 4-2; Table 4-2*) **Diagram A** represents hyposmotic overhydration, a state induced by drinking large amounts of water. The excessive ingestion of water leads to a dilution (i.e., a reduction in osmolality) of the ECF compartment and to an increase in the ECF volume. Water flows from a region of relatively low osmolality to one of higher osmolality and, in this case, moves from the ECF to the ICF compartment. This results in an expansion of the ICF compartment, a reduction in the osmolality of that compartment, and an increase in the total body water (TBW).

Diagram C represents isosmotic overhydration, a state induced by an infusion of isotonic NaCl. This infusion leads to an increase of TBW due to the expansion of the ECF volume. Since there is no change in the osmolality of the ECF compartment, there is no change in the ICF volume.

Diagram B represents hyperosmotic overhydration, a state induced by an infusion of hypertonic NaCl. This infusion increases the volume and the osmolality of the ECF compartment. The increase in plasma osmolality causes water to flow out of the ICF compartment, which eventually decreases the ICF volume, increases the osmolality of the ICF, and increases the ECF volume.

Diagram D represents hyposmotic dehydration. This state of NaCl deficiency results from any disorder that leads to loss of Na^+ salts in excess of water loss from the body. Hyposmotic dehydration is a clinically significant condition that can arise from profuse sweating, gastrointestinal Na^+ losses (e.g., due to vomiting), and renal Na^+ loss due to adrenal insufficiency. With adrenal insufficiency, the volume and osmolality of the ECF compartment are reduced, which causes water to flow into the ICF compartment. The ICF then increases in volume and decreases in osmolality. The protein concentration rises, however, in the ICF compartment as well as in the ECF compartment.

5
Acid-Base Physiology

I. THE HYDROGEN ION AND pH

A. FUNDAMENTAL CHEMISTRY (Table 5-1). The hydrogen ion (H^+) is a proton (i.e., a hydrogen atom without its orbital electron); H^+ in aqueous solution exists as a hydrated proton called the hydronium ion or H_3O^+. pH refers to the negative Briggsian logarithm of the H^+ concentration. The gain and loss of protons constitutes acid-base chemistry. The currently accepted model of acid-base relationships is that proposed by Brønsted.

1. An **acid** is a substance that acts as a proton donor, as illustrated by the dissociation of hydrochloric acid (HCl):

$$HCl \rightleftharpoons H^+ + Cl^-.$$

According to Brønsted, HCl is called an acid because it can generate H^+, and chloride ion (Cl^-) is called the **conjugate base** because at high H^+ concentration it can accept H^+. In aqueous systems, acids are classified as **strong** or **weak**. These terms should not be confused with acid **concentrations**. For example, a 10^{-5} mol/L HCl solution is a dilute solution of a strong acid, and a 5 mol/L acetic acid (CH_3COOH) solution is a concentrated solution of a weak acid.
   a. A **strong acid** completely or almost completely dissociates in aqueous solution. Thus, HCl is a strong acid because it ionizes virtually 100 percent into H^+ and Cl^-. A strong acid has a weak conjugate base.
   b. A **weak acid** only slightly dissociates in aqueous solution. Acetic acid is a weak acid because it ionizes only about 1 percent to yield H^+ and acetate ions.

2. A **base** is a substance that accepts protons (i.e., H^+) in solution. Thus, bicarbonate ion

Table 5-1. Examples of Acid Dissociation Reactions

Proton Donor (Conjugate Acid)	Proton (H^+)		Proton Acceptor (Conjugate Base)
Common Acids			
HCl	$\rightleftharpoons$ H^+	+	Cl^-
CH_3COOH	$\rightleftharpoons$ H^+	+	CH_3COO^-
HCO_3^-	$\rightleftharpoons$ H^+	+	CO_3^{2-}
NH_4^+	$\rightleftharpoons$ H^+	+	NH_3
HOH	$\rightleftharpoons$ H^+	+	OH^-
H_3O^+	$\rightleftharpoons$ H^+	+	H_2O
Important Buffer Acids at Physiologic [H^+]			
$(CO_2)H_2CO_3$	$\rightleftharpoons$ H^+	+	HCO_3^-
$H_2PO_4^-$	$\rightleftharpoons$ H^+	+	HPO_4^{2-}
H • Protein	$\rightleftharpoons$ H^+	+	Proteinate$^-$
$HHbO_2$	$\rightleftharpoons$ H^+	+	HbO_2^-
HHb	$\rightleftharpoons$ H^+	+	Hb^-

Note.—A conjugate acid can be an anion or a cation; a conjugate base usually is an anion; HCO_3^-, a conjugate base, also can be an acid.

(HCO_3^-), ammonia (NH_3), and acetate ion (CH_3COO^-) all are bases. A strong base has a weak **conjugate acid**. Sodium ion (Na^+) and potassium ion (K^+) are cations that constitute the metallic components of salts; Na^+ and K^+, therefore, are considered bases by this definition. However, to call these cations bases in the context of acid-base physiology does not lead to a logical development of the subject.

3. Some substances are more or less equally divided between the acidic and basic forms at the normal H^+ concentration of the body. For example, the imidazole side groups of hemoglobin undergo the following reaction:

$$HHb \rightleftharpoons H^+ + Hb^-.$$

The acid, deoxyhemoglobin (HHb), dissociates to form H^+ and the conjugate base, Hb^-. HHb and Hb^- occur in about equal concentrations in blood cells.

B. BLOOD SAMPLING PROCEDURES

1. **Arterial.** Arterial blood is the primary reference for all acid-base studies. Blood should be withdrawn slowly (over a period of about 1 minute) and only when the individual is relaxed and breathing in a normal pattern. Control of temperature in the measurement of pH is important. Arterial blood samples are required for clinical characterization of acid-base status. Blood-gas measurements do not indicate whether blood is arterial or venous. To be certain that arterial blood is obtained, the syringe must be filled without aspiration (i.e., by the pressure within the artery).

2. **Capillary.** Blood obtained by deep puncture from a vasodilated capillary bed (e.g., in the earlobe or finger) is nearly equivalent to arterial blood when capillary blood flow is not stagnant.

3. **Venous.** Venous blood generally is used for routine measurement of carbon dioxide (CO_2) content. Provided that the muscles of the part of the body from which the blood is withdrawn are relaxed, the CO_2 content is within 1–3 mmol/L of the CO_2 content of arterial blood. Properly collected venous blood is satisfactory for clinical estimation of arterial pH and arterial CO_2 tension ($PaCO_2$), if care is taken to minimize the effect of local metabolism. This is achieved with the following precautions.
   a. The individual should be warm and at rest in the supine position for at least 15 minutes before withdrawal of blood.
   b. A light tourniquet (at a pressure never higher than midway between systolic and diastolic blood pressures) may be placed on the arm but should not be released while blood is being withdrawn during an interval not to exceed 2 minutes.
   c. During withdrawal of blood, the individual should make no hand motions such as "hand pumping" and "fist clenching."

II. TERMINOLOGY, DEFINITIONS, AND FUNDAMENTAL MATHEMATICS OF ACID-BASE PHYSIOLOGY

A. CONCEPT OF pH AND H^+ CONCENTRATION

1. It is convenient to express H^+ concentration in two different ways, either directly as $[H^+]$ or indirectly as **pH**. (The symbol, $[H^+]$, refers to H^+ concentration in mol/L or Eq/L.) The relationship between $[H^+]$ and pH can be expressed as:
   a. $pH = \log_{10} \dfrac{1}{[H^+]}$
   b. $pH = -\log_{10} [H^+]$
   c. $[H^+] = 10^{-pH}$

2. The common textbook statement that pH values range from 0 (1 mol/L H^+ solution) to 14 (10^{-14} mol/L H^+ solution) has no basis in chemical reality but merely indicates the arbitrary scale of most pH meters. For instance, ordinary concentrated HCl has a pH of approximately -1.1.

3. pH is a **dimensionless** number and should be treated as such; it should not be referred to in concentration units or as "pH concentration." In fact, the quantity whose logarithm determines the pH is a volume per equivalent, which is the **inverse** of concentration. Thus, pH could be correctly conceptualized as a logarithmic expression of the volume required to contain 1 equivalent of H^+. In human plasma at pH 7.4 [i.e., $[H^+]$ of 40×10^{-9} mol (Eq)/L or 40 nmol (nEq)/L], that volume is 25 million L!

4. Because pH is the logarithmic expression of $[H^+]$, it permits a graphic representation of a wide range of $[H^+]$ values. (It is important to note that pH and $[H^+]$ are **inversely** related.)

Another advantage of the pH concept is that when the pK′ of a buffer system is known, it is immediately possible to determine the effective pH range of the buffer.*

5. A disadvantage of the pH system is that it both inverts and uses the logarithmic scale to express $[H^+]$. For example, it is not immediately apparent that a **decrease** in pH from 7.4 to 7.1 represents a **doubling** of the $[H^+]$ from 40 nmol/L to 80 nmol/L. (Beginning with pH 7.4 and a $[H^+]$ of 40 nmol/L, for every 0.1 **increase** in pH, $[H^+]$ is obtained by multiplying $[H^+]$ by 0.8; for every 0.1 **decrease** in pH, $[H^+]$ is calculated by multiplying by 1.25.)

B. H^+ CONCENTRATION OF BODY FLUIDS (Table 5-2)

1. **Blood and Plasma.** With regard to $[H^+]$, the body fluid compartment most studied is arterial blood plasma.

    a. In normal individuals, the $[H^+]$ is approximately 40 nmol (nEq)/L [4×10^{-8} mol (Eq)/L], which is equivalent to pH 7.4. The range of $[H^+]$ that is compatible with life is 20–126 nEq/L, which is equivalent to a pH range of 7.7–6.9, respectively.

    b. The term **blood** pH always refers to **plasma** pH (7.4), which is higher than the intracellular pH of the erythrocyte (7.2). The pH of whole blood and the pH of plasma are identical **when plasma is separated anaerobically at body temperature**; however, the measured pH of whole blood is about 0.01 pH units lower than the measured pH of plasma due to the effect of suspended erythrocytes at the reference electrode.

       (1) The CO_2 content of erythrocytes is considerably lower than that of plasma. Therefore, it is essential to specify whether whole blood or plasma is being analyzed, because the hematocrit affects the CO_2 content of whole blood.

           (a) In normal individuals at rest, the pH of mixed venous blood is 7.38 compared to 7.41 for arterial blood because of the uptake of CO_2 by blood as it perfuses the tissues.

           (b) To measure blood pH correctly, the blood must be well mixed just before introduction of the pH electrode.

       (2) The temperature at which plasma is separated from the erythrocytes has little effect on CO_2 content.

    c. The $[H^+]$ of plasma is very small compared to that of other ions found in plasma (e.g., the plasma Na^+ concentration ($[Na^+]$) is about 142 million nmol/L and the K^+ concentration ($[K^+]$) is about 4 million nmol/L).

    d. The pH of blood serum and the pH of plasma are identical when small amounts of heparin (i.e., < 100 units/ml of blood) are used to obtain plasma. Other anticoagulants such as edetic acid (EDTA), citrate, and oxalate may have a significant effect on blood pH and, therefore, should not be used.

Table 5-2. H^+ Concentration $[H^+]$ and pH of Biologic Fluids

Fluid	pH	$[H^+]$ (nEq/ or nmol/L)*	$[H^+]$ (Eq/L or mol/L)
Pure water	7.0	100	1×10^{-7}
Blood			
Normal mean	7.40	40	3.98×10^{-8}
Normal range	7.36–7.44	44–36	4.36×10^{-8}–3.63×10^{-8}
Acidosis (severe)	6.9	126	1.26×10^{-7}
Alkalosis (severe)	7.7	20	2.00×10^{-8}
CSF (normal range)	7.36–7.44	44–36	4.36×10^{-8}–3.63×10^{-8}
Pure gastric juice (normal)	1.0	100,000,000	1×10^{-1}
Urine			
Normal average	6.0	1000	1×10^{-6}
Maximum acidity	4.5	31,600	3.16×10^{-5}
Maximum alkalinity	8.0	10	1×10^{-8}
ICF (muscle)	6.8	158	1.58×10^{-7}

Note.—Adapted from: Regulation of hydrogen ion concentration in body fluids. In *Best and Taylor's Physiological Basis of Medical Practice*, 10th ed. Edited by Brobeck JR. Baltimore, Williams and Wilkins, 1979, p 5–13.
* n = nano- = 10^{-9}. Thus, 100 nEq/L = 100×10^{-9} Eq/L, where nEq = nmol of a monovalent ion.

---

*K = the ionization or dissociation constant; pK = the negative logarithm of K ($-\log$ K) and is equal to the pH at which half of the acid molecules are dissociated and half are undissociated.

2. Cerebrospinal Fluid (CSF). Generally, it is possible to characterize precisely the disturbances in terms of the **blood** acid-base data; however, the extracellular changes do not always reflect the intracellular changes. Also, acid-base alterations in arterial blood may produce similar or opposite changes in the CSF depending on whether the blood [H$^+$] changes are due to respiratory or metabolic abnormalities.

a. An increase in PaCO$_2$ leads to a comparable rise in the CSF PCO$_2$. This results in an increase in the [H$^+$] of both arterial blood and CSF.

b. Conversely, an increase in arterial [H$^+$] stimulates ventilation, which lowers the PCO$_2$ of the blood and CSF and leads to CSF alkalosis with blood acidosis. Similarly, metabolic alkalosis of arterial blood produces acidosis of the CSF.

C. THE HENDERSON-HASSELBALCH EQUATION

1. Significance. The acid-base balance is maintained primarily through the control of two organ systems; the lungs control the PCO$_2$ through the regulation of alveolar ventilation, and the kidneys control the HCO$_3^-$ concentration ([HCO$_3^-$]). The classic description of the acid-base state is based on the **Henderson-Hasselbalch equation**, which is an expression of three variables (pH, PCO$_2$, and [HCO$_3^-$]) and two constants (pK and S).

2. Definition of Parameters. The Henderson-Hasselbalch equation could functionally be written as:

$$pH = pK' + \log \frac{\text{kidneys}}{\text{lungs}} . \tag{1}$$

However, the equation is more usefully expressed as:

$$pH = pK' + \log \frac{[HCO_3^-]}{S \times PCO_2} , \tag{2}$$

or, using specific values for pK and S, as:

$$pH = 6.1 + \log \frac{[HCO_3^-]}{0.03 \times PCO_2} . \tag{3}$$

From equation (3) it is clear that the value of arterial pH depends on the **ratio** of [HCO$_3^-$] to S $\times$ PCO$_2$, not on the individual value of each variable. In clinical medicine, pH, PCO$_2$, and [HCO$_3^-$] can be measured directly. With equation (3), however, any one of the variables can be calculated if the other two are known.

a. pK is defined as the negative logarithm of the [H$^+$] at which half the acid molecules are undissociated and half are dissociated. When equimolar concentrations of weak acid and conjugate base exist, the pH value equals the pK (i.e., the log of 1 is 0).

(1) The **actual** dissociation constant **(K)** for carbonic acid (H$_2$CO$_3$) in dilute aqueous solution at 38° C is 1.6 $\times$ 10^{-4} mol/L (pK = 3.8). Thus, H$_2$CO$_3$ is almost completely dissociated in the body where [H$^+$] = 4 $\times$ 10^{-8} mol/L, and it exists in quantities that are too small to be analyzed (i.e., 2.4 $\times$ 10^{-3} mEq/L). The formation and dissociation of H$_2$CO$_3$ is expressed as:

$$\begin{array}{c} \text{carbonic} \\ \text{anhydrase} \\ CO_2 + H_2O \rightleftharpoons H_2CO_3 \rightleftharpoons H^+ + HCO_3^- . \\ 500{:}1 4000{:}1 \end{array} \tag{4}$$

At equilibrium there are approximately 500 mmol of CO$_2$ for every 1 mmol of H$_2$CO$_3$ and approximately 4000 mmol of H$_2$CO$_3$ for every 1 mmol of H$^+$. Because of the presence of carbonic anhydrase, equilibrium between CO$_2$ and H$_2$CO$_3$ is rapid and constant.*

(2) Because the denominator of equation (3) is increased by a factor of 500, the **apparent** dissociation constant **(K$'$)** for the CO$_2$/HCO$_3^-$ buffer system in **plasma** at 38° C is correspondingly smaller (8 $\times$ 10^{-7} mol/L or 800 nmol/L), and the apparent pK **(pK$'$)** for this same buffer pair is correspondingly larger (6.1). As a rule, the optimal buffer region of a buffer-pair system is within a range of $\pm$1 pH units of its pK value.

(3) It is important to note that because CO$_2$ increases [H$^+$], as shown in equation (4), it is considered an acid even though it is an acid **anhydride**. Thus, dissolved CO$_2$ is present

---

*Carbonic anhydrase is found in erythrocytes, gastric parietal cells, renal tubular cells, pancreatic and pulmonary tissue, bone, and the eye; it is **not** found in muscle, peripheral nerves, and skin.

Table 5-3. CO_2 Production under Basal Conditions*

Time Elapsed	Volume (L)	Molarity (mmol)	Molarity (mol)	Weight† (g)	Weight† (lb)
1 minute	0.2	9	9×10^{-3}	0.4	9×10^{-4}
1 hour	12.0	540	0.54	24	5.3×10^{-2}
1 day	300	13,500	13.5	593	1.3

* Data for a respiratory quotient of 1.0.
† The density of CO_2 gas at STPD = 1.976 g/L. (STPD = standard temperature, standard pressure, and dry; that is, at 0° C, 760 mm Hg, and containing no water vapor.)

as a **potential** H^+ donor, and its concentration is proportionate to the true donor, H_2CO_3. For these reasons, it is more meaningful to characterize acid-base disturbances in terms of the CO_2/HCO_3^- buffer system instead of the H_2CO_3/HCO_3^- system. For all practical purposes, the H_2CO_3/HCO_3^- buffer pair can be considered to be composed of HCO_3^- (conjugate base) and dissolved CO_2 (conjugate "acid").

b. S is defined as the solubility constant for CO_2 in plasma at 38° C and is equal to 0.03 mmol/L/mm Hg. S represents the proportionality constant between CO_2 and PCO_2. For blood plasma at 38° C, the amount of dissolved CO_2 is expressed as:

$$\text{dissolved } CO_2 = 0.03 \times PCO_2,$$

where CO_2 = the **millimoles** of dissolved CO_2 per liter of plasma. At a $PaCO_2$ of 40 mm Hg, the CO_2 concentration ($[CO_2]$) is more usefully expressed as:

$$0.03 \text{ mmol/L/mm Hg} \times 40 \text{ mm Hg} = 1.2 \text{ mmol/L}.$$

Multiplying the proportionality constant by the milliliters of CO_2 per millimole (22.3 ml/mmol) yields a solubility constant of 0.67 ml/L/mm Hg for CO_2 in plasma. At this solubility constant and at a PCO_2 of 40 mm Hg, the concentration of dissolved CO_2 is expressed as:

$$0.67 \text{ ml/L/mm Hg} \times 40 \text{ mm Hg} = 26.8 \text{ ml/L.}^*$$

c. [HCO_3^-] denotes the bicarbonate ion concentration, which is expressed in **millimolarity** rather than molarity. The normal value of $[HCO_3^-]$ in plasma is 24 mmol (mEq)/L.

III. GENERATION AND ELIMINATION OF H^+.
The greatest source of H^+ is the CO_2 produced as one of the end products of the oxidation of glucose and fatty acids during aerobic metabolism.

A. CARBONIC ACID (H_2CO_3) is an acid that forms a volatile end product upon dehydration and can, therefore, be excreted by the lungs.

1. The Hydration-Dehydration: Dissociation-Association Reaction.
a. HCO_3^- is formed through the hydration of CO_2 and the subsequent dissociation of H_2CO_3. A significant concentration of HCO_3^- cannot be formed by this reaction because the dissociation of H_2CO_3 forms **equal** numbers of H^+ ions and HCO_3^- ions. Therefore, the significant concentration of HCO_3^- can only be attained after the H^+ formed from H_2CO_3 is buffered by blood buffer anions (i.e., reduced hemoglobin, protein, and phosphate) as:

$$CO_2 + H_2O \rightarrow H_2CO_3 + Hb^- \rightarrow HHb + HCO_3^-,$$

where HHb/Hb^- = deoxyhemoglobin buffer system.
b. Thus, for every H_2CO_3 molecule that has its H^+ taken up by a buffer, one HCO_3^- appears in the blood. These HCO_3^- are distributed between the erythrocytes and the plasma. At any given pH, reduced blood (deoxyhemoglobin) contains more HCO_3^- than oxygenated blood.

2. CO_2 Production (Table 5-3). CO_2 is the chief product of metabolism and represents the greatest portion of the acid that is continuously eliminated from the body by the lungs.
a. Most CO_2 in the body is produced from the decarboxylation reactions of the tricarboxylic (citric) acid cycle.
b. HCO_3^- is an excellent buffer with respect to **noncarbonic** acid added to the blood by diet, metabolism, and disease.
c. Comparing CO_2 and HCO_3^-, there is more CO_2 than HCO_3^- produced metabolically in the body. However, the extracellular fluid (ECF) contains a preponderance of the HCO_3^- form.

*This amounts to about 5 percent of the total amount of CO_2 carried in the arterial blood.

   d. Under basal conditions, with a respiratory quotient (R.Q.) of 0.82, the average adult produces about 300 L (13 mol)* of CO_2 per day.
   e. The lungs and kidneys are the principal routes for the elimination of protons and for maintaining normal $[HCO_3^-]/S \times PCO_2$ ratios.
      (1) In the course of a day, the equivalent of 20–40 L of 1 N acid (20,000–40,000 mEq H^+)[†] are eliminated via the lungs, or, more correctly, the amount of CO_2 produced during a day in a normal individual is **potentially** capable of forming 20–40 Eq of H^+.
      (2) During a 24-hour period, the equivalent of 50–150 ml of 1 N acid (50–150 mEq H^+) is excreted via the kidneys.
   f. CO_2 must also be removed from the blood by the lungs at a rate of 200 ml/min during resting conditions.[‡] With a resting cardiac output of 5 L/min, 40 ml of CO_2 must be added to each liter of blood per minute (i.e., 5 L/min $\times$ 40 ml/L = 200 ml/min).

3. Buffering of H_2CO_3 is primarily by the intracellular buffers (i.e., proteins, organic and inorganic phosphates, and hemoglobin). Most of the buffering of H_2CO_3 occurs in the erythrocyte, where the hemoglobin buffer system is quantitatively the most important buffer.

B. NONCARBONIC ACIDS[§] are acids that do not form a volatile end product and must, therefore, be excreted by the kidneys.

1. Source.
   a. The primary source of noncarbonic acid in man is the metabolism of exogenous protein. The body produces 50–70 mEq of H^+ daily from the catabolism of proteins such as phosphoprotein and methionine, which are metabolically equivalent to phosphoric and sulfuric acids, respectively.
   b. Other noncarbonic acids are the products of intermediary metabolism.
      (1) Lactic acid, a product of anaerobic intermediary metabolism, accumulates in the ECF in large quantities during heavy exercise. This excess acid must be buffered until excreted or metabolized to CO_2 and water.
         (a) Although most organs generate lactic acid, skeletal muscle, erythrocytes, and skin produce the largest amount. (Liver, kidney, and muscle can convert lactic acid to HCO_3^-.)
         (b) Tissue hypoxia leads to **hyperlacticemia**, which reduces the plasma $[HCO_3^-]$ and increases the anion gap (see VII B 2 for discussion of anion gap). The increment in the anion gap in this situation is a measure of serum lactate. Clinical conditions associated with tissue hypoxia include cardiac arrest, shock, severe cardiac failure, and severe hypoxemia.
         (c) Alkalosis stimulates glycolysis and generates lactic acid mainly by activation of phosphofructokinase. Respiratory alkalosis is a more effective stimulus for glycolysis, because CO_2 crosses cellular membranes more readily.
      (2) Acetoacetic acid and β-hydroxybutyric acid also are products of intermediary metabolism. These acids are formed almost exclusively by the liver during fasting.
   c. Excess fluid intake (hydration) reduces the $[HCO_3^-]$, which causes a corresponding increase in $[H^+]$ if PCO_2 is constant. This type of noncarbonic acidosis is more rapidly corrected by the renal excretion of excess water than by renal secretion of the **apparent** excess of H^+. Dehydration has the opposite effect.

2. Excretion.
   a. Lactic acid production usually is transient; that is, after exercise is completed or the hypoxemic condition is removed, the lactic acid is metabolized to CO_2 and water. Lactic acid, therefore, can be eliminated by metabolic destruction of the acid rather than by renal excretion under normal conditions.
   b. Phosphoric and sulfuric acids cannot be further metabolized to CO_2, and the elimination of these acids and the H^+ that they yield can only be accomplished by the kidneys.
   c. Even noncarbonic acids that can be catabolized to CO_2 and water are excreted by the kidneys when present in large excess, as are β-hydroxybutyric and acetoacetic acids in diabetes mellitus. This condition is called **ketonuria**.
   d. Clinically more important is the acidosis of uncontrolled diabetes mellitus where the accumulation of acetoacetic and β-hydroxybutyric acids in the ECF may produce coma and death.

---

*1 mol of CO_2 = 22.26 L at STPD (standard conditions of temperature and pressure, dry).
[†]Depending on the level of physical activity; during **resting** metabolism, a normal individual is potentially capable of forming 13,000 mEq of H^+.
[‡]Equal to CO_2 production at rest.
[§]Carbonic and noncarbonic acids often are referred to as **volatile** and **fixed** acids, respectively.

3. Buffering of Noncarbonic Acids.

a. In contrast to the bicarbonate buffer system, which can only buffer noncarbonic acids, the nonbicarbonate buffer systems can buffer both noncarbonic and carbonic acids.

b. The plasma bicarbonate system is quantitatively the most important buffer in the ECF for noncarbonic acids.

c. When noncarbonic acids are being buffered in the erythrocytes, more than 60 percent of the buffering occurs by the hemoglobin and more than 30 percent by the bicarbonate system in the erythrocytes. Approximately 10 percent of the buffer capacity in erythrocytes is attributed to the organic phosphate esters.

d. The bodily response to mineral acids (e.g., HCl) is different in that lactic acid is distributed throughout the body water and is metabolized, whereas mineral acids are confined to the ECF and are buffered and excreted without being metabolized.

IV. BODY BUFFER SYSTEMS

A. INTRODUCTION

1. Blood Buffers (Table 5-4).

a. It is important to note that the blood buffers are not solely plasma buffers, but that hemoglobin, HCO_3^-, and phosphate, are found in erythrocytes and act as important blood buffers.

(1) In blood the chief H^+ acceptor is HCO_3^-, which exists in a concentration of 20–30 mEq/L.

(2) About 10 mEq/L of H^+ acceptor is available in the hemoglobin system, where reduced hemoglobin (deoxyhemoglobin) is a stronger base than oxyhemoglobin (i.e., deoxyhemoglobin has a stronger capacity to combine with H^+).

(3) The hemoglobin buffer system is quantitatively as important as the bicarbonate buffer system. The buffering capacity of 1 L of blood is as follows.

(a) Reduced hemoglobin = 27.5 mEq of H^+

(b) Plasma proteins = 4.2 mEq of H^+

b. The major buffer anions of whole blood, HCO_3^-, protein, and hemoglobin have a total concentration of approximately 48 mEq/L (see VII B 1 a).

c. With the bicarbonate and hemoglobin systems taken together, 5 L of blood of a normal adult have sufficient buffer capacity to combine with almost 150 mmol of protons (150 ml of 1 N HCl) before the pH of body fluids becomes dangerously acidic.

2. Tissue Buffers. The major buffer capacity of the body is not in the blood but in the H^+ acceptors found in other tissues, principally in the muscle, which constitutes about one-half of the cellular mass. These tissues can neutralize about five times as much acid as the blood buffers. Since skeletal muscle represents about half of the cellular mass, most intracellular buffering presumably occurs in muscle.

a. For any individual, the body HCO_3^- concentration averages 13 mEq/kg body weight. Muscle cells contain HCO_3^- at a concentration of about 12 mEq/L, and most other cells contain it at higher concentrations. The intracellular fluid (ICF) and ECF each contain about 50 percent of the total body HCO_3^-.

Table 5-4. Whole Blood Buffers

Buffer Type	Buffering Capacity of Whole Blood (%)
Bicarbonate	
Plasma	35
Erythrocyte	18
Total bicarbonate	53
Nonbicarbonate	
Hemoglobin and oxyhemoglobin	35
Plasma proteins	7
Organic phosphate	3
Inorganic phosphate	2
Total nonbicarbonate	47

Note.—Adapted from: Regulation of hydrogen ion concentration in body fluids. In *Best and Taylor's Physiological Basis of Medical Practice*, 10th ed. Edited by Brobeck JR. Baltimore, Williams and Wilkins, 1979, p 5–14.

b. The total body store of bone carbonate is about 50 times the amount of HCO_3^- in the ICF and ECF compartments. Bone carbonate appears to be the predominant source of base for neutralizing excess noncarbonic acid in the ECF. Indeed, it has long been recognized that chronic noncarbonic acidosis causes bone dissolution (resorption) through the loss of calcium carbonate ($CaCO_3$). The early carbonate release is in the form of sodium carbonate (Na_2CO_3).

c. It is important to remember that most of the body buffers are located in the ICF, the major exceptions being plasma proteins and the small amount of phosphate ion (HPO_4^{2-}) in the ECF. Hemoglobin, by definition, is an intracellular buffer.

B. DISTRIBUTION OF BODY BUFFER SYSTEMS IN THE THREE MAJOR FLUID COMPARTMENTS

1. Blood, in regard to its buffering activity, usually is considered as a whole rather than in terms of its components. Blood buffers are described here in terms of separate compartments (i.e., plasma and erythrocytes) for didactic purposes only.

a. General Considerations.

(1) Whole blood is an excellent buffering system for noncarbonic acids because of its nonbicarbonate buffers as well as its bicarbonate buffer system.

(2) More than 90 percent of the capacity of the blood to buffer carbonic acid is attributed to the hemoglobin buffer system. Thus, the nonbicarbonate buffer in the erythrocyte is quantitatively more important than the bicarbonate buffer in that compartment. The bicarbonate buffer system remains quantitatively important, however, in the erythrocyte (see Table 5-4).

b. Buffer Capacity of Blood Components.

(1) Plasma, which contains three buffer systems, has a considerable capacity for buffering noncarbonic acids but a much smaller capacity for buffering carbonic acid.

(a) Bicarbonate Buffer System (HCO_3^-/H_2CO_3 or HCO_3^-/CO_2). This buffer exists in a concentration of 24 mmol/L of plasma. When **noncarbonic acid** is added to normal plasma, more than 75 percent of the buffering capacity of plasma is due to the HCO_3^-/CO_2 system. Most of the remaining buffering of noncarbonic acid involves the plasma protein buffers and, to a small degree, the phosphate buffer system. Again, the HCO_3^-/CO_2 system plays no role in the buffering of carbonic acid.

(b) Nonbicarbonate Buffer Systems.

(i) Plasma Protein (Protein$^-$/H • Protein). Plasma is a salt solution containing 7 percent protein, which exists as polyanions at the pH of plasma. Plasma protein H^+ acceptors exist in a concentration of 1.5 mmol/L of plasma and account for less than one-sixth of the total buffering capacity of whole blood.

(ii) Inorganic Orthophosphate ($HPO_4^{2-}/H_2PO_4^-$). Because this buffer system exists in a concentration of only 0.66 mmol/L of plasma, it contributes little to the total buffering activity of plasma.* At a plasma pH of 7.4, the concentration ratio of $HPO_4^{2-}/H_2PO_4^-$ is 4:1. Therefore, 80 percent of the inorganic phosphate exists as disodium phosphate, and 20 percent is in the form of monosodium phosphate. The $HPO_4^{2-}/H_2PO_4^-$ system is a major elimination route for H^+ via the urine, which has a relatively high phosphate content.

(2) Erythrocytes. Although the blood contains other cells, the erythrocyte is the only important cellular component of the blood buffer system. The erythrocyte contains four buffer systems.

(a) Bicarbonate Buffer System. The HCO_3^-/CO_2 system exists in a concentration of 15 mmol/L of erythrocytes, compared to a concentration of 21 mmol/L and 25 mmol/L of whole blood and plasma, respectively.

(b) Nonbicarbonate Buffer Systems.

(i) Inorganic Orthophosphate. The $HPO_4^{2-}/H_2PO_4^-$ system exists in a concentration of 2 mmol/L of erythrocytes.

(ii) Organic Phosphate. Although the erythrocyte contains a significant amount of organic phosphate buffer, this amount is quantitatively small compared to the bicarbonate and hemoglobin buffer concentrations in the erythrocyte.

(iii) Hemoglobin Buffers (Hb^-/HHb and HbO_2^-/$HHbO_2$) [Table 5-5; see also V C]. One L of **erythrocytes** contains 334 g (5.1 mmol) of hemoglobin. One L of **whole blood** contains 150 g (2.3 mmol) of hemoglobin.

2. Interstitial Fluid (Including Lymph).

a. Bicarbonate Buffer System. The HCO_3^-/CO_2 system is quantitatively the most important

---

*The pK' of the acid form (i.e., $H_2PO_4^-$) is 6.8. With this pK', the $HPO_4^{2-}/H_2PO_4^-$ system would be a more effective buffer than the HCO_3^-/CO_2 system (pK' = 6.1) if it was present in an appreciable concentration. Many of the organic phosphate compounds found in the body have pK' values within one-half of a pH unit from 7.0.

buffer of noncarbonic acid. The $[HCO_3^-]$ of the interstitial fluid is 27 mmol/L, which is similar or about 5 percent higher than the $[HCO_3^-]$ of plasma. It is important to consider that in man the interstitial fluid volume is about three times that of plasma; therefore, the total capacity of the interstitial fluid to buffer **noncarbonic acid** is considerably greater than that of the total blood volume to buffer these acids. On a per unit of volume basis, however, the interstitial fluid has nearly the capacity of plasma to buffer noncarbonic acids.

    b. Nonbicarbonate Buffer Systems. The $HPO_4^{2-}/H_2PO_4^-$ system exists in a concentration of 0.7 mmol/L of interstitial fluid; therefore, this compartment has little capacity to buffer **carbonic acid**. The interstitial fluid is essentially free of protein.

  3. ICF (Excluding Erythrocytes).

    a. Bicarbonate Buffer System. The ICF contains only about 12 mmol of HCO_3^-/L in skeletal and cardiac muscle.

    b. Nonbicarbonate Buffer Systems. Protein and organic phosphate compounds exist in quantitatively significant amounts in the ICF, giving this compartment the capacity to buffer effectively both noncarbonic and carbonic acids as well as alkali. The ICF concentrations of these major buffer anions are:

    (1) $HPO_4^{2-}/H_2PO_4^-$ (skeletal muscle) = 6 mmol/L

    (2) Protein$^-$/H • Protein (skeletal muscle) = 6 mmol/L

    (3) Organic anions (skeletal muscle) = 84 mmol/L

C. HEMOGLOBIN: AN "EXTRACELLULAR" BUFFER (see Table 5-5)

  1. General Considerations. Although hemoglobin is found intracellularly within the erythrocyte, it is more conventionally regarded as extracellular and, therefore, part of the extracellular buffer system because:

    a. Hemoglobin is confined to the erythrocyte, which is a cellular component of the ECF.

    b. Hemoglobin is readily available for the buffering of extracellular acids.

  2. Chemical Forms of Hemoglobin. Four species of hemoglobin exist in the blood at any time, and the proportion of each depends on the O_2 tension (PO_2) and the pH.

    a. Hemoglobin is a mixture of **deoxyhemoglobin** (reduced hemoglobin) and **oxyhemoglobin** (oxygenated hemoglobin).

    (1) Oxyhemoglobin represents a mixture of protonated oxyhemoglobin (conjugate acid or **$HHbO_2$**) and unprotonated oxyhemoglobin (conjugate base or **HbO_2^-**).

    (2) Deoxyhemoglobin also represents a mixture of protonated deoxyhemoglobin (conjugate acid or **HHb**) and unprotonated deoxyhemoglobin (conjugate base or **Hb^-**).

    (3) At pH 7.4, the major oxygenated form of hemoglobin is HbO_2^- and the major deoxygenated form is HHb. HHb and $HHbO_2$ have lower O_2 affinities (i.e., higher O_2 dissociation constants) than do Hb^- and HbO_2^-.

    b. The **reduction of oxyhemoglobin** causes the hemoglobin to become a **weaker acid,**

Table 5-5. The Hemoglobin Buffer System

Buffer	Component	Description
Deoxyhemoglobin (HHb/Hb^-)	HHb	Weak conjugate acid; approximately two-thirds of deoxyhemoglobin delivered to the lungs in this protonated form
	Hb^-	Conjugate base (stronger than HbO_2^-); metabolically produced H^+ taken up by this unprotonated form
Oxyhemoglobin ($HHbO_2/HbO_2^-$)	$HHbO_2$	Weak conjugate acid (stronger than HHb)
	HbO_2^-	Conjugate base; most oxyhemoglobin transported to tissues in this unprotonated form

Note.—Oxyhemoglobin is treated as a separate buffer system from deoxyhemoglobin because there is a change in the pK' value for some of the histidine residues when hemoglobin changes from deoxy- to oxyhemoglobin.

whereas the **oxygenation of deoxyhemoglobin** causes the hemoglobin to become a **stronger acid**. The amount of acid formed following the oxygenation of human hemoglobin (i.e., in the pH range of 7.0–7.8) is a constant value of 0.3 mEq/10 g of hemoglobin.

   (1) As an acid (H^+ donor), $HHbO_2$ is a stronger acid than HHb.

   (2) As a base (H^+ acceptor), Hb^- is a stronger base than HbO_2^-.

  c. The buffer action of the hemoglobin system is relevant to the acid-base chemistry of respiration.

   (1) The deoxyhemoglobin produced after O_2 is delivered to the tissues is in the form of Hb^-, which can accept the H^+ from the dissociation of H_2CO_3 because Hb^- is a stronger base than HbO_2^-.

   (2) From the following dissociation reactions:

$$HHbO_2 \longleftrightarrow HbO_2^- + H^+; pK = 6.68$$

$$HHb \longleftrightarrow Hb^- + H^+; pK = 7.93$$

it is clear that oxyhemoglobin ($HHbO_2$) is a stronger acid than deoxyhemoglobin (HHb), and, consequently, Hb^- has a stronger affinity for protons (i.e., is a stronger base) than HbO_2^-.

3. Hemoglobin as a Buffer. Like all proteins, hemoglobin is a buffer. At pH 7.2 (i.e., the pH of normal arterial erythrocytes), the buffering action of hemoglobin is due mainly to the imidazole groups of the histidine residues.

  a. The titration curves of deoxyhemoglobin and oxyhemoglobin in Figure 5-1 illustrate the basis for the ability of hemoglobin to neutralize H^+ formed subsequent to the diffusion of CO_2 into the erythrocyte. Again, most of the H^+ are buffered by hemoglobin and most of the HCO_3^- diffuses into the plasma.*

  b. Oxyhemoglobin dissociates more completely than does deoxyhemoglobin, and as a result deoxyhemoglobin produces less H^+ at a given pH than does oxyhemoglobin, which is a stronger acid. Thus, hemoglobin becomes a more effective buffer when CO_2 and, hence, H^+ are added from the tissues. This is important because the diffusion of CO_2 from the

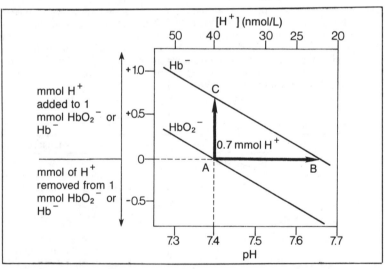

Figure 5-1. Titration curves of oxyhemoglobin (HbO_2^-) and deoxyhemoglobin (Hb^-), illustrating the importance of hemoglobin as a buffer. The complete deoxygenation of 1 mmol of HbO_2^- to liberate 1 mmol of O_2 results in the neutralization of 0.7 mmol of H^+ without a change in pH. *Line AC* indicates the amount of H^+ that can be added during the reduction of hemoglobin without causing a pH change. *Line AB* represents the pH change that would occur if the oxyhemoglobin at pH 7.4 was completely reduced. The reduction of HbO_2^- to Hb^- would cause a large increase in pH if CO_2 and, hence, H^+ were not added simultaneously to the system. Reduction of hemoglobin denotes the O_2-free state without a change in the valence of iron. (Adapted from White A, et al: Hemoglobin and the chemistry of respiration. In *Principles of Biochemistry*, 5th ed. New York, McGraw-Hill, 1973, p 843.)

---

*The buffer capacity of nonbicarbonate buffers is dependent primarily on the hemoglobin concentration.

tissues to the capillary blood is accompanied by the simultaneous reduction of oxy-hemoglobin.

   c. As the uptake of CO_2 depends on H^+ acceptors, this increase in H^+ acceptors in the form of deoxyhemoglobin facilitates the uptake and buffering of the H^+ generated by the hydration of CO_2 and the dissociation of H_2CO_3. As a result of these reactions, CO_2 is converted into HCO_3^- within the erythrocyte.

     (1) For each mmol of oxyhemoglobin that is reduced, about 0.7 mmol of H^+ can be taken up and, consequently, 0.7 mmol of CO_2 can enter the blood without a change in pH (see Fig. 5-1, *point A* to *point C*).

     (2) A reaction that causes no change in $[H^+]$ (or pH) is called **isohydric** buffering.

   d. Respiratory Quotient (R.Q.). When the body is at rest, the rate of O_2 consumption ($\dot{V}O_2$) under standard conditions is 250–350 ml/min, and the rate of CO_2 production ($\dot{V}CO_2$) is 200–250 ml/min. (The dot over the symbol V denotes **volume per unit time**.) R.Q. represents the ratio of $\dot{V}CO_2$ to $\dot{V}O_2$. Normally, on a mixed diet, the $\dot{V}O_2$ exceeds the $\dot{V}CO_2$, and the R.Q. is less than 1.

     (1) The metabolic R.Q. equals the **molar ratio** of CO_2 production rate to the corresponding O_2 consumption rate by metabolizing tissues.

     (2) If the R.Q. is 0.7, then for 1 mmol of O_2 consumed, 0.7 mmol of CO_2 is produced, which, when converted to H_2CO_3, yields 0.7 mmol of H^+ upon dissociation.

     (3) The complete deoxygenation of 1 mmol of oxyhemoglobin to liberate 1 mmol of O_2 results in the neutralization of 0.7 mmol of H^+ without a change in pH. Thus, all of the H^+ produced when the R.Q. is 0.7 can be buffered by deoxyhemoglobin with no change in pH (see Fig. 5-1).

   e. At pH 7.2, about 84 percent of the deoxyhemoglobin is in the form of HHb, whereas only about 23 percent of the oxyhemoglobin is in the form of $HHbO_2$. Of the total oxy-hemoglobin that causes O_2 to form deoxyhemoglobin:

     (1) 23 percent was already combined with H^+,

     (2) 16 percent will not combine with H^+, and

     (3) 61 percent will take up H^+ before a pH decrease occurs.

   f. CO_2 entering the erythrocytes rapidly undergoes two reactions.

     (1) CO_2 is hydrated to form H_2CO_3, a reaction that is catalyzed by carbonic anhydrase in the erythrocyte. The H_2CO_3 dissociates to form HCO_3^- and H^+, which is buffered primarily by the hemoglobin buffers. Much of the HCO_3^- formed within the erythrocyte diffuses into the plasma in exchange for Cl^-.

     (2) CO_2 combines with the amino groups of deoxyhemoglobin to form carbamino-hemoglobin. The carbaminohemoglobin dissociates to a carboxylate anion and H^+, which is buffered primarily by the hemoglobin buffers. In summary:

$$Hb \bullet NH_2 + CO_2 \rightleftharpoons Hb \bullet NHCOO^- + H^+.$$

   g. The unloading of O_2 from oxyhemoglobin to the tissues causes the formation of deoxy-hemoglobin that is better able to tie up the H^+ produced by the simultaneous uptake of CO_2. The loss of O_2 from hemoglobin facilitates the uptake of CO_2 in the form of carb-aminohemoglobin by the erythrocytes. Oxyhemoglobin contains about 0.1 mmol and deoxyhemoglobin about 0.3 mmol of carbaminohemoglobin per mmol.

V. RESPIRATORY REGULATION OF ACID-BASE BALANCE

A. BLOOD FORMS OF CO_2 (Table 5-6)

   1. It is not immediately apparent that the actual quantities of CO_2 and O_2 transported in the blood are far greater than the amounts of these gases in physical solution, because these gases are transported mainly in the form of chemical derivatives. Further, it is even less apparent that there is far **more** total CO_2 than O_2 in every liter of blood. At the level of the lung, the HCO_3^- combines with the potential protons (i.e., the protons associated with $HHbO_2$) to produce HbO_2^- and H_2CO_3 resulting in the elimination of the protons as CO_2 and water.* Thus, free H^+ never appear to any appreciable extent but are in the form of HHb or H_2CO_3. The entire respiratory cycle can be regarded as an exchange of a HCO_3^- for a HbO_2^-, with the HCO_3^- transported from the tissues to the lungs and the HbO_2^- transported to the tissues from the lungs (see Table 5-5).

   2. CO_2 is transported in three forms; however, CO_2 ultimately must be converted to a gas in order to be eliminated via the lungs.

   a. HCO_3^- accounts for 90 percent of the total CO_2 in plasma.

---

*$HHbO_2$ and HbO_2^- are forms of oxygenated hemoglobin and constitute a buffer pair where the protonated form, $HHbO_2$, is the conjugate acid and the unprotonated form, HbO_2^-, is the conjugate base (see Table 5-5).

b. Carbamino compounds (i.e., carbamates of hemoglobin and protein) represent the combination of CO_2 with free -NH_2 groups of blood proteins according to the equation:

$$R \bullet NH_2 + CO_2 \rightleftharpoons R \bullet NHCOO^- + H^+.$$

About 5 percent of the CO_2 normally carried in arterial blood is in the form of carbamino compounds (carbamates).

c. A small amount of CO_2 gas (5 percent) is physically dissolved in plasma and also hydrated as H_2CO_3. Also, approximately 500 mol of CO_2 exist for every 1 mol of H_2CO_3 [see II C 2 a (1)].

3. It is important to recognize that the term "bicarbonate" is used interchangeably with the term "total CO_2 combining power" or "total CO_2." This must never be confused with the term "partial pressure of CO_2," which is the PCO_2 of the arterial blood gas measurement.

4. In blood with an arterial PO_2 (PaO_2) of 100 mm Hg and a mixed venous PO_2 ($P\bar{v}O_2$) of 40 mm Hg, there are 200 ml O_2/L and 150 ml O_2/L of arterial and mixed venous blood, respectively. However, the $PaCO_2$ and mixed venous PCO_2 ($P\bar{v}CO_2$) are associated with 480 ml CO_2/L and 520 ml CO_2/L of arterial and mixed venous blood, respectively (see Table 5-6).

B. EXPRESSIONS FOR CO_2 (Table 5-7)

1. To express the CO_2 system in terms of the Henderson-Hasselbalch equation, the following modifications are made.

a. The numerator, [HCO_3^-], becomes [total CO_2] − [free CO_2], and the denominator, S × PCO_2, becomes [free CO_2], where [free CO_2] designates the combined concentrations of [H_2CO_3 + dissolved CO_2]. For practical purposes, [free CO_2] and S × PCO_2 can be equated with [dissolved CO_2].

b. Substituting these parameters in the Henderson-Hasselbalch equation (equation 3), the final modification of the equation is:

$$pH = 6.1 + \log \frac{[\text{total } CO_2] - [\text{free } CO_2]}{[\text{free } CO_2]},$$

where pH, [total CO_2], and [free CO_2] all are variables.

2. Blood Concentration of CO_2.

a. PCO_2 is equal to the partial pressure of CO_2 in mm Hg. For every 10 mm Hg rise in PCO_2 above 40 mm Hg, plasma [HCO_3^-] increases by about 1 mEq/L. PCO_2 is inversely related to pulmonary ventilation.

b. Physically dissolved CO_2 follows Henry's law; that is, the [CO_2] dissolved in blood plasma is proportional to the partial pressure of CO_2. [CO_2] can be determined in two ways with two different solubility constants.

(1) The equations are: [CO_2] = [dissolved CO_2] + [H_2CO_3] and, more practically, [CO_2] $\cong$ S × PCO_2.

Table 5-6. Normal Values for the Forms of CO_2 in the Blood of a Normal Individual at Rest

Form	Units	Arterial	Venous (Mixed)	% of [Total CO_2]
[Total CO_2]	mmol/L	21.5–22.4	23.3–24.2	100
	ml/dl	48–50	52–54	. . .
	ml/L	480–500	520–540	. . .
[Dissolved CO_2]*	mmol/L	1.20	1.34	5
	ml/dl	2.68	2.99	. . .
	ml/L	26.8	29.9	. . .
[HCO_3^-]	mmol/L	22–26 (24)†	24–28 (25.09)	90
[Carbamino-CO_2]‡	mmol/L	0.97	1.42	5
[H_2CO_3]	mEq/L	2.4×10^{-3}	. . .	Negligible
[CO_3^{2-}]	mEq/L	2.0×10^{-1}	. . .	Negligible
	mmol/L	1.0×10^{-1}	. . .	. . .
PCO_2	mm Hg	35–45 (40)	41–51 (46)	. . .
[HCO_3^-]/S × PCO_2	. . .	24/1.2 = 20/1	25.1/1.38 = 18/1	. . .
Base excess§	mEq/L	−2 to +2	0–4	. . .

* See II C 2 b.
† ( ) denote average value.
‡ [$HbNHCOO^-$], or carbaminohemoglobin, and carbamates of protein.
§ [HCO_3^-] or total CO_2 actually is used in acid-base problems as an indicator of base excess.

(2) The solubility constants are: S = 0.03, where $CO_2 = 0.03 \times PCO_2$ (in mmol/L), and S = 0.67, where $CO_2 = 0.67 \times PCO_2$ (in ml/L) [see II C 2 b].

(3) Alternatively, PCO_2 can be obtained from $[HCO_3^-]$ by the following empirical relationship: $PCO_2 = 1.5 \, [HCO_3^-] + 8$.

C. H_2CO_3 AND OTHER CO_2-FORMING ACIDS

1. Source.

a. The oxidation of glucose and triglyceride leads to the formation of CO_2, much of which is hydrated to H_2CO_3; H_2CO_3 in turn dissociates into H^+ and HCO_3^-.

b. CO_2 and water are the most abundant end products of metabolism.

(1) Lactic acid, a **noncarbonic acid**, normally is metabolized to CO_2 and water.

(2) The ketone bodies, β-hydroxybutyric acid and acetoacetic acid, also are noncarbonic acids, which are produced primarily by the liver during fasting. Upon reingestion, these acids that have accumulated are further catabolized to CO_2 and water by the extrahepatic tissues.

2. Elimination.

a. The respiratory system continuously excretes CO_2.

b. The lungs eliminate H_2CO_3 in the dehydrated (anhydrous) form of CO_2. Indeed, the unique property of the CO_2/HCO_3^- buffer system lies in the ability of the lungs to eliminate undissociated H_2CO_3 as nonionizable CO_2.

3. Buffering of H_2CO_3.

a. Because CO_2 readily penetrates cellular membranes, H_2CO_3 is buffered by the entire body.

b. Most of the buffering of H_2CO_3 occurs via the nonbicarbonate buffer systems within erythrocytes. The H^+ derived from the dissociation of H_2CO_3 is buffered primarily by the hemoglobin buffer system. More than 90 percent of the capacity of the blood to buffer H_2CO_3 resides in the hemoglobin buffer system.

c. The erythrocyte bicarbonate buffer and the plasma bicarbonate buffer play no role in the buffering of H_2CO_3.

VI. RENAL REGULATION OF ACID-BASE BALANCE

A. OVERVIEW

1. Acid-Base Conditions.

a. Type of diet is a major determinant of the daily acid-base conditions that must be regulated by the kidneys.

Table 5-7. The CO_2 System

[Total CO_2]	=	CO_2 content*
	=	$[H_2CO_3] + [CO_2] + [HCO_3^-] + [CO_3^{2-}] + [\text{carbamino } CO_2]$
	=	$[CO_2]^\dagger + [HCO_3^-]$
	=	$[\text{dissolved } CO_2] + [HCO_3^-]$
	=	$S \times PCO_2 + [HCO_3^-]$
$[CO_2]^\dagger$	=	Dissolved $CO_2 + H_2CO_3^{\ddagger}$
	=	H_2CO_3 "pool"
	=	$[\text{dissolved } CO_2] + [H_2CO_3] = [\text{dissolved } CO_2 + H_2CO_3]$
	=	$S \times PCO_2^{\S}$
$[HCO_3^-]^{\S}$	=	CO_2 content − content of dissolved $CO_2^{\parallel}$
	=	$[\text{total } CO_2] - [\text{dissolved } CO_2]$
	=	$[\text{total } CO_2] - S \times PCO_2$
[Carbamino CO_2]	=	Carbamates of hemoglobin and plasma protein. An appreciable fraction of [total CO_2] of blood consists of carbamino compounds of hemoglobin (about 1 mmol/L of blood).

* Total CO_2 content relates most closely to the $[HCO_3^-]$ of blood.
† $[CO_2]$ includes $[H_2CO_3]$, but $[\text{dissolved } CO_2]$ is used instead of $[H_2CO_3]$.
‡ Because $[H_2CO_3]$ is about 0.002 of $[\text{dissolved } CO_2]$, or 2.4×10^{-3} mEq/L, the inclusion of $[H_2CO_3]$ has no appreciable effect on the proportionality constant between $[CO_2]$ and PCO_2.
§ Within the usual range of PCO_2 (20–60 mm Hg), the $[\text{dissolved } CO_2]$ of plasma is 0.6–1.8 mmol/L. Therefore, no serious error would be introduced if an assumed value of 1 mmol/L for the $[\text{dissolved } CO_2]$ was to be subtracted from the $[\text{total } CO_2]$ to obtain the $[HCO_3^-]$.
‖ $[\text{Total } CO_2] - [H_2CO_2] + [CO_2]$ is called **actual HCO_3^-**, which corresponds to the alkali reserve.

(1) High protein diets contain large amounts of sulfur in the form of sulfhydryl groups. The sulfur is oxidized to sulfate ion (SO_4^{2-}), a process that tends to lead to metabolic acidosis.

(2) The most abundant of the weak acid waste products of metabolism is acid phosphate ($H_2PO_4^-$).

(3) Vegetarian diets are associated with large intakes of lactate and acetate, and the metabolites of these anions tend to lead to metabolic alkalosis.

b. Among the metabolically produced noncarbonic acids are sulfuric and phosphoric acids and smaller amounts of hydrochloric, lactic, uric, β-hydroxybutyric, and acetoacetic acids.

(1) Approximately one-half of these metabolically produced acids are neutralized by base in the diet.

(2) The other half of these noncarbonic acids must be neutralized by buffer anion systems of the body. Of those noncarbonic acids buffered in the ECF, 97–98 percent are buffered by reacting with HCO_3^-.

2. Kidney Function.

a. The primary role of the kidney in acid-base regulation is to conserve major cations and anions in the body fluids. To maintain the total quantities and concentrations of the major electrolytes within normal limits, these organs perform two major functions.

(1) The kidneys stabilize the standard HCO_3^- pool by obligatory reabsorption (mainly by the proximal tubule) and by controlled reabsorption of filtered HCO_3^- (by the distal and collecting tubules).

(2) The kidneys excrete a daily load of 50–100 mEq of metabolically produced noncarbonic acid.*

(a) In most cases, 25 percent of this noncarbonic acid (10–30 mEq/day) is excreted in the form of **titratable acid**.

(b) About 75 percent (30–50 mEq/day) is excreted in the form of acid combined with **ammonia** (i.e., NH_4^+).

(c) The normal urinary ratio of NH_4^+ to titratable acid is between 1 and 2.5.

b. The major sites of urine acidification are the distal and collecting tubules.

(1) Essentially all of the H^+ within the tubular lumen is from the tubular secretion of H^+ generated by metabolism. There is no significant contribution of H^+ from the glomerular filtrate, which accounts for less than 0.1 mmol of H^+ per day!

(2) Since the lowest pH attainable in urine is 4.4 (i.e., a $[H^+]$ of 40×10^{-6} Eq/L) and the plasma $[H^+]$ is 40×10^{-9} Eq/L, the kidney can cause a 1000-fold $[H^+]$ gradient between plasma and urine.

c. An important difference between the acidification of urine and free H^+ excretion is that the ability to reduce urinary pH (acidification) does not reveal a great deal about the amount of free H^+ excreted. This is because most of the H^+ that is excreted occurs in association with an anion (mostly as $H_2PO_4^-$) or in combination with ammonia (as NH_4^+).

d. The majority of secreted H^+ is used to bring about HCO_3^- reabsorption and, therefore, is not excreted.

B. BASIC ION EXCHANGE MECHANISMS (Figs. 5-2 and 5-3)

1. Entry of Na^+ into the luminal (adluminal) membrane of the tubular cell is operationally linked to the secretion of H^+ into the tubular lumen. (It is a cation-exchange process that maintains intracellular electroneutrality.)

a. Entry of Na^+ into the tubular cell is passive (i.e., by facilitated diffusion), because Na^+ proceeds down an electrochemical gradient.

b. Secretion of H^+ is active in that it occurs against an electrochemical gradient.

2. Subsequent to the passive Na^+ transport across the luminal membrane, Na^+ is actively reabsorbed across the basolateral (abluminal) membrane into the peritubular capillary in association with HCO_3^- that was formed within the cell. The secretion of 1 mol of H^+ leads to the reabsorption of 1 mol of Na^+ and 1 mol of HCO_3^- in the peritubular blood.

a. H^+ for secretion originates from the hydration-dissociation reaction within the tubular cell [see II C 2, equation (3)] and from the splitting of water, which liberates hydroxide ion (OH^-) within the cell; the OH^- then reacts with CO_2 to form HCO_3^-.

b. H^+ secretion exceeds H^+ excretion; the amount of H^+ excreted in the form of free H^+ is negligible.

c. The net effect of H^+ secretion is to facilitate the transfer of HCO_3^- from tubular lumen to tubular cell.

d. It is probable that H^+ secretion along the entire tubule is mediated by an active transport process.

---

*In normal individuals in Western countries, a net of 40-60 mEq of noncarbonic acid is excreted daily.

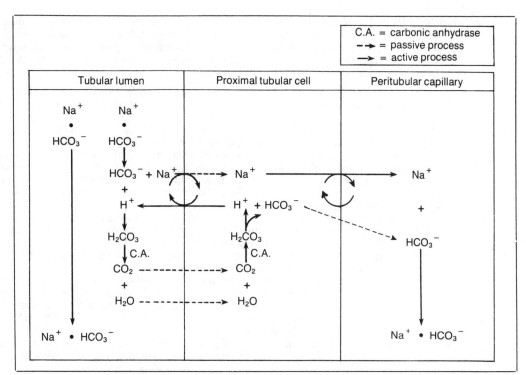

Figure 5-2. Mechanism of obligatory HCO_3^- reabsorption in the proximal tubule. The Na^+ is reabsorbed from the tubular lumen by a passive process and from the cell into the blood by an active process. Similarly, K^+ is reabsorbed as $KHCO_3$ or KCl. In contrast to Na^+, the transluminal reabsorption of K^+ is active, while transcellular K^+ transport is passive.

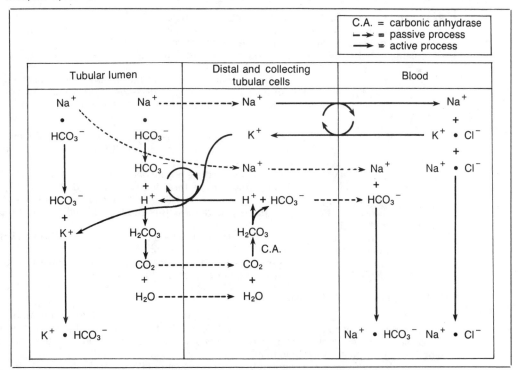

Figure 5-3. Mechanism of controlled HCO_3^- reabsorption in the distal and collecting tubules. Unlike the proximal tubule, the distal and collecting tubules lack luminal carbonic anhydrase. The exchange of K^+ for Na^+ leads to the reabsorption of 1 mol of $NaCl$ for every 1 mol of $KHCO_3$ excreted. Note that **intracellular** carbonic anhydrase is found throughout the nephron, including the distal and collecting tubules.

C. REABSORPTION OF HCO$_3^-$ (see Figs. 5-2 and 5-3). It is important to recognize that the filtered HCO$_3^-$ is not directly reabsorbed by the renal tubules; instead, it must first be converted to H$_2$CO$_3$, with the H$^+$ secreted into the tubular lumen in exchange for the Na$^+$ that is transported from the lumen. The HCO$_3^-$ is reabsorbed **indirectly** by conversion to CO$_2$ within the tubular lumen, and most of the H$^+$ is reabsorbed following its conversion to water within the lumen.

    1. Proximal Reabsorption. Approximately 90 percent of the filtered HCO$_3^-$ is reabsorbed into the proximal tubules as a result of the Na$^+$–H$^+$ exchange and represents more than 4000 mEq/day.

        a. The presence of carbonic anhydrase on the microvilli of the luminal border (brush border) of the proximal tubules catalyzes the rapid dehydration of H$_2$CO$_3$ to form CO$_2$ and water.

        b. CO$_2$ diffuses back into the proximal tubular cell where it is rehydrated by intracellular carbonic anhydrase into H$_2$CO$_3$. The HCO$_3^-$ formed by the dissociation of H$_2$CO$_3$ is passively reabsorbed into the peritubular blood along with equimolar amounts of Na$^+$, which is actively transported into the peritubular blood. The H$^+$ formed by the dissociation of H$_2$CO$_3$ serves as a source for another H$^+$ to be secreted.

        c. The entry of Na$^+$ into the tubular cell is balanced electrochemically by two mechanisms.

            (1) Cl$^-$, which is the only quantitatively important reabsorbable anion in the filtrate, is transported from lumen to cell by passive diffusion.

            (2) In the presence of carbonic anhydrase and with OH$^-$ as its substrate, HCO$_3^-$ is regenerated from the cellular CO$_2$ and water.

            (3) The net result of either the Cl$^-$ diffusion into the cell or the regeneration of HCO$_3^-$ is a net reabsorption of sodium chloride (NaCl) or sodium bicarbonate (NaHCO$_3$), respectively, into the peritubular capillaries. It is important to note that the Na$^+$ delivered to the peritubular capillaries comes from the filtrate in the lumen, but the HCO$_3^-$ transported into the capillaries is synthesized within the proximal tubular cell.

        d. Any secreted H$^+$ that combines with HCO$_3^-$ in the lumen to bring about HCO$_3^-$ reabsorption DOES NOT contribute to the urinary excretion of acid. Thus, the majority of the secreted H$^+$ is used to accomplish HCO$_3^-$ reabsorption.

        e. Referring to this process as HCO$_3^-$ reabsorption, then, seems inaccurate, since the HCO$_3^-$ that appears in the peritubular capillaries is not the same HCO$_3^-$ that was filtered. Moreover, most of the H$^+$ that was secreted into the lumen is not excreted in the urine but is incorporated into water and reabsorbed.

    2. Reabsorption from the Distal and Collecting Tubules.

        a. The remaining 10–15 percent of the filtered HCO$_3^-$ is reabsorbed by the distal and collecting tubules via a mechanism that involves the exchange of Na$^+$ for K$^+$ or H$^+$. As in the proximal tubules, every H$^+$ secreted into the distal tubular lumen leaves a HCO$_3^-$ within the tubular cell.

        b. Except for the Na$^+$–H$^+$ or Na$^+$–K$^+$ exchange pump at the luminal border of the distal cellular membranes and the absence of carbonic anhydrase on the distal luminal border, the reabsorption pathways for HCO$_3^-$ are analogous to those described for the proximal tubule.*

    3. Addition of New HCO$_3^-$. In addition to the renal conservation of HCO$_3^-$, the kidneys add newly synthesized HCO$_3^-$ to the plasma so that the quantity of HCO$_3^-$ in the renal vein exceeds the amount that entered the kidneys.

        a. The addition of new HCO$_3^-$ to the plasma does not involve the HCO$_3^-$ reabsorbed into the tubule but the HCO$_3^-$ generated within the tubular cell via the hydration of CO$_2$ and dissociation of H$_2$CO$_3$. This process is similar to the scheme for the reabsorption of filtered HCO$_3^-$; however, **the HCO$_3^-$ that is generated within the tubular cell does not represent filtered HCO$_3^-$.**

        b. The renal contribution of new HCO$_3^-$ is accompanied by the excretion of an equivalent amount of acid in the urine in the form of titratable acid, NH$_4^+$, or both.

        c. The amount of new HCO$_3^-$ formed per day (approximately 50 mEq) is much less than the quantity of filtered HCO$_3^-$ reabsorbed per day (more than 4300 mEq).

D. EXCRETION OF TITRATABLE ACID (Fig. 5-4)

    1. Titratable acid denotes that portion of H$^+$ bound to filtered buffers and equals the amount of alkali, in the form of sodium hydroxide (NaOH), required to titrate urine back to the normal pH of blood (i.e., the number of milliequivalents of H$^+$ added to the tubular fluid which combined with phosphate or organic buffers).

        a. Titratable acid is largely attributed to the conversion of HPO$_4^{2-}$ to H$_2$PO$_4^-$.

        b. Titratable acid is a poor measure of the total amount of H$^+$ secreted by the tubules, since

---

*Intracellular carbonic anhydrase is found throughout the renal tubule, including the distal convoluted tubule and collecting duct. In the proximal convoluted tubule, carbonic anhydrase also is located in the luminal cell membranes.

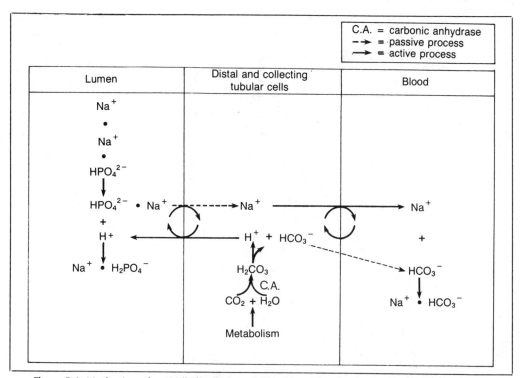

Figure 5-4. Mechanism of controlled excretion of titratable acid in the distal and collecting tubules.

most of the acid produced by this secretion is H_2CO_3, which disappears from the urine as CO_2; however, the H^+ that is trapped by anions of noncarbonic acids remains in the final urine in the form of titratable acid.

    c. Titratable acid is a measure of the content of weak acids. It is **not** a measure of the H^+ that combines with ammonia to yield NH_4^+, because the pK′ of the NH_3/NH_4^+ buffer system is high (9.2).

  2. The exchange of H^+ for Na^+ converts dibasic sodium phosphate (Na_2HPO_4) in the glomerular filtrate into dihydrogen phosphate (NaH_2PO_4), which is excreted in the urine as titratable acid. The H^+ secreted into the tubules, therefore, can react with filtered HPO_4^{2-} rather than the filtered HCO_3^-.

    a. Besides HCO_3^-, HPO_4^{2-} represents a major filtered conjugate base.* Furthermore, the $HPO_4^{2-}/H_2PO_4^-$ buffer pair provides an excellent buffer system because it has a pK′ of 6.8.

    b. The blood $HPO_4^{2-}/H_2PO_4^-$ molar ratio is 4:1. The H^+ secretory system converts much of the HPO_4^{2-} to $H_2PO_4^-$ in the tubular fluid, resulting in a urinary $HPO_4^{2-}/H_2PO_4^-$ molar ratio of 1:4. The acidification of the phosphate buffer system occurs significantly in the proximal tubule, and much of the H^+ generated by the formation of H_2SO_4 and H_3PO_4 during protein and phospholipid metabolism is excreted in this way.

    c. Each H^+ secreted into the lumen that reacts with HCO_3^- is not excreted, whereas each H^+ secreted into the lumen that reacts with a nonbicarbonate buffer remains in the tubular fluid and is excreted.

E. EXCRETION OF AMMONIA (Fig. 5-5)

  1. Unlike phosphate, ammonia enters the tubular lumen not by filtration but by tubular synthesis and secretion, which normally are confined to the distal and collecting tubules.

  2. Ammonia is synthesized mainly by the deamidization and deamination of glutamine in the presence of glutaminase.

  3. The NH_3/NH_4^+ system has a very high pK′ (about 9.2), which means that at the usual urine pH, virtually all the nonpolar ammonia that enters the tubular lumen immediately combines

---

*Since about 75 percent of the filtered HPO_4^{2-} is reabsorbed, only about 25 percent of the filtered HPO_4^{2-} is available for buffering.

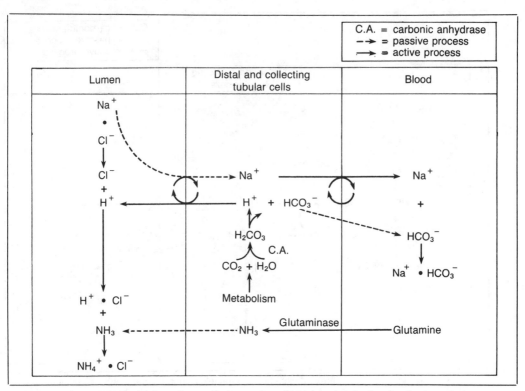

Figure 5-5. Mechanism of controlled excretion of NH_4^+ in the distal and collecting tubules.

with H^+ to form NH_4^+, which is nondiffusible because it is lipid insoluble. The renal excretion of NH_4^+ causes the addition of HCO_3^- to the plasma.

a. The important physiologic characteristic of the NH_3/NH_4^+ buffer pair is that, as H^+ is combined with intraluminal buffer (NH_3), H^+ is excreted in the urine as NH_4^+, a substance that does **not** cause the pH of urine to fall.

b. Under most conditions, the excretion of NH_4^+ is quantitatively more important to the acid-base balance of the body than the excretion of titratable acid. The normal kidney excretes almost twice as much acid combined with ammonia (30–50 mEq/day) as titratable acid (10–30 mEq/day).

    (1) In chronic severe metabolic acidosis, ammonia serves as the major urinary buffer, and NH_4^+ excretion can increase from a normal value of 30 mEq/day to 500 mEq/day. However, in chronic acidosis, the excretion of $H_2PO_4^-$ may increase by only 20–40 mEq/day.

    (2) In diabetic ketoacidosis, the excretion of titratable acid may reach 75–250 mEq/day, while the excretion of acid combined with ammonia may reach amounts of 300–500 mEq/day. However, the ratio of NH_4^+ to titratable acid remains within the normal range of 1 to 2.5.

4. The sum of titratable acid and urinary NH_4^+ excretion represents the net gain of HCO_3^- for the body fluids.

a. The net amount of fixed acid excreted (in mEq/day) equals the sum of titratable acid and NH_4^+ minus the urinary $[HCO_3^-]$.

b. Normally, the urine is free of HCO_3^- because all the HCO_3^- remaining in the distal and collecting tubules has combined with secreted H^+ and been reabsorbed.

F. $PaCO_2$ AND RENAL INTRACELLULAR pH (Fig. 5-6)

1. The single most important determinant of the rate of tubular acid secretion is $PaCO_2$.

2. An increase in $PaCO_2$ causes, by CO_2 diffusion, an equivalent increase in the PCO_2 within the tubular cells. This causes an increased rate of H_2CO_3 formation and, in turn, an elevated $[H^+]$ following the dissociation of H_2CO_3. (Presumably, it is this change that directly stimulates the rate of H^+ secretion; thus, the ultimate stimulus for H^+ secretion is not $PaCO_2$, per se, but the decreased intracellular pH).

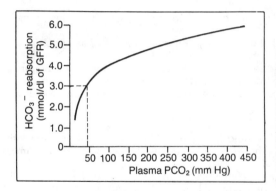

Figure 5-6. The relationship between arterial CO_2 tension (PaCO_2) and reabsorption of HCO_3^-. The most important determinant of tubular acid secretion is PaCO_2. The curve shows that the rate of H^+ secretion (as manifested by the HCO_3^- reabsorption rate) is directly related to PaCO_2. The renal response to increasing PaCO_2 leads to a greater increase in plasma $[HCO_3^-]$ than that produced by buffering of H_2CO_3 by blood and body buffers alone. (Reprinted with permission from Koushanpour E: Renal regulation of acid-base balance. In *Renal Physiology: Principles and Functions.* Philadelphia, WB Saunders, 1976, p 289.)

3. The renal mechanism that produces an increase in $[HCO_3^-]$ in response to increasing PaCO_2 has several important characteristics.
   a. Maximal changes in HCO_3^- reabsorption produced by increases in PaCO_2 require 48–72 hours.
   b. The major increase in $[HCO_3^-]$ involves the plasma and ECF, not the ICF, because HCO_3^-, unlike CO_2, crosses cellular membranes slowly.
   c. An elevation in the PaCO_2 enhances H^+ excretion (as NH_4^+), resulting in an increase in renal HCO_3^- reabsorption and an increase in plasma $[HCO_3^-]$. Plasma $[HCO_3^-]$ increases an average of 1 mmol/L (acute respiratory acidosis) or 3.5 mmol/L (chronic respiratory acidosis) for every 10 mm Hg increment in the PaCO_2.
      (1) Thus, of the total increase in $[HCO_3^-]$ evoked by increases in PaCO_2, approximately 65 percent occurs as a result of renal tubular activity.
      (2) Only about 35 percent of the total increase in plasma $[HCO_3^-]$ is related to the buffering of H_2CO_3 by the nonbicarbonate buffers of the blood and body fluids.

VII. ACID-BASE ABNORMALITIES

A. DEFINITIONS (Tables 5-8 and 5-9)

1. Acidosis and Alkalosis.
   a. Acidosis or **acidemia** is an abnormal clinical condition or process caused by the accumulation of acid (or the loss of base) sufficient to decrease pH below 7.36 or to increase the $[H^+]$ above 43.6 nEq/L of blood in the absence of compensatory (secondary) changes.
   b. Alkalosis or **alkalemia** is an abnormal condition or process caused by the accumulation of base (or the loss of acid) sufficient to raise pH above 7.44 or to decrease the $[H^+]$ below 36.3 nEq/L of blood in the absence of compensatory changes.

2. Respiratory and Metabolic.
   a. The adjective **respiratory** denotes that the primary abnormality involves impairment in alveolar ventilation, which results in an abnormally high or low [total CO_2] of the ECF.*
      (1) Respiratory acidosis refers to a condition of abnormally high PaCO_2, which is termed **hypercapnia** (hypercarbia).

Table 5-8. Characteristics of the Uncompensated Acid-Base Abnormalities

Acid-Base Abnormality	Primary Disturbance	Effect on:			Compensatory Response
		$[HCO_3^-]/S \times PCO_2$	$[H^+]$	pH	
Acidosis					
Respiratory	↑ PCO_2	< 20	↑	↓	↑ $[HCO_3^-]$
Metabolic	↓ $[HCO_3^-]$	< 20	↑	↓	↓ PCO_2
Alkalosis					
Respiratory	↓ PCO_2	> 20	↓	↑	↓ $[HCO_3^-]$
Metabolic	↑ $[HCO_3^-]$	> 20	↓	↑	↑ PCO_2

Note.—These conditions are simple disturbances (i.e., they represent the effects of one primary etiologic factor) and, thus, are termed uncompensated or **pure** acid-base abnormalites. The compensatory response always occurs in the same direction as the primary disturbance.

*[Total CO_2] refers to the dissolved CO_2 ($S \times PCO_2$) plus the $[HCO_3^-]$.

    (2) Respiratory alkalosis refers to a condition of abnormally low $PaCO_2$, which is called **hypocapnia** (hypocarbia).

  b. The adjective **metabolic** denotes that the primary abnormality involves an abnormal gain or loss of noncarbonic acid by the ECF, which affects $[HCO_3^-]$.

    (1) Metabolic acidosis refers to a disturbance that leads to the accumulation of noncarbonic acid in the ECF or to the loss of HCO_3^- from the ECF.

    (2) Metabolic alkalosis refers to an imbalance characterized by a loss of noncarbonic acid or a gain of HCO_3^- by the ECF.

3. Primary and Secondary Factors.

  a. The factor in the $[HCO_3^-]/S \times PCO_2$ ratio (i.e., $[HCO_3^-]$ or PCO_2) that undergoes the greater degree of displacement (i.e., the larger proportional change) indicates the **primary abnormality** (see Tables 5-8 and 5-9).

    (1) If the $[HCO_3^-]/S \times PCO_2$ ratio becomes less than 20/1, by either decreasing the numerator or increasing the denominator, pH falls ($\uparrow[H^+]$) and acidosis occurs.

    (2) If the $[HCO_3^-]/S \times PCO_2$ ratio becomes more than 20/1, by either increasing the numerator or decreasing the denominator, pH rises ($\downarrow[H^+]$) and alkalosis occurs.

  b. A **secondary response** in the alternate variable, which occurs in the same direction as the primary abnormality, **counteracts** the effect of the primary abnormality on the $[HCO_3^-]/S \times PCO_2$ ratio.

4. Simple and Mixed Acid-Base Disturbances.

  a. Acid-base imbalances that are caused by one primary factor are described as **simple** disturbances.

  b. Acid-base imbalances that are caused by more than the primary factor are described as **mixed** disturbances.

    (1) Mixed-type disturbances are not uncommon. Sometimes a primary abnormality of one type is superimposed on a primary abnormality of another type.

      (a) A patient with respiratory acidosis from pulmonary emphysema may develop metabolic acidosis from uncontrolled diabetes or metabolic alkalosis from large doses of corticosteroids used in the treatment of an attack of status asthmaticus.

      (b) A patient with a metabolic acid-base disturbance, in turn, may develop a respiratory acid-base abnormality.

    (2) The changes in the variables may be difficult to interpret in mixed acid-base disturbances, because manifestations of one primary abnormality may be either cancelled out or augmented by those of the other primary abnormality. Therefore, a normal pH may not necessarily mean a normal compensation in mixed acid-base imbalances.

5. Plasma $[HCO_3^-]$ and $[Total\ CO_2]$. For an adequate analysis of any acid-base disorder, it is necessary to measure pH, PCO_2, and the total CO_2 content of blood.

  a. An acid-base disorder cannot be diagnosed with certainly from the plasma $[HCO_3^-]$ alone. Although a reduction in the plasma $[HCO_3^-]$ may be due to metabolic acidosis, it also can indicate a renal compensation for respiratory alkalosis. Similarly, an elevated $[HCO_3^-]$ can result from a metabolic alkalosis or the secondary response to respiratory acidosis. Since

Table 5-9. Primary and Secondary* Changes in the CO_2/HCO_3^- System

	Acidosis		Alkalosis	
Respiratory	$\dfrac{[HCO_3^-]\ \uparrow}{S \times PCO_2\ \Uparrow}$	< 20:1	$\dfrac{[HCO_3^-]\ \downarrow}{S \times PCO_2\ \Downarrow}$	> 20:1
	$[total\ CO_2]\ \uparrow$		$[total\ CO_2]^\dagger\ \downarrow$	
Metabolic	$\dfrac{[HCO_3^-]\ \Downarrow}{S \times PCO_2\ \downarrow}$	< 20:1	$\dfrac{[HCO_3^-]\ \Uparrow}{S \times PCO_2\ \uparrow}$	> 20:1
	$[total\ CO_2]\ \Downarrow$		$[total\ CO_2]\ \Uparrow$	

Note.—*Dashed arrows* denote direction of **compensatory** responses. *Thin solid arrows* show changes in $[total\ CO_2]$ during respiratory imbalances. *Wide arrows* depict direction of change of **primary** acid-base imbalance and direction of change in $[total\ CO_2]$. (Adapted from Christensen HN: *Diagnostic Biochemistry*. New York, Oxford University Press, 1959, p 106.)

* A compensatory (secondary) response in the **alternate** variable occurs in the same direction and counteracts the effect of the primary disturbance by returning the $[HCO_3^-]/S \times PCO_2$ ratio toward normal (20:1).

† Total CO_2 does not always increase in alkalosis or decrease in acidosis. Dissolved CO_2 contributes little to the $[total\ CO_2]$. $[Total\ CO_2]$ provides little information about pulmonary function.

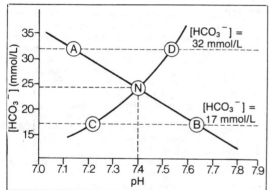

Figure 5-7. A pH-[HCO_3^-] diagram illustrating that acidosis (*points A and C*) and alkalosis (*points B and D*) cannot be differentiated solely on the basis of [HCO_3^-] or [total CO_2]. The [HCO_3^-] pertains to plasma. A = respiratory acidosis; B = respiratory alkalosis; C = metabolic acidosis; D = metabolic alkalosis; and N = normal point. The slope of *line ANB* is a measure of the nonbicarbonate buffer capacity of whole blood. (After Davenport HW: *The ABC of Acid-Base Chemistry*, 5th ed. Chicago, University of Chicago Press, 1969, p 65.)

the aim of therapy is to bring about acid-base balance and not to normalize [HCO_3^-], it is necessary to measure the pH (i.e., [H^+]) when acid-base imbalance is suspected.

b. [Total CO_2] does not always increase in alkalosis or decrease in acidosis (Fig. 5-7; see also Tables 5-6, 5-7, and 5-9). About 90 percent of the [total CO_2] of plasma is contributed by [HCO_3^-], and 5 percent is contributed by dissolved CO_2 and H_2CO_3; therefore, dissolved CO_2 contributes little to the [total CO_2]. The [total CO_2] gives little information about the functional status of the lungs.

c. The physician accepts [total CO_2] and [HCO_3^-] as essentially interchangeable terms. The [total CO_2] normally exceeds the [HCO_3^-] by 1.2 mmol. Most laboratories measure the [total CO_2] and not the [HCO_3^-].

B. CLINICAL EXPRESSIONS FOR EVALUATION OF ACID-BASE STATUS

1. Base Excess (BE) [see Table 5-6; Fig. 5-8 and Table 5-10]

a. BE is the **change** in the concentration of buffer base [BB] from its normal value; that is, BE equals the **observed** [BB] minus the **normal** [BB] in mEq/L of whole blood. The range of normality for BE is − 2 to + 2 mEq/L in arterial whole blood.*

(1) The [BB] in normal whole blood is 48 mEq/L, which is arbitrarily set to 0, and deviations (i.e., ±BE) are measured from this reference point.

(2) The normal [BB] of 48 mEq/L equals the sum of **all** the conjugate bases in 1 L of arterial whole blood. These bases and their concentrations are:

 (a) [HCO_3^-] = 24 mEq/L

 (b) [$Protein^-$] = 15 mEq/L

 (c) [Hb^-/HbO_2^-] = 9 mEq/L

b. BE refers principally to the [HCO_3^-] but also to other bases in the blood (mainly plasma protein and hemoglobin). However, the [HCO_3^-] or [total CO_2] is used in acid-base problems as the indicator of BE.

(1) Because [HCO_3^-] and BE are influenced only by metabolic processes, there are only two acid-base conditions associated with abnormalities in [HCO_3^-] and BE.

 (a) Metabolic Acidosis. When a metabolic process leads to the accumulation of noncarbonic acid in the body or the loss of HCO_3^-, the [HCO_3^-] falls and the BE value becomes negative. Metabolic acidosis is associated with a BE below − 5 mEq/L.

 (b) Metabolic Alkalosis. When a metabolic process causes a loss of acid or an accumulation of HCO_3^-, the [HCO_3^-] rises above normal and the BE value becomes positive. Metabolic alkalosis is associated with a BE above + 5 mEq/L.

Table 5-10. Base Excess and Metabolic Acid-Base Abnormalities

[HCO_3^-]	Base Excess	Metabolic Abnormality	Characteristics
↑	+	Metabolic alkalosis	Noncarbonic acid is lost; HCO_3^- is gained
↓	−	Metabolic acidosis	Noncarbonic acid is gained; HCO_3^- is lost

Note.—Adapted from Broughton JO: *Understanding Blood Gases*. Madison, Wisconsin, Ohio Medical Products, form no. 456, 1979, p 8. (Reprinted with permission of The BOC Group, c1669, The BOC Group, Inc.)

*Because BE is a negative or positive value, the term "base deficit" is avoided.

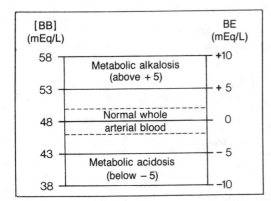

Figure 5-8. Diagram illustrating the relationship between base excess (*BE*) and the acid-base status of normal whole arterial blood. BE is the base concentration (in mEq/L), as measured by the titration with strong acid to pH 7.4, at PCO_2 of 40 mm Hg, and at 37° C. For negative values of BE, the titration must be carried out with strong base. BE measures the **change** in the concentration of the buffer base (*[BB]*) from its normal value, which is 48 mEq/L in normal whole arterial blood. (BB is the sum of the buffer anions of blood or plasma.) When observed [BB] is less than normal [BB], BE is a negative value; when observed [BB] is greater than normal [BB], BE is a positive value; and when observed [BB] equals normal [BB], BE equals zero. (After Winters RW, et al: *Acid-Base Physiology in Medicine.* Westlake, Ohio, London Co, 1967, p 45.)

(2) The BE value serves only as a rough guide for therapy for alkalosis and acidosis.

2. Anion Gap (Fig. 5-9).

  a. Definition.

    (1) Routine serum electrolyte determinations measure virtually all cations but only a fraction of the anions. This **apparent** disparity between the total cation concentration and the total anion concentration is termed the anion gap.

    (2) The anion gap, which has a normal value of 12 ± 4 mEq/L, reflects the concentrations of those anions actually present but routinely undetermined (i.e., other than $[Cl^-]$ and $[HCO_3^-]$), such as:

      (a) Polyanionic plasma proteins

      (b) Phosphates

      (c) Sulfate

      (d) Ions of organic acids (e.g., lactic, β-hydroxybutyric, and acetoacetic acids)

    (3) When the anion gap is increased, unmeasured anions fill this gap. (Most of these anions usually are the products of metabolic processes that generate H^+.)

  b. Determination of Anion Gap.

    (1) The anion gap (AG) equals 16 mEq/L of anions other than HCO_3^- and Cl^- when determined by the following equation:

$$\begin{aligned} AG &= ([Na^+] + [K^+]) - ([HCO_3^-] + [Cl^-]) \\ &= (142\ \text{mEq/L} + 4\ \text{mEq/L}) - (25\ \text{mEq/L} + 105\ \text{mEq/L}) \\ &= 146\ \text{mEq/L} - 130\ \text{mEq/L} \\ &= 16\ \text{mEq/L}. \end{aligned}$$

    (2) Often the anion gap is calculated with Na^+ as the major cation, as:

$$\begin{aligned} AG &= [Na^+] - ([HCO_3^-] + [Cl^-]) \\ &= 142\ \text{mEq/L} - (25\ \text{mEq/L} + 105\ \text{mEq/L}) \\ &= 142\ \text{mEq/L} - 130\ \text{mEq/L} \\ &= 12\ \text{mEq/L}. \end{aligned}$$

    (3) It is important to recognize that the plasma proteins contribute a significant amount of negative charges and, thus, anionic equivalence (16 mEq/L). The plasma proteins account, almost stoichiometrically, for the difference between the $[Na^+]$ and the sum of $[HCO_3^-]$ and $[Cl^-]$.

  c. Acidification with Acids Other than HCl.

    (1) Organic acids increase the anion gap. In this case, the lost (neutralized) HCO_3^- is *not* replaced by the routinely unmeasured anions of noncarbonic acids such as lactic acid and the ketoacids. Thus, the anion gap is increased by all metabolic acidoses *except* the hyperchloremic acidoses. With the accumulation of acid, there is rapid extracellular buffering by HCO_3^-. If the acid is HCl, then:

$$HCl + NaHCO_3 \rightleftharpoons NaCl + H_2CO_3.$$

    The **net** effect is the milliequivalent-for-milliequivalent replacement of extracellular HCO_3^- by Cl^-. Since the **sum** of $[Cl^-]$ and $[HCO_3^-]$ remains constant, the anion gap is unchanged in hyperchloremic acidosis.

    (2) It is apparent that the **decrease** in plasma $[HCO_3^-]$ equals the **increase** in the anion gap (see Fig. 5-9). The presence of an increased anion gap usually indicates an excess of H^+ derived from noncarbonic acid. Whatever the size of the anion gap, **it is the retention of H^+—not the particular anion—that is responsible for the acidosis.**

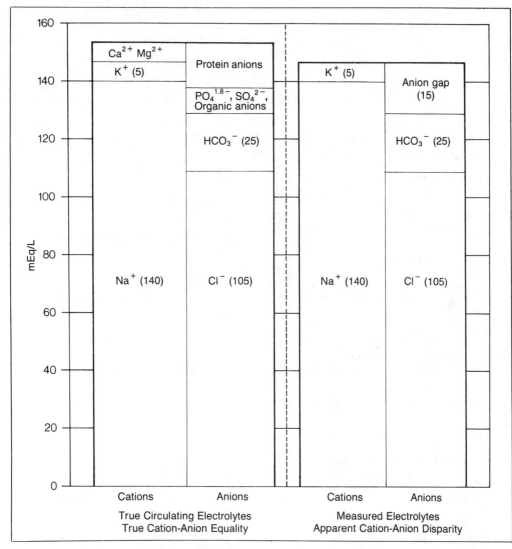

Figure 5-9. Circulating cations always counterbalance anions, thereby maintaining electroneutrality (*left*). Routine electrolyte assays only measure a portion of circulating anions, and an **apparent** cation-anion disparity exists (*right*). This difference, in mEq/L, is termed the anion gap. *Parentheses* indicate concentrations in mEq/L. [Reprinted with permission from Narins RG, et al: Lactic acidosis and elevated anion gap (I). *Hosp Pract* 15:125–136, 1980. (Original drawing was by Albert Miller.)]

    (a) An increased anion gap due to excess H^+ derived from noncarbonic acid occurs in **metabolic acidosis**; this also may result from a compensatory increase in lactic acid production in response to respiratory alkalosis via activation of the phosphofructo-kinase step in glycolysis.

    (b) In **respiratory acidosis** the anion gap is not increased, because excess H^+ is de-rived from the H_2CO_3 pool, not the noncarbonic acid pool.

  d. Conditions that increase the anion gap include diabetic and alcoholic ketoacidosis, intoxi-cant and lactic acidoses, and renal failure (Table 5-11).

  e. Conditions that cause metabolic acidosis without an increase in the anion gap are associ-ated with a high serum [Cl^-] and include diarrhea; pancreatic drainage; ureterosigmoidos-tomy; ileal loop conduit; treatment with acetazolamide (Diamox), ammonium chloride, or arginine-HCl; renal tubular acidosis; and, rarely, intravenous hyperalimentation (see Table 5-11). (The metabolic conditions that decrease [HCO_3^-] and BE with and without an in-crease in anion gap are summarized in Table 5-11 to facilitate differential diagnosis of met-abolic acidosis.)

  f. **Osmolar gap** refers to the disparity between the **measured** serum (plasma) osmolality and the **calculated** serum osmolality. Osmolar gap measurement provides a reasonably good screening procedure for toxins. Other causes of metabolic acidosis do not affect the osmolar gap, since the metabolic acid simply replaces the HCO_3^- with another anion and HCO_3^- is lost as CO_2.

(1) Serum osmolality is estimated by dividing blood urea nitrogen (BUN) and glucose concentrations (in mg/L) by their molecular weights,* thereby converting them to milliosmoles (mOsm)/L. Doubling the serum $[Na^+]$ (in mEq/L) provides an estimate of the serum **ion** concentration. With a BUN of 14 mg/dl, plasma glucose of 90 mg/dl,

Table 5-11. Causes of Metabolic Acidosis (↓ $[HCO_3^-]$ and ↓ Base Excess)

Without an increase in anion gap*

Gastrointestinal loss of HCO_3^-
Diarrhea
Draining of gastrointestinal fistulas (e.g., small bowel and, particularly, pancreatic fistulas)
Ureterosigmoidostomy (ureteroenterostomy)
Obstructed ileal loop (ileal loop conduit)
Anion exchange resins (e.g., cholestyramine, which can exchange Cl^- for HCO_3^-)

Renal loss of HCO_3^-
Renal tubular acidosis (proximal or distal)
Treatment with carbonic anhydrase inhibitors (e.g., acetazolamide)
Treatment with NH_4Cl and other chlorides (e.g., $CaCl_2$, arginine HCl, and excess NaCl)
Renal insufficiency

Renal dysfunction
Pyelonephritis and obstructive uropathy
Low plasma renin and aldosterone
Treatment with aldosterone antagonists (e.g., spironolactone)

Endocrine dysfunction
Adrenal insufficiency (e.g., hypoaldosteronism)

Miscellaneous
Intravenous hyperalimentation (rarely a cause)
Dilutional (expansion) acidosis
Addition of HCl and its congeners

With an increase in anion gap

Increased acid production
Diabetic ketoacidosis
Alcoholic ketoacidosis
Lactic acidosis
Inborn errors of metabolism
Azotemia

Ingestion of intoxicants
Paraldehyde (rarely a cause)
Ethylene glycol
Methanol
Salicylate overdose
Isopropanol
Formic acid
Boric acid
Oxalic acid

Failure of acid excretion
Acute renal failure
Chronic renal failure

Note.—After Broughton JO: *Understanding Blood Gases.* Madison, Wisconsin, Ohio Medical Products, form no. 456, 1979, p 9.
* Conditions that cause a metabolic acidosis without an increase in unmeasured anions are associated with a high serum $[Cl^-]$ or **hyperchloremia**.

*The molecular weight of urea (60) is not used because it is blood urea **nitrogen** that is measured, and urea contains two nitrogen atoms ($2 \times 14 = 28$).

and a plasma $[Na^+]$ of 142 mEq/L, serum (plasma) osmolality (P_{osm}) is estimated as follows:

$$P_{osm} = 2\ [Na^+] + [BUN\ (mg/L)/28] + [glucose\ (mg/L)/180]$$
$$= 2\ (142) + 140/28 + 900/180$$
$$= 286 + 5 + 5 = 296\ mOsm/kg.$$

(2) Circulating intoxicants increase measured serum osmolality without altering serum $[Na^+]$. Therefore, the measured serum osmolality exceeds the calculated serum osmolality, and the difference closely reflects the osmolar concentration of the circulating toxins.

C. GRAPHIC EVALUATION OF ACID-BASE STATUS: THE pH-[HCO₃⁻] DIAGRAM (Figs. 5-10 and 5-11)

1. **Value of Diagram.** When acid-base data obtained from blood are plotted on the pH-$[HCO_3^-]$ diagram, the physician has a tool that can be used to:
   a. Diagnose the primary cause of an acid-base abnormality
   b. Determine the degree of compensatory response of the kidneys and of the lungs by monitoring the $[HCO_3^-]$ and PCO_2, respectively
   c. Estimate the concentration of noncarbonic and carbonic acids in the ECF
   d. Aid in the choice of therapy to correct an acid-base imbalance
   e. Measure the nonbicarbonate buffer power of whole blood by the slope of the buffer line
   f. Indicate not only the total buffer activity but also the distribution of the buffering activity between the bicarbonate and nonbicarbonate buffer systems

2. **Construction of Diagram: Effects of H⁺ and CO₂ on pH** (see Figs. 5-10 and 5-11). The basis for plotting pH and $[HCO_3^-]$ on cartesian coordinates is best understood from a consideration of the CO_2/HCO_3^- buffer system:

$$CO_2 + H_2O \rightleftharpoons H_2CO_3 \rightleftharpoons H^+ + HCO_3^-. \tag{5}$$

   a. Effect of Fixed Acid or Base on pH.
      (1) Addition of H⁺ at a Constant PCO₂.
         (a) When H^+, in the form of fixed (noncarbonic) acid, are added to the system depicted in equation (5), most of them combine with HCO_3^- to form H_2CO_3, which dehydrates to form CO_2 and water. Thus, the series of reactions in equation (5) is driven to the **left**.
         (b) The decrease in $[HCO_3^-]$ closely approximates the amount of H^+ added **only if the solution has no other nonbicarbonate buffer substances**. Otherwise, the amount of acid added would be greater than the decrease in $[HCO_3^-]$.
      (2) Addition of Base at a Constant PCO₂.
         (a) When a base is added to the system depicted in equation (5), some of the base combines with H^+, causing more H_2CO_3 to dissociate into H^+ and HCO_3^-. Here the series of reactions in equation (5) proceeds to the **right**.
         (b) The increase in $[HCO_3^-]$ determines the amount of base added.
      (3) Principles (see Fig. 5-10).
         (a) When fixed acid or base is added under the conditions described, there is an inverse relationship between $[H^+]$ and $[HCO_3^-]$, and the primary acid-base distur-

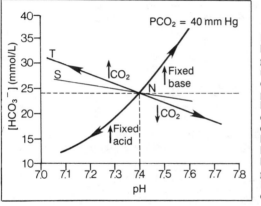

Figure 5-10. The pH-$[HCO_3^-]$ diagram showing the buffer curves of separated plasma (*S*) and true oxygenated plasma (*T*). *Point N* denotes the intercept for a normal plasma $[HCO_3^-]$ of 24 mmol/L and a normal pH of 7.4. The slope of the normal in vitro buffer line of true plasma is a function of the hemoglobin content of blood. *Lines S and T* are called **nonbicarbonate buffer curves**, and the steepness of the slopes of these lines is a quantitative assessment of the amount of nonbicarbonate buffer present in the system. These same two lines represent CO_2-titration curves. The steeper the buffer line, the smaller the pH change resulting from a given increase or decrease in PCO_2. (Adapted from Davenport HW: *The ABC of Acid-Base Chemistry*, 5th ed. Chicago, University of Chicago Press, 1969, p 48.)

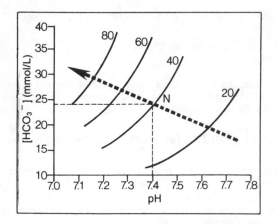

Figure 5-11. The pH-[HCO_3^-] diagram with arterial CO_2 (PaCO_2) isobars for 20, 40, 60, and 80 mm Hg. Each PaCO_2 isobar is the titration curve of a HCO_3^--H_2CO_3 solution with PaCO_2 held constant. *Point N* denotes the intercept for a normal plasma [HCO_3^-] of 24 mmol/L and a normal pH of 7.4. The *dashed line* represents the buffer line. (After Davenport HW: *The ABC of Acid-Base Chemistry*, 5th ed. Chicago, University of Chicago Press, 1969, p 45.)

bance is **metabolic**. Thus, when fixed acid or base is added, the [H^+] and [HCO_3^-] change in **opposite** directions.

   (b) Or, if the direction of change in [HCO_3^-] is the same as that for pH, the primary acid-base abnormality is metabolic.

b. Effect of CO_2 on pH.

   (1) Addition of CO_2 (see Fig. 5-11).

   (a) An increase in PaCO_2 titrates the blood (and ECF) in the acid direction. This is clear from equation (5), which shows that an increase in PCO_2 causes more CO_2 to hydrate to form H_2CO_3; the H_2CO_3 in turn dissociates into H^+ and HCO_3^-, moving the reactions to the **right**.

   (b) In the system shown in equation (5), every molecule of CO_2 that hydrates and dissociates forms one H^+ and one HCO_3^-; therefore, the changes in [H^+] and [HCO_3^-] are exactly equal.

   (c) The final [H^+] depends not only on the change in PCO_2 but on the buffers in the system. For every H_2CO_3 molecule that has its H^+ taken up by a buffer ion, one HCO_3^- appears in the solution. For example, the reaction with hemoglobin is:

$$H_2CO_3 + Hb^- \rightleftharpoons HHb + HCO_3^-.$$

   (d) If CO_2 is added to **whole blood** (which contains many buffer systems), the H^+, rather than remaining in solution, are mostly bound to protein buffers (Pr$^-$) as:

$$CO_2 + H_2O \rightleftharpoons H_2CO_3 \rightleftharpoons HCO_3^- + H^+$$
$$+$$
$$Pr^-$$
$$\Updownarrow$$
$$H \bullet Pr.$$

Therefore, the CO_2/HCO_3^- system alone is not effective in buffering the changes in pH that are induced by the addition of CO_2.

   (2) Removal of CO_2.

   (a) When CO_2 is removed from the system shown in equation (5), fewer CO_2 molecules are available for combination with water to form HCO_3^-.

   (b) Thus, not only is CO_2 decreased but [H^+] and [HCO_3^-] also are decreased and the reactions shift to the **left**.

   (3) Principles. The concentration of acid added to a buffer solution by a change in PCO_2 is equal to the change in [HCO_3^-].*

   (a) If CO_2 is added or removed, the [H^+] and [HCO_3^-] change in the **same** direction, and the primary acid-base disturbance is **respiratory**.

   (b) Or, if the direction of change in [HCO_3^-] is opposite to that for pH, the primary acid-base disturbance is respiratory.

   (c) As PCO_2 is increased or decreased, the values for pH and [HCO_3^-] form coordinates that define a nearly straight line termed the **buffer line** or the CO_2-titration curve (see Fig. 5-11).

---

*In man, the serum [HCO_3^-] is 24×10^{-3} Eq/L (24×10^{-3} mol/L), whereas the serum [H^+] is 40×10^{-9} Eq/L (40×10^{-9} mol/L). Therefore, the [HCO_3^-] is 6×10^5 greater than [H^+], and any shift (left or right) in equilibrium has a much greater proportionate effect on [H^+] than on [HCO_3^-].

3. **Properties of PCO₂ Isobars** (see Figs. 5-10 and 5-11).
   a. At constant [HCO₃⁻], PCO₂ is proportional to [H⁺]. This property can be illustrated using the following form of the Henderson-Hasselbalch equation to solve for PCO₂:

$$[H^+] = 24 \ \frac{PCO_2}{[HCO_3^-]}$$

$$PCO_2 = \frac{[H^+][HCO_3^-]}{24}.$$

   (1) **Example 1.** At [H⁺] = 40 nmol/L and [HCO₃⁻] = 24 mmol/L:

$$PCO_2 = \frac{40(24)}{24} = 40 \ mm \ Hg.$$

   (2) **Example 2.** At [H⁺] = 80 nmol/L and [HCO₃⁻] = 24 mmol/L:

$$PCO_2 = \frac{80(24)}{24} = 80 \ mm \ Hg.$$

   b. At constant pH (i.e., along any vertical line in Figure 5-11), PCO₂ is proportional to [HCO₃⁻].*
   (1) **Example 1.** At [HCO₃⁻] = 24 mmol/L and pH = 7.4:

$$PCO_2 = 5/3 \ [HCO_3^-]$$
$$= 5/3 \ (24) = 40 \ mm \ Hg.$$

   (2) **Example 2.** At [HCO₃⁻] = 20 mmol/L and pH = 7.32:

$$PCO_2 = 2 \ [HCO_3^-]$$
$$= 2 \ (20) = 40 \ mm \ Hg.$$

   c. Any PCO₂ isobar is separated from any other PCO₂ isobar by a constant number of pH units.
   (1) **Example 1.** The 90 mm Hg PCO₂ isobar always is 0.3 pH unit to the left of the 40 mm Hg PCO₂ isobar, and the 20 mm Hg PCO₂ isobar always is 0.3 pH unit to the right of the 40 mm Hg PCO₂ isobar.
   (2) **Example 2.** The 60 mm Hg PCO₂ isobar always is 0.18 pH unit to the left of the 40 mm Hg PCO₂ isobar, and the 20 mm Hg PCO₂ isobar always is 0.18 pH unit to the right of the 40 mm Hg PCO₂ isobar.

4. **True Plasma and Separated Plasma** (see Fig. 5-10). The major contributor to the buffering activity of blood is in the erythrocytes. Because plasma alone is a poor buffer, a distinction is made between true plasma and separated plasma.
   a. True plasma is plasma equilibrated with CO₂ in the presence of erythrocytes prior to the anaerobic removal of plasma for analysis. When these data are plotted on the pH-[HCO₃⁻] diagram, they show the buffering power of plasma plus erythrocytes. For this reason, the blood buffer line exhibits an increased slope.
   b. Separated plasma is plasma removed from the erythrocytes prior to titration with CO₂ followed by analysis. Thus, the pH-[HCO₃⁻] diagram of separated plasma does not reflect the buffering activity of hemoglobin. The buffer line for separated plasma has a slope that is less steep than that of the true plasma buffer line, indicating that plasma alone is a less efficient buffer.
   c. The buffer capacities of whole blood and true plasma are nearly directly proportionate to the concentration of hemoglobin in blood. Because hemoglobin concentrations vary among patients, a direct measurement of the hemoglobin concentration of blood or of the buffer value of blood (true plasma) is advisable in acid-base disorders.

5. **Alternative Plots of the Henderson-Hasselbalch Equation** (Fig. 5-12).
   a. The plot of [HCO₃⁻] on the ordinate and of PCO₂ on the abscissa represents graphically the [HCO₃⁻]/S × PCO₂ ratio.

---

*The equation for the calculation of PCO₂ is shown above.

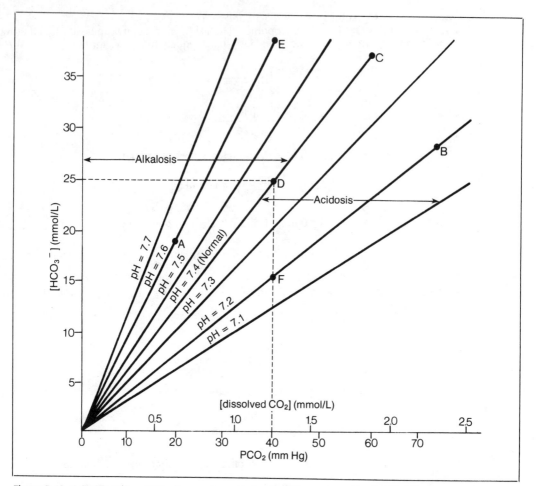

Figure 5-12. A [HCO$_3^-$]/S × PCO$_2$ diagram for analysis of acid-base status. The *[HCO$_3^-$]* pertains to plasma, and the *radiating lines* are iso-pH lines. Note that [dissolved CO$_2$] is equal to S × PCO$_2$. *A* = respiratory alkalosis (uncompensated); *B* = respiratory acidosis (uncompensated); *C* = respiratory acidosis (fully compensated); *D* = normal point; *E* = metabolic alkalosis (uncompensated); and *F* = metabolic acidosis (uncompensated). *Point C*, which is associated with a normal pH, represents fully compensated respiratory acidosis because, of the the two variables in the [HCO$_3^-$]/S × PCO$_2$ ratio, the PCO$_2$ undergoes the greater degree of change (increase) and, therefore, indicates the primary event. The increase in [HCO$_3^-$] represents the compensatory response.

    b. At constant temperature (which keeps the pK′ constant), the [HCO$_3^-$]/S × PCO$_2$ ratio is the pH with iso-pH lines radiating from zero.
    c. Table 5-12 lists examples of the [HCO$_3^-$]/S × PCO$_2$ plot, using *points A* through *F* of Figure 5-12.

Table 5-12. Interrelationships among pH, [HCO$_3^-$], and PCO$_2$ of Plasma

Point*	[HCO$_3^-$] (mmol/L)	PCO$_2$ (mm Hg)	S × PCO$_2$ (mmol/L)	$\dfrac{[\text{HCO}_3^-]}{\text{S} \times \text{PCO}_2}$	pH	Acid-Base State
A	19	20	0.6	31.7	7.6	Respiratory alkalosis
B	28	72	2.2	12.7	7.2	Respiratory acidosis
C	37	60	1.8	20.6	7.4	Respiratory acidosis (fully compensated)
D	25	40	1.2	20.8	7.4	Normal
E	39	40	1.2	32.5	7.6	Metabolic alkalosis
F	15	40	1.2	12.5	7.2	Metabolic acidosis

*Data from [HCO$_3^-$]/S × PCO$_2$ diagram in Figure 5-12.

D. INTERPRETATION OF ACID-BASE ABNORMALITIES USING THE pH-[HCO$_3^-$] DIAGRAM

(Fig. 5-13). Acid-base imbalances are determined graphically with reference to the intercept of the PCO$_2$ isobar of 40 mm Hg and the normal buffer line. The intercept of these two curves marks the point of normality (*N*), which is associated with a pH of 7.4 (the *abscissa*) and a [HCO$_3^-$] of 24 mmol/L (the *ordinate*). Point *N* is the triple intercept that defines the pH, [HCO$_3^-$], and PCO$_2$ of true arterial plasma of a normal individual.

1. **Points to the left** of *point N* (*points A, C, E,* and *F*) indicate acidosis. **Points to the right** of *point N* (*points B, D, G,* and *H*) indicate alkalosis.
    a. *Line NA* represents the direction of an individual's response to an increase in PCO$_2$. Points to the left of the normal PCO$_2$ isobar (*points A, E,* and *F*) indicate **respiratory acidosis**.
        (1) *Point A* denotes a condition of uncompensated respiratory acidosis, which is characterized by a low pH, high [HCO$_3^-$], and high PCO$_2$.
        (2) *Point E* denotes respiratory acidosis ($\uparrow$ PCO$_2$) with metabolic acidosis ($\downarrow$ [HCO$_3^-$]).
        (3) *Point F* denotes respiratory acidosis ($\uparrow$ PCO$_2$) with metabolic alkalosis ($\uparrow$ [HCO$_3^-$]).
    b. *Line NB* represents the direction of an individual's response to a decrease in PCO$_2$. Points to the right of the normal PCO$_2$ isobar (*points B, G,* and *H*) indicate **respiratory alkalosis**.
        (1) *Point B* denotes a condition of uncompensated respiratory alkalosis, which is characterized by high pH, low [HCO$_3^-$], and low PCO$_2$.
        (2) *Point G* denotes respiratory alkalosis ($\downarrow$ PCO$_2$) with metabolic acidosis ($\downarrow$ [HCO$_3^-$]).
        (3) *Point H* denotes respiratory alkalosis ($\downarrow$ PCO$_2$) with metabolic alkalosis ($\uparrow$ [HCO$_3^-$]).

2. **Points below** the normal buffer line (*points C, E,* and *G*) indicate conditions with a component of metabolic acidosis ($\downarrow$ [HCO$_3^-$]). **Points above** the normal buffer lines (*points D, F,* and *H*) indicate conditions with a component of metabolic alkalosis ($\uparrow$ [HCO$_3^-$]).
    a. *Line NC* represents the direction of the development of metabolic acidosis in an individual; the [HCO$_3^-$] of this individual moves down the normal PCO$_2$ isobar.
        (1) *Point C* denotes a condition of uncompensated metabolic acidosis, which is characterized by low pH, low [HCO$_3^-$], and normal PCO$_2$.
        (2) *Point E* denotes metabolic acidosis ($\downarrow$ [HCO$_3^-$]) with respiratory acidosis ($\uparrow$ PCO$_2$).
        (3) *Point G* denotes metabolic acidosis ($\downarrow$ [HCO$_3^-$]) with respiratory alkalosis ($\downarrow$ PCO$_2$).
    b. *Line ND* represents the direction of the development of metabolic alkalosis in an individual; the [HCO$_3^-$] of this individual moves up the normal PCO$_2$ isobar.
        (1) *Point D* denotes a condition of uncompensated metabolic alkalosis, which is characterized by high pH, high [HCO$_3^-$], and normal PCO$_2$.
        (2) *Point F* denotes metabolic alkalosis ($\uparrow$ [HCO$_3^-$]) with respiratory acidosis ($\uparrow$ PCO$_2$).
        (3) *Point H* denotes metabolic alkalosis ($\uparrow$ [HCO$_3^-$]) with respiratory alkalosis ($\downarrow$ PCO$_2$).

3. **Mixed acid-base disturbances**, in which the primary state of acidosis or alkalosis has an additional abnormality with respect to PCO$_2$ and [HCO$_3^-$], are denoted by *points E, F, G,* and *H*.

VIII. COMPENSATORY MECHANISMS FOR PRIMARY ACID-BASE DISORDERS (see Tables 5-8 and 5-9)

A. TERMINOLOGY

1. **Compensation** is the secondary physiologic process occurring in response to a primary acid-base disturbance by which the deviation of blood pH is ameliorated. Thus, abnormal pH is returned toward normal by **altering the component that is not primarily affected.**

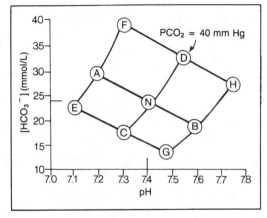

Figure 5-13. Pathways of acid-base imbalance. The [HCO$_3^-$] pertains to plasma. *Line ANB* represents the normal blood buffer line, and *line CND* represents the normal PCO$_2$ isobar (40 mm Hg). *N* = normal point; *A* = respiratory acidosis (uncompensated); *B* = respiratory alkalosis (uncompensated); *C* = metabolic acidosis (uncompensated); *D* = metabolic alkalosis (uncompensated); *E* = respiratory acidosis + metabolic acidosis; *F* = respiratory acidosis + metabolic alkalosis; *G* = respiratory alkalosis + metabolic acidosis; and *H* = respiratory alkalosis + metabolic alkalosis. (After Davenport HW: *The ABC of Acid-Base Chemistry*, 5th ed. Chicago, University of Chicago Press, 1969, p 65.)

    a. The terms secondary and compensatory may be used to describe a change in the composition of the blood or to describe a process.

    b. The terms secondary and compensatory may be used to describe a process, but they should not be used to describe acidosis or alkalosis.

  2. Correction denotes the **therapeutic** course of action that is directed toward counteracting or altering the factor that is primarily affected. Correction involves the amelioration of the primary underlying abnormality in $[H^+]$, $[HCO_3^-]$, or PCO_2. In correction, as in compensation, the pH is returned toward normal.

B. BASIC MECHANISMS

  1. Buffers of the ECF represent the body's first defense mechanism for neutralizing noncarbonic acid. H^+ is transferred across tissue cell membranes in a direction that normalizes blood pH.

  2. A deviation in plasma pH acts on the respiratory center to change alveolar ventilation, which, in turn, alters the alveolar PCO_2 (P_ACO_2) and $PaCO_2$ so as to counteract the change in plasma pH. This is called **respiratory compensation** of a primary noncarbonic acid excess or deficit; respiratory compensation usually is not complete (i.e., does not restore pH to normal).

  3. Carbonate ion is released from bone, which is the predominant source of alkali for neutralizing excess noncarbonic acid added to the ECF.

  4. Renal excretion of noncarbonic acid or base increases in order to eliminate an excess of noncarbonic acid or base or to compensate for the pH changes due to an abnormal $PaCO_2$. This is called **renal compensation** of a primary hyper- or hypocapnia.

C. PRINCIPLES OF RESPIRATORY AND RENAL COMPENSATORY RESPONSES

  1. General Considerations.

    a. Physiologic compensation for major acid-base abnormalities rarely is complete. Therefore, it must be emphasized that the singular concern of clinical diagnosis is to differentiate the primary cause of the imbalance from the secondary (compensatory) response.

    b. It also must be understood that most data indicate that the pH ($[H^+]$) is the most important factor in determining the bodily response to acid-base imbalances. (In both acid and base disturbances, the abnormal pH is returned toward normal.) In spite of its critical biological significance, however, a change in pH does not supply all the information needed for quantitative assessment and, thus, for planning therapy.

      (1) The key determinants of the cause and compensation of acid-base abnormalities are PCO_2 and $[HCO_3^-]$, not pH.

      (2) The most important factor in determining the efficacy of the bodily responses to deviations from the normal acid-base status is pH or $[H^+]$.

  2. Respiratory and Renal Processes.

    a. Both respiratory and renal compensatory responses tend to restore the abnormal $[HCO_3^-]/S \times PCO_2$ ratio toward its normal value of 20/1.

      (1) Primary acid-base disturbances of **metabolic origin** lead to secondary adjustment of the PCO_2 by changes in the rate of alveolar ventilation.

      (2) Primary acid-base disturbances of **respiratory origin** lead to secondary changes in blood $[HCO_3^-]$ by appropriate adjustment of the rate of H^+ secretion/excretion from the renal tubular cell into the tubular lumen.

    b. These two processes interact to control the $[H^+]$ of the ECF, and this integration is clearly understood using the mathematical relationship of the Henderson equation:

$$[H^+] = 24 \ \frac{PCO_2}{[HCO_3^-]} \ .$$

    From this equation it is apparent that:

      (1) An increase in the $[H^+]$, regardless of cause, can be reduced toward normal by a decrease in PCO_2, an increase in plasma $[HCO_3^-]$, or by both changes.

      (2) A decrease in the $[H^+]$, regardless of cause, can be increased toward normal by an increase in PCO_2, a decrease in plasma $[HCO_3^-]$, or by both changes.

D. PATHWAYS FOR COMPENSATION OF PRIMARY ACID-BASE DISTURBANCES: THE pH-$[HCO_3^-]$ RELATIONSHIP (Fig. 5-14)

  1. Respiratory Acidosis (*point A*). The system at fault in this condition is the respiratory system, and compensation occurs through metabolic processes.

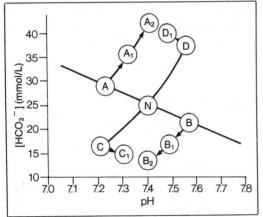

Figure 5-14. Effects of renal and respiratory compensation on the pH and plasma [HCO_3^-]. N = normal point; A = respiratory acidosis; A_1 = respiratory acidosis with partial renal compensation; A_2 = respiratory acidosis with complete renal compensation; B = respiratory alkalosis; B_1 = respiratory alkalosis with partial renal compensation; B_2 = respiratory alkalosis with complete renal compensation; C = metabolic acidosis; C_1 = metabolic acidosis with partial respiratory compensation; D = metabolic alkalosis; and D_1 = metabolic alkalosis with partial renal compensation. (After Davenport HW: *The ABC of Acid-Base Chemistry*, 5th ed. Chicago, University of Chicago Press, 1969, p 65.)

    a. The kidneys excrete more acid and reabsorb more HCO_3^-, returning the [HCO_3^-]/S × PCO_2 ratio toward 20/1 and, therefore, returning pH toward normal (*point A_1*).

    b. If the PCO_2 is elevated but the pH is normal, the kidneys had time to retain HCO_3^- to compensate for the elevated PCO_2 and the process is not acute; that is, the condition has existed at least a few days to give the kidneys time to compensate (*point A_2*).

    c. Usually, the body does not fully compensate for respiratory acidosis. The renal compensation for acute and chronic cases of this condition is as follows:

        (1) Acute: 1 mEq/L increase in plasma [HCO_3^-] for each 10 mm Hg increase in PCO_2

        (2) Chronic: 3.5 mEq/L increase in plasma [HCO_3^-] for each 10 mm Hg increase in PCO_2

2. Respiratory Alkalosis (*point B*). In this condition, the acid-base abnormality again is respiratory, and compensation again occurs through metabolic means.

    a. The kidneys compensate by excreting HCO_3^-, thus returning the [HCO_3^-]/S × PCO_2 ratio toward 20/1; this compensation takes 2–3 days (*point B_1*).

    b. Of the four acid-base abnormalities, only in respiratory alkalosis is the body able to compensate fully; the [HCO_3^-]/S × PCO_2 ratio and pH, therefore, return entirely to normal (*point B_2*).

    c. Renal compensation for acute and chronic cases of this condition is as follows:

        (1) Acute: 2 mEq/L fall in plasma [HCO_3^-] for each 10 mm Hg decrease in PCO_2

        (2) Chronic: 5 mEq/L reduction in plasma [HCO_3^-] for each 10 mm Hg decrease in PCO_2

3. Metabolic Acidosis (*point C*). In this condition, the major abnormality is low [HCO_3^-] or negative BE, and the compensation is a respiratory process.

    a. By hyperventilation, the PCO_2 is lowered so that the [HCO_3^-]/S × PCO_2 ratio is returned toward 20/1 (*point C_1*).

    b. Since the compensatory system is the lungs, compensation can occur rapidly. If the metabolic acidosis is severe, however, the lungs may not be able to expel sufficient CO_2 to compensate fully.

    c. Respiratory compensation for metabolic acidosis is a 1.2 mm Hg fall in PCO_2 for each 1 mEq/L decrease in plasma [HCO_3^-].

4. Metabolic Alkalosis (*point D*). In this condition, the major abnormality is a high [HCO_3^-], and the compensation is a respiratory mechanism.

    a. By hypoventilation, the PCO_2 is elevated so that the [HCO_3^-]/S × PCO_2 ratio is increased toward normal (*point D_1*).

    b. The body usually cannot compensate fully for metabolic alkalosis.

    c. Respiratory compensation for this condition is a 0.6 mm Hg increase in PCO_2 for each 1 mEq/L increase in plasma [HCO_3^-].

IX. THE UNCOMPENSATED ACID-BASE ABNORMALITIES: A COMPARISON (see Fig. 5-13 and Tables 5-8, 5-9, and 5-13)

    A. GENERAL CONSIDERATIONS. The four uncompensated acid-base disorders are compared below on the basis of the following criteria.

        1. Reference Ponts.

            a. As discussed earlier, *point N* in Figure 5-13 represents the intercept for the pH, [HCO_3^-], and PCO_2 of true arterial plasma of a normal individual. Each uncompensated acid-base abnormality is represented by a point that has a unique relation to *point N*.

   b. The mechanism of buffer action for each acid-base disorder can be defined in terms of the hydration-dehydration:dissociation-association of the CO_2/HCO_3 system:

$$CO_2 + H_2O \rightleftharpoons H_2CO_3 \rightleftharpoons H^+ + HCO_3^-. \tag{5}$$

  2. Compensatory Response. The responses to metabolic and respiratory acid-base imbalances differ in that there is no extracellular buffer for respiratory acid-base abnormalities. HCO_3^- is not an effective buffer for H_2CO_3, because the association of H^+ with HCO_3^- results in the regeneration of H_2CO_3 as:

$$H_2CO_3 + HCO_3^- \rightleftharpoons HCO_3^- + H_2CO_3.$$

Consequently, H_2CO_3 is buffered primarily by the intracellular buffers (i.e., hemoglobin, protein, and organic/inorganic phosphates) in respiratory acid-base disorders.

   a. Metabolic acid-base abnormalities are compensated by the respiratory regulator. Sequential responses to increases in PCO_2 include:

    (1) Intracellular buffering within 10 to 30 minutes

    (2) Increased renal H^+ excretion within hours to days

   b. Respiratory acid-base abnormalities are compensated by the renal regulator. Sequential responses to a H^+ load that lead to increased renal H^+ excretion include:

    (1) Immediate extracellular buffering by HCO_3^-

    (2) Respiratory buffering (by decreasing PCO_2) within minutes to hours

    (3) Intracellular buffering within 2 to 4 hours

    (4) Increased renal H^+ excretion within hours to days

B. MAJOR DISTINGUISHING FEATURES

  1. Respiratory Acidosis. CO_2 retention is the primary cause of respiratory acidosis. The $[HCO_3^-]/S \times PCO_2$ ratio is decreased below 20/1, because the denominator is increased by pulmonary retention of CO_2, the immediate effect of which is a lowering of pH ($\uparrow [H^+]$).

   a. Reference Points.

    (1) *Point A* in Figure 5-13 represents a patient with pulmonary hypoventilation or hypercapnia. This condition causes increases in $PACO_2$ and $PaCO_2$, which lead to respiratory acidosis. *Point A* lies on the normal buffer line to the left of pH = 7.4; therefore, this patient has uncompensated respiratory acidosis.

    (2) From equation (5), increasing PCO_2 causes CO_2 molecules to move to the **right** through hydration and dissociation, forming more HCO_3^- and H^+. Therefore, uncompensated respiratory acidosis is associated with above normal concentrations of CO_2, HCO_3^-, and $[H^+]$ ($\downarrow$ pH).

   b. Compensation for respiratory acidosis is by the kidney, which secretes H^+ and conserves HCO_3^-. In addition, elevation of PCO_2 results in new HCO_3^- generation in the kidneys. Thus, the compensatory response for respiratory acidosis is to increase the numerator of the $[HCO_3^-]/S \times PCO_2$ ratio.

  2. Respiratory Alkalosis. The excessive elimination of CO_2 is the primary cause of respiratory

Table 5-13. Responses of the Renal pH Regulator to Various Conditions

Condition	Acid-Base Abnormality	Effect on:		
		Plasma pH	Renal Cellular pH*	Urine pH
NH_4Cl ingestion	Metabolic acidosis	$\downarrow$	$\downarrow$	Aciduria
CO_2 inhalation (acute)	Respiratory acidosis	$\downarrow$	$\downarrow$	Aciduria
K^+ depletion	Hypokalemic alkalosis	$\uparrow$	$\downarrow$	Aciduria
$NaHCO_3$ ingestion	Metabolic alkalosis	$\uparrow$	$\uparrow$	Alkaluria
Voluntary hyperventilation	Respiratory alkalosis	$\uparrow$	$\uparrow$	Alkaluria
K^+ injection	Hyperkalemic acidosis	$\downarrow$	$\uparrow$	Alkaluria

Note.—Reprinted with permission from Koushanpour E: Renal regulation of acid-base balance. In *Renal Physiology: Principles and Functions*. Philadelphia, WB Saunders, 1976, p 299.
* The renal cellular pH is the key factor determining the renal excretion of acids and bases. In general, alterations in plasma pH, whether due to changes in the plasma $[HCO_3^-]$ or the PCO_2, result in parallel changes in the intracellular pH. Note that acidosis favors H^+ excretion (aciduria), except in hyperkalemic acidosis. Most patients with metabolic alkalosis have an increase in HCO_3^- reabsorption (see XIII D 1).

alkalosis. The $[HCO_3^-]/S \times PCO_2$ ratio is increased above 20/1, because the denominator is decreased by primary pulmonary loss of CO_2.

a. Reference Points.

    (1) *Point B* in Figure 5-13 represents a patient with alveolar hyperventilation or hypocapnia. This condition causes decreases in $PACO_2$ and $PaCO_2$, which lead to respiratory alkalosis. *Point B* lies on the normal buffer line to the right of pH $= 7.4$; therefore, this patient has uncompensated respiratory alkalosis.

    (2) From equation (5), the initiating event in respiratory alkalosis is a decrease in PCO_2 and a net loss of H_2CO_3, causing HCO_3^- and H^+ to associate and move to the **left**. Uncompensated respiratory alkalosis is associated with below normal concentrations of CO_2, HCO_3^-, and H^+ ($\uparrow$ pH).

b. Compensation for respiratory alkalosis is by the kidney, which conserves H^+ by reducing the excretion of NH_4^+ and titratable acid. Initially the $[HCO_3^-]$ is low, because lowering PCO_2 converts more HCO_3^- to H_2CO_3 and then to CO_2, which is exhaled. However, this conversion reaction lags behind the rapid ventilation, and the low PCO_2 diminishes H^+ secretion and HCO_3^- reabsorption. Thus, the renal compensatory response to respiratory alkalosis is to decrease the numerator of the $[HCO_3^-]/S \times PCO_2$ ratio.

3. Metabolic Acidosis. A primary cause of metabolic acidosis is an increase in $[H^+]$ brought about by the addition of a noncarbonic acid (e.g., HCl, lactic acid, acetoacetic acid, and β-hydroxybutyric acid) to the ECF. The $[HCO_3^-]/S \times PCO_2$ ratio is less than 20/1, because the numerator is decreased by the addition of H^+; the H^+, in turn, reacts with HCO_3^- to produce a primary decrease in $[HCO_3^-]$.

a. Reference Points.

    (1) *Point C* in Figure 5-13 represents a patient with either diabetes mellitus, which leads to ketoacidosis, or diarrhea, which leads to HCO_3^- loss. In going from *point N* to *point C*, the $[HCO_3^-]$ of this patient moves from the normal buffer line to another buffer line below and parallel to the normal one. If the PCO_2 is not changed as a result of acidosis, *point C* will lie on the normal PCO_2 isobar to the left of pH $= 7.4$. This position defines uncompensated metabolic acidosis.*

    (2) Referring to equation (5), many of the added H^+ of metabolic acidosis react with HCO_3^- (moving the series of reactions to the **left**) and are quickly eliminated as CO_2 from the lungs. Therefore, uncompensated metabolic acidosis is associated with a decreased concentration of HCO_3^-, an increased concentration of H^+, and a normal PCO_2.

b. Compensation. The increase in $[H^+]$ stimulates alveolar ventilation, which reduces PCO_2 and causes a smaller, secondary decrease in H_2CO_3. Thus, the compensatory response for metabolic acidosis is to decrease the denominator of the $[HCO_3^-]/S \times PCO_2$ ratio.

4. Metabolic Alkalosis. The initiating events that cause metabolic alkalosis are the loss of noncarbonic acid from the ECF (e.g., the loss of the mineral acid HCl via persistent vomiting) or the addition of HCO_3^- to the body fluids through alkali ingestion. Either of these initiating events causes the other to occur, and the end result is increased concentrations of CO_2 and HCO_3^- and a decreased concentration of H^+. The $[HCO_3^-]/S \times PCO_2$ ratio rises above 20/1, because the numerator is increased by the accumulation of HCO_3^-, which increases pH.

a. Reference Points.

    (1) The pH, PCO_2, and $[HCO_3^-]$ values at *point D* in Figure 5-13 represent a patient who either has had persistent vomiting or has ingested an excessive amount of antacids; both conditions lead to an excessive loss of noncarbonic acid. In moving from *point N* to *point D*, the $[HCO_3^-]$ of this patient moves from the normal buffer line to one that is higher and parallel to the normal one. If there is no respiratory response to the change in pH ($\downarrow [H^+]$) [i.e., if PCO_2 is not changed as a result of alkalosis], *point D* will lie on the normal PCO_2 isobar to the right of pH $= 7.4$. This position defines the condition of uncompensated metabolic alkalosis.

    (2) Referring to equation (5), the H^+ lost in metabolic alkalosis causes CO_2 to move to the **right** through a series of reactions that lead to an elevation in $[HCO_3^-]$. If HCO_3^- is added, then some of the H^+ reacts with the HCO_3^-, causing a decrease in $[H^+]$ ($\uparrow$ pH). The end result is increased concentrations of CO_2 and HCO_3^- and a decreased concentration of H^+.

b. Compensation for metabolic alkalosis is by the lung with a decrease in minute ventilation leading to an increase in PCO_2, but this latter effect is not pronounced. Thus, CO_2 retention may or may not be measurable. The usual compensatory response for metabolic

---

*A metabolic acidosis rarely is "pure" because the elevated $[H^+]$ stimulates the respiratory drive, and resultant hyperventilation lowers the PCO_2. Only a patient with an impaired ventilatory response would present with uncompensated metabolic acidosis (e.g., a patient with respiratory muscular paralysis of a neurologic disorder or a patient on mechanical ventilation).

alkalosis is an increase in the denominator of the $[HCO_3^-]/S \times PCO_2$ ratio. The renal compensation consists of increased HCO_3^- excretion.

C. OTHER DISTINGUISHING FEATURES

1. The **compensatory changes in [H$^+$] and [HCO$_3^-$]** that occur in metabolic acid-base abnormalities contrast sharply to the changes that occur in respiratory acid-base abnormalities.
   a. In metabolic acidosis the [H$^+$] increases and the [HCO$_3^-$] decreases, whereas in respiratory acidosis the [HCO$_3^-$] and [H$^+$] increase together.
   b. In metabolic alkalosis the [H$^+$] decreases and the [HCO$_3^-$] increases, whereas in respiratory alkalosis the [HCO$_3^-$] and [H$^+$] decrease together.

2. **CO$_2$ Content** (see Table 5-9).
   a. The overall direction of the change of CO_2 content is downward in both metabolic acidosis and respiratory alkalosis.
   b. The overall direction of the change of CO_2 content is upward in both metabolic alkalosis and respiratory acidosis.

X. ETIOLOGY OF ACID-BASE ABNORMALITIES

A. RESPIRATORY ACIDOSIS is caused by hypoventilation.

1. **Acute** cases of respiratory acidosis are caused by the following agents and conditions:
   a. General anesthesia
   b. Drugs, such as anesthetic agents, narcotics, sedatives, barbiturates, succinylcholine, curare, nerve gases, and botulinus toxin
   c. Cardiac arrest
   d. Pneumothorax and pleural accumulation of fluid or blood
   e. Pulmonary edema
   f. Severe pneumonia
   g. Bronchospasm and laryngospasm
   h. Aspiration of a foreign body
   i. Mechanical ventilators
   j. Tetanus (i.e., thoracic muscle spasm)

2. **Chronic** cases of respiratory acidosis are attributed to:
   a. Obstructive pulmonary diseases, such as chronic bronchitis, emphysema, bronchial asthma, tracheal stenosis, and tumor
   b. Restrictive pulmonary diseases, such as infiltrative tumors, atelectasis, bronchiectasis, mucoviscidosis, and interstitial fibrosis (late)
   c. Brain tumor, stroke, cerebral ischemic disease, trauma, and increased CSF pressure
   d. Respiratory nerve damage, such as spinal cord lesions, poliomyelitis, amyotropic lateral sclerosis, peripheral neuritis, myasthenia gravis, and Guillain-Barré syndrome
   e. Primary respiratory myopathy that restricts thoracic movement
   f. Restrictive diseases of the thorax, such as scleroderma and arthritis
   g. Prolonged pneumonia
   h. Obesity, such as Pickwickian syndrome

B. RESPIRATORY ALKALOSIS results from:

1. Anxiety and hysteria

2. Fever

3. Salicylate intoxication (early)

4. CNS diseases, such as cerebrovascular accident, trauma, infection (e.g., encephalitis and meningitis), and tumor

5. Congestive heart failure

6. Pneumonia

7. Pulmonary emboli

8. Hypoxia due to lowered ambient pressure (at high altitude), ventilation-perfusion imbalance, or severe anemia

9. Hepatic insufficiency, such as cirrhosis, hepatic coma, and hepatic failure

10. Gram-negative septicemia

11. Mechanical respirators

12. Pregnancy

13. Drugs, such as analeptics, and hormones, such as epinephrine and progesterone

14. Thyrotoxicosis

C. **METABOLIC ACIDOSIS** (see also Table 5-11). Etiologic agents of metabolic acidosis include:

1. Ketoacidosis from diabetes, starvation, or alcoholism

2. Lactic acidosis, as seen in severe hypoxemia, which may be spontaneous or secondary to shock, hypoxia, or diabetic nonketotic acidosis (see also X D)

3. Acid metabolized from exogenous substances, such as:
   a. Ammonium chloride → urea + H^+
   b. Methyl alcohol → oxalic acid
   c. Ethylene glycol → oxalic acid
   d. Sulfur → sulfide + H^+
   e. Paraldehyde (rarely)
   f. Salicylate poisoning by ingestion of acetylsalicylic acid (aspirin), methyl salicylate (oil of wintergreen), phenyl salicylate (salol), salicylic acid, or sodium salicylate or by the use of salicylic acid ointments

4. Treatment with acetazolamide (Diamox), a carbonic anhydrase inhibitor

5. Treatment with infusion of arginine HCl or ingestion of lysine HCl

6. Renal acidosis due to:
   a. Reduced nephron mass
   b. Renal tubular acidosis [type I (distal) and type II (proximal), including Fanconi's syndrome and its variants]
   c. Diminished NH_3 production due to renal failure

7. Exogenous acid ingestion or administration [e.g., mineral acid ingestion and rhubarb poisoning (oxalic acid)]

8. Extrarenal loss of HCO_3^- due to:
   a. Diarrhea
   b. Pancreatic fistula (drainage)
   c. Ureterosigmoidostomy
   d. Obstructive ileal loop
   e. Hyperalimentation fluids (i.e., cationic amino acids), such as arginine, lysine, and histidine
   f. Neoaminosol
   g. Cholestyramine

9. Dilution acidosis (hydration)

10. Endocrine disorders, such as hypoaldosteronism, Addison's disease, and hyperparathyroidism

D. **LACTIC ACIDOSIS** is caused by:

1. Increased generation of lactate as a result of:
   a. Decreased tissue perfusion due to shock or cardiac arrest
   b. Increased skeletal muscle activity, such as convulsive states
   c. Large tumor masses, such as leukemias, lymphomas, and visceral metastatic diseases
   d. Cyanide poisoning
   e. Carbon monoxide poisoning

2. Decreased utilization of lactate as a result of:
   a. Phenformin (DBI) intoxication*
   b. Hepatic failure due to reduced hepatic perfusion, hepatocyte failure, or hepatocyte replacement by tumor
   c. Diabetes mellitus
   d. Alcoholism (possible cause)

3. Lactic acidosis X

4. Factitial lactic acidosis

---

*This hypoglycemic agent is no longer in use.

E. **METABOLIC ALKALOSIS** results from:

1. Excessive ingestion of antacids

2. Extrarenal loss of acid due to:
   a. Persistent vomiting
   b. Nasogastric suction or fistula
   c. Congenital alkalosis with diarrhea

3. Treatment with diuretics, such as mercurials, ethacrynic acid, furosemide, and thiazides

4. Rapid correction of chronic hypercapnia (posthypercapnic metabolic alkalosis)

5. Rapid contraction of ECF ("contraction alkalosis")

6. Excessive licorice ingestion

7. Metabolic conversion of organic acids (e.g., lactic acid) to bicarbonate ("alkaline overshoot")

8. Renal tubular alkalosis due to:
   a. Cl^- deficiency
   b. K^+ deficiency
   c. Increased Na^+ delivery to distal tubules
   d. Hyperaldosteronism
   e. Cushing's syndrome
   f. Treatment with corticosteroids (e.g., prednisone and cortisone) or adrenocorticotropic hormone (ACTH)
   g. Bartter's syndrome

9. Nonparathyroid hypercalcemia

10. Transfusion of large volumes of blood*

XI. CLINICAL MANIFESTATIONS OF ACID-BASE ABNORMALITIES

A. RESPIRATORY ACIDOSIS

1. The clinical features of respiratory acidosis are collectively referred to as **CO_2 narcosis**. (Since the CSF is relatively poorly buffered[†] and CO_2 readily diffuses across the blood-brain barrier, the increase in PCO_2 due to respiratory acidosis is associated with a rapidly increased $[H^+]$ ($\downarrow$ pH) of the CSF.)

2. The early stages of CO_2 narcosis are characterized by cyanosis with hyperventilation at rest, fatigue, weakness, blurred vision, and headache.

3. The excess CO_2 can lead to severe depression of mentation, impairment of consciousness, twitching fingers at rest (tremors), asterixis, and finally to delirium, somnolence, convulsions, increased CSF pressure, and coma.

B. RESPIRATORY ALKALOSIS

1. The clinical characteristics of this condition are predominantly those of an increased excitability of the central and peripheral nervous systems and include lightheadedness, altered consciousness, paresthesia of the extremities and circumoral area, and carpopedal spasm.

2. Respiratory alkalosis frequently is associated with paresthesia and a tetany that is indistinguishable from hypocalcemic tetany.

3. Epileptogenic foci become more active in respiratory alkalosis.

C. METABOLIC ACIDOSIS. As in most or all clinical situations, the clinical features of metabolic acidosis are difficult to define because the associated biochemical agent of the underlying disease (e.g., chronic renal failure and diabetes mellitus) cannot be isolated from the effect of the increased $[H^+]$ on cellular function.

1. A mild degree of metabolic acidosis may be symptomless or may be accompanied by anorexia, listlessness, nausea, headache, and lethargy.

2. Metabolic acidosis can predispose toward potentially fatal ventricular arrhythmias and can reduce both cardiac contractility and the inotropic response to catecholamines.

---

*Most bank blood is anticoagulated with acid-citrate-dextrose (ACD), each unit (500 ml) containing 16.8 mEq/of citrate, which can be metabolized to HCO_3^-. More than 8 units of blood must be given acutely to produce a significant increase in arterial pH (alkalosis).

†The protein concentration of CSF is approximately 20 mg/dl (200 mg/L).

3. Severe degrees of metabolic acidosis may be associated with other clinical features, some of which are directly attributable to elevated [H^+]. Examples include:

   a. Air hunger (Kussmaul's respiration), which is characterized by an initial increase in tidal volume followed by an increase in respiratory rate

   b. Lethargy and impairment of consciousness, with progression to stupor and coma

   c. Peripheral vasodilation. Initially, cardiac output may be increased, but with severe acidosis it falls and hypotension may be pronounced.

   d. Skeletal decalcification. Over a protracted period of time (e.g., with renal failure), a loss of bone mineral mass may occur as part of the whole body buffering mechanism to counteract the H^+ retention.

D. METABOLIC ALKALOSIS

1. Symptoms of metabolic alkalosis are primarily related to volume depletion (e.g., polyuria, polydipsia, and muscular weakness) rather than to the alkalosis itself.

2. The predominant effect of metabolic alkalosis is an increased irritability of the central and peripheral nervous systems manifested by numbness, paresthesia progressing to mental confusion, and tetany. Other common findings include slow, shallow respiratory activity, nausea, vomiting, drowsiness, and coma.

3. The clinical characteristics of this disturbance can occur in other clinical settings not involving metabolic alkalosis (i.e., in hypokalemia alone).*

XII. DIAGNOSIS OF ACID-BASE ABNORMALITIES (Fig. 5-15)

A. GENERAL CONSIDERATIONS. A salient point regarding plasma [HCO_3^-] is to be reiterated. An acid-base disorder cannot be diagnosed with certainty from the plasma [HCO_3^-].

1. Although a reduction in [HCO_3^-] may be due to metabolic acidosis, it also can reflect the renal compensation to respiratory alkalosis.

2. Similarly a high plasma [HCO_3^-] can represent a metabolic alkalosis or the renal compensatory response to respiratory acidosis.

B. pH ALTERATIONS. Since the aim of therapy is to correct the acid-base imbalance, it is important to establish the correct diagnosis. This requires the measurement of pH to determine whether the arterial pH is low (< 7.35), high (> 7.45), or normal (between 7.35 and 7.45).

1. If the pH is low, two abnormalities are possible (see Fig. 5-15*A*).

   a. In respiratory acidosis, the PCO_2 and [HCO_3^-] are high.

   (1) With respiratory acidosis, if the [HCO_3^-] is less than normal (i.e., < 25 mEq/L or not appropriately high), the patient must have a complicating metabolic acidosis in addition to respiratory acidosis.

   (2) Some patients with a high PCO_2 develop complicating metabolic alkalosis and still have low pH values.

   b. In metabolic acidosis the PCO_2 and [HCO_3^-] are low.

   (1) For patients with metabolic acidosis, if the PCO_2 is not appropriately low, complicating respiratory acidosis must be present.

   (2) This complication is the most common additional acid-base abnormality in such patients.

2. If the pH is high, two abnormalities are possible (see Fig. 5-15*B*).

   a. In acute respiratory alkalosis, the PCO_2 and [HCO_3^-] are low.

   (1) If the [HCO_3^-] is not appropriately low, the patient has a complicating metabolic alkalosis.

   (2) This complication is the most common acid-base abnormality in this group of patients.

   b. The presence of metabolic alkalosis is indicated by a high PCO_2 and a high [HCO_3^-].

   (1) Some patients do not have substantial elevations in PCO_2.

   (2) Some patients with a combination of respiratory acidosis and complicating metabolic alkalosis may have elevated pH values.

3. If the pH is normal, there are four possible diagnoses (see Fig. 5-15*C*).

   a. The patient has no acid-base disturbance, which is indicated by a normal PCO_2 and a normal [HCO_3^-].

   b. The patient has chronic respiratory alkalosis, which is demonstrated by a low PCO_2 and a low [HCO_3^-].

---

*The clinical features of the consequent hypokalemia are identical to those from any other cause and include weakness, paralytic ileus, tachycardia, cardiac arrhythmias, and increased susceptibility to digitalis intoxication.

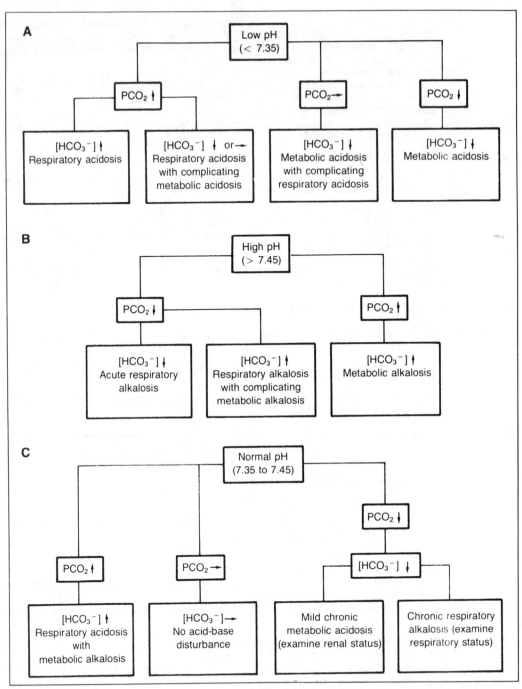

Figure 5-15. Diagnostic approach to common acid-base abnormalities. Changes in PCO_2 and $[HCO_3^-]$ and the associated abnormalities observed at different pH values are shown. Some patients with a high PCO_2 and complicating metabolic alkalosis may have a normal pH (*diagram A*). (Reprinted with permission from Robin ED: Disorders of acid-base metabolism. In *Scientific American: Medicine*, vol 2, sect 14. New York, Scientific American, 1984, p 15.)

  c. Mild chronic metabolic acidosis also may be associated with a normal pH, a low PCO_2, and a low $[HCO_3^-]$. To determine which of these two abnormalities is present (chronic respiratory alkalosis or mild chronic metabolic acidosis), the physician should check:
    (1) The patient's renal status, to establish the possibility of metabolic acidosis
    (2) The patient's respiratory status, to establish the possibility of chronic respiratory alkalosis

d. A normal pH also may be present in patients with respiratory acidosis with metabolic alkalosis. In such patients both the PCO_2 and the $[HCO_3^-]$ are high. Some patients with high PCO_2 and high $[HCO_3^-]$ may have low **or** high pH values (see Fig. 5-15A and B, respectively).

XIII. TREATMENT OF ACID-BASE ABNORMALITIES

A. RESPIRATORY ACIDOSIS

1. Therapy for respiratory acidosis should be directed toward improvement of ventilation and should include treatment of pulmonary infection, clearing of secretions, and, when necessary, assistance with ventilation.

2. High-flow O_2 should not be used when PCO_2 is very high, because in such cases respiratory sensitivity to CO_2 is lost and low PO_2 is the only stimulus to respiratory drive. Elevation of PO_2 quickly removes respiratory drive, and tidal volume and depth of respiration may **decrease** ("oxygen paradox"), further elevating PCO_2 and leading to CO_2 narcosis.

3. Although bronchodilators (e.g., aminophylline and epinephrine) may be effective in reversing the bronchoconstrictive component, corticosteroids and, infrequently, intubation and mechanical ventilation may be necessary to control the PCO_2 by providing adequate ventilation and removing thick mucus plugs.

B. RESPIRATORY ALKALOSIS

1. Therapy for respiratory alkalosis should be directed toward decreasing the rate of pulmonary gas exchange, and treatment of the primary disease causing hyperventilation is paramount. When hypoxemia is the driving force behind the hyperventilation, O_2 therapy is effective.

2. Increasing alveolar PCO_2 can be accomplished with the simple paper bag technique or by having the patient breathe 5–10 percent CO_2.

3. There is no rationale for the use of respiratory depressants or the administration of acid in the treatment of respiratory alkalosis.

C. METABOLIC ACIDOSIS

1. Unless the arterial pH is markedly depressed (< 7.2), it is seldom necessary to use alkali once proper treatment of the primary disease is begun.
   a. If the pH is below 7.25 and the total CO_2 content is less than 15 mEq/L, alkali may be a useful adjunct.
   b. In acute metabolic acidosis, alkali ($NaHCO_3$) is the mainstay of therapy, because profound acidosis may depress myocardial function, antagonize insulin action, inhibit glycolysis, and increase hepatic and renal gluconeogenesis.
   c. In **lactic acidosis**, continuing production of excess lactate often requires alkali until the primary cause is treated.

2. When intravenous $NaHCO_3$ is required, a number of factors should be considered.
   a. When HCO_3^- is added to the ECF, approximately equal amounts of Cl^- are excreted.
   b. Whenever alkali is used, there must be careful monitoring of serum $[K^+]$, because sudden increases in $[HCO_3^-]$ can drive large amounts of K^+ into the ICF, resulting in cardiac arrhythmia, especially in patients receiving digitalis.
   c. Giving large doses of $NaHCO_3$ can give the patient a large osmotic load, which may be more detrimental than the acidosis. Therefore, metabolic acidosis is not usually treated with $NaHCO_3$ unless the pH is below 7.25.

3. Lactate can be used; however, in the presence of lactic acidosis the possibility exists that lactate cannot be metabolized to HCO_3^- (and therefore cannot act as an alkali).

4. When chronic metabolic acidosis requires therapy (e.g., in chronic renal disease and renal tubular acidosis), one of two preparations may be employed.
   a. Oral $NaHCO_3$ (tablets) may be given, unless gastric upset occurs.
   b. Shohl's solution may be administered. This solution contains 140 g citric acid plus 98 g of the hydrated crystalline salt of sodium citrate in 1 L of water; sodium citrate is metabolized to $NaHCO_3$. Each milliliter of Shohl's solution contains 1 mEq of Na^+.
   c. In chronic renal disease, the load of Na^+ in the alkalinizing solution (either as HCO_3^- or citrate$^-$) may exceed the excretory capacity of the diseased kidney(s) and lead to edema. In this case, the acidosis may have to be treated with dialysis.

5. Patients with diarrhea lose alkaline, bicarbonate-containing intestinal (pancreatic) secretions

that contain more Na^+ than Cl^-. In the usual case of diarrhea, dehydration occurs with significant loss of salt and water in addition to the decreased $[HCO_3^-]$. As a result, the administration of Na^+ with Cl^- is required. [In this case, the disorder is not corrected by administering NH_4HCO_3, because the metabolic products (i.e., urea, CO_2, and water) are neutral.]

a. Extra provision of HCO_3^- is not necessary in milder cases, when renal function is normal and when HCO_3^- can be formed endogenously by the reaction:

$$CO_2 + H_2O$$
$$Na^+ + Cl^- + H_2CO_3 \rightarrow Na^+ + HCO_3^- + H^+ + Cl^-$$
$$\text{(retained)} \qquad \qquad \qquad NH_3$$
$$NH_4^+ + Cl^-$$
$$\text{(excreted)}.$$

b. The metabolic error is corrected by Na^+ (with HCO_3^-) retention and by excretion of Cl^- (with H^+). In patients with metabolic acidosis, NH_4^+ excretion can exceed 250 mEq/day in contrast to a limited ability to increase titratable acid excretion.

c. With deficiencies of both Na^+ and HCO_3^-, infusion of $NaHCO_3$ is useful.

d. Organic anions such as sodium lactate, sodium citrate, sodium acetate, and Tris buffer have been used because they are rapidly metabolized to HCO_3^-, but they offer no particular advantage over $NaHCO_3$. For example:

$$CH_3CHOHCOO^- \text{ (lactate)} + 3O_2 \rightarrow 2CO_2 + 2H_2O + HCO_3^-.$$

e. In the presence of insulin, the anions of the ketoacids can also be metabolized to HCO_3^-.

D. METABOLIC ALKALOSIS

1. The common type of metabolic alkalosis is associated with varying degrees of hypokalemia and volume (salt) depletion. Increased K^+ excretion in volume-depleted patients may be due to increased aldosterone secretion. [It is important to remember that volume depletion refers to **effective circulating volume**. Thus, effective volume depletion can be produced not only by actual volume loss (due to vomiting or diuretics) but also in edematous states caused by a reduction in cardiac function (congestive heart failure), ascites (cirrhosis), or movement of fluid into the interstitium (nephrotic syndrome). In these conditions there is a reduction in effective circulating volume, even though the **total** extracellular volume is expanded.]

a. Metabolic alkalosis can be corrected most easily by the renal excretion of the excess HCO_3^-. This must be induced, because HCO_3^- absorption (by H^+ secretion) is **increased** in response to volume depletion, K^+ depletion, or both.

b. The aim of the therapy is to restore volume and K^+ balance, which will reduce HCO_3^- reabsorption and allow excess HCO_3^- to be excreted. This requires the administration of Cl^-. (In mild hypokalemia, correction of hypovolemia and excess HCO_3^- can be controlled by the administration of NaCl without K^+ supplementation. The provision of filtered Cl^- and correction of volume depletion are the cornerstones of treatment for metabolic alkalosis along with the correction of hypokalemia, which usually exists.)

c. In metabolic alkalosis caused by vomiting, an essential requirement is to resupply H^+ with Cl^-, both of which can be made available by neutral isotonic NaCl administration by the reaction:

$$CO_2 + H_2O$$
$$Na^+ + Cl^- + H_2CO_3 \rightarrow H^+ + Cl^- + Na^+ + HCO_3^-.$$
$$\text{(retained)} \qquad \text{(excreted)}$$

The missing acid is recovered concomitantly with the excretion of $NaHCO_3$. The test for efficacy of corrective treatment is the formation of an alkaline urine.

d. Although NaCl corrects alkalosis, KCl corrects the alkalosis via the excretion of $KHCO_3$ and also provides supplemental K^+.

(1) The administration of K^+ with any anion other than Cl^- results in an increase in H^+ secretion preventing the correction of alkalosis.

(2) This is important clinically, since many of the K^+ supplements contain HCO_3^- or acetate, citrate, phosphate, or gluconate ions. Only the chloride of potassium is effective.

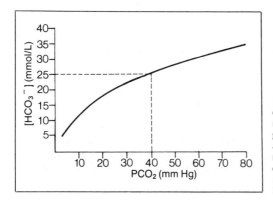

Figure 5-16. Empirical evaluation of the effect of elevated PCO_2 on the $[HCO_3^-]$ of the ECF. The curve represents data for CO_2 absorption of normal oxygenated true plasma. The change in plasma $[HCO_3^-]$ is approximately 1–1.2 mEq/L for a change in PCO_2 of 10 mm Hg above 40 mm Hg. For example, a PCO_2 increase from 40 to 60 mm Hg is associated with a $[HCO_3^-]$ increase from 25 to 27 mmol/L (see Section VI F 3 c).

2. **Cl^-** can be administered as NaCl, KCl, HCl or NH_4Cl.
   a. Intravenous injection of K^+ is potentially dangerous because a sudden increase in the plasma K^+ can depolarize the heart. This hazard is avoided with K^+ treatment by the oral route, since absorption is gradual.
   b. HCl can be given as HCl or as NH_4Cl, which is metabolized in the liver to HCl and ammonia; the ammonia then is converted to urea. NH_4Cl is easier to administer because it can be given orally or intravenously. Because of the accumulation of ammonia and urea, the clinical use of NH_4Cl is contraindicated in the presence of renal or hepatic disease.
   c. HCl can be infused, but because of its corrosive property must be given slowly and only through a central venous catheter. An isotonic HCl solution (0.15 mol/L or 150 mEq each of H^+ and Cl^- in 1 L of distilled water) can be administered over a period of 8–24 hours.
   d. An alternative to HCl is peritoneal dialysis or hemodialysis to remove the excess HCO_3^-.
   e. Administration of acetazolamide (Diamox) can be used to allow the kidneys to excrete more HCO_3^-.

3. It is important to understand that therapy for metabolic alkalosis in which H^+ and Cl^- are lost in gastric secretions cannot be directed to the loss of H^+ alone. This disorder cannot be corrected by restoring H^+ in the form of organic acids (e.g., lactic acid), because the acid is metabolized to HCO_3^-.

E. **RECOMMENDED APPROACH FOR THERAPY: SUMMARY**

1. **Step 1: Correct the $[HCO_3^-]$** for the effect of an abnormal PCO_2 (Fig. 5-16).
   a. For each 10 mm Hg increase in PCO_2 **subtract** 1.0 mEq/L.*
   b. For each 10 mm Hg decrease in PCO_2 **add** 1.2 mEq/L.

2. **Step 2: Identify the Primary Cause.** The variable that undergoes the greater degree of alteration (i.e., the larger proportional change) indicates the primary event. Attention is directed toward PCO_2 and $[HCO_3^-]$, which are the key determinants of cause and of compensation (even though pH or $[H^+]$ is the most important factor in determining the bodily response to acid-base imbalances).

3. **Step 3: Assess the Extent.** The excess amount of acid in patients can be estimated by multiplying the decrease in $[HCO_3^-]$ by the total volume of body water.
   a. The plasma $[HCO_3^-]$ is the primary index of extracellular buffering.
   b. In respiratory disease, the extent of the disorder is measured by the deviation of PCO_2 from normal, and the correction is effected by altering ventilation to restore the PCO_2 to normal.

*As the PCO_2 **increases**, the H_2CO_3 is buffered by the **nonbicarbonate buffers**; therefore, the plasma $[HCO_3^-]$ rises when the PCO_2 rises.

STUDY QUESTIONS

Directions: Each question below contains five suggested answers. Choose the **one best** response to each question.

1. Most of the CO_2 carried in whole blood is in the form of

(A) dissolved CO_2
(B) carbaminohemoglobin
(C) methemoglobin
(D) HCO_3^-
(E) protein carbamates

Questions 2–4

The following pH and blood gas findings were determined in a 30-year-old man while he was breathing room air.

$$pH = 7.30$$
$$PCO_2 = 60 \text{ mm Hg}$$
$$PO_2 = 38 \text{ mm Hg}$$

2. Prior to treatment, the patient described above would have a plasma $[HCO_3^-]$ (in mmol/L) of approximately

(A) 48
(B) 38
(C) 28
(D) 18
(E) 8

3. The total plasma CO_2 content (in mmol/L) of this patient would be approximately

(A) 50
(B) 40
(C) 30
(D) 20
(E) 10

4. Prior to correction of the acid-base abnormality, the urine of this patient would show

(A) a high pH
(B) glycosuria
(C) an increased inulin clearance
(D) hypernatriuria
(E) a very low HCO_3^- concentration

Questions 5–7

A 62-year-old woman who is brought to the emergency room is in a state of confusion, is unable to answer questions coherently, and exhibits tachypnea. The laboratory results are listed below.

Plasma Findings	Arterial Blood Findings
$[Na^+] = 144 \text{ mEq/L}$	pH = 7.17
$[K^+] = 4.4 \text{ mEq/L}$	$PCO_2 = 25 \text{ mm Hg}$
$[Cl^-] = 107 \text{ mEq/L}$	
$[HCO_3^-] = 9 \text{ mEq/L}$	
Ketonemia not present	

5. The laboratory findings for the patient described above are consistent with which of the following conditions?

(A) Respiratory acidosis
(B) Metabolic acidosis
(C) Insulin-dependent diabetes mellitus
(D) Decreased anion gap
(E) Diarrhea

6. This patient's condition is most likely attributed to

(A) salicylate poisoning
(B) increased fat catabolism
(C) myasthenia gravis
(D) decreased insulin secretion
(E) severe vomiting

7. What is the most likely treatment for this patient's condition?

(A) A mechanical respirator
(B) Insulin
(C) KCl
(D) $NaHCO_3$
(E) Rebreathing into a paper bag

8. Which of the following conditions most likely leads to acidosis combined with marked dehydration?

(A) Chronic vomiting
(B) Excessive ingestion of citrus fruits
(C) Severe diarrhea
(D) Drinking sodium lactate solution
(E) Water deprivation for 1 day

9. A patient has a plasma $[HCO_3^-]$ of 15 mmol/L and PCO_2 of 50 mm Hg. Based on these data, the most likely diagnosis is

(A) compensated respiratory alkalosis
(B) compensated respiratory acidosis
(C) compensated metabolic acidosis
(D) combined respiratory and metabolic acidosis
(E) combined respiratory alkalosis and metabolic acidosis

Directions: Each question below contains four suggested answers of which **one or more** is correct. Choose the answer

A if **1, 2, and 3** are correct
B if **1 and 3** are correct
C if **2 and 4** are correct
D if **4** is correct
E if **1, 2, 3, and 4** are correct

10. Compensatory responses to metabolic acidosis include

(1) hyperventilation
(2) an increase in titratable acid excretion
(3) an increase in NH_4^+ excretion
(4) an increase in glycolysis by H^+ activation of phosphofructokinase

Questions 11–13

A 75-year-old man with peptic ulcer disease has severe vomiting for 1 week and is admitted to the hospital. During the previous 6 months this man has experienced recurrent epigastric pain, which is relieved by tablets of aluminum hydroxide gel or by ingestion of a meal. The patient now reports that for the last 6 weeks he has had a decreased appetite, an early sense of fullness when eating a meal, and vomiting. The following laboratory data are reported.

Serum Findings	Urine Findings
$[Na^+]$ = 140 mEq/L	$[Na^+]$ = 10 mEq/L (N = 50–130 mEq/L)
$[K^+]$ = 2.2 mEq/L	$[K^+]$ = 21 mEq/L (N = 20–70 mEq/L)
$[Cl^-]$ = 86 mEq/L	$[Cl^-]$ = 3 mEq/L (N = 50–130 mEq/L)
$[HCO_3^-]$ = 42 mEq/L	pH = 5.0 (N = 5–7)
PCO_2 = 53 mm Hg	
PO_2 = 68 mm Hg	
pH = 7.53	

11. The patient described above has alkalosis; likely causes of this alkalosis include

(1) primary hyperventilation due to abdominal pain
(2) excessive intake of aluminum hydroxide
(3) hypokalemia due to loss of K^+ in the vomitus
(4) substantial loss of HCl due to chronic vomiting

12. This patient is likely to demonstrate which of the following physiologic and chemical responses?

(1) Excretion of alkaline urine
(2) Compensatory hypoventilation
(3) Decreased reabsorption of HCO_3^-
(4) Intracellular alkalosis

13. Corrective procedures for this alkalotic patient include the administration of

(1) citric acid
(2) KCl
(3) 1 N solution of HCl
(4) neutral isotonic NaCl

14. "Titratable acid" represents a number of urinary organic and inorganic substances. The components of titratable acid include

(1) NH_4^+
(2) ketoacids
(3) NH_3
(4) monobasic $H_2PO_4^-$

15. In the treatment of hypokalemic alkalosis, it is necessary to

(1) restore vascular volume
(2) reduce renal HCO_3^- reabsorption
(3) administer Cl^-
(4) excrete an alkaline urine

Directions: The groups of questions below consist of lettered choices followed by several numbered items. For each numbered item select the **one** lettered choice with which it is **most** closely associated. Each lettered choice may be used once, more than once, or not at all.

Questions 16–19

Points A–D on the pH-$[HCO_3^-]$ diagram below indicate states of acid-base imbalance; point N indicates a normal acid-base state. Match each of the following conditions to the appropriate point (A–D) on the pH-$[HCO_3^-]$ diagram.

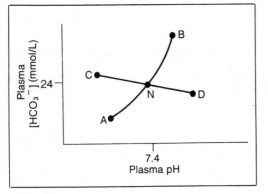

16. Hypocapnia
17. Hypercapnia
18. Ketoacidosis
19. NaHCO₃ ingestion

Questions 20–24

The table below shows arterial blood acid-base data for five individuals who are denoted by the letters A-E. For each of the following descriptions of acid-base status, choose the individual with the appropriate acid-base data.

	PCO₂ (mm Hg)	[HCO₃⁻] (mmol /L)	pH	[H⁺] (nmol /L)
(A)	25	15.0	7.40	40.0
(B)	40	25.0	7.41	38.4
(C)	45	35.0	7.51	30.8
(D)	54	27.5	7.37	47.1
(E)	60	15.0	7.02	96.0

20. Normal

21. Partially compensated metabolic alkalosis

22. Fully compensated respiratory alkalosis

23. Uncompensated respiratory acidosis

24. Combined respiratory and metabolic acidosis

Questions 25–29

The table below shows blood data for five patients who are designated by the letters A–E. For each of the following patients, select the blood data that define his or her condition.

	pH	PCO_2 (mm Hg)	$[HCO_3^-]$ (mmol/L)
(A)	7.22	60	26
(B)	7.35	60	32
(C)	7.37	65	37
(D)	7.47	20	14
(E)	7.53	50	40

25. A 60-year-old man with chronic bronchitis who develops diarrhea

26. An obese 24-year-old man

27. A 14-year-old girl with an acute attack of asthma

28. A 45-year-old woman with cirrhosis and ascites who develops acute gastrointestinal bleeding due to esophageal varices and is treated with 19 units of blood

29. A 30-year-old man who makes an ascent by mountain climbing

ANSWERS AND EXPLANATIONS

1. The answer is D. (*V A; Table 5-6*) About 90 percent of the total CO_2 content of blood is carried in the form of HCO_3^-. This ion is formed within the erythrocyte in a two-step process. First, CO_2 is hydrated via carbonic anhydrase to form H_2CO_3. The H_2CO_3 then is dissociated into H^+ and HCO_3^-. The H^+ formed within the erythrocyte is buffered primarily by the hemoglobin buffer system, while most of the HCO_3^- generated within the erythrocyte diffuses into the plasma in exchange for the inward diffusion of Cl^-.

2. The answer is C. (*II C*) The Henderson-Hasselbalch equation contains three variables (pH, $[H^+]$, and PCO_2), and two must be known in order to solve the equation, as:

$$pH = pK' + \log \frac{[HCO_3^-]}{S \times PCO_2}$$
$$pH = pK' + \log [HCO_3^-] - \log S \times PCO_2$$

then: $\qquad \log [HCO_3^-] = pH - pK' + \log S \times PCO_2,$

therefore: $\qquad [HCO_3^-] = 10^{(pH-pK')} (S \times PCO_2)$
$$= (\text{antilog } pH - 6.1) (S \times PCO_2).$$

In the example:
$$[HCO_3^-] = (\text{antilog } 7.3 - 6.1) (0.03 \, PCO_2)$$
$$= (\text{antilog } 1.2) (1.8)$$
$$= (15.8) (1.8)$$
$$= 28.5 \text{ mmol/L}.$$

Or, using the antilogarithmic equation, which is simpler:

$$[HCO_3^-] = \frac{24}{[H^+]} \, PCO_2,$$

where: $[H^+] = \text{antilog} - pH$. In the example, then:

$$[H^+] = \text{antilog} - 7.3$$
$$= \text{antilog} (-8 + 0.7)$$
$$= \text{antilog} (0.7 - 8)$$
$$= 5.01 \times 10^{-8}$$
$$= 50.1 \text{ nmol /L}.$$

Substituting and solving for $[HCO_3^-]$:

$$[HCO^-] = \frac{24}{[H^+]} \, PCO_2$$

$$= \frac{24}{50.1} \, 60$$

$$= 28.7 \text{ mmol/L}.$$

3. The answer is C. (*V A; Tables 5-6, 5-7, and 5-9*) The total CO_2 content is essentially the sum of $[HCO_3^-]$ and $S \times PCO_2$. Using the $[HCO_3^-]$ value determined above, then:

$$[\text{total } CO_2] = [HCO_3^-] + S \times PCO_2$$
$$= 28.7 + 1.8$$
$$= 30.5 \text{ mmol/L}.$$

4. The answer is E. (*III A; VI C, F; VII C 2 b; VIII C, D; IX B; Tables 5-8, 5-9, and 5-13*) The immediate compensatory response to respiratory acidosis ($\uparrow PCO_2$) is to increase the numerator (i.e., $[HCO_3^-]$). This increase in plasma $[HCO_3^-]$ is brought about by the buffering of H_2CO_3 by cell buffers (hemoglobin and proteins) as:

$$CO_2 + H_2O \rightarrow H_2CO_3 + Hb^- \rightarrow HHb + HCO_3^-,$$

where Hb^- = buffer anion; and HHb = buffered H^+. The second response is a slower increase in extracellular $[HCO_3^-]$ brought about by renal reabsorption of HCO_3^- with a reciprocal excretion of Cl^-, as:

$$Na^+ + Cl^- + H_2CO_3 \rightarrow Na^+ + HCO_3^- + H^+ + Cl^-$$
$$\text{(reabsorbed)} \quad \qquad \llcorner NH_3$$
$$NH_4^+ + Cl^-.$$
$$\text{(excreted)}$$

Thus, there is accelerated H^+ secretion and increased reabsorption of HCO_3^- in response to respiratory acidosis.

5. The answer is B. [*VII B 1 b (1) (a), B 2 b; VIII D 3; IX B 3, C; Fig. 5-9; Tables 5-8, 5-9, and 5-13*] Diabetes mellitus is a major cause of metabolic acidosis, and insulin-dependent diabetes mellitus is the most common cause of ketoacidosis. The absence of ketonemia indicates that this acidosis is not due to the lack of insulin and, therefore, rules out insulin-dependent diabetes mellitus. The patient has metabolic acidosis with a high anion gap without ketoacidosis. Diarrhea is a cause of metabolic acidosis but is also characterized by the gastrointestinal loss of K^+ (hypokalemia), which is not seen in this patient.

6. The answer is A. (*X C*) This patient has a high anion gap metabolic acidosis without ketoacidosis; these symptoms are consistent with salicylate toxicity or the ingestion of toxins (e.g., methanol, ethylene glycol, and paraldehyde). Decreased insulin secretion leads to fat catabolism and ketoacidosis, which are not present in this patient. Moreover, insulin deficiency is associated with hyperkalemia, which also is absent. Severe vomiting is a cause of metabolic alkalosis, and myasthenia gravis is a cause of respiratory acidosis. These two conditions are characterized by elevated $[HCO_3^-]$ and elevated PCO_2, respectively, which are not observed in this patient.

7. The answer is D. (*XIII C*) In most clinical situations, $NaHCO_3$ is the mainstay of therapy for metabolic acidosis. After $NaHCO_3$ administration, the plasma $[HCO_3^-]$, the arterial pH, and the PCO_2 will increase. The rise in PCO_2 is due to a decrease in alveolar ventilation induced by the inhibitory effect of the increased pH on the peripheral chemoreceptors. Since this patient is not diabetic, insulin is not needed. Also, the patient does not exhibit hypokalemia or hypochloremia; therefore, KCl administration is not part of the therapy. Lastly, rebreathing into a bag to increase the PCO_2 does not correct the underlying metabolic problem. The PCO_2 is low as a secondary response to the increased H^+, which is the primary acid-base imbalance.

8. The answer is C. (*III B 1 c; IX B 4*) In contrast to gastric secretions, intestinal secretions usually are relatively alkaline, containing a high concentration of HCO_3^-. Severe diarrhea leads to the loss of HCO_3^- and, therefore, is a cause of metabolic acidosis. Metabolic alkalosis is caused by excessive ingestion of citrus fruits and by chronic vomiting; the latter condition also causes dehydration. Drinking sodium lactate solution can cause acidosis and overhydration, and water deprivation can lead to dehydration and contraction alkalosis.

9. The answer is D. (*XII A–B; Fig. 5-15*) The condition of combined respiratory and metabolic acidosis is characterized by a high PCO_2, a low plasma $[HCO_3^-]$, and a low plasma pH. The arterial plasma pH of this patient is calculated as follows:

$$pH = pK + \log ([HCO_3^-]/S \times PCO_2)$$
$$= 6.1 + \log (15/1.5)$$
$$= 6.1 + \log 10$$
$$= 6.1 + 1.0 = 7.1$$

Respiratory acidosis is characterized by a high PCO_2 and a variable increase in plasma $[HCO_3^-]$. The latter characteristic represents the compensatory change in the alternate variable, which is in the same direction as the primary disturbance. In acute respiratory acidosis, the plasma $[HCO_3^-]$ should rise 1 mmol/L for each 10 mm Hg increase in PCO_2. The acute elevation of PCO_2 to 50 mm Hg should increase the plasma $[HCO_3^-]$ to 25 mmol/L (pH of 7.3), if this patient has acute respiratory acidosis. The low plasma $[HCO_3^-]$ of this patient, however, indicates a condition of combined respiratory and metabolic acidosis.

Metabolic acidosis is characterized by a decreased plasma $[HCO_3^-]$ and compensatory hyperventilation, which reduces the PCO_2. Respiratory alkalosis is associated with a low PCO_2 and a variable decrease in plasma $[HCO_3^-]$. The condition of combined respiratory alkalosis and metabolic acidosis is characterized by a low PCO_2 and a low plasma $[HCO_3^-]$.

10. The answer is A (1, 2, 3). *(VI D, E; VIII C, D 3; IX B 3; Fig. 5-14; Tables 5-8 and 5-9)* Acidosis stimulates the chemoreceptors that control respiration, causing an increase in alveolar ventilation. PCO_2 consequently declines in a patient with metabolic acidosis, and pH returns toward normal. This hyperventilatory response is characterized more by an increase in tidal volume than by an increase in respiratory rate (i.e., Kussmaul's respiration).

The renal response to increased H^+ load is increased H^+ excretion, which occurs primarily by augmented cellular NH_3 production and, consequently, NH_4^+ excretion. In contrast, there usually is a limited ability to enhance titratable acid secretion. In metabolic acidotic states, the **total H^+ secretion** is reduced because of the decrease in plasma $[HCO_3^-]$ and, therefore, in the quantity of filtered HCO_3^- that must be reabsorbed; however, the **net H^+ excretion** (which refers to the amount of H^+ eliminated in the urine and does not include the secreted H^+ used for HCO_3^- reabsorption) is increased and results in the addition of new HCO_3^- to the body.

There is an inverse relationship between $[H^+]$ and the activity of several glycolytic enzymes, particularly phosphofructokinase.

11. The answer is D (4). *[VII B 1 b (1) (b); VIII D 4; Tables 5-8, 5-9, and 5-13]* Vomiting is the primary cause of this patient's metabolic alkalosis. If this vomiting persists, K^+ depletion may result because vomiting (like diarrhea and continuous aspiration of intestinal fluids during surgery) causes rapid K^+ deficiency. The transcellular fluids that enter the gastrointestinal tract contain K^+ in concentrations that are 2 to 5 times higher than that of the ECF. It is important to understand that this alkalosis is due to the loss of Cl^- from the plasma and not the loss of H^+ from the stomach.

Respiratory alkalosis is also characterized by high pH ($\downarrow [H^+]$). In **contrast** to metabolic alkalosis, respiratory alkalosis is characterized by a low PCO_2 (hypocapnia) and a variable reduction in the plasma $[HCO_3^-]$. One way to establish the correct diagnosis is to treat the metabolic alkalosis and monitor the PCO_2. If the PCO_2 remains elevated, then the patient has an underlying respiratory acidosis.

Aluminum hydroxide $[Al (OH)_3]$ is classified as a nonsystemic gastric antacid because the cationic portion (Al) is not readily absorbed from the intestine. In the acid medium of the stomach, aluminum chloride ($Al Cl_3$) is formed; however, in the alkaline fluid of the intestine, $Al (OH)_3$ is reformed and the Cl^- is absorbed. Consequently, no alteration in systemic acid-base balance occurs with the ingestion of $Al (OH)_3$.

The primary cause of this patient's metabolic alkalosis is not hypokalemia, although hypokalemia and metabolic alkalosis commonly occur together. In alkalotic states, H^+ migrate from the ICF compartment into the ECF compartment in an attempt to restore the $[H^+]$ of the ECF. To preserve electroneutrality, K^+ moves into the ICF and lowers the plasma $[K^+]$. The increase in intracellular $[K^+]$ favors increased renal K^+ secretion.

12. The answer is C (2, 4). *(VIII D 4; IX D 4; Fig. 5-14; Tables 5-8, 5-9, and 5-13)* The pulmonary compensatory response to metabolic alkalosis is hypoventilation, which leads to increases in PCO_2, H_2CO_3, $[H^+]$, and $[HCO_3^-]$. It is well known that as PCO_2 rises, the H_2CO_3 is buffered by nonbicarbonate buffer; thus, the $[HCO_3^-]$ rises when the PCO_2 rises. As the PCO_2 increases, even more HCO_3^- is added to the ECF, and the increased H^+ decreases the pH toward normal. The limitation on this pulmonary response is the PaO_2, because hypoxemia is a potent stimulus of the respiratory centers through the mediation of the peripheral chemoreceptors.

This patient does not have an uncomplicated alkalosis as occurs with excess HCO_3^- ingestion. The renal response to this type of alkalosis is increased urinary excretion of HCO_3^- (alkaluria). Rather, this patient has a complicated metabolic alkalosis caused by the loss of HCl together with a "contraction alkalosis" caused by depletion of the ECF volume. The loss of HCl and water are caused by chronic vomiting. The resultant hypovolemia stimulates the secretion of aldosterone, which promotes the renal reabsorption of Na^+ and the secretion of H^+ (i.e., Na^+–H^+ cation exchange). In turn, the increased H^+ secretion promotes the indirect reabsorption of HCO_3^- via the formation of H_2CO_3 within the tubule and the dehydration of H_2CO_3 to form CO_2 and water, which are reabsorbed. The increased reabsorption of HCO_3^- exacerbates the metabolic alkalosis and, together with the increased secretion and excretion of H^+, account for the paradoxical finding of an acid urine in the presence of an extracellular alkalosis.

In metabolic alkalosis, the increase in plasma $[HCO_3^-]$ is associated with a decrease in the plasma $[Cl^-]$ due, in this case, to vomiting. Thus, less Cl^- is available for reabsorption with Na^+. As a result, Na^+ reabsorption requires an increase in tubular H^+ secretion and, therefore, in HCO_3^- reabsorption. Since K^+ is present in significant concentrations in gastrointestinal fluids, whole body K^+ deficiencies may occur with vomiting.

Extracellular alkalosis also tends to produce hypokalemia partly as a result of K^+ loss in the vomitus and mainly as a result of a larger K^+ depletion via urinary excretion. This latter effect is caused by movement of H^+ from the ICF to the plasma and the concomitant cellular uptake of K^+ to maintain electroneutrality. This increase in the intracellular $[K^+]$ promotes the net renal secretion of K^+. Also, the cellular efflux of H^+ leads to an intracellular alkalosis. Thus, cellular stores of K^+ are depleted via the gastrointestinal tract and via the urine, and these combined losses account for the hypokalemia found in this patient.

Patients with renal K^+ wasting should excrete more K^+ than observed in this patient, whose urine $[K^+]$ is 21 mEq/L. However, high values of K^+ do not necessarily imply renal K^+ wasting, since the urine ion **concentration** must be correlated with urine **volume**. For example, this urine $[K^+]$ of 21 mEq/L would not be excessive with a daily urine volume of 500 ml. K^+ excretion also is increased as a result of enhanced aldosterone secretion. It also should be emphasized that hypokalemia promotes renal HCO_3^- reabsorption and contributes to the continuation of metabolic alkalosis. Hypokalemia is a frequent finding in patients with metabolic alkalosis.

13. The answer is C (2, 4). (*VIII D; IX B 4 b*) When ingested in large amounts, Na^+ and K^+ salts of organic acids (e.g., lactate and citrate) are metabolized to CO_2 and water and lead to a further elevation of serum $[HCO_3^-]$. Therefore, these agents are not effective in treating alkalosis. Indeed, the major source of alkali is the oxidation of organic anions. For example:

$$citrate^- + O_2 \rightarrow CO_2 + H_2O + HCO_3^-.$$

Other organic ions that rapidly metabolize to HCO_3^- include acetate, lactate, and, in the presence of insulin, the anions of ketoacids.

The aim of the therapy is to restore volume and K^+ balance, which will reduce HCO_3^- reabsorption. The restoration of circulating blood volume inhibits the secretion of aldosterone and permits the excess HCO_3^- to be excreted in the urine as $NaHCO_3$. An essential requirement is to resupply H^+ with Cl^-. The restoration of the circulating blood volume and repletion of Cl^- are accomplished by the oral or intravenous administration of NaCl and water as half-isotonic or isotonic saline. The test for effective corrective treatment is the excretion of an alkaline urine. These reactions are summarized as:

$$\underset{\substack{\\ \text{(retained)}}}{Na^+ + Cl^- + \overset{\displaystyle CO_2 + H_2O}{\overset{\curlyvee}{H_2CO_3}}} \rightarrow \underset{\substack{\\ \text{(excreted)}}}{H^+ + Cl^- + Na^+ + HCO_3^-.}$$

The oral administration of K^+ is required to restore K^+ balance. The correction of hypokalemia by the intravenous administration of K^+ is potentially dangerous, because it may produce a sudden increase in the extracellular K^+ concentration and thereby depolarize cardiac muscle. This hazard is avoided by the **oral** administration of K^+, since absorption from the gastrointestinal tract is gradual. Soups, fruit juices, and bananas are particularly rich in K^+. In mild K^+ depletion, the contraction of the ECF volume may be the underlying cause of alkalosis, and HCO_3^- reabsorption can be decreased by administering isotonic saline with supplemental K^+. In patients with severe hypokalemia and in those who cannot eat (due to vomiting), K^+ must be administered slowly and intravenously as a KCl solution.

14. The answer is C (2, 4). (*VI D–E*) Titratable acid is determined by the amount of NaOH that must be added to a 24-hour urine collection to bring it to a pH of 7.4. H^+ cannot be excreted as free H^+ but must be buffered in the urine to minimize the fall in urine pH. The amount of H^+ excreted in the form of free H^+ is negligible. Only weak acids with pK values of approximately 4.6 or above are involved. These include primarily organic acids (i.e., ketoacids and uric acid) and phosphate.

The pK of the NH_3/NH_4^+ system is 9.3. The very low concentration of NH_3 in urine makes it an ineffective buffer. NH_4^+ is not a fraction of titratable acid. It is important to differentiate urinary acidification from H^+ excretion. The ability to reduce the pH of urine is acidification and under normal conditions represents primarily $H_2PO_4^-$ (titratable acidity). However, the amount of H^+ being excreted represents titratable acid **and** NH_4^+. Thus, the ability to reduce urinary pH does not necessarily reveal much about the amount of H^+ being excreted.

15. The answer is E (all). (*XIII D*) The administration of filtered Cl^- and the correction of volume depletion are the cornerstones for the treatment of metabolic alkalosis; a third crucial step is the correction of hypokalemia, which usually exists. The renal compensation for metabolic alkalosis depends more on concomitant levels of Cl^-, extracellular fluid volume, and K^+ than it does on the magnitude of noncarbonic acid deficit or alkali excess.

The deficient Cl^- is provided by the administration of neutral isotonic saline with the concomitant excretion of Na^+ and HCO_3^-. If renal function is normal, this compensation occurs without difficulty. The test for efficacy of treatment of a hypokalemic alkalotic patient is the production of an alkaline urine.

16–19. The answers are: 16-D, 17-C, 18-A, 19-B. (*VII A 5, B 1 b (1), C 2 a (2), (3), D; VIII D; IX B 2–4; Figs. 5-8, 5-13, and 5-14; Tables 5-6, 5-8, 5-9, 5-10, and 5-12*) In analyzing these four acid-base disturbances, it is important to consider the PCO_2 and $[HCO_3^-]$ rather than the pH, as PCO_2 and $[HCO_3^-]$ are the key determinants of the cause of, and compensation for, these disturbances.

Point D represents a patient with an increased pH ($\downarrow [H^+]$) brought about by respiratory alkalosis. This condition is characterized by a primary decrease in PCO_2 (hypocapnia or hypocarbia) and a variable secondary decrease in the plasma [HCO_3^-]. Metabolic acidosis also is characterized by declines in these two variables, but the pH is decreased ($\uparrow$ [H^+]) as well. Respiratory alkalosis is defined as alveolar ventilation in excess of the existing need of the body to eliminate CO_2. This excess in alveolar ventilation, called hyperventilation, results in a reduced arterial PCO_2. Hyperpnea is the general term used to describe any increase in ventilatory effort. With respiratory alkalosis there is a decline in the [total CO_2].

Point C represents a patient with decreased pH ($\uparrow$ [H^+]) caused by respiratory acidosis. This clinical disorder is characterized by a primary elevation in the PCO_2 (hypercapnia or hypercarbia) and a variable secondary increase in the plasma [HCO_3^-]. The common denominator in respiratory acidosis is hypoventilation, which is defined as alveolar ventilation insufficient to excrete CO_2 rapidly enough to meet the existing needs of the body. With respiratory acidosis there is a relatively small increment in the [total CO_2], because the major fraction of the CO_2 content is comprised of HCO_3^-.

Point A represents a patient with diabetes mellitus, which is the most common cause of ketoacidosis. This overproduction of ketoacids is caused by a deficiency of insulin, which leads to: (1) increased lipolysis and an increased delivery of free fatty acids to the liver and (2) the preferential conversion of free fatty acids to ketoacids rather than to triglycerides. Thus, metabolic acidosis is characterized by a low arterial pH ($\uparrow$ [H^+]), a reduced [HCO_3^-], and a compensatory hyperventilation resulting in hypocapnia. The renal compensatory response for respiratory alkalosis also diminishes the plasma [HCO_3^-], but the pH in that disorder is elevated ($\downarrow$ [H^+]).

Overproduction of β-hydroxybutyric and acetoacetic acids causes acidosis by two mechanisms: (1) a decrease in plasma [HCO_3^-] with an increase in the anion gap and (2) overloading of the renal capacity to excrete H^+ resulting in a loss of Na^+ and a failure to recover $NaHCO_3$. In metabolic acidosis, there is a decline in the [total CO_2]. Furthermore, ketoacidosis, like lactic acidosis, differs from other forms of metabolic acidosis in that the anion associated with H^+ can be metabolized back to HCO_3^-, as:

$$\beta\text{-hydroxybutyrate}^- + O_2 \rightarrow CO_2 + H_2O + HCO_3^-.$$

β-hydroxybutyrate represents about 75 percent of the circulating ketoacids in diabetic ketoacidosis. It can be seen from the above chemical equation that the metabolism of the β-hydroxybutyrate anion results in the regeneration of the HCO_3^- that was neutralized in buffering the H^+. Since the HCO_3^- is replaced by an anion that is metabolized back to HCO_3^-, there is no actual loss of HCO_3^- from the body in ketoacidosis (or lactic acidosis). Insulin administration decreases the accumulation of β-hydroxybutyric acid and allows the metabolism of the acid ions back to HCO_3^-.

Point B represents a patient with metabolic alkalosis. Excessive ingestion of $NaHCO_3$ can result in metabolic alkalosis and an increase of pH ($\downarrow$ [H^+]). Metabolic alkalosis is characterized by an increase in the plasma [HCO_3^-] and a compensatory increase in the PCO_2 produced by a decline in alveolar ventilation. Since elevation of plasma [HCO_3^-] can be due to the renal compensation for chronic respiratory acidosis, the diagnosis of metabolic alkalosis cannot be made without measuring the pH. Metabolic alkalosis is associated with a large increase in the [total CO_2].

20–24. The answers are 20-B, 21-C, 22-A, 23-D, 24-E. [*VII A 2–4, B 1 b (1) (b); VIII D 1, 2, 4; IX B 1, 2, 4; XII B 1; Figs. 5-12, 5-13, 5-14, and 5-15; Tables 5-2, 5-6, 5-8, 5-9, and 5-12*] Metabolic alkalosis is characterized by a high pH ($\downarrow$ [H^+]), a high [HCO_3^-], and a compensatory increase in PCO_2. The individual designated by (**C**) shows a change in the [H^+] that is opposite in direction to the change in [HCO_3^-], which indicates that the primary disturbance is metabolic. The plasma pH of this individual (7.5) indicates alkalosis. The increment in PCO_2 shows that partial respiratory compensation has occurred.

Respiratory alkalosis is characterized by a low [H^+] ($\uparrow$ pH), a low PCO_2 (hypocapnia), and a variable decrease in plasma [HCO_3^-]. The individual denoted by (**A**) shows a compensatory change in the alternate variable ([HCO_3^-]) that is in the same direction as the change in the primary variable (PCO_2); both variables decrease by 25 percent. This characteristic denotes a respiratory disturbance. In this individual, the pH of 7.4 together with hypocapnia and decreased [HCO_3^-] indicates full compensation for a respiratory alkalosis with a [HCO_3^-]/$S \times PCO_2$ ratio of 20:1.

The individual designated by (**D**) has uncompensated respiratory acidosis as indicated by a 35 percent increase in PCO_2 (hypercapnia) with only a 12 percent compensatory increase in the [HCO_3^-].

The condition of respiratory acidosis with a complicating metabolic acidosis is characterized by an acid pH ($\uparrow$ [H^+]), high PCO_2 (hypercapnia), and low [HCO_3^-]. In acute respiratory acidosis, the plasma [HCO_3^-] should rise 1 mmol/L for each 10 mm Hg increase in PCO_2. The acute increase of the PCO_2 to 60 mm Hg, shown by individual (**E**), should increase the [HCO_3^-] to 26 mmol/L (pH of 7.26) in a case of acute respiratory acidosis. The low [HCO_3^-] of this individual, however, indicates that his or her condition is a combined respiratory and metabolic acidosis.

25–29. The answers are: 25-A, 26-B, 27-A, 28-E, 29-D. (*X A 1, 2, B 8, E 1; XII A–B; Fig. 5-15*) To diagnose these patients' conditions, first determine the acid-base disorders defined by the three variables given in the table.

At a PCO_2 of 60 mm Hg, the plasma $[HCO_3^-]$ should be approximately 26 mmol/L in acute respiratory acidosis (1 mmol/L for each 10 mm Hg increase in PCO_2) and 31 mmol/L in chronic respiratory acidosis (3.5 mmol/L for each 10 mm Hg increase in PCO_2). Therefore, the data designated by (**A**) are consistent with either acute respiratory acidosis (due to bronchitis) or metabolic acidosis (lowering of the $[HCO_3^-]$ from 31 to 26 mmol/L, due to diarrhea) superimposed on chronic respiratory acidosis.

The data designated by (**B**) are consistent with chronic respiratory acidosis (hypercapnia due to hypoventilation) or metabolic alkalosis (raising of the $[HCO_3^-]$ from 24 to 32 mmol/L) superimposed on acute respiratory acidosis. Chronic hypercapnia can be seen in extremely obese patients (Pickwickian syndrome), in when the increased weight of the chest wall increases the work of breathing.

Acute respiratory acidosis is caused by asthma. An acute increase in PCO_2 to 60 mm Hg should cause a 1 mmol/L rise in plasma $[HCO_3^-]$ for each 10 mm Hg increase in PCO_2 or a $[HCO_3^-]$ of 26 mmol/L. Such an increase in PCO_2 also should lower the pH to 7.22, if there was no buffering and the $[HCO_3^-]$ remained at 24 mmol/L. However, the increase in $[HCO_3^-]$ from 24 to 26 mmol/L indicates an increase in buffering capacity with the pH remaining at 7.22. Therefore, the data designated by (**A**) appear to be consistent with a respiratory acidosis or respiratory acidosis superimposed on a metabolic acidosis. It should be noted again that an increase in plasma $[HCO_3^-]$ can represent either a metabolic alkalosis or the compensatory response to respiratory acidosis, and, therefore, the acid-base disorder cannot be diagnosed with certainty from the plasma $[HCO_3^-]$ alone.

The data designated by (**E**) are consistent with the acute metabolic alkalosis of the cirrhotic, ascitic patient. This woman's metabolic acidosis is due to the citrate load associated with the blood transfusions. Since cirrhosis with ascites is associated with the depletion of the effective circulating blood volume, the urinary $[Na^+]$ of this woman should be less than 10 mEq/L. The Na^+ conservation requires an increase in H^+ secretion and, therefore, an increase in HCO_3^- reabsorption (acid urine). The net effect is that volume is maintained at the expense of extracellular pH, the correction of which would require the loss of sodium bicarbonate $(NaHCO_3)$ in the urine and would result in further volume depletion.

Because of the ascites, volume expansion with saline is not indicated. Acetazolamide (Diamox) inhibits carbonic anhydrase, which plays an important role in the proximal reabsorption of HCO_3^-. Acetazolamide produces a $NaHCO_3$ diuresis, which tends to restore acid-base balance.

The data designated by (**D**) [i.e., alkaline pH and hypocapnia] are diagnostic of respiratory alkalosis. With a PCO_2 of 20 mm Hg, the plasma $[HCO_3^-]$ should be about 19 mEq/L in acute respiratory alkalosis (a decrease of 2 mEq/L per 10 mm Hg fall in PCO_2) and 14 mEq/L in chronic respiratory alkalosis (a reduction of 5 mEq/L per 10 mm Hg decline in PCO_2). In the mountain climber, the plasma $[HCO_3^-]$ of 14 mEq/L is consistent with uncomplicated chronic respiratory alkalosis.

6
Gastroenterology

I. ORAL CAVITY, PHARYNX, AND ESOPHAGUS

A. MASTICATION

1. Control. Mastication, the first (although not absolutely essential) step in digestion, has both voluntary and involuntary components. The involuntary component is represented by the **chewing reflex.**
   a. When food touches buccal receptors, the masseter, medial pterygoid, and temporal muscles, which hold the jaws closed, are relaxed or inhibited by reflex. The digastric and lateral pterygoid muscles, which open the jaws, are contracted.
   b. As the lower jaw drops, a stretch reflex in the muscles that close the jaw leads to rebound contraction.
   c. When the jaw closes and the food bolus is again pressed against the buccal surface, the cycle is repeated.
   d. Cycle length is variable, averaging 1 second.
   e. Pressures generated by the jaw muscles reach 11–25 kg on the incisors and 29–90 kg on the molars.

2. Functions.
   a. Mastication has several mechanical purposes:
      (1) To crush and break food into a size suitable for swallowing
      (2) To break apart otherwise undigestible cellulose coatings on plant foods, allowing later digestion of the nutrients within
      (3) To increase the surface areas of all food products that will later be exposed to digestive enzymes
   b. In addition, chewing mixes food with saliva, which serves several purposes (See Section I C 5) and permits:
      (1) Dissolution of readily soluble components of food
      (2) Initiation of digestion of starch by salivary amylase
      (3) Lubrication and softening of the bolus, which aids in swallowing
   c. Mastication stimulates the taste buds and releases food odors, which stimulate the olfactory receptors. These serve both to increase the pleasure of eating and to initiate gastric secretion.

B. DEGLUTITION

1. There are **three phases** of swallowing.
   a. Oral (Voluntary) Phase. After the bolus is formed, the anterior tongue is pressed against the hard palate with the jaws closed and the soft palate elevated, and the bolus is propelled into the oropharynx.
   b. Pharyngeal Phase.
      (1) Receptors in the oropharynx are stimulated to initiate an involuntary and precisely sequenced reflex response as afferent impulses are relayed to the **swallowing center** in the medulla and lower pons.
      (2) Contraction of the palatopharyngeal muscle pulls the soft palate upward to close the posterior nares, preventing regurgitation of food into the nasal cavities.
      (3) The palatopharyngeal folds on either side of the pharynx are pulled medially to form a sagittal slit or pharyngeal gate, which allows properly masticated food to pass rapidly to the posterior pharynx while impeding the passage of large food fragments. Since this process lasts only 1 second, larger objects usually are not permitted to pass into the esophagus.

(4) The larynx is raised, the vocal cords are closed, and the epiglottis swings over the superior laryngeal opening.

(5) The swallowing center directly inhibits the respiratory center for the duration of this phase (a maximum of 2 seconds).

(6) Pharyngeal pressure rises sharply from atmospheric pressure at rest to 100 mm Hg for 0.5 second as the bolus passes.

c. Esophageal Phase.

(1) Movement of the Bolus. Two esophageal sphincters (constrictors) move the bolus from the pharynx to the stomach.

(a) The pharyngoesophageal (hypopharyngeal) sphincter is formed by the cricopharyngeal muscle.

(b) The gastroesophageal sphincter [lower esophageal sphincter (LES)] is a 4–6 cm zone, the location of which varies with respiration. It is located 3 cm above and 3 cm below the diaphragm at the end of expiration and 1 cm above and 3 cm below the diaphragm at the end of the inspiration.

(2) Three forces guide the bolus to its destination.

(i) Momentum is imparted by a fast persistaltic wave that begins in the pharynx and pushes the bolus into the esophagus.

(ii) Force is exerted by esophageal persistalsis.

(iii) Gravity alone is an important force: In the upright position the bolus literally falls toward the stomach.

(3) Sequence.

(a) At rest, the pressure of the hypopharyngeal sphincter normally is higher than the intrathoracic pressure of the midesophagus in order to prevent air entry. As the pharyngeal phase of swallowing proceeds, the brief peak of pharyngeal pressure combines with upward motion of the larynx and cricoid to open the sphincter and allow passage of the bolus. The sphincter then closes rapidly, increasing pressure to 90–100 mm Hg to prevent reflux. As the bolus reaches the clavicular level, the larynx descends, the glottis opens, and respiration resumes.

(b) As hypopharyngeal pressure peaks, distention produced by the bolus initiates esophageal peristalsis. Unlike peristalsis elsewhere in the gut, the process in both the striated muscle of the upper third and the smooth muscle of the lower two-thirds of the esophagus is not propagated myogenically by intrinsic plexuses. It is, rather, controlled by the swallowing center via impulses conducted by the vagus nerve. Traveling at 2–4 cm/sec and generating pressures of 30–120 mm Hg, the peristaltic wave reaches the stomach in 9 seconds. If liquids are swallowed repeatedly, however, the esophagus remains relaxed until the final swallow, which is followed by a contraction wave.

(c) The LES, which at rest maintains a pressure 5–10 mm Hg greater than intragastric pressure, relaxes within 1.3 seconds as the relaxation wave that precedes the wave of peristaltic contraction sweeps over it and passes onto the fundus. Following passage of the wave, the sphincter again contracts.

2. Disorders of Swallowing.

a. Esophageal Reflux.

(1) Factors Normally Preventing Reflux.

(a) Despite the fact that during normal quiet breathing gastric pressure is approximately 10 mm Hg higher than esophageal pressure, LES pressure exceeds this higher figure by 5 mm Hg, which prevents reflux. The LES resists distention, yielding only to pressures of 25 mm Hg or greater in normal individuals.

(b) Gastrin, which is released by entry of food into the stomach, increases LES pressure, thus preventing reflux when the stomach is full.

(c) Since the LES is partly below the diaphragm, anything that increases intra-abdominal pressure, such as abdominal compression or forced expiration against a closed glottis, also increases LES pressure, thus maintaining the pressure gradient.

(d) In addition, mucosal folds in the gastric cardia may act as a valve.

(2) Clinical Considerations.

(a) Reflux of gastric acid may be associated with **heartburn (pyrosis),** with the esophageal pH dropping as low as 2; however, in some individuals a drop to 4 can produce symptoms.

(b) Motor abnormalities associated with reflux include:

(i) Nonpropulsive synchronous esophageal contractions (diffuse esophageal spasm)

(ii) Prolonged peristalsis

(iii) Increased intraluminal pressure

    (c) Physiologic conditions may predispose to reflux.
      (i) In pregnancy, the enlarged uterus displaces the distal esophagus, which normally is located intra-abdominally, into the thorax, decreasing LES resistance. Approximately 50 percent of women suffer heartburn until the uterus descends in the last few weeks of pregnancy.
      (ii) In the neonatal period, reflux occurs because controls of the swallowing mechanism are not fully developed. Not only does the LES fail to close between swallows, but peristalsis does not occur in the lower esophagus, and the hypopharyngeal sphincter does not close tightly.
  b. Belching (Eructation). Following a heavy meal or the ingestion of large amounts of gas (as in the form of carbonated beverages), the gastric gas bubble is displaced from its usual location in the fundus to the cardia. When the LES relaxes during the swallowing process, gas enters the esophagus and is regurgitated.
  c. Achalasia is a neuromuscular disorder of the lower two-thirds of the esophagus, which leads to absence of peristalsis and failure of the LES to relax. Food accumulates above the LES, taking hours to enter the stomach and dilating the esophagus. In its extreme form this dilatation produces the condition of **megaesophagus.**

C. SALIVARY SECRETION

  1. Secretory structures include the three paired major salivary glands: the **parotid** glands, which are histologically of the serous type, and the **submaxillary** (submandibular) and **sublingual glands,** which are both of the mixed serous and mucous type. The many small buccal and labial glands are also of the mixed type.

  2. Composition of Saliva.
    a. Daily volume of the saliva is 1000–1500 ml.
    b. Although **pH** ranges from 6.0–7.4, it is generally near 7.0.
    c. Ionic Composition. Electrolyte concentration and osmolality both vary with secretory flow rates, but generally, in comparison to plasma, saliva contains higher concentrations of potassium (K^+) and bicarbonate (HCO_3^-) and lower concentrations of sodium (Na^+) and chloride (Cl^-) [see Section I C 3 d].
    d. Protein Composition. Two major types of protein are found in saliva.
      (1) Salivary amylase (α-amylase or ptyalin), the major protein of the serous secretions, is a digestive enzyme that initiates the breakdown of starches within the mouth. This digestion continues within the stomach for a variable period of time until the enzyme is inactivated by gastric pH.
      (2) Mucin, the major protein of the mucous secretions, is a glycoprotein that serves the lubricating function of saliva.

  3. Mechanism of Secretion.
    a. A typical salivary gland (of the mixed type) is composed of both **acini** and **salivary ducts;** each is important in one of the two steps of the secretory process.
      (1) The **acini produce a primary secretion,** which resembles an ultrafiltrate of plasma, although it is actually a product of active transport and contains α-amylase, mucin, or both.
      (2) In the salivary ducts, the primary secretion is modified by the secretion and absorption of electrolytes. **First,** in the ducts near their junction with the acini, HCO_3^- is actively secreted, a process catalyzed by carbonic anhydrase, which also necessarily entails passive Cl^- reabsorption. **Second,** further along the ducts, Na^+ is actively absorbed and K^+ is actively secreted.
    b. Secretory flow rates determine the composition of saliva. At rest, Na^+ and Cl^- concentrations are low in comparison to plasma, while those of HCO_3^- and K^+ are high. When flow rates increase, less time is available for active processes in the ducts, and saliva resembles the primary secretion.
    c. Exchange of substances between capillary blood and salivary fluid is facilitated by **countercurrent blood flow** (i.e., blood reaches the salivary ducts before it reaches the acini).
    d. Aldosterone increases Na^+ and Cl^- reabsorption and K^+ secretion.
    e. Salivary glands, like the thyroid, function as an iodide trap—that is, they concentrate or extract iodide from the plasma.

  4. Control of Salivary Secretion.
    a. The control of salivary secretion is exclusively neural. Each gland is supplied by both parasympathetic and sympathetic nerves.
    b. The salivary glands are controlled by the **superior and inferior salivatory nuclei,** which are found in the brain stem near the junction of the medulla and pons.

    c. Parasympathetic signals from these nuclei to the salivary glands produce vasodilation and profuse watery secretion, which is high in osmolality but low in protein.

    d. Excitatory impulses reach the salivatory nuclei from receptors in the tongue and mouth in response to taste and tactile stimuli. Central nervous system (CNS) centers also influence the nuclei, which is obvious when it is considered that the smell and sight of food can induce salivation, as can even the memory or "thought" of food.

    e. Salivatory reflexes produce secretion when food reaches several points in the alimentary tract, including the mouth, stomach, and upper small intestine.

    f. Sympathetic stimuli from the superior cervical ganglion cause vasoconstriction and, at least from the submaxillaries, secretion of a reduced amount of saliva, which is high in protein content but low in osmolality.

    g. Atropine and other cholinergic blockers decrease salivation.

5. Functions.

    a. Saliva facilitates chewing, swallowing, and speech by moistening and lubrication.

    b. It dissolves readily soluble components of food, thus permitting stimulation of taste buds.

    c. It cleanses and helps to maintain healthy oral tissues through its content of proteolytic enzymes (including lysozyme), antibodies, and other antibacterial factors.

    d. Saliva initiates starch digestion via salivary amylase.

    e. It aids in excretion of heavy metals such as lead and mercury as well as other substances such as iodide and thiocyanate.

II. STOMACH

A. MOTOR ACTIVITY

1. General Considerations.

    a. Anatomy.

        (1) The three functional parts of the stomach are the **fundus, corpus** (body), and **antrum.**

        (2) Gastric contents are isolated from other parts of the gastrointestinal tract by the LES proximally and by the pylorus (pyloric sphincter) distally.

        (3) The antrum and pylorus are anatomically continuous and respond to nervous control as a unit.

        (4) As elsewhere in the gut, each muscle layer in the stomach forms a functional syncytium and therefore acts as a unit. In the fundus, where the layers are relatively thin, strength of contraction is weak; in the antrum, where the muscle layers are thick, strength of contraction is greater.

        (5) The stomach and duodenum are divided by a hypomuscular segment; only a few longitudinal muscle bundles traverse this region.

        (6) Once food has been mixed with gastric secretions, it is referred to as **chyme.**

    b. Innervation.

        (1) Intrinsic. The interconnected myenteric plexus **(Auerbach's plexus)** and submucosal plexus **(Meissner's plexus)** within the stomach wall comprise the intrinsic innervation of the stomach, as they do elsewhere in the gut. They are **directly responsible for peristalsis** and other contractions. Because this system is continuous between the stomach and duodenum, peristalsis in the antrum influences the duodenal bulb.

        (2) Extrinsic. Autonomic innervation is dual—both **sympathetic,** via the celiac plexus, and **parasympathetic,** via the vagus nerve. Sympathetic innervation inhibits motility, and parasympathetic innervation stimulates motility. Together the two systems modify the coordinated motor activity that arises independently in the intrinsic system.

    c. The basic electrical rhythm (BER) is an excitatory wave originating in a pacemaker within the longitudinal muscle layer high on the greater curvature of the corpus. It is responsible for the rhythm and frequency of gastric contraction. The BER has a **frequency** of 3/min and a velocity of 1 cm/sec when it sweeps over the corpus, increasing to 3–4 cm/sec in the antrum.

2. Stomach Movements and Pressures.

    a. Classification of Pressure Waves. Three types of wave forms have been recognized using balloons and catheters and are classified as:

        (1) Type I, which are monophasic, 3–10 mm Hg waves that last 5–20 seconds

        (2) Type II, which are monophasic, 8–40 mm Hg waves that last 12–60 minutes

        (3) Type III, which are complex, 30–60 mm Hg waves with a duration that varies from seconds to minutes

    b. The Empty Stomach.

        (1) The fasting volume of the stomach is approximately 50 ml. Intraluminal pressure equals intra-abdominal pressure since there is little or no wall tension. No movement occurs in the corpus; only the antrum contracts in response to the BER at a rate of 3/min.

(2) As fasting continues, before hunger is noted, type I pressure waves are seen 25 percent of the time, type II are seen 15 percent, and the stomach is quiescent for the remaining 60 percent.

(3) When hunger appears, type I contractions disappear and type II contractions with an elevated baseline occur. Although these waves may be associated with hunger, they are not unique to that state and thus are not true "hunger contractions." This increase in motility is produced by a falling blood sugar level, which in turn produces increased vagal output.

c. The Full Stomach.

(1) Pressure. Receptive relaxation precedes the peristaltic entry of a bolus into the stomach. As filling begins, pressure rises a small amount (2–7 mm Hg), but as filling proceeds further, there is no corresponding pressure increase until after a very large volume change.

(2) Peristalsis proceeds at a rate that is dictated by the BER. At any given time, two to three waves are present.

    (a) During the first hour after a meal, type I waves are seen, and peristalsis is weak.

    (b) During the second postprandial hour, type II waves, corresponding to stronger contractions, are evident and gastric contents are moved toward the antrum.

(3) Terminal antral contraction, a simultaneous nonperistaltic contraction of the prepyloric antrum, follows most peristaltic waves. This raises antral pressure over pyloric 60 percent of the time; despite this, no more than 1 percent of chyme enters the duodenum since the pyloric sphincter is wholly or partially closed. The terminal antrum and pylorus then act as a pump to deliver chyme to the duodenal bulb.

(4) Chyme, after passing the pylorus, **enters the duodenal bulb.** Because the BERs of the stomach and duodenum differ (3/min and 12/min, respectively), it is in this region that coordination of the two must occur. It is believed that coordination occurs as the antral BER is conducted into the duodenum by longitudinal muscle fibers traversing the hypomuscular segment. The result is one antral contraction for every five duodenal contractions. Distention of the bulb by entering chyme promotes reflex contraction.

(5) Control of the rate of gastric emptying depends on the physical and chemical properties of chyme when it reaches the upper small intestine.

    (a) Particle Size. Smaller particles empty rapidly, while the larger are returned to the proximal antrum.

    (b) Volume. The greater the initial volume, the faster the emptying.

    (c) Osmotic Pressure. The higher the osmolar concentration (above 200 mOsm/L), the slower the emptying. This response is mediated by duodenal osmoreceptors, although the mechanism is unknown.

    (d) Acidity. The lower the pH (< 3.5), the slower the emptying.

    (e) Chemical Composition. Fats in the duodenum strongly inhibit gastric motility, whereas the inhibitory effect of proteins and carbohydrates is weaker.

3. Mechanisms Regulating Gastric Motility.

a. Neural.

(1) Sympathetic.

    (a) Efferent nerves reduce the frequency and velocity of the BER, decrease the force of contraction of circular muscle through their influence on the intrinsic plexuses, reduce blood flow, and, most importantly, counteract vagally induced increases in motility. In addition, they mediate the antral inhibition resulting from distention of the duodenum and jejunum.

    (b) Afferent nerves serve a sensory function, mediating pain, which is produced chiefly by increased tension in the muscular wall of the stomach.

(2) Parasympathetic.

    (a) Vagal efferent nerves carry both excitatory and inhibitory stimuli.

        (i) Excitatory influences predominate, serving to increase the frequency and velocity of the BER. After bilateral vagotomy, peristalsis is weak and emptying is delayed. Administration of an anticholinesterase such as neostigmine enhances motility, while an anticholinergic drug such as atropine inhibits motility.

        (ii) Inhibitory influences are responsible for receptive relaxation of the fundus during swallowing, relaxation of the stomach following esophageal distention, and the slowing of gastric motility, which is produced by the presence of fats on the duodenum.

    (b) Vagal afferent nerves from the stomach are stimulated by changes in gastric pH (<3 or >8) and gastric distension.

(3) The **enterogastric reflex** depresses gastric motility and relaxes the pylorus in response to irritation of the duodenum, produced, for example, by high acidity (pH < 3.5), hypertonicity, increased intraluminal pressure, and high alcohol concentration (> 10 percent). This reflex is reduced by vagotomy but abolished by destruction of the celiac plexus.

(4) The CNS has definite influences on gastric motility and secretion, seen, for example, in conjunction with pain, as well as anxiety, sadness, anger, and other emotional states. These emotions, however, are variable and, hence, unpredictable.

b. Hormonal.

(1) **Excitatory effects** are produced by **gastrin,** which increases LES pressure, increases the frequency and velocity of the BER, and increases antral motility.

(2) **Inhibitory Hormones.**

(a) **Enterogasterone** is a poorly defined hormone, thought to be produced by the small intestine; it may be responsible for the inhibitory effect of high acidity and high fat concentration on gastric motility.

(b) **Secretin,** another product of the mucosa of the small intestine, counters the effects of gastrin (e.g., by lowering LES pressure). It accounts for the inhibitory effect of high pH on gastric motility.

(c) **Cholecystokinin (CCK or pancreozymin),** another small intestinal hormone, contributes to the inhibitory effect of fats on gastric motility and emptying.

(d) **Systemic catecholamines** decrease motility.

4. Vomiting (Emesis).

a. Definitions. Vomiting or emesis describes a reflex loss of upper gastrointestinal contents through the mouth. **Nausea** describes the disagreeable sensation that precedes vomiting, which is accompanied by secretion of copious amounts of thin mucoid saliva. (An individual so afflicted is said to be **nauseated;** a stimulus capable of producing such a sensation is described as **nauseous.)** **Retching** involves all of the involuntary motions of vomiting but without the production of vomitus.

b. Stimuli include:

(1) Unusual stimulation of the gastrointestinal tract, including mechanical stimulation of the pharynx and tonsilar fauces, mechanical or inflammatory disturbances of the intestine (e.g., obstruction, strangulation, and appendicitis), and the presence of irritating substances (e.g., bacterial enterotoxins)

(2) Abnormal stimulation of nongastrointestinal organs, such as the heart, kidney, uterus, and testes

(3) Intense emotions, including fright

(4) Strong sensations, including many odors and tastes

(5) Excessive vestibular stimuli, which produce states such as vertigo and various types of motion sickness

(6) Pregnancy, which produces morning sickness, probably in response to hormonal and metabolic changes

c. Vomiting is controlled by the **vomiting center,** which is located in the dorsolateral reticular formation of the medulla, closely related to the tractus solitarius, at about the level of the dorsal motor nucleus of the vagus nerve.

(1) Afferent stimuli are transmitted via vagal and sympathetic pathways.

(2) Efferent motor impulses are carried through cranial nerves V, VII, IX, and XII to the upper gastrointestinal tract as well as through spinal nerves supplying the diaphragm and abdominal muscles.

(3) Direct stimulation of the vomiting center produces the phenomenon of central vomiting, which is distinguished by its projectile nature and minimal nausea. This may be achieved pharmacologically (e.g., through administration of apomorphine, ipecac, or picrotoxin) and also is seen in any condition producing increased intracranial pressure (e.g., intracranial hemorrhage and tumor) or meningeal irritation (e.g., meningitis).

d. Mechanical Sequence of Vomiting.

(1) Relaxation of the gastric corpus and fundus allows entry of chyme, which has been pushed there by sequential contraction of the upper small intestine, pylorus, and antrum.

(2) The LES, esophagus, and upper esophageal sphincter relax.

(3) Following a deep inspiration, the glottis closes, inspiration is inhibited, and the abdominal muscles contract, squeezing the stomach between the diaphragm and other intra-abdominal organs, thus raising intragastric pressure and expelling gastric contents upward and out of the mouth.

(4) During the ejection period, the esophagus remains relaxed, the glottis closed, and inspiration inhibited. The larynx and hyoid are drawn and rigidly held upward and forward, enlarging the throat and giving free passage to the vomitus. The soft palate is raised to prevent entry of vomitus into the nasal passages.

B. GASTRIC SECRETION

1. General Considerations.

a. There are **three phases of gastric secretion.**

    (1) Cephalic Phase. The presence of food in the mouth produces reflex stimulation of gastric secretion mediated by vagal pathways.

    (2) Gastric Phase. This phase is initiated when food reaches the stomach both by the distension that the food produces and by the secretagogues (e.g., protein) that it contains.

    (3) Intestinal Phase. Further secretion is stimulated by the presence of chyme in the upper small intestine.

b. Secretory Gastric Cells.

    (1) Cardiac glands are found in a narrow band surrounding the esophagogastric junction and are composed of tubular epithelia, which only secrete mucus.

    (2) Gastric glands are structures that contain three types of cells.

        (a) The mucous neck cells secrete mucus.

        (b) The oxyntic (parietal) cells are responsible for both hydrochloric acid (HCl) and intrinsic factor secretion.

        (c) The chief cells secrete pepsinogen, the inactive precursor of the proteolytic enzyme, pepsin.

    (3) Pyloric glands contain not only mucus-secreting cells but also the G cells responsible for gastrin secretion.

2. Components of Gastric Secretion.

a. Hydrochloric Acid.

    (1) The functions of HCl are proteolytic and bacteriostatic; it also provides an optimal pH for pepsin action.

    (2) The Secretory Mechanism.

        (a) The primary secretory process involves active concentration of hydrogen ion (H^+) from 0.00005 mEq/L in plasma to 140–170 mEq/L in gastric juice. Each liter of gastric juice produced requires 1532 kcal supplied by adenosine triphosphate (ATP) derived from oxidative metabolism and glycolysis.

        (b) Proposed Steps in Secretion.

            (i) H^+ is secreted into gastric juice by an active process, labeled the "proton pump," in the parietal cell. Hydroxide ion (OH^-) remains in the cell, increasing intracellular pH.

            (ii) Carbon dioxide (CO_2), as a local metabolic byproduct or a derivative of plasma CO_2, reacts with water under the influence of carbonic anhydrase to yield H^+ and HCO_3^- within the cell. The production of H^+ neutralizes the OH^- produced in the proton pumping process; the HCO_3^- enters the systemic circulation and is responsible for the alkaline tide associated with gastric secretion.

            (iii) Cl^- is secreted actively against an electrochemical gradient by a pump believed to be separate from but coupled with the proton pump.

    (3) Control of Gastric Acid Secretion.

        (a) Neural.

            (i) Vagal stimulation increases both pepsin and acid output directly as well as indirectly by causing secretion of gastrin. This pathway mediates the effects of stress and emotional states on gastric secretion.

            (ii) Because acetylcholine (ACh), which is released by the postganglionic parasympathetic neuron, is responsible for the effects of vagal stimulation, administration of anticholinergic drugs can be used to decrease gastric acid output.

        (b) Hormonal.

            (i) Gastrin, produced within the stomach itself, is the most important hormonal influence on gastric acid secretion (see Section II B 2 b).

            (ii) Small intestinal hormones inhibit gastric acid production in response to the presence of fatty or hypertonic chyme. Secretin, CCK, gastric inhibitory peptide (GIP), vasoactive inhibitory peptide (VIP), and the putative hormone enterogastrone are among the hormones known to have this effect.

b. Gastrin

    (1) Gastrin is not a single hormone but a family of at least six structurally related polypeptides, which are produced by the G cells of the antral mucosa. They fall into two groups: **little gastrin,** G-17, 17 amino acids in length; and **big gastrin,** G-34, 34 amino acids in length.

        (a) Gastrin I and gastrin II, both G-17 and differing only in the sulfation of the tyrosine moiety in gastrin II, are the predominant forms found in the antral mucosa.

        (b) G-34 predominates in the circulation; this is attributed to its higher half-life in the circulation than that of G-17. It is probable that this form is converted to the more active G-17.

    (c) Physiologic activity resides in the C-terminal tetrapeptide common to all forms of gastrin. **Pentagastrin,** a synthetic relative used clinically, consists of this C-terminal tetrapeptide attached to a substituted alanine. Its physiologic properties are identical to those of the larger, naturally occurring molecules.

    (d) The active terminal portion of gastrin is very similar to that of CCK, a polypeptide hormone of the small intestine. Differences are great enough, however, to give the two substances different activities.

  (2) The **functions** of gastrin include the following.

    (a) It stimulates acid and pepsin secretion and increases the total volume of gastric secretion.

    (b) It increases gastric and intestinal motility.

    (c) It increases pressure in the LES while decreasing that of the ileocecal sphincter.

    (d) It increases HCO_3^- secretion by the pancreas and liver.

  (3) Control.

    (a) Entry of food into the stomach produces gastrin release by two mechanisms—antral distension triggers local vagal reflexes, and digestion releases potent **secretagogues** (e.g., protein degradation products). Food may also contain substances like caffeine and alcohol, which further increase secretion.

    (b) Vagal stimulation directly increases gastrin output as mentioned previously.

    (c) Gastrin release is regulated to a large extent by feedback inhibition, the controlling stimulus being pH. In the periods between meals when the stomach is empty, pH is low and gastric output is small. The buffer action of food, when it enters the stomach, produces a rise in pH and gastric secretion.

    (d) Gastrin effects on parietal cell secretion may be mediated by histamine, which is produced locally by mast cells in the gastric mucosa. This theory is supported by the clinical success of histamine analogs such as cimetidine and ranitidine, which bind to and block histamine H_2 receptors on the parietal cell, reduce gastric acid output, and promote healing of peptic ulcers.

c. Enzymes.

  (1) Pepsins, a family of related protein-degrading enzymes, are secreted in an inactive form as pepsinogens by the chief cells. Activation occurs rapidly in the presence of high concentrations of HCl. The optimal pH for pepsin activity is 1 to 3; pepsins are inactive at a pH of 7.

  (2) Minor enzyme components of the gastric secretion include a **gelatinase** and **gastric lipase.** The latter digests tributyrin, found in butterfat, but has no activity on dietary triglycerides.

d. Intrinsic Factor.

  (1) Intrinsic factor is a glycoprotein secreted by the parietal cells of the gastric mucosa, chiefly by those in the fundus.

  (2) In the stomach, digestion liberates vitamin B_{12} from ingested food, and the vitamin is then bound to intrinsic factor.

  (3) The intrinsic factor–vitamin B_{12} complex is carried to specific mucosal receptors located only in the terminal ileum, where the vitamin is absorbed. It is uncertain whether intrinsic factor enters the muscosal cell or is left behind in the intestine; it does not, however, appear in the plasma.

3. The **gastric mucosal barrier** is the property that protects the gastric lining cells from damage by intraluminal HCl or autodigestion. Its chief component is a thick viscous alkaline mucous layer, which is over 1 mm thick and is secreted by the mucous cells. The mucous cells are the most numerous cell type of the gastric mucosa, covering the surface between the various gastric glands.

  a. The turnover rate of the gastric mucosa is extremely high: 5×10^5 mucosal cells are shed each minute, and the entire mucosa is replaced in 1–3 days.

  b. Mild injury results in increased mucous secretion and surface desquamation followed by regeneration.

  c. More serious, repeated injury denudes the mucosal surface, forming an ulcer, and produces bleeding. Ulceration results when damage to the mucosa (e.g., due to a highly concentrated HCl, 10 percent ethanol, salicylic acid, or acetyl-salicylic acid) allows acid to penetrate the mucosal barrier and destroy mucosal cells. This liberates histamine, which increases acid secretion and produces increased capillary permeability and vasodilation. The latter two effects lead to edema. Exposure of mucosal capillaries to the digestive process is what leads to bleeding.

  d. The rate of repair of mucosal injury depends on the extent of injury, from as little as 48 hours for restricted desquamation to up to 3 to 5 months if damage has left only the deepest portions of the gastric pits intact.

C. GASTRIC DIGESTION AND ABSORPTION

1. Digestion.

a. **Carbohydrate digestion** depends on the action of salivary amylase, which remains active in the full stomach until halted by falling pH. This action may be substantial: up to 45 percent of existing starch can be hydrolyzed within 40 minutes.

b. **Protein digestion** is minimal: only 10 percent of protein absorbed is broken down fully. However, pepsin action does destroy cellular integrity and reduce meat particles.

c. **Fat digestion** is minimal due to the restriction of gastric lipase activity to triglycerides containing short-chain (< 10 carbons) fatty acids. Acid and pepsin break emulsions so that fats coalesce into droplets, which float and empty last.

2. Gastric Absorption.

a. Very little absorption of nutrients takes place in the stomach. The only substances absorbed to any appreciable extent are highly lipid soluble substances (e.g., the unionized triglycerides of acetic, propionic, and butyric acids). Aspirin at gastric pH is unionized and fat soluble; after absorption, it ionizes intracellularly, damaging mucosal cells and ultimately producing bleeding. Ethanol is rapidly absorbed in proportion to its concentration.

b. Water moves in both directions across the mucosa. It does not, however, follow osmotic gradients. Water-soluble substances, including Na^+, K^+, glucose, and amino acids, are absorbed in insignificant amounts.

III. SMALL INTESTINE

A. MOTILITY

1. General Considerations.

a. Functions of the small intestine include **digestion** and **absorption** as well as **delivery of unabsorbed residue to the colon.**

b. In keeping with these functions, chyme moves slowly through this organ, with one meal leaving the ileum at the same time a second enters the stomach. Movements affect both mixing and propulsion.

2. Intestinal BER.
The intestinal BER is faster than the gastric BER and progressively decreases along the length of the intestine from 12/min in the duodenum to 9/min in the ileum. These slow waves originate in a pacemaker found in the longitudinal muscle near the entry of the bile duct.

3. Types of Movements.

a. **Rhythmic segmentation** mixes chyme with digestive juices and exposes it to the absorptive surfaces. An intestinal segment undergoing this process takes on the appearance of a string of sausages as adjacent areas alternately contract and relax, pushing chyme back and forth from one relaxed segment to the other. The frequency of segmentation corresponds to the BER; because of this gradient down the intestine, segmentation contractions are weakly propulsive.

b. **Short, propulsive movements** are annular contractions, which at a velocity of 1.2 cm/sec, move the very short distance of 0.15 cm. Their importance to the movement of chyme toward the colon is slight.

c. **Peristalsis** is the major propulsive force in the small intestine. It arises as a local reflex governed by intrinsic plexuses; it is not affected by extrinsic denervation. Stimulation of a segment of the intestine, that produced by distension, induces contraction in the segment proximal to it and relaxation distally. Contraction occurs first in the longitudinal muscle layer, second in the circular layer, 90° out of phase; reciprocal inhibitory neurons in the myenteric plexus ensure that the two layers never contract simultaneously.

d. **Movements of the mucosa and villi** are not well organized but contribute nonetheless to the absorptive process. Spontaneous contractions of the muscularis mucosa beneath the complexly folded mucosal surface shift the folding patterns between the longitudinal and circular layers and back again. These contractions are modified by chyme and mechanical stimuli. Villi move both with swaying and shortening motions to an irregular rhythm, which is greatest in the proximal small intestine and in the fed state. Their motility may be increased by a hypothetical hormone, villikinin.

4.
The **ileocecal sphincter,** separating the terminal ileum and colon, is an area that is 4 cm in length, is normally closed, and has a resting pressure that exceeds colonic pressure by 20 mm Hg. It acts as a valve, functioning to prevent colonic reflux. The ileocecal sphincter is not absolutely essential, however, for if removed, significant reflux does not occur.

a. At rest, the ileocecal sphincter is kept closed through a myenteric reflex pathway from the cecum. When the cecum is full, ileocecal pressure is raised, further delaying passage of chyme through this area.

b. The sphincter relaxes as a wave of relaxation precedes each propulsive peristaltic wave that sweeps over the terminal ileum. Distension of the terminal ileum also produces reflex relaxation of the sphincter. Finally, gastrin, in contrast to its action on the LES, also stimulates relaxation of the ileocecal sphincter.

5. Control of Intestinal Motility.
  a. Neural.
    (1) Intrinsic reflex pathways, as mentioned earlier, establish the basic mixing and propulsive movements of the small intestine.
    (2) Extrinsic Neural Influence.
      (a) Autonomic influences are diffuse rather than localized; they modulate the intrinsic motility. Vagal stimulation enhances the strength of intestinal contraction, while adrenergic stimulation inhibits motility.
      (b) Long intestinal reflexes play an uncertain role in normal functioning; to what extent the observed effects are hormonally rather than neurally mediated is not known.
        (i) The gastroileal (gastroenteric) reflex intensifies peristalsis in the ileum, emptying chyme through the ileocecal valve into the cecum as soon as a second meal is eaten.
        (ii) The ileogastric reflex slows gastric emptying when the ileum is distended.
        (iii) The intestino-intestinal reflex results in cessation of intestinal movement [i.e., adynamic (paralytic) ileus] whenever the intestine is injured (e.g., in surgery) or distended. This reflex is mediated through a spinal reflex arc, with both afferent and efferent arms carried in the splanchnic (adrenergic) nerves; it is initiated by stretch receptors in the longitudinal muscle layer.
  b. Hormonal interactions with the nervous system in the control of intestinal motility are not well-defined.
    (1) Gastrin and CCK increase motility of the upper small intestine. Serotonin, the neurotransmitter between the sensory and motor neurons in the peristaltic reflex, also stimulates intestinal motility.
    (2) Glucagon, secretin, and epinephrine inhibit intestinal motility.

6. Intestinal Blood Flow.
  a. Intestinal blood flow is **autoregulated** between perfusion pressures of 90–270 mm Hg. This means that constant flow is maintained within this pressure range through local modulation of arteriolar constriction. For example, reduced arterial pressure leads to a drop in arterial resistance and an increase in venous constriction in order to maintain capillary pressure supplying the mucosa, submucosa, and musculature.
  b. The mucosa and submucosa receive greater flow than the muscularis; control mechanisms for these two regions are separate.
  c. Increased motor activity in the intestine increases blood flow; the rhythmic contractions act as a muscle pump. Hormonal influences on intestinal blood flow are poorly understood.
  d. Vasodilatation in the mucosa can be caused by ACh, bradykinin, CCK, gastrin, secretin, and histamine, while vasoconstriction is the effect of angiotensin II, norepinephrine, and serotonin.
  e. Extrinsic neural control of blood flow to the intestine is an autonomic function. Sympathetic stimulation produces vasoconstriction, whereas parasympathetic stimulation increases blood flow (although this effect may be due more to increased motility than to vasodilatation).

B. PANCREATIC SECRETION

1. General Considerations. The pancreas daily secretes about 1200–1500 ml of clear, colorless, odorless alkaline (pH 7.6–8.2) fluid, which is isosmotic with plasma. Its viscosity varies with the stimulus to its production; specific gravity is 1.01–1.02. It contains HCO_3^-, which neutralizes gastric acid and regulates the pH of the upper intestine, as well as enzymes to digest the three major types of food: protein, fat, and carbohydrates.

2. Composition.
  a. Electrolytes.
    (1) Na^+ and K^+ concentrations in the pancreas are the same as those in plasma water (154 and 4.8 mmol/kg of water, respectively).
    (2) Calcium ion (Ca^{2+}), secreted together with enzymes, is in a lower concentration than in plasma (1.7 mmol/kg H_2O).

(3) The pancreatic Cl^- and HCO_3^- concentrations change reciprocally so that the sum of these concentrations balances the total cation concentrations of the pancreas. As the secretory rate rises, HCO_3^- concentration rises and Cl^- concentration declines. The exact mechanism by which HCO_3^-, other electrolytes, and water are secreted by the pancreas is disputed. It is known that the cells responsible are the cells of the intercalated ducts and the centroacinar cells. It is these cells and not the acinar cells that contain the high levels of carbonic anhydrase necessary for HCO_3^- formation from the CO_2 derived from respiration. According to one theory, as isotonic sodium bicarbonate ($NaHCO_3$) solution passes through the duct, it is modified as a passive exchange of HCO_3^- for Cl^- occurs.

(4) Other ions present in small amounts include magnesium (Mg^{2+}), zinc (Zn^{2+}), phosphate (HPO_4^-), and sulfate (SO_4^{2-}).

   b. Proteins. At least 10 proteins, mostly enzymes, comprise 0.1–10 percent of the fluid, depending on the secretory state. The pancreas synthesizes and secretes more protein than all other glands in the body except the lactating mammary gland.

(1) **Trypsinogen** is an inactive proenzyme, which is split either by enterokinase (enteropeptidase) or trypsin itself (in autocatalysis) to yield trypsin, the active enzyme. Trypsin functions as an endopeptidase.

(2) **Chymotrypsinogens** are a family of proenzymes, cleaved by trypsin to their active forms, chymotrypsins. These enzymes all are endopeptidases.

(3) **Procarboxypeptidases A and B, proelastase,** and **procollagenase** are also proenzymes cleaved by trypsin to their active proteolytic forms.

(4) **Ribonuclease** and **deoxyribonuclease** serve to split RNA and DNA, respectively.

(5) **Pancreatic amylase,** secreted in the active form, hydrolyzes glycogen, starch, and most other complex carbohydrates except cellulose to form disaccharides.

(6) **Pancreatic lipase,** also secreted in the active form, acts at the lipid-water interface in the presence of both Ca^{2+} and bile salts to hydrolyze triglycerides. Other lipolytic pancreatic enzymes include **cholesterol esterase** and **phospholipases A and B.** The first of these hydrolyzes cholesterol esters, while the second two, after activation by trypsin, act sequentially to hydrolyze lecithin to glycerylphosphoryl choline.

(7) **Trypsin inhibitor** is a substance secreted by the same cells that produce the pancreatic proenzymes simultaneously with these proenzymes. It protects the pancreas from autodigestion, which occurs in acute pancreatitis when trypsin is activated either in the secretory cells or within the acini and ducts of the pancreas.

3. Control of Pancreatic Secretion.

   a. Dietary adaptation describes the effect of dietary composition on both pancreatic enzyme content and water and electrolyte secretion. This is a long-term effect, however, for amylolytic, lipolytic, and proteolytic enzymes all increase in activity only after diets high in their respective substrates are consumed for many months.

   b. Neural Control.

(1) Sympathetic influence may not be exerted directly but indirectly through its effect on vascular tone. Vasoconstriction is associated with decreased secretion.

(2) Parasympathetic stimulation via vagal pathways increases enzyme output but has no effect on water or HCO_3^- secretion. This is the mechanism behind the gastropancreatic reflex, by which distension of the stomach leads to increased enzyme output. Atropine blocks these effects.

   c. Hormonal Control.

(1) **Secretin** is a polypeptide hormone released by the mucosa of the upper small intestine in response to the entry of HCl, lipids, protein-breakdown products, and other constituents of chyme into this area from the stomach. It stimulates a high output of fluid and HCO_3^-.

(2) **CCK** is a polypeptide hormone that is identical to pancreozymin, which was thought to be a separate hormone. Also produced by the upper small intestine mucosa, CCK stimulates pancreatic enzyme release.

C. BILIARY SECRETION

1. General Considerations.

   a. Between 250 and 1100 ml of bile are secreted daily by the liver.

   b. Although it is secreted continuously, bile is stored in the gallbladder and is released into the duodenum only after chyme has triggered the release of CCK, which then produces gallbladder contraction and emptying.

   c. Bile contains no digestive enzymes and is dependent on **bile salts** to function in digestion. Bile salts emulsify fats into small droplets, which permits access to lipase, and then carry the breakdown products in soluble complexes (i.e., **micelles**) to the sites of absorption.

d. Bile also serves as a vehicle for the excretion of such water-soluble substances as cholesterol, drugs, steroid hormones, and bilirubin.

2. **Composition.**
   a. **Bile Salts (0.7 Percent of Bile).**
      (1) There are two **primary bile salts** formed in the liver from cholesterol: **cholic acid** and **chenodeoxycholic acid.**
      (2) The two **secondary bile salts** are formed in the colon by bacterial action on the primary bile acids. By this means, **deoxycholic acid** is derived from cholic acid, and **lithocholic acid** is derived from chenodeoxycholic acid.
   b. **Bile Pigments (0.2 Percent of Bile).**
      (1) Bilirubin and biliverdin, the two principal bile pigments, are metabolites of hemoglobin formed in the liver and conjugated as glucuronides for excretion. They are responsible for the golden yellow color of bile.
      (2) Intestinal bacteria metabolize bilirubin further to urobilin, which is responsible for the brown color of stool.
   c. The **electrolyte composition** of bile is similar to that of pancreatic juice and plasma.

3. **Control of Biliary Secretion.**
   a. **Neural.**
      (1) Parasympathetic stimulation via the vagus nerve enhances bile secretion.
      (2) Sympathetic stimulation via the splanchnic nerve indirectly decreases bile secretion by producing vasoconstriction.
   b. **Hormonal.**
      (1) **Secretin** increases the volume and HCO_3^- concentration of bile as it does the volume of pancreatic secretion, but it has no effect on bile salt concentration.
      (2) **Gastrin** has an effect similar to the effect of secretin but weaker.
      (3) The major action of **CCK** is to increase bile output; however, it has only a minor effect.
   c. Choleresis (increased bile flow) results when the amount of bile salts reaching the liver increase, for water and electrolytes passively follow the secretion of bile salts.
   d. Hepatic blood flow also determines bile flow, since secretion is an active, oxygen-dependent process.

4. **Enterohepatic Circulation.**
   a. Ninety to ninety-five percent of all bile salts that enter the small intestine are actively reabsorbed in the lower ileum. As no absorption takes place from the duodenum or jejunum, high bile salt concentrations are available at the major sites of fat digestion and absorption. The liver extracts all bile salts from the portal circulation, secreting them once again into the bile.
   b. The total **circulating pool of bile salts** is approximately 3.6. g. As minimal secretion after a fatty meal is 4–8 g, it follows that the total pool must circulate twice during digestion of each meal, which is, therefore, 6–8 times daily.
   c. The **rate of bile salt synthesis** is determined by the rate of return to the liver. A normal rate is 0.2–0.4 g/day, which replaces normal fecal losses. Maximally it can reach 3–6 g/day; if loses exceed this rate, the pool size decreases.
   d. As bile salts are necessary to proper digestion and absorption of fats, any condition that disrupts the enterohepatic circulation (as occurs following ileal resection or in small intestinal diseases such as sprue and Crohn's disease) leads to a decreased bile acid pool and malabsorption of fat and fat-soluble vitamins, the clinical manifestations of which are **steatorrhea** and nutritional deficiency. If fecal losses of bile salts increase but the bile salt pool remains adequate for fat digestion, watery diarrhea results as bile salts inhibit water and Na^+ absorption in the colon.

5. **Gallbladder.**
   a. **Functions.**
      (1) The chief function of the gallbladder is **storage.** During interdigestive periods, the sphincter of Oddi is closed and bile flows into the relaxed gallbladder. Total capacity is about 30 ml.
      (2) The gallbladder also concentrates bile, as rapid reabsorption of water and electrolytes increases bile salt concentration five- to tenfold. If emptying is delayed, this process slows the increase in intrabiliary pressure.
   b. **Control.**
      (1) **Neural.** Parasympathetic stimuli via the vagus nerve cause contraction of the gallbladder and relaxation of the sphincter of Oddi.
      (2) **Hormonal.** When chyme reaches the upper small intestine, fatty acids and protein digestive products cause the release of CCK, which is the main stimulus to gallbladder contraction.

c. **Effects of Cholecystectomy.** Bile, not the gallbladder, is essential to digestion. After removal of the gallbladder, bile empties slowly but continuously into the intestine, allowing digestion of fats sufficient to maintain good health and nutrition. Only meals high in fat need to be avoided.

d. **Gallstones** form in an estimated 10–30 percent of the population, although only a fraction, perhaps 20 percent of these ever produce symptoms. In the Western societies, about 85 percent of gallstones are composed chiefly of cholesterol; the remainder are pigment stones, composed chiefly of calcium bilirubinate. The former are radiolucent; the latter are radiopaque.

   (1) **Cholesterol,** a highly water-insoluble substance representing only 0.06 percent of bile at the time of secretion, is kept in solution in bile through the formation of micelles with bile salts and lecithin. When the proportions of lecithin, cholesterol, and bile salts are altered, cholesterol crystallizes, leading to stone formation.

   (2) **Calcium bilirubinate stones** can form when infection of the biliary tree leads to bacterial deconjugation of conjugated bilirubin. This, however, is not the only mechanism for formation. Unconjugated bilirubin, which is insoluble in bile, then precipitates to begin the stone-forming process.

6. **Bilirubin Metabolism.**

a. Bilirubin is formed during the catabolism of the heme moiety of hemoglobin, which is released during red cell destruction, and other heme-containing proteins (e.g., myoglobin and catalase). Bilirubin is carried to the liver combined with plasma albumin. This combination is necessary, for unbound, bilirubin is poorly soluble in water and therefore in blood.

b. Within the hepatic cell, bilirubin is conjugated to glucuronic acid by the enzyme **glucuronyl transferase.** Thus conjugated, bilirubin becomes soluble in bile and is excreted actively into the bile canaliculi.

c. **Jaundice** (or icterus) describes the yellow coloration that appears whenever either free or conjugated bilirubin reaches a level high enough to stain tissue. Usually it is visible in the skin mucous membranes and conjunctiva when the plasma bilirubin rises above 2 mg/dl. Causes include:

   (1) Excess production, as occurs in hemolytic anemias

   (2) Hepatic disease associated with decreased hepatocellular uptake of bilirubin, failure of intracellular protein binding, or impaired secretion of bilirubin into the canaliculi

   (3) Intra- or extrahepatic obstruction of the biliary tree

D. INTESTINAL SECRETION

1. **Mucus Production.**

a. **Brunner's glands** are found only in the first part of the duodenum, between the pylorus and the papilla of Vater.

   (1) Histologically, they are compound mucous glands; their secretion is a thick, alkaline mucus, which protects the duodenal mucosa from the acidic chyme entering from the stomach. Also present in their secretion is enterokinase, which is important in activating trypsin produced by the pancreas.

   (2) Secretion is partially under hormonal control, increasing in response to gastrin, secretin, CCK, and glucagon, and is partially under neural control, increasing with vagal stimuli and decreasing with sympathetic stimuli. Secretion also increases in response to direct tactile stimulation and mucosal irritation.

b. Mucous also is secreted along the entire length of the small intestine by **goblet cells** scattered among the cells of the mucosal epithelium. This mucus also serves a protective function. Secretion increases as the passage of chyme supplies direct tactile and chemical stimulation. Similar goblet cells are found in the intestinal glands **(crypts of Lieberkühn).**

2. **Intestinal Juice.**

a. The intestinal glands or crypts of Lieberkühn secrete an intestinal juice (also known as the **succus entericus),** which is isosmotic with plasma, almost identical in composition with the extracellular fluid, and has a pH in the neutral range of 6.5–7.5. Total output is 1.5–2 L/day. Intestinal juice is virtually enzyme free, containing the only two enzymes that are thought to have any physiologic role within the lumen, enterokinase and intestinal amylase. Small amounts of secretory immunoglobulin A (IgA) protect the mucosa.

b. This fluid functions mainly **to assist digestion** by providing an aqueous medium for hydrolytic reactions and **to enhance absorption** by serving as a solvent for the products of digestion and as an aqueous phase for suspension and emulsification of lipids and their by-products.

c. Intestinal secretion is chiefly **under the control of local neural reflexes** triggered by the presence of chyme. Whether vagal stimulation has any effect is controversial; however, secretion is increased by the parasympathetic drugs, pilocarpine and physostigmine, and is decreased by the anticholinergic, atropine. Hormonal influences are poorly defined, although CCK, secretin, and vasoactive intestinal peptide (VIP) are able to increase secretion.

d. The intestinal glands are **found along the entire length of the intestine** with the exception of the first few centimeters of the duodenum, in which are found Brunner's glands. The epithelial cells are continually dividing within these sample tubular glands, with the newly formed cells migrating up out of the crypts to the tips of the villi from which they are shed. Turnover is rapid; the life span of an intestinal epithelial cell is but 3–5 days, which defines the length of time necessary to replace the entire intestinal epithelium.

e. In addition to goblet cells and epithelial cells in various stages of maturation, the crypts of Lieberkühn also contain two other types of secretory cells, the **Paneth's** cells and **argentaffin cells.** The products of the first are not known, but the products of the second are believed to be hormones.

E. INTESTINAL DIGESTION AND ABSORPTION

1. The Absorptive Surface.
   a. The small intestine is a cylindrical structure approximately 285 cm in length from the pylorus to the ileocecal valve in the living human; however, in death, as its muscles relax, it stretches to over 700 cm.
   b. Surface area is increased by the presence of many valve-like folds in the mucosa, the **valvulae conniventes.**
   c. The mucosal surface is covered with minute finger-like projections called villi. Each villus is 0.5–1.0 mm in length with a density of 20–40/mm^2. The intestinal epithelial cells are on the surface of the villi, and the intestinal glands are at the base.
   d. Each of the intestinal mucosal cells has at its luminal surface many finger-like projections of its cell membrane, termed **microvilli,** which comprise the **brush border.**
   e. In all, the valvulae conniventes, villi, and microvilli increase the absorptive surface of the small intestine to 2×10^6 cm^2, which is 600 times greater than the surface would be if the intestine was a simple cylinder with a smooth lining.

2. Intestinal Enzymes.
   a. While the pancreas is the major source of digestive enzymes, these enzymes alone are not sufficient to digest chyme completely enough to allow easy and thorough absorption. Completion of digestion and absorption requires enzymes formed by intestinal epithelial cells.
   b. Intestinal enzymes act both extracellularly, although they are bound to the plasma membranes of the microvilli, and within the epithelial cell cytoplasm. They are not secreted freely into the intestinal juice. Each enzyme will be considered as the digestion of its substrate is discussed.

3. Carbohydrates.
   a. General Considerations.
      (1) Dietary intake of carbohydrates equals 250–800 g/day, which represents 50 to 60 percent of the diet.
      (2) Carbohydrates may be categorized as mono-, di-, and polysaccharides.
         (a) Monosaccharides are the only forms that may be absorbed directly.
         (b) Disaccharides require the action of oligosaccharidases, which are found in the intestinal brush border, for reduction to monosaccharides before they can be absorbed.
         (c) Polysaccharides must be hydrolyzed by amylase before they can be acted on by intestinal surface enzymes and absorbed.
      (3) The **three major carbohydrates** in the human diet are the disaccharides, **sucrose** (cane sugar) and **lactose** (milk sugar), as well as the polysaccharide starches (which may be in either the straight chain form, **amylose,** or the branched chain form, **amylopectin**). **Cellulose,** another plant polysaccharide, is present in the diet in large amounts, but no enzymes in the human digestive tract can digest it, and it is excreted unused.
   b. Digestion.
      (1) Starch digestion begins in the mouth with the action of salivary amylase. However, this action is short-lived, and the bulk of starch digestion takes place in the small intestine through the action of pancreatic amylase, with a minor contribution from intestinal amylase. This process yields **glucose** as well as three oligosaccharides—**maltose**, a disaccharide, **maltotriose,** a trisaccharide, and α-limit dextrins, which contain eight glucose moieties on the average.

(2) Oligosaccharides, whether ingested directly or produced by the action of amylase on polysaccharides, must be hydrolyzed to monosaccharides before absorption. This is carried out by several oligosaccharidases found on the epithelial brush border, including **lactase, maltase, sucrase,** and **isomaltase** (or α-dextrinase).

(3) Some oligosaccharidase activity may be increased by increasing the proportion of the respective substrate in the diet. This phenomenon is termed **induction.** Sucrase may be induced, for example, while lactase may not.

(4) Carbohydrates that are unable to be digested by the intestinal and pancreatic enzymes enter the lower gastrointestinal tract unchanged. **Bacteria** are present in the lower ileum and colon; many of these organisms are able to metabolize the undigestible carbohydrates. Products of this bacterial metabolism include gases, such as hydrogen (H_2), methane (CH_4), and CO_2, as well as intestinal irritants, such as some short-chain fatty acids. Consequences of this process may include increased fluid secretion and intestinal motility, manifest as diarrhea, cramps ("gas pains"), bloating, and flatulence.

(5) **Oligosaccharide deficiencies** are relatively common, resulting from genetic defects, aging, and mucosal injury, resulting either from disease or surgery. The most common deficiency is that of lactase, leading to **lactose or milk intolerance.** In most races intestinal lactase activity is high only in the neonatal period, declining during childhood to a low level maintained throughout adult life. In Western Europeans and their American descendants, however, high lactase levels usually persist in adulthood, while in most Americans of different extraction, including blacks, age-related decline is the norm. Lactase deficiency, then, can hardly be considered a disease.

c. Mechanisms of Absorption.

(1) Glucose and other monosaccharides are hydrophilic, but in order to be absorbed, they must pass through the hydrophobic lipid cell membrane of the intestinal epithelial cell. This is accomplished predominantly through transportation by a **carrier molecule** in an active energy-consuming process. Each monosaccharide is believed to have its own selective transport system.

(2) Glucose and galactose, for example, share a carrier transport system and compete for access to it. This system is Na^+-dependent, and, for this reason, has been termed the Na^+ **co-transport mechanism.**

(3) Fructose is absorbed along a concentration gradient. Although this mechanism does require a specific carrier, it does not consume energy and, thus, is termed **facilitated diffusion.** Fructose absorption occurs readily due to the maintenance of a low intracellular fructose concentration by the rapid conversion of the entering fructose into glucose and lactate.

4. Proteins.

a. General Considerations.

(1) The daily dietary protein requirement for adults is 0.5–0.7 g/kg of body weight, and for children 1–3 years of age it is 4 g/kg.

(2) The protein that is found in the intestines comes from two sources:

(a) **Endogenous proteins,** totaling 180 g/day, are secretory proteins as well as the protein components of desquamated cells.

(b) **Exogenous proteins** are dietary proteins,, which total at least 75–100 g daily in the average American diet.

(3) Proteins, except in the neonatal period, cannot be absorbed intact; first they must be broken down into their constituent amino acids. Enzymes that bring this about fall into two categories.

(a) **Endopeptidases** attack interval peptide bonds and are responsible for the initial breakdown of polypeptides into smaller fragments. This group of enzymes includes gastric pepsin, pancreatic trypsin, chymotrypsin, and elastase.

(b) **Exopeptidases** attack only the terminal peptide bond of polypeptides and oligopeptides, which are formed by polypeptide breakdown by the endopeptidases, liberating single amino acids. Among these enzymes are pancreatic carboxypeptidases A and B and small intestinal aminopeptidase and dipeptidase.

b. Digestion.

(1) Digestion of protein begins in the stomach with the action of pepsin. This yields large peptide fragments but only a very few free amino acids, which serves not only to increase the sites available to enzyme attack in the small intestine but also to create secretagogues, which stimulate the production and release of the pancreatic enzymes that carry on the bulk of protein digestion.

(2) Pancreatic endo- and exopeptidases together continue protein degradation until only dipeptides, oligopeptides, and free amino acids remain.

(3) Since 60 percent of protein residues remain in di- and oligopeptide forms, which cannot be absorbed, final digestion is dependent on the small intestinal enzymes.

    (a) Aminopeptidase and dipeptidase, which are bound to the surface of the microvilli, yield amino acids plus di- and tripeptides, which can be absorbed by the epithelial cells.

    (b) Within the epithelial cell, cytoplasmic dipeptidases complete hydrolysis of the small peptides; thus, postprandial portal blood almost exclusively contains only free amino acids.

  (4) Protein digestion and absorption are highly efficient processes; only 2–5 percent of protein in the small intestine is not digested and absorbed. Six to twelve g of protein are lost in the stool daily, but most of this does not come from endogenous and exogenous sources but from bacteria in the lower gut and cellular debris.

c. Mechanism of Absorption.

  (1) Amino acid absorption occurs through active, **carrier-mediated transport.**

  (2) There are at least five carrier-transport systems, each selective for the transport of different subsets of amino acids. An individual system transports:

    (a) Neutral amino acids, including alanine, valine, leucine, methionine, phenylalanine, tyrosine, and isoleucine.

    (b) Cystine plus the basic amino acids, lysine and arginine.

    (c) Glycine plus the amino acids, proline and hydroxyproline

    (d) Acidic amino acids, including aspartic acid and glutanic acid

    (e) β-amino acids, β-alanine and taurine

  (3) Like glucose transport, amino acid transport is Na^+-dependent.

  (4) Vitamin B_6 or pyridoxine (in the form of pyridoxal phosphate) is required for transport of many amino acids.

  (5) Di- and tripeptides are absorbed into mucosal cells via a separate transport system.

  (6) The small intestine in the fetus and neonate has the capacity to absorb intact polypeptides and proteins. This occurs through **pinocytosis** (endocytosis). This ability is important, less for nutrition than for the uptake of maternal antibodies secreted in breast milk. In the adult the extent to which intact protein absorption occurs is unknown. Endocytosis is felt, however, to be the mechanism by which intrinsic factor complexed with vitamin B_{12} is absorbed in the terminal ileum.

  (7) Congenital defects in any of the amino acid transport systems produce disease. For example, **Hartnup disease** is produced by an inborn defect in the transport mechanism for neutral amino acids.

5. Fats.

a. General Considerations.

  (1) Daily dietary fat intake varies widely from 25–160 g.

  (2) Triglycerides comprise more than 90 percent of dietary fat, while the remainder consists of phospholipids, cholesterol, cholesterol esters, free fatty acids, and minute quantities of the fat-soluble vitamins (A, D, E, and K).

b. Digestion.

  (1) Fat digestion effectively begins in the duodenum with the action of pancreatic lipase since gastric lipase has no effect on the triglycerides.

  (2) Fats must be made accessible to the water-soluble digestive enzymes before their digestion can begin. This is accomplished by the process of **emulsification,** which occurs through intestinal motor activity and the detergent action of bile salts, lecithin, lysolecithin, monoglycerides, and fatty acids. Emulsification creates many small (200–5000 nm in diameter) fat globules, thus increasing the surface area available to pancreatic lipase, an enzyme that is only active at the lipid-water interface.

  (3) The principal products of lipase action are free fatty acids, and 2-monoglycerides (β-monoglycerides).

  (4) Bile salts, through their detergent action in high concentrations interact with the emulsified lipids to form **micelles,** spherical structures 3–10 nm in diameter. In each micelle, the molecules arrange themselves so that the polar (hydrophilic) groups face outward into the aqueous medium and nonpolar (hydrophobic) groups face inward where monoglycerides, fatty acids, lecithin, and lysolecithin are found dissolved in the core.

  (5) Micelles are able to dissolve in the aqueous digestive juices. In this soluble form, they serve to transport the products of lipid digestion to the site of absorption at the brush border of the intestinal epithelial cells. Once the products of fat digestion have been released at the epithelial cell surface, the bile salts, which are not absorbed until they reach the ileum, are free to repeat the transfer process with additional quantities of lipid breakdown products.

c. Mechanism of Absorption. Fat absorption is most rapid in the duodenum and proximal jejunum where bile salt concentrations are high. It is a highly efficient process. Only 2–4 g of fat appear daily in the stool; most of this is not dietary but is derived from desquamated cells and colonic bacteria.

(1) Uptake of lipids and lipid breakdown products by the intestinal epithelial cell is brought about by **passive diffusion through the plasma membrane.**

(2) The intracellular fate of absorbed fatty acids depends on their size.

    (a) Medium-chain fatty acids (6–12 carbon atoms) are absorbed as they are water soluble.

    (b) Long-chain fatty acids (> 12 carbon atoms) are resynthesized into triglycerides with glycerol for their process derived from absorbed 2-monoglycerides.

(3) The resynthesized triglycerides are collected in large lipid globules within the epithelial cells. After the addition of phospholipids, apolipoproteins, and cholesterol, chylomicrons are formed from the fat globules. Chylomicrons are then released into the lymphatic system by exocytosis.

(4) **Cholesterol Digestion and Absorption.**

    (a) Cholesterol present in the small intestine is derived from two sources. Exogenous or dietary cholesterol totals 0.5 mg/day; endogenous cholesterol, derived from biliary excretion and desquamated cells, totals 2–3 g/day.

    (b) Cholesterol esters are converted to free cholesterol by the pancreatic enzyme, cholesterol esterase.

    (c) Cholesterol enters micelles along with lipid breakdown products and is absorbed into the epithelial cells in the same manner.

6. Water and Electrolytes.

  a. Water.

    (1) Water in the small intestine represents the 1500–2500 ml ingested daily plus the 7000–10,000 ml secreted by the gut and its associated glands and organs. By the time the colon is reached, the small intestine has absorbed all but 1000–1500 ml. Fluid loss in the stool is 150–200 ml per day.

    (2) Water undergoes isosmotic absorption; that is, it moves passively by diffusion in both directions across the intestinal mucosa, following the osmotic gradients between the luminal contents and plasma. Thus, for example, as nutrients are absorbed, reducing the osmotic pressure of the luminal contents, water follows the gradient into the epithelial cell and is absorbed as well, leaving the intraluminal space isosmotic again. Although the duodenal contents may be either hypo- or hyperosmolar, depending on the contents of a meal, they are isotonic throughout the rest of the intestine.

  b. Na^+.

    (1) Na^+ enters the epithelial cell down both electrical and chemical gradients by a process of diffusion; however, it is uncertain whether this is simple diffusion, carrier-mediated, or a combination of the the two. It is known, however, that glucose and amino acid absorption across the intestinal epithelium is coupled to Na^+ transport, using a common carrier in each case.

    (2) Na^+ is actively transported out of the mucosal cell into the extracellular space by Na^+ pumps located in the lateral and basal cell membranes. Water follows this movement passively.

  c. K^+ is secreted to some extent by the intestinal cells, chiefly within mucus, but on the whole, its movement through the intestinal mucosa occurs through simple diffusion.

  d. Cl^- in the upper small intestine is absorbed by passive diffusion as it follows Na^+. In the ileum, as in the colon, an active transport mechanism exists, which absorbs Cl^- in exchange for excreting HCO_3^-.

7. Vitamins and Minerals.

  a. Water-soluble vitamins are rapidly absorbed in the upper small intestine. The fat-soluble vitamins, A, D, E, and K, are absorbed in the same manner as the lipids also in the upper small intestine. Vitamin B_{12} absorption, as mentioned earlier, depends on the presence of gastric intrinsic factor and occurs in the ileum.

  b. Calcium intake averages 1000 mg daily. Of this, 25–80 percent is absorbed, chiefly in the duodenum.

    (1) Absorption is strictly controlled by the body's needs. For example, absorption is increased when supplies are low (e.g., in deficiency states such as rickets) and when requirements are high (e.g., in growing children and lactating women).

    (2) Ca^{2+} is absorbed primarily by an active process, although some passive diffusion does occur.

    (3) The vitamin D metabolite, 1,25-dihydroxycholecalciferol (DHCC), increases absorption. DHCC is produced in the kidney under the influence of parathormone, which is secreted in response to plasma Ca^{2+} levels.

(4) Absorption is blocked by phosphate and oxalates, which form insoluble salts with Ca^{2+} in the intestinal lumen. Similarly, fatty acids form soaps with Ca^{2+}, blocking absorption.

c. Iron (Fe) intake averages 15–20 mg daily; normally only 3–6 percent of this is absorbed or enough to replace daily losses of 0.6 mg in a man and 1.3 mg in a menstruating woman
   (1) Fe absorption is facilitated by the conversion of dietary Fe from the ferric (Fe^{3+}) state, in which most of it exists, to the ferrous (Fe^{2+}) form. This is accomplished chiefly by the low gastric pH, which favors reduction, but it is facilitated by the presence of reducing substances such as vitamin C (ascorbic acid).
   (2) Fe absorption is blocked by phosphates, oxalates, and phytic acid (found in cereal grains), which form insoluble compounds in the intestinal lumen.
   (3) The absorption of Fe across the intestinal mucosa is an active process, occurring most readily in the upper small intestine. Some of the absorbed Fe is passed to the bloodstream, where it is carried bound to the protein, **transferrin,** but most is bound within the cell to the protein, **apoferritin,** to form **ferritin,** the major storage form of Fe in the body. Fe in ferritin is in equilibrium with plasma Fe bound to transferrin.
   (4) Fe is released from ferritin in the mucosal cell according to the body's need as expressed plasma Fe or the degree of transferrin saturation. Any Fe not so released during the short life span of the mucosal cell is lost with that cell as it is desquamated at the end of its life span. Thus, in Fe deficiency the ferritin stores are drawn down and more Fe enters the plasma, which decreases mucosal cell Fe and enhances transport of Fe into the cell. In Fe overload ferritin stores build to maximal levels, resulting in increasing excretion and decreasing absorption. This function of the intestine in regulating Fe absorption has been labeled the **mucosal block.**

F. DISORDERS OF SMALL INTESTINAL FUNCTION

1. Maldigestion, failure of the intestine to break down food sufficiently to allow adequate absorption of nutrients, occurs most commonly when pancreatic enzymes do not enter the intestinal lumen.
   a. Pancreatic insufficiency is a common sequela of **chronic pancreatitis.**
   b. In **acute pancreatitis,** which most commonly results from chronic alcohol abuse or obstruction of the common bile duct by a stone, pancreatic enzymes are activated while still within the ducts and acini of the pancreas. This leads to destruction of some portion of the pancreatic parenchyma. If enough of the pancreas is destroyed, which usually requires repeated acute episodes, exocrine insufficiency and even endocrine insufficiency result.
   c. Without the pancreatic enzymes, digestion of fats, proteins, and carbohydrates is significantly impaired. Digestion of fats is most affected; thus **steatorrhea** (the production of fatty stools) is a hallmark of this disorder.
   d. Treatment requires the administration of enzyme-rich pancreatic extract.

2. Malabsorption is inadequate nutrient absorption despite adequate digestion, which occurs most often after damage to the intestinal absorptive surface either from surgery or disease.
   a. Several diseases impair absorption by the intestinal mucosa. One of the most common is **idiopathic** or **nontropical sprue,** which is also known as celiac disease and, more descriptively, as gluten-sensitive enteropathy. This disease results from a toxic effect of the wheat protein **gluten** on the mucosa. In its most severe form, damage is so extensive that the villi are blunted or entirely destroyed.
   b. In mild cases, only steatorrhea may be evident; however, in the fully developed disease nutritional deficiency becomes apparent due to failure to absorb proteins, carbohydrates, calcium, vitamins including K and B_{12}, and folate.

IV. COLON

A. MOTILITY

1. Types of Movement.
   a. Haustral shuttling describes the random segmental contractions of circular muscle that divide the large bowel into sacs called haustra. As the haustra are not fixed in location, being alternately formed, obliterated, and reformed elsewhere, the luminal contents are displaced first in one direction and then in the opposite direction in a nonprogressive shuttling movement, which aids the absorptive process.
   b. Segmental propulsion moves haustral contents toward the rectum but only involves a short colonic segment.
   c. Systolic multihaustral propulsion involves coordinated contraction of several adjacent segments and obliteration of haustral folds in recipient segments, with the result that intestinal contents are moved aborally.

    d. Peristalsis, here as elsewhere in the gut, is a progressive contractile wave preceded by a wave of relaxation.

    e. Mass movements occur most often from the transverse colon to the sigmoid. They represent simultaneous forceful contractions of a long segment of the large bowel. These occur only three to four times daily, initiated, at least partially, by the gastrocolic and duodenocolic reflexes. These reflexes cause emptying of the bowel after a meal when food distends the stomach and duodenum, respectively.

2. Rates of Movement.

    a. The net **rate of aboral movement** of large intestinal contents is 5 cm/hr, rising to 10 cm/hr after a meal. However, this rate is highly variable. In patients suffering constipation, it may be no more than 1 cm/hr, while in a normal individual, maximal parasympathetic stimulation, as occurs with a carbachol injection, may increase the rate to 20 cm/hr.

    b. Transit time describes the period between ingestion of a meal and excretion of its residues. It is influenced to some extent by the contents of the meal; for example, high-residue (high-fiber) diets have a shorter transit time than low-residue (low-fiber) diets. However, it must be realized that meals are not digested, absorbed, and excreted in the discrete units in which they usually are eaten. Residues within the colon are not completely evacuated at any time so those from several meals or even several days of meals are mixed. Thus, while laboratory studies show the first appearance of the residue from a test meal in the stool as soon as 4–10 hours after ingestion, final excretion of the last residues of that meal may take 58 to 165 hours.

3. Defecation.

    a. The anal sphincter has **two major components:** the internal and external sphincters.

        (1) The **internal sphincter** is the thick terminus of the circular smooth muscle layer of the rectum. It is not under voluntary control, relaxing only when the rectum or rectosigmoid colon is distended. Should distention be prolonged, it returns to the contracted state.

        (2) The **external sphincter** consists of striated muscle, which not only surrounds the internal sphincter but extends distal to it. Control is voluntary, modifying the reflex contraction maintained at the rest by receptors within the muscle, which act like muscle spindles. Increased intra-abdominal pressure as well as distension of the sigmoid and rectum reflexly produce contraction in the external sphincter, preventing incontinence; this is, in fact, the major mechanism by which continence is maintained. Gross distension, as occurs during defecation causes relaxation; relaxation also occurs during micturition.

    b. When, as follows a mass movement, the rectum is distended by fecal material, peristaltic contraction of the distal colon, sigmoid, and rectum begins eliminating the angles between them. As contraction raises rectal pressure, both the internal and external sphincters open and feces are expelled, emptying the colon as high as the splenic flexure. The process ends when the external sphincter closes. Overall, defecation is an intrinsic reflex arising in the myenteric plexus within the colonic musculature.

        (1) Defecation is assisted by any maneuver that increases intra-abdominal and thoracic pressure as, for example, straining (forced expiration against the closed glottis with contraction of the abdominal muscle) and squatting.

        (2) There also is a spinal defecation reflex under parasympathetic control. This begins as afferent fibers in the rectum are stimulated by distension. Efferent pathways from the spinal cord, via the **nervi erigentes,** reach the descending colon, sigmoid, rectum, and anus, serving to strengthen the peristaltic contractions produced by the intrinsic reflex.

        (3) Despite the defecation reflexes, defecation in all but infants, does not occur without conscious, voluntary inhibition of the reflex contraction produced in the external sphincter by sudden distention of the rectum. Voluntary effort can strengthen external sphincter contraction if the time is not appropriate for defecation. If this occurs, the reflexes die out and with them the desire to defecate; they do not return until another mass movement distends the rectum, again triggering the reflexes and desire.

B. ABSORPTION, SECRETION, AND GAS PRODUCTION

1. Colonic function is restricted to **absorption of water, Na$^+$**, and other minerals. Of the 500–2000 ml of chyme that enter the colon daily, usually less than 100 ml of fluid are excreted in the feces as 90 percent of the fluid is absorbed.

2. Colonic absorption of water and electrolytes is most active proximally in the ascending and transverse colon.

    a. Na$^+$ is actively absorbed here as it is in the small intestine.

    b. Cl$^-$ is absorbed as it is in the ileum, through active exchange with HCO_3^-, which is secreted into the lumen.

    c. Water is absorbed passively by diffusion, following the osmotic gradient established by Na^+ and Cl^- absorption.

    d. K^+ is passively absorbed along its electrochemical gradient.

3. Colonic **secretion** consists only of an **alkaline mucus,** which not only lubricates and protects the mucosa but also binds together the fecal mass. No digestive enzymes are secreted in the colon.

4. Intestinal Gas.

    a. There are three sources of gas in the gastrointestinal tract.

        (1) Swallowed air, including air released from food and carbonated beverages, enters the stomach from which it is removed by eructation or passed into the intestines with chyme.

        (2) Gas is formed by bacterial action in the ileum and large intestine.

        (3) Some gases diffuse into the gastrointestinal tract from the bloodstream.

    b. While the little amount of gas in the small bowel originates as swallowed air, colonic gas or flatus is produced in large volumes chiefly through bacterial breakdown of the undigested nutrients that reach it. The total volume of gases entering and forming in the colon may be as high as 7–10 L/day, including chiefly CO_2, CH_4, H_2, and nitrogen gas (N_2). As these gases, with the exception of N_2, diffuse readily through the intestinal mucosa, the final volume of the flatus expelled is highly reduced to perhaps 600 ml daily.

C. DISORDERS OF COLONIC FUNCTION

1. Constipation is a difficult disorder to define given the wide range of stool frequency in normal individuals, which ranges from three or four per day, to once daily, and even to once every 2–3 days.

    a. The two most **common causes** are not diseases of the colon but are **chronic voluntary suppression** of defecation and **chronic laxative abuse.** With time, these conditions progressively weaken the defecation reflexes, resulting in an **atonic colon.**

    b. The only true **symptoms** of constipation are slight **anorexia,** mild **abdominal discomfort,** and **distension.** These are due solely to the unpleasant mechanical effects of a distended colon and rectum. The symptoms are not due to "autointoxication" or the absorption of toxic substances from the feces. Once the distension is relieved, symptoms vanish. Potentially toxic substances (e.g., ammonia and histamine) are absorbed normally in small quantities from the colon into the portal circulation, where they are detoxified by the liver.

    c. Constipation is best treated by prevention, that is, by establishment of regular bowel habits and a diet high in dietary bulk (i.e., residue and fiber). Dietary bulk stimulates motility, decreases the frequency of diverticular disease, and may reduce the frequency of colonic neoplasia.

2. Megacolon develops in patients with a congenital absence of the myenteric and submucosal plexuses in a segment of the colon. The condition is also known as **Hirschsprung's disease** and is analogous to achalasia. Massive amounts of feces may eventually collect proximal to the affected segment, producing remarkable distension of the colon.

3. Diarrhea is the opposite of constipation and is best defined as an increase in stool volume, frequency, or both. The comparison is not made with some absolute norm but rather with a patient's customary pattern of defecation. Four categories of diarrhea may be distinguished.

    a. Osmotic diarrhea results when a substance enters the intestinal tract and is not absorbed but is osmotically active. Such substances include the saline cathartic group of laxatives (e.g., magnesium sulfate). This type of diarrhea also is seen in oligosaccharidase deficiency. For example, in lactose intolerance undigested carbohydrates enter the colon with a relatively high volume of water, which is retained because of its osmotic activity. In the colon, bacterial fermentation products such as lactic acid and volatile fatty acids cause additional fluid secretion by further increasing luminal osmolality and inhibiting absorption.

    b. Secretory diarrhea results from secretion of fluid and electrolytes in excess of the colon's ability to absorb fluid. The most common cause is abuse of contact cathartics such as phenolphthalein and bisacodyl. Infection with an enterotoxin-producing bacterium (e.g., cholera) also may produce a secretory diarrhea. In cholera, the organism, *Vibrio cholerae,* adheres to the intestinal mucosa and releases a toxin that induces an outpouring of fluid and electrolytes from the small intestine.

    c. Motor diarrhea results when intestinal contents are moved at abnormally high rates through the intestines. One such variety is the **neurogenic diarrhea associated with stress.** In this condition, excessive stimulation of the parasympathetic nervous system causes increased motility and secretion of mucus from the colon.

d. Defective ion absorption produces diarrhea because the gut fails to absorb a single ionic species actively but the intestinal contents remain isosmotic. It is a minor cause of diarrhea, associated with only some obscure congenital conditions, such as congenital chloridorrhea.

V. HUNGER, APPETITE, AND SATIETY

A. DEFINITIONS

1. **Hunger** is a conscious sensation of the physical need for food.

2. **Appetite** generally refers to psychic or emotional desire to eat without reference to the need for food.

3. **Satiety** is the conscious sensation of food sufficiency, and it leads to cessation of eating. Both eating and its cessation are voluntary acts.

B. CENTRAL FACTORS. Hunger and satiety are inherent in the organization of the brain; the hypothalamus regulates food intake.

1. A **satiety center** is believed to be located in the ventromedial nuclei of the hypothalamus. In experimental studies with animals, lesions in this area lead to voracious eating as long as food is available, ultimately producing morbid obesity. Stimulation, on the other hand, produces a refusal to eat, even in the presence of highly appetizing food after deprivation.

2. **Feeding centers** are thought to be located in the lateral areas of the hypothalamus. Lesions here produce a failure to eat, with resulting emaciation and death by starvation. Stimulation produces voracious eating.

3. The feeding center is thought to be continuously active unless it is inhibited by the satiety center. Activity of the satiety center, in turn, is regulated by the blood levels of glucose and glucagon. For example, a rise in the glucose level as occurs postprandially stimulates the satiety center, which then inhibits the feeding center.

C. PERIPHERAL FACTORS

1. **Gastrointestinal mechanisms** for influencing hunger and satiety are presumed to exist, acting by stimulation of sensory nerve endings. This would involve, for example, stimulation of oropharyngeal receptors and gastric stretch receptors during eating.

2. **Metabolic mechanisms** would include those means by which the nutrient status of the body might activate central sensory mechanisms. According to the **glucostatic theory,** for example, a high arteriovenous (A-V) glucose ratio, not an absolutely high glucose level, inhibits the contractions in the gut that are associated with hunger. The A-V glucose ratio also suppresses appetite and food intake when the glucostatic center in the hypothalamus is stimulated.

STUDY QUESTIONS

Directions: Each question below contains five suggested answers. Choose the **one best** response to each question.

1. Gastric acid secretion is increased by all of the following substances EXCEPT

(A) gastrin
(B) caffeine
(C) protein
(D) histamine
(E) secretin

2. Sectioning of all vagal innervation of the gastrointestinal tract would produce all of the following conditions EXCEPT

(A) decreased salivation
(B) increased gastric motility
(C) decreased gastrin release
(D) decreased transit time
(E) increased protein absorption

3. A patient who has undergone resection of the terminal ileum may suffer from all of the following conditions EXCEPT

(A) steatorrhea
(B) iron deficiency anemia
(C) megaloblastic anemia due to vitamin B_{12} deficiency
(D) night blindness due to vitamin A deficiency
(E) bleeding tendency due to vitamin K deficiency

4. A child with Hartnup disease, a congenital disorder characterized by a defect in one of the specific amino acid transport systems, would be expected to show impaired intestinal absorption of all the amino acids listed EXCEPT

(A) alanine
(B) isoleucine
(C) phenylalanine
(D) valine
(E) lysine

5. In a patient complaining of heartburn, swallowing might be associated with all of the following conditions EXCEPT

(A) prolonged esophageal peristalsis
(B) increased intraluminal esophageal pressure
(C) failure of the lower esophageal sphincter to relax
(D) a drop in esophageal pH to 3
(E) nonpropulsive synchronous esophageal contractions

6. Elevated gastrin levels are characteristic of the Zollinger-Ellison syndrome, which is associated with pancreatic tumors known as gastrinomas. All of the following conditions would be expected in this syndrome EXCEPT

(A) diarrhea
(B) peptic ulcer
(C) esophageal reflux
(D) increased hepatic secretion of HCO_3^-
(E) decreased intestinal transit time

Directions: The question below consists of lettered choices followed by several numbered items. For each numbered item select the **one** lettered choice with which it is **most** closely associated. Each lettered choice may be used once, more than once, or not at all.

Questions 7–12

For each area of gastrointestinal activity, select the location in which this activity is most likely to occur.

(A) Stomach
(B) Duodenum
(C) Jejunum
(D) Ileum
(E) Colon

7. The principal area of bile salt uptake

8. The area from which the least absorption occurs

9. The chief site of iron absorption

10. The chief site of vitamin B_{12} absorption

11. The only area in which cells do not secrete or produce digestive enzymes

12. The area in which absorption is restricted to water and electrolytes

ANSWERS AND EXPLANATIONS

1. The answer is E. [*II A 3 b (2) (b), B 2 b (2) (a), (d), (3) (a), c*] Caffeine and protein in the diet both serve to increase gastrin production upon entry into the stomach. Gastrin, a hormone produced in the gastric antrum, increases acid production by the gastric parietal cells. Histamine, found in the mast cells of the antral mucosa, also is a potent stimulus to acid secretion, although its precise role in this process is controversial. Secretin, a polypeptide hormone product of the small intestinal mucosa, acts to decrease gastric acid secretion, although its principal action is to stimulate the pancreas to produce water and HCO_3^-.

2. The answer is B. [*I C 4; II A 3 a, B 2 b (3) (c); III E 4 c; IV A 2*] Parasympathetic influences conveyed by the vagal nerves on the gastrointestinal system enhance secretion and motility and, in general, stimulate the entire digestive process. Thus, loss of these influences would be associated with decreased saliva, acid, and enzyme production and decreased gastric, small intestinal, and colonic motility, all of which would increase transit time. Absorption is not directly affected by nervous stimuli; therefore, sectioning of vagal innervation would not produce any change in protein absorption.

3. The answer is B. (*III C 4 d*) Bile salts, which are essential for fat absorption, are secreted by the liver into the bile, absorbed in the distal ileum, and carried back to the liver in the portal vein. Disruption of this enterohepatic circulation leads to fat malabsorption. Manifestations of this include steatorrhea and deficiencies of the fat-soluble vitamins A, D, E, and K. Iron is absorbed throughout the small intestine but most actively in the upper small intestine. Vitamin B_{12} is absorbed only from the terminal ileum.

4. The answer is E. (*III E 4 c*) Amino acid absorption occurs through the active process of carrier-mediated transport. There are at least five transport systems, each of which preferentially carries amino acids of a similar chemical nature. Thus, the neutral amino acids (i.e., alanine, valine, leucine, isoleucine, methionine, phenylalanine, and tyrosine) share one carrier, while cystine and the basic amino acids, lysine and arginine, share another.

5. The answer is C. [*I B 2 a (2) (a), (b)*] The occurrence of esophageal reflux often is associated with heartburn (pyrosis). Esophageal motor abnormalities seen in reflux include prolonged peristalsis, increased intraluminal pressure, and nonpropulsive synchronous esophageal contractions (diffuse esophageal spasm). Measurement of intraluminal esophageal pH often shows a drop to a pH of 2 associated with reflux of gastric acid, although heartburn may appear when the pH drops only to 4. Failure of the lower esophageal sphincter to relax during swallowing is associated with achalasia, not esophageal reflux.

6. The answer is C. [*II B 2 b (2)*] Manifestations of the Zollinger-Ellison syndrome result from the high levels of gastrin that characterize this disorder. To understand the manifestations, it is necessary only to understand the normal functions of gastrin. Gastrin stimulates gastric acid and pepsin secretion, increasing the total volume of gastric secretion. Peptic ulcer disease in both the duodenum and stomach results from the increased acid production. Gastrin also increases both gastric and intestinal motility. It is this increase in motility as well as the increase in the volume and acidity of gastric secretion that lead to the diarrhea and shortened intestinal transit time seen in Zollinger-Ellison syndrome. In addition, gastrin is responsible for increases in HCO_3^- in both hepatic and pancreatic secretions, although in Zollinger-Ellison syndrome these fail to compensate for the increased acid production. Esophageal reflux is not characteristic of Zollinger-Ellison syndrome and would not be expected since gastrin serves to increase pressure in the lower esophageal sphincter.

7–12. The answers are: 7-D, 8-A, 9-B, 10-D, 11-E, 12-E. (*II C 2 a; III C 4 a, E 7 a, c; IV B 1, 3*) Little absorption of any kind occurs in the stomach; only alcohol and highly lipophilic molecules such as aspirin are absorbed there to any appreciable extent. Iron (Fe) is absorbed principally in the duodenum and upper small intestine. Bile salt absorption occurs chiefly in the ileum as does absorption of the vitamin B_{12} intrinsic factor complex. The colonic mucosa absorbs only water and electrolytes; its cells secrete no digestive enzymes. The jejunum is the site of very active digestion and absorption of fats, carbohydrates, and proteins.

I. GENERAL PRINCIPLES OF ENDOCRINOLOGY AND INFORMATION TRANSFER

A. DEFINITION AND CHARACTERISTICS OF A HORMONE

1. Definition. Hormones are secretory products of ductless (endocrine) glands released into the extracellular (interstitial) space, absorbed into the bloodstream, and transported by the circulatory system to **target cells** (organs). Hormones interact with their target cells via **receptors**, which are large protein molecules with specific binding sites for specific hormones. Generally, there are 2000–100,000 receptor molecules per target cell.

  a. Melatonin, the major hormone produced by the pineal gland, is found in both the cerebrospinal fluid (CSF) and the blood. Since the melatonin levels are lower in the CSF than in the blood of adult humans, melatonin probably enters the CSF from the blood.

  b. The hormone **angiotensin II** differs from other hormones in that this octapeptide is formed in the bloodstream from a precursor hormone or **prohormone** decapeptide called angiotensin I by the catalytic action of pulmonary and renal converting enzyme. The protein **angiotensinogen** secreted by the liver could be called the prohormone of angiotensin I and, therefore, the **pro-prohormone** for angiotensin II.

  c. Some hormones are converted from prohormones to more active forms within the target tissues. A special characteristic of **testosterone** is that it serves as a circulating prohormone for the formation of a more potent subcellular androgen called **dihydrotestosterone** in many peripheral (extrahepatic) target organs including the skin, seminiferous tubules, epididymis, seminal vesicles, prostate gland, and brain.

2. Characteristics.

  a. Function. Hormones regulate existing fundamental bodily processes but do not initiate cellular reactions de novo.

    (1) As regulators, hormones stimulate or inhibit the rate and magnitude of biochemical reactions by their control of enzymes and thereby cause morphologic, biochemical, and functional changes in target tissues. Although they are not used as energy sources in biochemical reactions, hormones modulate energy-producing processes and regulate the circulating levels of energy-yielding substrates such as glucose and fatty acids.

    (2) Hormones regulate such bodily processes as growth, maturation, differentiation, regeneration, reproduction, pigmentation, behavior, metabolism, and chemical homeostasis. Regulation of the slower processes (e.g., growth, reproduction, and metabolism) requires longer periods of continual hormone stimulation in contrast to rapid coordinations of the body, which are regulated by the nervous system.

    (3) In contrast to vitamins, hormones do not serve a nutritive role in responsive tissues, and they are not incorporated as a structural moiety into another molecule.

  b. Concentration. Hormones usually are secreted into the circulation in extremely low concentrations.

    (1) Plasma concentrations of steroid and thyroid hormones are between 10^{-6} mol/L and 10^{-9} mol/L.

    (2) Plasma concentrations of peptide hormones are between 10^{-10} mol/L and 10^{-12} mol/L.

  c. Latent Period. Hormones have a much longer latent period than that associated with neurons following their stimulation.

    (1) Following the administration of **oxytocin,** milk ejection occurs in a few seconds.

    (2) The metabolic response to **thyroxine** can take as long as 3 days.

B. CHEMISTRY OF HORMONES.
There are three major classes of hormones: steroids, proteins and polypeptides, and amino acid derivatives. No polysaccharides or nucleic acids are known to function as hormones. The majority of hormones are water soluble.

1. **Steroids** comprise a group of biologically active substances including androgens (C-19), estrogens (C-18), progesterone (C-21), glucocorticoids (C-21), and mineralocorticoids (C-21). 25-Hydroxycholecalciferol (25-hydroxyvitamin D_3) and 1,25-dihydroxycholecalciferol (1,25-dihydroxyvitamin D_3) are modified steroids (C-27) called **secosteroids**.
   a. **Structure.**
      (1) Steroids are hydrophobic, lipid-soluble substances consisting of three cyclohexyl rings and one cyclopentyl ring combined into a single structure.
      (2) The steroids are derivatives of the cyclopentanoperhydrophenanthrene nucleus consisting of a fully hydrogenated phenanthrene (rings A, B, and C), to which is attached a cyclopentane ring. This fully saturated, four-ring structure consisting of 17 carbon atoms is the hypothetical parent compound called **gonane** or **sterane**.
   b. **Synthesis, Secretion, and Circulation.**
      (1) The immediate precursor of steroids is stored cholesterol (esters), which is either synthesized intracellularly or absorbed from the plasma. For the biosynthesis of all steroids, the initial step is conversion of cholesterol to pregnenolone, which is the rate-limiting step in steroidogenesis.
      (2) Steroids are synthesized and secreted by the adrenal cortex, testis, and ovary, which are endocrine organs derived from mesoderm. The placenta also synthesizes and secretes steroids.
      (3) After being released into the general circulation by diffusion, steroids circulate in the plasma, bound to transport proteins such as cortisol-binding globulin and sex steroid–binding globulin.
      (4) The half-life of steroids varies from 60 minutes for testosterone to 100 minutes for cortisol.

2. **Proteins and Polypeptides.**
   a. **Structure.** The protein and polypeptide hormones generally are water soluble and circulate unbound in plasma.
   b. **Synthesis, Secretion, and Circulation.**
      (1) Many of the protein-type hormones are synthesized as prohormones, which must be modified after synthesis by the secretory cells to produce biologically active hormones.
      (2) Protein and polypeptide hormones are secreted by endocrine organs derived from ectoderm (e.g., the anterior and posterior lobes of the pituitary gland,* tuberoinfundibular nuclei, supraoptic and paraventricular nuclei, and placenta) as well as those derived from endoderm (e.g., the pancreatic islets of Langerhans, thyroid gland, and parathyroid glands).
      (3) Protein and polypeptide hormones probably are stored exclusively in subcellular membrane-bound secretory granules within the cytoplasm of the endocrine cells.
      (4) These hormones are released into the blood by exocytosis, which involves the fusion of the secretory granule and cellular membrane followed by the extrusion of the granular contents into the bloodstream.
      (5) The half-life of protein and polypeptide hormones varies from 5–6 minutes for adrenocorticotropic hormone (ACTH), to 20 minutes for growth hormone (GH), to 60 minutes or more for gonadotropins. In the circulation, parathormone splits into two fragments, of which only one, the amino terminal portion, has significant biologic activity.

3. **Amino Acid Derivatives.** These derivatives of the amino acid tyrosine sometimes are called **phenolic derivatives** and include epinephrine, norepinephrine, triiodothyronine (T_3), and tetraiodothyronine (thyroxine or T_4).
   a. **Secretion.** These hormones are secreted by endocrine tissues derived from ectoderm (adrenal medulla) and from tissues derived from endoderm (thyroid gland).
   b. **Circulation.**
      (1) Most of the thyroid hormones (i.e., T_4 and T_3) are bound to throxine-binding globulin. T_4 has a biologic half-life of about 7–9 days, and T_3 has a plasma half-life of about 2 days.
      (2) Epinephrine and norepinephrine exist in plasma either in the free form or in conjugation with sulfate or glucuronide. Most of the circulating epinephrine is bound to blood proteins (mainly albumin), while norepinephrine does not bind to blood proteins to any appreciable extent. The circulatory half-life of epinephrine and norepinephrine is about 1–3 minutes.

---

*The anterior and posterior lobes of the pituitary gland are derivatives of buccal ectoderm and neural ectoderm, respectively.

C. MECHANISMS OF HORMONE ACTION

1. Overview.

   a. Hormones produce their effects by first combining with specific cell components called receptors. Only cells with receptors for a specific hormone respond; those cells lacking the specific receptors are unaffected. Hormone receptors found on the target cell membrane are termed **external receptors,** and those within the cytoplasm and nucleus are termed **internal receptors.**

   b. Hormones and receptors interact in the following ways to affect intracellular metabolism.

     (1) Polypeptide hormones bind to a **fixed** receptor at the outer cell surface.

     (2) Steroids bind to a specific **mobile** receptor in the cytoplasm.

     (3) Thyroid hormones combine with a **nuclear** receptor.

   c. Hormone-sensitive cells respond to high concentrations of certain hormones by reducing the number of receptors on the cell surface. For example, elevated ambient insulin concentrations cause a loss or inactivation of insulin receptors in liver cells, fat cells, and white blood cells.

   d. Polypeptide hormones that have been found to enter cells include insulin, prolactin, parathormone, and gonadotropins. Thyroid hormone, an amino acid derivative, also can enter cells.

     (1) Human chorionic gonadotropin (HCG) remains bound to its receptor when it enters ovarian cells.

     (2) Since a receptor may enter cells with the hormone, it is possible that the biologically active component is the receptor and not the hormone. An example is **insulin antibody**, which causes the insulin receptor to change its configuration and allows the receptors to exert insulin effects on cells.

2. Hormone-Receptor Interaction. For many hormones, two mechanisms for hormone-receptor coupling operate. One mechanism involves **cyclic adenosine 3′,5′-monophosphate (cAMP)** as a second messenger that changes enzyme activities. Cyclic AMP is the intracellular nucleotide messenger that mediates the effects of a variety of hormones on subcellular processes, which lead to many physiologic responses. The other mechanism involves **transcription and translation** effects that change enzyme activities.

   a. cAMP-Mediated Hormone Activity.

     (1) Most polypeptide hormones, many biogenic amines, and some prostaglandins activate their target cells by stimulating the synthesis of the second messenger, cAMP, which causes enzyme phosphorylation, which usually is associated with enzyme activation. The first messenger is the hormone that binds to the membrane receptor and leads to the activation of the membrane-bound enzyme **adenyl cyclase**, which converts adenosine triphosphate (ATP) to cAMP in the presence of magnesium ion (Mg^{2+}).

     (2) Cyclic AMP exerts biologic activity via the phosphorylation of **cAMP-dependent protein kinases**.

       (a) Kinases are a family of enzymes that phosphorylate their substrate. Protein kinases, a subgroup of the kinases, can be soluble or membrane bound. Protein kinases transfer a phosphate group from ATP to the hydroxyl group of the substrate. Only the protein kinases that are regulated by cAMP are called cAMP-dependent protein kinases.

       (b) The phosphorylation of a cAMP-dependent protein kinase can lead to the activation (e.g., via phosphorylase kinase, phosphorylase, and triglyceride lipase) or the inactivation (e.g., via glycogen synthetase and pyruvate dehydrogenase) of the substrate (enzyme).

       (c) The effects of the cAMP-dependent protein kinases on their substrates (enzymes) are reversed by a group of enzymes called phosphoprotein phosphatases, which remove the phosphate group from the protein enzyme by hydrolysis.

       (d) The enzyme that inactivates cAMP is called cyclic nucleotide phosphodiesterase. Most of the phosphodiesterases are soluble enzymes.

     (3) An example of cAMP-mediated hormone activity is the hormone regulation of glycogen metabolism. Glucagon and epinephrine stimulate glycogenolysis and also inhibit glycogen synthesis.* Insulin has the opposite effect on both processes.

       (a) The enzymes that promote glycogenolysis are active in their phospho- form and inactive in their dephospho- form. The reverse is true for glycogen synthetase, which is the major enzyme in glycogen synthesis.

---

*The glycogenolytic effect of epinephrine in human liver probably occurs via the activation of a cAMP-independent phosphorylase.

(b) In muscle, a cAMP-dependent protein kinase is activated either by catecholamines (via their β-adrenergic receptors) or by glucagon (via its receptor in the liver).

(c) This activated protein kinase phosphorylates another protein kinase, phosphorylase kinase, thereby activating it.

(d) The activated phosphorylase kinase activates (phosphorylates) the enzyme, phosphorylase, which initiates glycogen breakdown. Phosphorylase kinase also phosphorylates glycogen synthetase, thereby inactivating it.

b. Transcription and Translation Effects. Steroid hormones and thyroid hormones modulate transcription in specific areas of the nuclear chromatin by interacting with DNA molecules in the chromatin to cause **enzyme induction**. Steroid-receptor complexes are translocated through the cytosol and enter the nucleus. Thyroid hormones enter the nucleus in the free state and then combine with nuclear receptors.

(1) Chromatin consists of DNA, histone proteins, and nonhistone (acidic) proteins. The specificity of nuclear binding is a property of a particular acidic protein.

(2) As a result of the interaction of steroid hormones and thyroid hormones with the chromatin, transcription is stimulated and specific messenger RNA (mRNA) synthesis increases.

(3) The specific mRNAs enter the cytoplasm, where they direct the synthesis (translation) of specific proteins. These proteins may be enzymes, structural proteins, receptor proteins, or secretory proteins.

c. Other Mechanisms of Hormone Action.

(1) Hormonal control of membrane permeability is well documented for ions and for metabolites such as glucose and amino acids.

(a) For example, the increased entry of glucose into cells, which is mediated by a special transport system, is the major way in which insulin controls glucose utilization by muscle.

(b) Other peptide hormones that change membrane permeability include GH, ACTH, calcitonin, epinephrine, glucagon, antidiuretic hormone (ADH), parathormone, thyroid-stimulating hormone (TSH or thyrotropin), and thyroxine.

(c) Steroid hormones that alter membrane permeability include mineralocorticoids, glucocorticoids, estrogens, and androgens.

(2) Angiotensin and Aldosterone.

(a) Vascular smooth muscle contracts directly in response to angiotensin II and indirectly in response to aldosterone via an increase in the blood volume. Both hormones cause the elevation of blood pressure.

(b) During sodium deprivation or pregnancy, patients produce abnormally high amounts of angiotensin II and aldosterone without the usual high blood pressure response.

(i) Angiotensin II increases the number of its receptors on adrenocortical cells.

(ii) Angiotensin II decreases the number of its receptors in vascular smooth muscle.

(3) Catecholamines and Thyroid Hormone Receptors.

(a) Catecholamines exert their effects via plasma membrane receptors.

(b) Iodothyronines (thyroid hormones) are lipid soluble and have receptors in the nucleus of the target cell.

(c) Excess thyroid hormone leads to an increased number of catecholamine receptors in the myocardium of experimental animals. This may explain the catecholamine-like cardiac effects that are noted in hyperthyroid patients, even though these patients produce normal amounts of catecholamines.

(i) Hyperthyroid patients have rapidly beating hearts and palpitations, effects also observed with excessive amounts of catecholamines (pheochromocytoma).

(ii) The cardiac symptoms of hyperthyroidism can be ameliorated by the administration of a β-blocker such as propranolol.

(iii) The increase in cardiac β-adrenergic receptors in hyperthyroidism makes the heart more responsive to catecholamines (e.g., epinephrine and norepinephrine).

D. OTHER MODES OF INFORMATION TRANSFER (Fig. 7-1)

1. Paracrine communication involves the local diffusion of a peptide or other regulatory molecule to its target cell through the extracellular space.

a. Cells affected by a paracrine regulatory substance are in the vicinity of the cell that releases the messenger (see Fig. 7-1B).

b. Although paracrine substances may diffuse into the blood (as do hormones), it is not necessary for them to do so.

c. An example of a paracrine substance is the hormone **somatostatin**, which is enzymatically

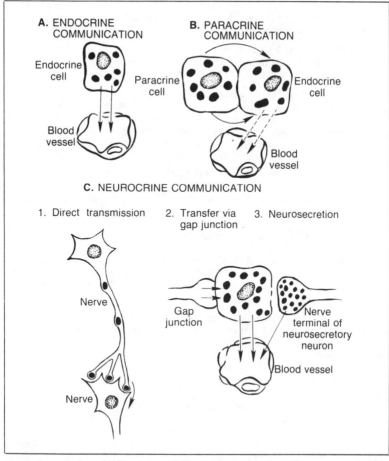

Figure 7-1. Modes of transmission of chemical messengers. It is now recognized in modern endocrinology that a hormone can be transmitted by routes other than the bloodstream (*A*). Hormones can act in close proximity (*B*) as well as distant to their site of release (*C3*). Also shown are types of neural transmission (*C1* and *C2*). *Solid arrows* indicate the major route of hormonal secretion, and *dashed arrows* denote a minor route of hormonal diffusion. (Adapted from Smith PH and Madson KL: Interactions between autonomic nerves and endocrine cells of the gastroenteropancreatic system. *Diabetologia* 20:314–324, 1981.)

degraded in the blood. This process protects distant cells from a paracrine substance should one enter the bloodstream. Somatostatin secretion is stimulated by glucose, glucagon, and gut hormones.

2. **Neurocrine communication** involves the release of chemical messengers from nerve terminals. Neurocrine substances may reach their target cells via one of three routes (see Fig. 7-1 C).

   a. The neurotransmitter can be released directly into the intercellular space, cross the synaptic junction, and inhibit or activate the postsynaptic cell (see Fig.7-1 *C1*).

      (1) Neurocrine substances are inactivated by degrading enzymes and by reuptake of the substances by neurons.

      (2) Examples of neurocrine substances secreted by this route are acetylcholine (ACh) and norepinephrine.

   b. A neural signal also can be transferred to a target cell via a **gap junction**, which is a membrane specialization between nerve cells, between nerve terminals and endocrine cells, and between endocrine cells. Gap junctions allow the movement of small molecules and electric signals from one cell to another, creating a functional **syncytium** (see Fig. 7-1 *C2*).

   c. The third potential route for the transmission of a neural signal is identical to the classic neurosecretory mechanism, which involves the release of a peptide from or neurohormone from a **neurosecretory neuron** into the blood followed by the interaction of this neurohormone with specific receptors on distant target cells (see Section II C and D). Examples of such neurocrine substances are oxytocin and ADH (see Fig. 7-1 C3). The effector sites of neurohormones are not always endocrine cells.

3. **Autacoid** is a term used to designate a compound that is synthesized at or close to its site of action. This is in contrast to the circulating hormones, which act on tissues distant from their site of synthesis. Prostaglandins are autacoids.

4. Signal transmission also might occur through the release of peptides into the lumen of the gastrointestinal tract, where these peptides interact with endocrine cells to cause the release of a second endocrine or paracrine messenger. It is known that gastrin, somatostatin, and substance P are released into the gut lumen following nerve stimulation. It is possible that these substances are released as precursor molecules and are activated by digestive enzymes. Gastrin also is secreted into the bloodstream.

5. Peptides that mediate (via endocrine, paracrine, or neurocrine mechanisms) the regulatory effects in the brain and gut include substance P, somatostatin, secretin, bombesin, vasoactive intestinal peptide (VIP), cholecystokinin (CCK),* endorphins, enkephalins, gastrin, and neurotensin.

II. BASIC CONCEPTS OF ENDOCRINE CONTROL

A. **HOMEOSTASIS AND STEADY STATE.** One of the major functions of the endocrine system is to maintain the homeostasis of the **milieu intérieur**. This condition of relative constancy in the concentration of dissolved substances, in temperature, and in pH is a basic requirement for the normal function of cells.

1. The concept of **homeostasis** as a straight line function must be modified because many regulated organismic processes are not constant but conform to a persistent endogenous or exogenous **rhythm**.
   a. A circadian pattern in the levels of plasma 17-hydroxycorticosteroids has been demonstrated in man. Such 24-hour cycles are not solely a response to fluctuating environmental stimuli but also are due to internal endogenous oscillators whose phases are influenced by environmental stimuli.
   b. In man, certain corticosteroids (e.g., cortisol) have a rhythmic pattern of secretion, with secretory rates highest early in the morning and lowest late at night. Accordingly, plasma cortisol concentration is at a peak between 6 A.M. and 8 A.M. and at a nadir between midnight and 2 A.M. This rhythmic secretory pattern persists but shifts to correspond with a change in sleeping habit (e.g., during illness, night work, changes in longitude, and total bed rest or confinement). For this reason, treatment of patients with exogenous corticosteroids is on an alternate-day dosage regimen whereby the entire dose is given in the morning of every other day. This dosage schedule simulates the normal adrenocortical secretory rhythm.
   c. The rhythmic pattern of corticosteroid secretion occurs in isolated adrenal glands and even in single adrenocortical cells.

2. The term **steady state** indicates that a function or a system does not vary with time, but that the system is not in true equilibrium. The system is said to be in a **dynamic equilibrium** because matter and energy flow into the system at a rate equal to that at which matter and energy flow out of the system.

B. **THE HYPOTHALAMIC-HYPOPHYSIAL AXES AND FEEDBACK CONTROL** (Fig. 7-2)

1. **Overview.** The hypothalamus has **neural control** over hormone secretion by the posterior lobe of the pituitary gland. The secretory activity of the anterior lobe is controlled by **hypothalamic hormones**, which are secreted into the **hypothalamic-hypophysial portal system** (the hypophysial portal system).

2. **Hypophysial Portal System.** The median eminence has a poorly developed blood-brain barrier, and there is relatively little **arterial** blood perfusing the cells of the anterior lobe. The blood supply of the anterior lobe is derived from branches of the internal carotid arteries (mainly the superior hypophysial artery).
   a. The posterior lobe derives its blood from a capillary plexus emanating from the inferior hypophysial artery. This capillary plexus drains into the dural sinus. The neural tissue of the upper infundibular stem (neural stalk) and of the median eminence is supplied largely by branches of the superior hypophysial artery. (The median eminence is the specialized area of the hypothalamus located beneath the inferior portion of the third ventricle.)

---

*Also called pancreozymin-cholecystokinin or PZ-CCK.

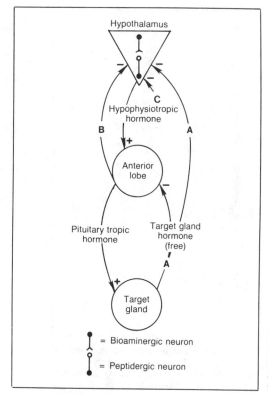

Figure 7-2. Diagram showing the three levels of feedback mechanisms for controlling hormone synthesis and secretion; long-loop feedback (*A*), short-loop feedback (*B*), and ultrashort-loop feedback (*C*). *Plus signs* indicate stimulation, and *minus signs* indicate negative feedback.

    b. The **primary capillary plexus**, which emanates from the superior hypophysial artery, forms a set of long portal veins that carry blood downward into the anterior lobe.
        (1) The portal veins, which give rise to the **secondary capillary plexus**, constitute about 90 percent of the blood supply to the cells of the anterior lobe. The secondary capillary plexus drains into the dural sinus.
        (2) The anterior lobe receives its remaining blood from the short portal veins, which originate in the capillary plexus of the inferior hypophysial artery at the base of the neural stalk (lower infundibular stem).

  3. Feedback Control. An important concept of hormone control systems is that of the **feedback mechanism** for the control of hormone synthesis and secretion. Negative feedback control occurs on three levels (see Fig. 7-2).
    a. Long-Loop Feedback (see Fig. 7-2A). Peripheral gland hormones and substrates arising from tissue metabolism can exert what is called long-loop feedback control on both the hypothalamus and the anterior lobe of the pituitary gland. Long-loop feedback usually is negative but occasionally can be positive and is particularly important in the control of thyroidal, adrenocortical, and gonadal secretions.
    b. Short-Loop Feedback (see Fig. 7-2B). Negative feedback also can be exerted by the anterior pituitary tropic hormones on the synthesis or release of the hypothalamic releasing or inhibiting hormones, which collectively are called **hypophysiotropic hormones**.
    c. Ultrashort-Loop Feedback (see Fig. 7-2C). Evidence suggests that the hypophysiotropic hormones may inhibit their own synthesis and secretion via a control system referred to as ultrashort-loop feedback.
 C. NEUROSECRETORY NEURONS: THE CONCEPT OF NEUROSECRETION (Fig. 7-3). Neural control of the pituitary gland is exerted through neurohumoral secretions that arise from specialized neurosecretory neurons (peptidergic neurons) and are carried by the bloodstream to a target site. A neuroendocrine transducer is an endocrine gland that secretes a hormone in response to a neural stimulus.

  1. Structure and Function.
    a. Neurosecretory neurons are glandular, unmyelinated neurosecretory cells that have a dual function.
        (1) They function as typical neurons, in that they conduct action potentials.
        (2) They also function as endocrine glands, in that they synthesize and release neurohormones either directly into the general circulation (as in the case of the neurosecretory

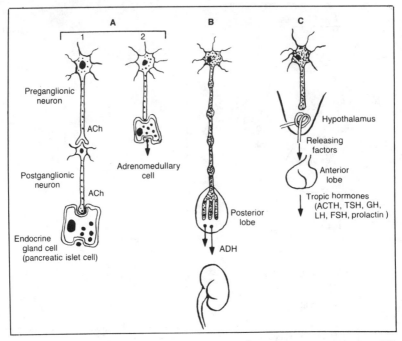

Figure 7-3. The three types of neuroendocrine transducers: secretomotor neurons (*A*), magnocellular neurosecretory neurons (*B*), and parvicellular neurosecretory neurons (*C*). *A* represents the control of endocrine glands by the direct innervation of autonomic fibers. The adrenal medulla is the only autonomic neuroeffector that is innervated by preganglionic neurons. *A1* represents the parasympathetic nervous system, and *A2* shows the sympathoadrenomedullary axis. Postganglionic innervation by the sympathetic nervous system is not shown in this figure. *B* and *C* represent the control of the anterior and posterior lobes by the neurosecretory neurons. *ACh* = acetylcholine; *ADH* = antidiuretic hormone; *ACTH* = adrenocorticotropic hormone; *TSH* = thyroid-stimulating hormone; *GH* = growth hormone; *LH* = luteinizing hormone; and *FSH* = follicle-stimulating hormone. (Reprinted with permission from Martin JB, et al: Neuroendocrine transducers and neurosecretion. In *Neuroendocrinology*. Philadelphia, FA Davis, 1977, p 5.)

neurons of the pars nervosa) or into a portal system (as in the case of the hypophysiotropic neurons, which release their neurohormones into the primary plexus of the hypophysial portal system).

b. A neurosecretory cell system consists of axons that terminate directly on or near blood vessels. This differentiates these cells from typical neurons, which release neurotransmitters at localized synaptic regions. The functional complex of a neurosecretory neuron together with a blood vessel (hemocoele) is called a **neurohemal organ.**

2. Classification. Neurosecretory neurons in humans and other mammals are restricted to the hypothalamus, where they occur as two distinct populations of cells that secrete neurohormones (see Fig. 7-3; Table 7-1).

a. The **magnocellular neurosecretory system** refers to the neurosecretory neurons of the supraoptic and paraventricular nuclei, which together form the supraopticohypophysial tract. Magnocellular neurons synthesize and secrete the neurohomones ADH and oxytocin.

b. The **parvicellular** (also **parvocellular**) neurosecretory system refers to the neurosecretory neurons of the tuberoinfundibular (tuberohypophysial) tract.

   (1) These neurons of the medial basal hypothalamus have axons that terminate directly on the capillaries of the portal vessels in the median eminence, and they form a final common pathway for neuroendocrine function.

   (2) The parvicellular neurosecretory neurons mainly are peptidergic. An important exception, however, is the dopaminergic neurosecretory neurons that form and secrete prolactin-inhibiting factor (PIF).

c. Neural information is transmitted to the parvicellular and magnocellular neurosecretory cells by monoaminergic neurons. Most of the cell bodies of the monoaminergic neurons are located in the mesencephalon and lower brain stem.

Table 7-1. Neuroendocrine Transducer Systems

Neuroendocrine System	Hormone
Magnocellular neurosecretory neurons— posterior lobe	
Supraoptic	Antidiuretic hormone
Paraventricular	Oxytocin
Parvicellular neurosecretory neurons— median eminence	Hypophysiotropic hormones
Preganglionic fibers— adrenal medulla	Epinephrine*
Postganglionic fibers— pineal gland	Melatonin†
juxtaglomerular apparatus	Renin†

Note.—After Martin JB, et al: Neuroendocrine transducers and neurosecretion. In *Clinical Endocrinology.* Philadelphia, FA Davis, 1977, p 4.
*Acetylcholine is the neurotransmitter preceding epinephrine release.
†Norepinephrine precedes the release of both melatonin and renin.

    (1) The monoaminergic neurons that innervate the parvicellular neurons produce and secrete biogenic amines (i.e., norepinephrine, dopamine, and serotonin), which modulate the hypothalamic release of the hypophysiotropic hormones.
    (2) The function of the magnocellular neurosecretory neurons is controlled by cholinergic and noradrenergic neurotransmitters.
      (a) ACh stimulates the release of ADH and oxytocin.
      (b) Norepinephrine inhibits the secretion of ADH and oxytocin.
    (3) Since the secretion of the parvicellular and magnocellular peptidergic neurons is regulated by biogenic amines, the neurosecretory neurons can correctly be viewed as *neuroeffector* cells.

D. NEUROENDOCRINE TRANSDUCERS are endocrine glands that convert neural signals into hormonal signals. There are three established types of neural control of endocrine tissues (see Fig. 7-3).
  1. Direct Innervation By Autonomic Secretomotor Neurons.
    a. The Pancreatic Islets of Langerhans.
      (1) The islets of Langerhans have a postganglionic parasympathetic innervation. Increased vagal activity to the beta cells stimulates insulin release only during periods of elevated blood sugar.
      (2) The islets of Langerhans also have a postganglionic sympathetic innervation. When the sympathetic nerves to the beta cells are stimulated or when norepinephrine or epinephrine is infused, the predominant effect is inhibition of insulin secretion.
    b. The Pineal Gland. This endocrine structure of the diencephalon is classified as a periventricular (circumventricular) organ because it borders on the third ventricle.
      (1) The pinealocytes are innervated by the **postganglionic** adrenergic (sympathetic) fibers, which originate in the superior cervical ganglia of the sympathetic chain.
      (2) When the neurotransmitter norepinephrine is released by the autonomic fibers, it stimulates the synthesis and release of melatonin and other indoleamine hormones.
      (3) The pineal gland, like other periventricular organs (e.g., the median eminence), has a poorly developed blood-brain barrier.
    c. The Juxtaglomerular Cells. These granular cells of the juxtaglomerular apparatus receive a postganglionic input, which, when stimulated, leads to the release of the proteolytic enzyme renin. Renin can be classified, according to the neuroendocrine transduction concept, as a hormone.
    d. The adrenal medulla, which is composed of chromaffin cells, is innervated by **preganglionic** (cholinergic) sympathetic fibers, which, when stimulated, cause the release of the adrenomedullary hormones epinephrine and norepinephrine. The release of ACh at the synapses causes the secretion of these catecholamines.
  2. The Magnocellular Neurosecretory Regulation of the Posterior (Neural) Lobe of the Pituitary Gland (see Fig. 7-3).

    a. Depolarization of the magnocellular neurosecretory cells by ACh released at synapses on the cell bodies of these neurons causes the release of ADH and oxytocin. The axons of these neurons terminate directly on the blood vessels of the posterior lobe.

    b. The neural input to the cell bodies of the magnocellular neurons is cholinergic, and the hormonal output consists of peptidergic hormones.

3. The Parvicellular Neurosecretory Regulation of the Anterior Lobe of the Pituitary Gland (see Fig. 7-3).

    a. The anterior lobe lacks a direct nerve supply, but the pituitary gland does possess an innervation. The neurons present in the anterior lobe are exclusively postganglionic sympathetic, which are **vasomotor fibers** and not secretomotor fibers.

        (1) The hypothalamic regulation of the anterior lobe is achieved through the tuberohypophysial neurons of the medial basal hypothalamus. These peptidergic neurons synthesize and secrete specific hypothalamic (hypophysiotropic) hormones, which enter the hypophysial portal system and stimulate or inhibit the secretion of anterior pituitary hormones.

        (2) The arcuate nucleus (nucleus infundibularis) is the main site of origin of the fine unmyelinated axons of the tuberoinfundibular pathway; however, tuberohypophysial neurons are located throughout the hypophysiotropic area, including the ventromedial nuclei and the periventricular area. The arcuate nucleus serves as the final neural link in the neurovascular connection between the hypothalamus and the anterior lobe.

    b. The cells of the intermediate lobe, unlike those of the anterior lobe, receive a direct bioaminergic secretomotor supply from the hypothalamus and are not perfused directly by the hypophysial portal system.

III. ENDOCRINE CELLS OF THE GASTROENTEROPANCREATIC (GEP) SYSTEM

A. NEUROPEPTIDES OF THE GEP SYSTEM

1. The endocrine cells of the gut produce peptides that also are found in a variety of other tissues, including the pancreas, pituitary gland, and central and peripheral nerves. The same peptide might have dual functions as both gut hormone and neurotransmitter. The pancreatic hormones insulin, glucagon, and somatostatin are considered to be part of the GEP system.

2. The cells of the GEP system share with certain neural tissues the capacity of **a**mine **p**recursor **u**ptake and **d**ecarboxylation and, therefore, are called **APUD cells**. APUD cells are located in the gastrointestinal tract and pancreatic islets.

    a. The endocrine cells of the gastrointestinal tract are found singly or in small groups and are dispersed among the other epithelial cells of the mucosa.

    b. The GEP cells usually are situated at or near the base of the intestinal glands.

B. APUD CELLS

1. APUD cells represent a group of neurons and endocrine cells, which take up amino acids and modify them into amines and peptides. APUD cells synthesize and secrete all of the hormones of the body except the steroids.

2. Most APUD cells are of ectodermal origin. Those in the gut and pancreas, however, originate from endoderm. Some APUD cells, such as those of the adrenal medulla and neurons of sympathetic ganglia, are derived from the neuroectoderm component called the **neural crest**.

3. Many APUD endocrine cells contain dopamine, serotonin, and histamine as well as polypeptide hormones, and these substances tend to be released together.

    a. In some cases, enough hormone is secreted to enter the plasma and be transported to other tissues in the body and, therefore, an **endocrine** function is performed.

    b. In other instances, little hormone is bound to distant receptors (due to too few receptors, too little available hormone to bind them, or both) and, therefore, a **paracrine** action results (see Fig. 7-1).

IV. THE ENDOGENOUS OPIOID PEPTIDES

A. TERMINOLOGY

1. Opioid denotes opiate-like in terms of a functional similarity with a chemical dissimilarity.

    a. Opioid peptides refer to endogenous or synthetic compounds that have a spectrum of pharmacologic activity similar to that of morphine.

    b. Since the opioid peptides bind with morphine receptors in the central nervous system (CNS), they are called **morphinomimetic peptides**.

2. The two chemical groups of the neuroactive opioids or opioid peptides are called **endorphins** and **enkephalins**.

  a. The term endorphin originally was used to designate all opiate-like componds occurring in the brain. This term now is restricted to the specific endogenous morphine-like substances, α-, β-, and γ-endorphins.

  b. The enkephalins include met-enkephalin, leu-enkephalin, dynorphin, and α-neo-endorphin.

B. ENDORPHINS

1. Chemistry. Endorphins are structural derivatives of β-lipotropic hormone (β-LPH). The lipotropin-related peptides (β-LPH and β-endorphin) and ACTH have a common glycoprotein precursor molecule (molecular weight of about 31,000 daltons), which is called pro-opiocortin or pro-opiomelanocortin (**POMC**).

  a. POMC is activated by proteolytic cleavage to yield 10 individual peptides that have been classified into four groups.

  (1) ACTH and corticotropin-like intermediate lobe peptide (CLIP)

  (2) Lipotropic hormones (lipotropins), represented by β- and γ-LPH

  (3) Melanocyte-stimulating hormones (MSHs) represented by α-, β, and γ-MSH

  (4) Opioid peptides or endorphins, represented by α, β, and γ-endorphins

  b. POMC consists of three major chemical moieties.

  (1) The N-terminal fragment (16 K) is cleaved to form γ-MSH.

  (2) ACTH consists of the fragment containing amino acid residues 1–39.

  (a) ACTH is cleaved to produce α-MSH, which consists of the fragment containing amino acid residues 1–13. In man, α-MSH normally is not synthesized in significant quantities.

  (b) ACTH also is cleaved into CLIP, which consists of the fragment containing amino acid residues 18–39.

  (3) β-LPH consists of the fragment containing amino acid residues 1–91. β-LPH is the precursor of the endogenous opioids, γ-LPH and β-endorphin.

  (a) γ-LPH consists of the β-LPH fragment containing amino acid residues 1–58.

  (i) In nonhuman species, γ-LPH is converted by proteolytic cleavage to β-MSH, which is the fragment containing amino acid residues 41–58.

  (ii) β-MSH and α-MSH are not formed in humans in significant quantities because the intermediate lobe is vestigial in adult humans. β-MSH does not exist as such in normal human plasma.*

  (b) β-Endorphin consists of the β-LPH fragment containing amino acid residues 61–91.

  (i) β-Endorphin is cleaved into γ-endorphin, which is the fragment containing amino acid residues 61–77.

  (ii) γ-Endorphin forms α-endorphin, which is the fragment containing amino acid residues 61–76.

2. Distribution. The endogenous opioids (endorphins) do not cross the blood-brain barrier.

  a. POMC is synthesized in the anterior and intermediate lobes of the pituitary gland, in the hypothalamus and other areas of the brain, and in several peripheral tissues including the placenta, gastrointestinal tract, and lung.

  (1) In the human pituitary gland, which has a vestigial intermediate lobe, POMC is localized principally in the anterior lobe, where it is synthesized by basophils. POMC is the precursor of ACTH and β-LPH, which are secreted together from the basophils.

  (a) ACTH and β-LPH occur within the same pituitary cell and possibly within the same secretory granule.

  (b) The intermediate lobe does not produce ACTH and β-LPH as final secretory products.

  (2) In the brain, the concentrations of ACTH and β-LPH are much lower than in the anterior lobe, and all neural cells or fibers that contain ACTH also contain β-LPH. The highest extrapituitary concentrations of ACTH, α-MSH, β-LPH, γ-LPH, and β-endorphin exist in the hypothalamus followed by the limbic system.

  b. β-Endorphin.

  (1) β-Endorphin is the principal opioid peptide in the pituitary gland, and it exists in the highest concentration in the intermediate lobe of experimental animals. In the human pituitary, which has a vestigial intermediate lobe, β-endorphin generally is confined to the cells of the anterior lobe. In response to acute stress, the pituitary gland secretes concomitantly ACTH and β-endorphin.

---

*β-MSH has been isolated from the human pituitary gland, but the MSH activity in human plasma probably is due largely to the MSH-related peptides, β-LPH and γ-LPH.

(2) β-Endorphin has been found in the human pancreas, placenta, semen, and in the male reproductive tract.

(3) Immunoassayable β-endorphin also exists in the plasma and CSF.

c. CNS Endorphins. The endorphin system exists in the CNS and is characterized by long fiber projection systems from the arcuate region of the hypothalamus to the periventricular region.

(1) Only in the arcuate area do ACTH, β-LPH, and the endorphins occur within **cell bodies** of neurons. Therefore, endorphin cell bodies exist only in the ventral hypothalamus.

(2) Outside the arcuate area, the endorphins are found within the **fibers** of neurons, which project mainly to the mesencephalic periaqueductal gray area. Other areas of projection include the periventricular thalamus, medial amygdala, locus ceruleus, and the zona incerta.

(3) Brain β-endorphin content is unaltered by hypophysectomy, suggesting that the pituitary gland is not the source of CNS endorphins.

3. Physiologic Effects.

a. Opiate-Receptor Binding. Endorphins bind to opiate receptors in the brain to cause analgesia, sedation, and respiratory depression.

(1) The intraventricular administration of β-endorphin in humans relieves intractable pain.

(a) It is believed that this analgesic effect involves the pituitary release of endorphins because the effect is abolished with hypophysectomy.

(b) The characteristic effect of opiates in humans is less a specific blunting of pain than it is an induced state of indifference or emotional detachment from the experience of suffering.

(c) Endorphins play a central role in controlling affective states and may be involved in controlling the drive for food, water, and sex, all of which are associated with the limbic nervous system.

(2) The intracerebral administration of endorphins produces a profound sedation and immobilization (catatonia).

b. Stimulation of the Hypothalamus. The intracisternal or intracerebral injection of β-endorphin elicits a hypophysiotropic effect via the stimulation of the hypothalamus.

(1) This hypothalamic effect on pituitary function causes the secretion of GH, prolactin, ACTH, and ADH.

(2) In addition, there is a diminished pituitary secretion of TSH and the gonadotropic hormones, luteinizing hormone (LH) and follicle-stimulating hormone (FSH).

c. β-LPH is a more potent stimulator of **aldosterone synthesis** than is angiotensin II, and β-LPH is thought to be the stimulator of the zona glomerulosa in primary aldosteronism.

d. β-LPH, γ-LPH, and MSH have a lipolytic effect resulting in **fat mobilization**; however, these compounds do not have significant regulatory effects on fat metabolism in humans.

(1) β-MSH has been isolated from the human pituitary gland.

(2) MSH activity in human plasma is due largely to the MSH-related peptides, β-LPH and γ-LPH.

e. β-LPH, γ-LPH, and ACTH have a **weak MSH activity,** which explains the hyperpigmentation associated with increased ACTH secretion.

C. ENKEPHALINS

1. Chemistry.

a. Enkephalins, represented by met-enkephalin and leu-enkephalin, are pentapeptide opioids, which are not breakdown products of POMC. The precursor substance for met- and leu-enkephalin is **proenkephalin A.**

b. Proenkephalin B contains the sequences of α-neo-endorphin, dynorphins A and B, and leu-enkephalin. Although α-neo-endorphin and dynorphin contain the leu-enkephalin sequence, neither is considered a precursor for leu-enkephalin.

2. Distribution.

a. The pituitary gland is virtually devoid of enkephalins.

(1) Of the pituitary enkephalins, including dynorphin (amino acid residues 1–13), the greatest concentrations exist in the pars nervosa.

(2) Brain enkephalins are not depleted following hypophysectomy, suggesting a neural origin for these peptides.

b. The CNS. Enkephalins usually are localized in neuronal processes and terminals. The enkephalin-containing neurons have the widest distribution throughout the CNS and are characterized by short axon projections.

(1) The highest concentrations of met-enkephalin are in the basal ganglia (globus pallidus) and substantia nigra.*

---

*The substantia nigra is classified by some neuroanatomists as a component of the basal ganglia.

> > **(2)** In the spinal cord, enkephalins exist in highest concentration in the dorsal gray matter, which corresponds to the synapses of primary sensory nerve endings. Enkephalins also are found in other areas of the spinal cord, notably the substantia gelatinosa, which is known to be involved in the transmission of pain impulses.
> >
> > **(3)** Many limbic structures have relatively high enkephalin levels, including the lateral septal nucleus, the interstitial nucleus of the striae terminalis, the prefornical area, and the central amygdala.
> >
> > **(4)** Dynorphin also is distributed widely in the CNS, with the highest concentrations in the hypothalamus, medulla-pons, midbrain, and spinal cord.
>
> > **c.** Enkephalin-like material has been demonstrated in the CSF and has been isolated from the adrenal medulla, where it exists in high concentrations both in axon terminals of the splanchnic nerve and in adrenomedullary chromaffin cells.
> >
> > > **(1)** Met-enkephalin in the circulation originates from the adrenal medulla.
> > >
> > > **(2)** High plasma concentrations of met-enkephalin are found in patients with pheochromocytoma.
> >
> > **d.** Enkephalin immunoreactivity has been reported in the human gastrointestinal tract (myenteric plexus and mucosa), gallbladder, and pancreas.
> >
> > **e.** Ectopic production of enkephalins has been reported in carcinoid tumors of the lung and thymus.
>
> **3. Physiologic Effects.** Enkephalins are most appropriately considered to be neurotransmitters or neuromodulators. A neuromodulator is a substance that modifies (positively or negatively) the action of a neurotransmitter at a presynaptic site (by modulating the release of a neurotransmitter) or at a postsynaptic site (by modulating the neurotransmitter action).
>
> > **a.** The enkephalins in the dorsal gray matter of the spinal cord function to suppress substance P–containing nerve endings and provide an analgesic (antinociceptive) role.
> >
> > **b.** The vagal nuclear localization of enkephalins corresponds to the emetic actions and antitussive properties of morphine.
> >
> > **c.** Morphine and enkephalins inhibit the firing of the noradrenergic neurons of the locus ceruleus, which is the origin of the ascending noradrenergic fibers.
> >
> > > **(1)** The localization of enkephalins in the locus ceruleus may account for the euphoria-producing actions of morphine.
> > >
> > > **(2)** The amygdala is considered the prime site for morphine-generated euphoria.
> > >
> > > **(3)** The enkephalin tracts and opiate receptors in the limbic system may explain the euphoric effects of opiates.
> > >
> > > **(4)** Respiratory depression, which accounts for the lethal effects of opiates, may involve receptors in the nucleus solitarius of the brain stem, which regulates visceral reflexes including respiration.
> >
> > **d.** Electric stimulation of the periaqueductal gray area of the spinal cord of humans, which causes analgesia, is associated with an increase in the concentration of enkephalins in the CSF (lumbar spinal fluid).
> >
> > > **(1)** Opioids are believed to modulate pain impulses by acting to reduce sensitivity to pain peripherally as well as centrally by activating opiate receptors in the periaqueductal gray area.
> > >
> > > **(2)** Acupuncture is associated with the release of enkephalins into the CSF.

V. THE PITUITARY GLAND

> **A. EMBRYOLOGY.** The pituitary gland is in close anatomic relation to the hypothalamus. This relationship has both embryologic and functional significance.
>
> > **1.** The anterior lobe of the pituitary gland (also known as the **adenohypophysis)** is derived from the primitive gut by an upward extension (Rathke's pouch) of the epithelium of the primitive mouth cavity (stomodeum). The adenohypophysis is a derivative of oral (buccal) ectoderm.
> >
> > **2.** The neural or posterior lobe (also known as the **neurohypophysis)** develops as a downward evagination of the neural tube at the base of the hypothalamus (infundibulum) and, therefore, represents a true extension of the brain. Neuroregulation of this structure is achieved by direct neural connections. The neurohypophysis is a derivative of neural ectoderm.

> **B. MORPHOLOGY**
>
> > **1. Gross Anatomy.**
> >
> > > **a.** The pituitary gland lies in a bony walled cavity, the **sella turcica,** in the sphenoid bone at the base of the skull.
> > >
> > > **b.** The sella turcica is separated superiorly from the cranial cavity by the dura mater, which is one of the three meninges that envelop the entire brain. This tough fibrous connective tissue membrane forms a shelf known as the **diaphragma sellae**, which extends over most of the top of the gland. The dura mater completely lines the sella turcica and nearly surrounds the pituitary gland.

c. The pituitary (hypophysial) stalk and its blood vessels reach the main body of the gland through the diaphragma sellae. The pituitary stalk consists of the neural stalk and the adenohypophysial tissue that is contiguous with the neural stalk.

d. The **adenohypophysis** is divided into three parts.

(1) The **pars distalis** represents the bulk of the anterior lobe in man and receives most of its blood supply from the superior (anterior) hypophysial artery, which gives rise to the hypophysial portal system

(2) The **pars intermedia** lies between the pars distalis and the neural lobe and is a vestigial structure in humans. It is relatively avascular and is considered almost nonexistent in humans.

(3) The **pars tuberalis** is an elongated collection of secretory cells, which superficially envelops the neural stalk and extends upward as far as the basal hypothalamus. It is the most vascular portion of the anterior lobe.

e. The **neurohypophysis** also consists of three components.

(1) The **median eminence**, located beneath the third ventricle, is a small, highly vascular protrusion of the dome-shaped base of the hypothalamus, which is designated grossly as the tuber cinereum. The floor of the third ventricle is designated as the infundibulum because of its resemblance to a funnel.

(2) The **neural stalk** of the posterior lobe arises in the median eminence.

(3) The **pars nervosa** (neural lobe) retains its neural connection with the ventral diencephalon.

(4) The dominant features of the posterior lobe are the neurosecretory neurons of the supraopticohypophysial and paraventriculohypophysial tracts, which form the magnocellular neurosecretory system. These unmyelinated nerve tracts arise from the supraoptic and paraventricular nuclei within the ventral diencephalon (hypothalamus) and descend through the infundibulum and neural stalk to terminate in the posterior lobe. The posterior lobe is a storage site for hormones and, therefore, is not correctly termed an endocrine gland.

2. Histology.

a. Cells. Two major cell types are found in equal numbers in the anterior lobe.

(1) Chromophils (granular secretory cells) exist in two forms.

(a) Acidophils (eosinophils) account for about 80 percent of the chromophils. The acidophils are the cellular source of prolactin and GH.

(b) Basophils comprise about 20 percent of the chromophils. The anterior pituitary hormones that are secreted by the basophils are TSH, ACTH, LH, FSH, and β-LPH.

(2) Chromophobes (agranular cells) are not precursors of the chromophils and are now known to have an active secretory function. It is likely that most of these cells are degranulated secretory cells

b. The **neurons** found in the anterior lobe are almost exclusively postganglionic sympathetic fibers that innervate blood vessels.

c. The **nerve fibers** of the neurohypophysial system terminate mostly in the pars nervosa. Interspersed between these neurosecretory fibers are numerous glial cells called **pituicytes**, whose function, other than structural support, remains unknown.

d. Vascular Supply (see detailed description of the hypophysial portal system in Section II B 2). A basic tenet of the neurovascular hypothesis is that the concentration of the hypothalamic hypophysiotropic hormones is greater in hypophysial portal blood than at any other site in the vasculature.

(1) In humans, the capillaries at the base of the hypothalamus are formed directly from branches of the superior hypophysial arteries, which arise from the internal carotid arteries. There are few vascular anastomoses between the hypothalamic artery and the superior hypophysial artery. The crucial regulatory connection between the hypothalamus and the anterior lobe is via the hypophysial portal vessels.

(2) The intermediate lobe is not perfused directly by the hypophysial portal system but is regulated by bioaminergic secretomotor fibers originating in the hypothalamus.

(3) The blood supply to the posterior lobe is largely separate from that of the anterior lobe. The blood supply to the median eminence is greater than that to the entire pituitary gland.

C. HORMONES OF THE POSTERIOR LOBE: ADH AND OXYTOCIN. The physiologic aspects of ADH are described in Section VIII of Chapter 4, "Renal Physiology." The physiologic aspects of oxytocin are described below.

1. Synthesis and Storage.

a. Like ADH, oxytocin is a nonapeptide* that is synthesized within the cell bodies of the pep-

---

*If the two cysteine residues are counted together as a single cystine residue, ADH and oxytocin are correctly classified as octapeptides.

tidergic neurons of the magnocellular neurosecretory system or the hypothalamic-neurohypophysial neural tract (supraopticohypophysial tract).
   b. This polypeptide is synthesized mainly in the paraventricular nuclei of the hypothalamus and, like ADH, is stored in the posterior lobe of the pituitary gland.

  2. Stimuli for Release.
    a. Oxytocin secretion is brought about by stimulation of cholinergic nerve fibers.
    b. Stimulation of the tactile receptors in the areolar region of the breast during suckling activates somesthetic neural pathways, which transmit this signal to the hypothalamus. This leads to the reflex secretion of oxytocin into the bloodstream and to milk release following a latent period of 30–60 seconds. This reflex is called the **milk let-down** or **milk ejection reflex**.
      (1) Oxytocin causes milk release in lactating women by contraction of the myoepithelial cells, which cover the stromal surface of the epithelium of the alveoli, ducts, and cisternae of the mammary gland.
      (2) Oxytocin secretion can be conditioned so that the physical stimulation of the nipple no longer is required. Thus, lactating women can experience milk release in response to the sight and sound of a baby.
      (3) Oxytocin is not required for successful nursing in humans.
    c. Oxytocin secretion also can occur in response to genital tract stimulation, such as that which occurs during coitus and parturition.
    d. Oxytocin is produced in men and also is released during genital tract stimulation. The role of this neurohypophysial hormone in men is unknown.

  3. Inhibition of Release.
    a. Milk let-down can be inhibited by emotional stress and psychic factors such as fright.
    b. Excitation of adrenergic fibers to the hypothalamus inhibits peptide release. Activation of the sympathetic neurons with the concomitant release of norepinephrine and epinephrine inhibits oxytocin secretion.
    c. Ethanol inhibits endogenous oxytocin release, resulting in reduced myometrial contractility.
    d. Enkephalins also inhibit oxytocin release.

  4. Physiologic Effects.
    a. Oxytocin stimulates contraction of the smooth muscle (myoepithelium) of the lactating mammary gland.
    b. It also stimulates contraction of the smooth muscle of the uterus (myometrium).
      (1) The sensitivity of the myometrium to exogenous oxytocin during pregnancy increases as pregnancy advances.
      (2) Oxytocin plays a role in labor and has been shown to be a useful therapeutic agent in the induction of labor.

D. HORMONES OF THE ANTERIOR LOBE. The principal hormones of the anterior lobe of the pituitary gland can be classified conceptually into two groups: hormones that stimulate other endocrine glands to secrete hormones and hormones that have a direct effect on nonendocrine target tissues. Only the latter group is described here, using GH and prolactin as examples.

  1. GH also is known as human growth hormone (HGH), somatotropic hormone (STH), somatotropin, and somatocrinin.
    a. Synthesis, Chemistry, and General Characteristics.
      (1) GH is synthesized by the acidophils of the anterior lobe and is stored in very large amounts in the human pituitary gland. GH represents approximately 4–10 percent of the wet weight of the pituitary gland, which is equivalent to 5–15 mg.
      (2) GH is a single, unbranched polypeptide chain containing 191 amino acid residues. It has been synthesized in bacteria using recombinant DNA techniques.
      (3) Humans exhibit a **species specificity** for GH, and only human and monkey GH preparations have biologic activity in humans.
      (4) Like all other pituitary hormones, GH is secreted episodically in periods of 20–30 minutes. The large diurnal fluctuations represent integrations of many small secretory episodes.
        (a) A regular nocturnal peak in GH secretion occurs 1–2 hours after the onset of deep sleep.
        (b) The nocturnal peak of GH secretion correlates with stage 3 or stage 4 slow-wave sleep, as indicated on the electroencephalogram (EEG).
      (5) The plasma GH concentration in the growing child is not significantly higher than that in the adult whose growth has ceased.
    b. Control of Secretion. The release of GH is primarily under the control of two hypothalamic (hypophysiotropic) hormones.
      (1) Stimuli for Release. Somatotropin releasing factor (SRF) is the putative releasing hor-

mone for GH. SRF has not been identified or chemically synthesized. However, several pharmacologic, physiologic, and psychic agents are known to stimulate GH release.

(a) Release of GH is mediated by monoaminergic and serotoninergic pathways; thus, α-adrenergic (norepinephrine), dopaminergic, and serotoninergic agonists as well as β-adrenergic antagonists all stimulate GH release in humans.

(b) Bromocriptine (2-bromo-α-ergocryptine, a dopamine agonist), enkephalins, endorphins (β-endorphin), and opiates stimulate GH secretion.

(c) Insulin-induced hypoglycemia is a potent stimulus of GH secretion as are pharmacologic doses of glucagon and vasopressin. In addition, GH release is evoked in experimental animals by infusing 2-deoxy-D-glucose, which is a competitive inhibitor of glucose transport. This substance produces systemic hyperglycemia but intracellular glycopenia.

(d) Physiologic stimuli include hypoglycemia, increased plasma concentrations of amino acids (arginine, leucine, lysine, tryptophan, and 5-hydroxytryptophan), and decreased free fatty acid concentrations. In addition, estrogens stimulate GH synthesis and secretion.

(e) GH secretion is stimulated by moderate to vigorous exercise, emotional stress, and stress due to fever, surgery, anesthesia, trauma, pyrogen administration, and repeated venipuncture. Fasting leads to elevated GH secretion after 2 or 3 days.

(2) Inhibitors of Secretion.

(a) GH can inhibit its own secretion via a short-feedback loop mechanism that operates between the anterior lobe and the median eminence. Somatotropin-inhibiting hormone (SIH or somatostatin) inhibits the synthesis and release of GH.

  (i) Somatostatin is a tetradecapeptide (14 amino acid residues) that has been chemically synthesized.

  (ii) It is a product of the parvicellular neurosecretory neurons that terminate in the median eminence.

  (iii) Somatostatin also is found in other parts of the brain, in the gastrointestinal tract, and in the delta cells of the pancreatic islets of Langerhans.

  (iv) In addition to inhibiting GH secretion, somatostatin blocks the secretion of insulin, glucagon, and gastrin and inhibits the intestinal absorption of glucose. These effects of somatostatin produce a state of hypoglycemia.

(b) The secretion of GH in response to the aforementioned stimuli often is blunted in obese individuals.

(c) Glucocorticoids (cortisol) decrease GH secretion, but their predominant effect is the interference with the metabolic actions of GH.

(d) A decline in GH secretion is observed in late pregnancy, despite the presence of high estrogen levels.

  (i) Impairment of glucose tolerance is common, and clinical diabetes occurs frequently, despite the above normal insulin secretion in response to a glucose load during pregnancy.

  (ii) Pregnancy regularly antagonizes the action of insulin and increases the pancreatic secretory capacity of both normal and diabetic individuals.

  (iii) The development of gestational diabetes probably is due to a greater degree of insulin antagonism caused by normal plasma concentrations of human placental lactogen (HPL or chorionic somatomammotropin).

c. Physiologic and Metabolic Effects.

(1) Stimulation of Growth of Bone, Cartilage, and Connective Tissue.

(a) The effects of GH on skeletal growth are mediated by a family of polypeptides called **somatomedins,** which are synthesized mainly in the liver. (Thyroid hormone and insulin also are necessary for normal osteogenesis).

(b) It seems likely that somatomedin may be produced in nonhepatic tissues as well, as somatomedin activity has been found in the serum, liver, kidney, and muscle tissue.

  (i) Receptors for somatomedin exist in chondrocytes, hepatocytes, adipocytes, and muscle cells.

  (ii) Somatomedin has insulin-like effects on tissues, including lipolysis, increased glucose oxidation in fat, and increased glucose and amino acid transport by muscle.

(c) GH, through somatomedin, stimulates proliferation of chondrocytes and the appearance of osteoblasts. The increase in the thickness of the epiphysial (cartilagenous) end-plate accounts for the increase in linear skeletal growth.

(d) After epiphysial closure (fusion), bone length can no longer be increased by GH, but bone thickening can occur through periosteal growth. It is this growth that accounts for the changes seen in hypersomatotropism (acromegaly).

    (e) These reactions are the biochemical correlates of protein synthesis in general body growth and also account for the hyperplasia and hypertrophy associated with increased tissue mass.

  (2) Protein Metabolism.

    (a) GH has predominantly anabolic effects on skeletal and cardiac muscle, where it stimulates the synthesis of protein, RNA, and DNA.

    (b) GH causes the reduction of circulating levels of amino acids and urea (i.e., the promotion of nitrogen retention), which accounts for the term **positive nitrogen balance**. Urinary urea concentration also is decreased.

    (c) GH promotes amino acid transport and incorporation into proteins.

  (3) Fat Metabolism.

    (a) GH has an overall catabolic effect on adipose tissue. It stimulates the mobilization of fatty acids from adipose tissue, leading to a decreased triglyceride content of fatty tissue and increased plasma levels of free fatty acids and glycerol (a gluconeogenic substance).

    (b) GH has a ketogenic effect in that it increases hepatic oxidation of fatty acids to the ketone bodies, acetoacetate and β-hydroxybutyrate.

  (4) Carbohydrate Metabolism. GH is a diabetogenic hormone. Because of its anti-insulin effect, GH has the tendency to cause hyperglycemia.

    (a) GH can produce an insulin-resistant diabetes mellitus primarily because of its lipolytic effect.

      (i) Free fatty acids can antagonize the effect of insulin to promote glucose uptake by skeletal muscle and adipose tissue.

      (ii) Free fatty acids can stimulate gluconeogenesis.

      (iii) Excess acetyl coenzyme A (acetyl-CoA) production favors gluconeogenesis because pyruvate carboxylase requires acetyl-CoA to form oxaloacetate from pyruvate. Oxaloacetate is the rate-limiting factor in gluconeogenesis.

      (iv) Acetyl-CoA also inhibits glycolysis by inhibition of pyruvate kinase.

      (v) Free fatty acids stimulate hepatic glucose synthesis mainly via the stimulation of fructose diphosphatase. At the same time, pyruvate kinase and phosphofructokinase both are inhibited by free fatty acids, which block glycolysis and favor gluconeogenesis. Citrate also blocks glycolysis at the phosphofructokinase step.

    (b) GH induces an elevation in basal plasma insulin levels.

    (c) Because of its anti-insulin effect, GH inhibits glucose transport in adipose tissue. Since adipose tissue requires glucose for triglyceride synthesis, GH antagonizes insulin-stimulated lipogenesis.

  (5) Mineral Metabolism. GH promotes urinary retention of Ca^{2+}, phosphorus, and Na^+

d. Endocrinopathies.

  (1) Growth retardation can occur when GH levels are increased and somatomedin levels are depressed (e.g., in kwashiorkor).

  (2) In the African pygmy, who is resistant to the action of GH, both GH and somatomedin levels are normal. This condition is due to a decrease in cellular receptors.

  (3) Overproduction of GH during **adolescence** results in **giantism** (gigantism), which is characterized by excessive growth of the long bones. Patients may grow to heights of as much as 8 feet.

  (4) Excessive secretion of GH during **adulthood**, after the epiphysial plates of the long bones have fused, causes growth in those areas where cartilage persists. This clinical condition, called **acromegaly,** has several characteristics, including:

    (a) Elongation and widening of the mandible (prognathism), resulting in an underbite and increased interdental spaces

    (b) Enlargement of the frontal, mastoid, ethmoid, and maxillary sinuses, causing a prominent brow

    (c) Thickening of the skin and coarsening of the facial features, which is due mainly to the proliferation of connective tissue and leads to edema

    (d) Periosteal growth of the vertebrae and the phalanges of the hands (metacarpals) and feet (metatarsals)

    (e) Hypertrophy of the soft tissues such as the heart (cardiomegaly), liver (hepatomegaly), kidney (renomegaly), intestine, spleen (splenomegaly), tongue, and muscles

  (5) Deficiency of GH secretion in immature individuals leads to a stunting of growth or **dwarfism**, which is accompanied by sexual immaturity, hypothyroidism, and adrenal insufficiency.

  (6) Deficiency of GH may be part of an overall lack of anterior pituitary hormones (panhypopituitarism) or from an isolated genetic deficiency. In the adult, in whom GH deficiency alone is rare, clinical manifestations may include impaired hair growth and a tendency toward fasting hypoglycemia.

(7) Treatment.
    (a) Conditions associated with GH deficiencies can be treated with HGH.
    (b) The treatment of choice for hypersomatotropism is the selective surgical extirpation of the tumor without damage to other pituitary functions. Additional forms of therapy include external irradiation with cobalt and internal implantation of radioactive yttrium.
    (c) Somatostatin lowers GH levels in most patients with acromegaly, although usually not to normal levels. Somatostatin is not a treatment of choice because it also suppresses secretion of insulin, glucagon, and TSH.
    (d) Bromocriptine is effective in suppressing, but not normalizing, GH in most patients with acromegaly. This substance tends to stimulate GH secretion in normal individuals.

2. Prolactin also is known as lactogenic hormone, mammotropic hormone, and galatopoietic hormone.
    a. Synthesis, Chemistry, and General Considerations.
        (1) Prolactin is synthesized in the pituitary acidophils.
        (2) Human prolactin is a single peptide chain containing 198 amino acid residues.
        (3) It does not regulate the function of a secondary endocrine gland in humans.
    b. Control of Secretion. Two hypothalamic neurosecretory substances have been implicated in prolactin secretion.
        (1) Stimuli for Release.
            (a) Prolactin releasing factor (PRF) is the putative releasing hormone for prolactin. PRF has not been identified or chemically synthesized; however, one of these releasing factors is thyrotropin releasing hormone (TRH), which causes the release of thyrotropin (TSH).
            (b) Prolactin secretion increases about 1 hour after the onset of sleep, and this increase continues throughout the sleep period. The nocturnal peak occurs later than that for GH.
            (c) Prolactin secretion is enhanced by exercise and by stresses such as surgery under general anesthesia, myocardial infarction, and repeated venipuncture.
            (d) Plasma prolactin levels begin to increase by the eighth week of pregnancy and usually reach peak concentrations by the thirty-eighth week.
            (e) Nursing and breast stimulation are known to stimulate prolactin release. Oxytocin, the hormone that stimulates milk release, does not stimulate prolactin secretion.
            (f) Serum prolactin levels are elevated in those patients with primary hypothyroidism who are believed to have high TRH levels in the hypophysial portal circulation.
            (g) Dopamine antagonists (phenothiazine and tranquilizers), adrenergic blockers, and serotonin agonists stimulate prolactin secretion.
            (h) Pituitary stalk section and lesions that interfere with the portal circulation to the pituitary gland also cause prolactin release.
        (2) Inhibitors of Release.
            (a) Normally, the control of prolactin secretion is a constant tonic inhibition via PIF. Dopamine is secreted into the hypophysial portal vessels and may be the PIF. It is physiologically the most important PIF.
            (b) Serotonin antagonists and dopamine agonists (bromocriptine) block the secretion of prolactin. Bromocriptine administered during the postpartum period reduces prolactin secretion to nonlactating levels and terminates lactation.
    c. Physiologic Effects.
        (1) Prolactin does not have an important role in maintaining the secretory function of the corpus luteum and, therefore, is not a gonadotropic hormone in women.
        (2) Prolactin plays an important role in the development of the mammary gland and in milk synthesis.
            (a) During pregnancy, the mammary duct gives rise to lobules of alveoli, which are the secretory structures of this tissue. This differentiation requires prolactin, estrogens, and progestogens. Once the lobuloalveolar system is developed, the role of prolactin and corticosteroids in milk production, although essential, becomes miminal. GH and thyroid hormone enhance milk secretion.
            (b) Immediately following pregnancy, prolactin stimulates galactosyltransferase activity, leading to the synthesis of lactose.
            (c) In women, high serum levels of prolactin are associated with suppressed LH secretion and anovulation, which account for amenorrhea during postpartum lactation.
                (i) With continued nursing, FSH levels rise, but LH levels remain low.
                (ii) In the early postpartum period, both the FSH and LH levels are low and account for the antireproductive and antigonadal effects of prolactin.
    d. Endocrinopathies. Hyperprolactinemia is not a rare condition but frequently is undiagnosed because galactorrhea occurs in only about 30 percent of cases.

(1) In women, elevated serum prolactin manifests as infertility and menstrual abnormalities (amenorrhea).

(2) In men, hyperprolactinemia is a cause of decreased libido and impotence.

(3) Treatment of prolactin hypersecretion includes:

    (a) Surgical removal of tumors

    (b) Administration of bromocriptine, which lowers prolactin levels and usually restores normal gonadal function

VI. THE ADRENAL MEDULLA

A. EMBRYOLOGY

1. The neural crest gives rise to neuroblasts, which eventually give rise to the autonomic postganglionic neurons, the adrenal medulla, and the spinal ganglia.

2. The adrenal medulla consists of chromaffin cells (pheochromocytes), which are neuroectodermal derivatives and the functional analogs of the sympathetic postganglionic fibers of the autonomic nervous system.

3. In early **fetal** life, the adrenal medulla contains only norepinephrine.

B. MORPHOLOGY

1. Gross Anatomy.

    a. The adrenal medulla represents essentially an enlarged and specialized sympathetic ganglion and is called a **neuroendocrine transducer** because a neural signal to this organ evokes hormonal secretion.

    b. The adrenal medulla is the only autonomic neuroeffector organ without a two-neuron motor innervation. It is innervated by long sympathetic preganglionic, cholinergic neurons that form synaptic connections with the chromaffin cells.

    c. Small clumps of chromaffin cells also can be found outside the adrenal medulla, along the aorta and the chain of sympathetic ganglia.

2. Histology.

    a. There are two types of adrenomedullary chromaffin cells. Individual cells contain either **norepinephrine** or **epinephrine**, which is stored largely in subcellular particles called chromaffin granules. These granules are osmiophilic, electron-dense, membrane-bound secretory vesicles.

        (1) Approximately 80 percent of the chromaffin granules in the human adrenal medulla synthesize epinephrine (adrenalin). The remaining 20 percent synthesize norepinephrine (noradrenalin).

        (2) The chromaffin granules contain catecholamines, protein, lipids, and adenine nucleotides (mainly ATP).

            (a) One of the proteins localized in the particulate fraction is the enzyme, dopamine-β-hydroxylase.

            (b) Soluble acidic proteins found in the granules are called **chromogranins.**

    b. The preganglionic sympathetic fibers that innervate the adrenal medulla traverse the splanchnic nerve, which contains myelinated (type B) secretomotor fibers emanating mainly from lower thoracic segments (T-5 to T-9) of the ipsilateral intermediolateral gray column of the spinal cord.

    c. The cell bodies of the chromaffin cells do not have axons.

3. Vascular Supply (Fig. 7-4).

    a. The arterial blood supply to the adrenal gland reaches the outer capsule from branches of the renal and phrenic arteries, with a less important arterial input directly from the aorta. The adrenal medulla is perfused by blood vessels in two ways.

        (1) A type of **portal circulation** exists in the adrenal gland where the cortex and medulla are in juxtaposition. From the capillary plexus on the outer adrenal capsule most of the blood enters venous sinuses, which drain into and supply the medullary tissue. Thus, most of the blood perfusing the adrenal medulla is derived from the portal system and is partly deoxygenated.

        (2) There also exists a direct arterial blood supply to the medulla via the medullary arteries, which traverse the cortex.

    b. Venous blood drains via a single central vein, composed almost entirely of bundles of longitudinal smooth muscle fibers, which passes along the longitudinal axis of the gland.

C. ADRENOMEDULLARY HORMONES: MONOAMINES

1. The adrenal medulla synthesizes and secretes biogenic amines. These dihydroxylated

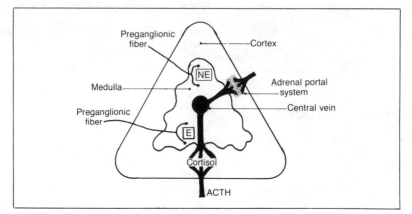

Figure 7-4. The anatomic relationship of an inner adrenal medulla surrounded by a cortex is phylogenetically new, in that it appears only in mammals. The adrenal portal vascular system, which constitutes a functional connection, contains a high concentration of cortisol. Most of the blood perfusing the adrenal medulla is partially deoxygenated. *NE* = norepinephrine; *E* = epinephrine; and *ACTH* = adrenocorticotropic hormone. (Adapted from Pohorecky LA and Wurtman RJ: Adrenocortical control of epinephrine synthesis. *Pharmacol Rev* 23(1):1–35, 1971.)

phenolic amines or **catecholamines** are epinephrine and norepinephrine.

    a. Epinephrine is produced almost exclusively in the adrenal medulla, with smaller amounts synthesized in the brain. Essentially all of the circulating epinephrine is derived from the adrenal medulla.

    b. Norepinephrine is widely distributed in neural tissues, including the adrenal medulla, sympathetic postganglionic fibers, and the CNS. In the brain, the concentration of norepinephrine is the highest in the hypothalamus. The norepinephrine content of a tissue reflects the extent or density of its sympathetic innervation. Norepinephrine has been demonstrated in almost all tissues except the placenta, which is devoid of nerve fibers.

2. Bilaterally adrenalectomized human patients excrete practically no epinephrine in the urine. Urinary levels of norepinephrine remain within normal limits, however, indicating that the norepinephrine comes from extra-adrenal sources (i.e., the terminals of the postganglionic sympathetic fibers and the brain).

3. Most of the met-enkephalin in the circulation originates in the adrenal medulla. Enkephalins are pentapeptides that belong to the class of opiate-like neuropeptides (also called endogenous opioid peptides; see Section IV A and C). Enkephalins function as neurotransmitters or neuromodulators, which normally are localized in neuronal processes and terminals.

D. CONTROL OF CATECHOLAMINE SYNTHESIS

    1. The biosynthetic pathway originates with L-tyrosine, which is derived from the diet or from the hepatic hydroxylation of L-phenylalanine. Tyrosine is hydroxylated in the cytoplasm by tyrosine hydroxylase to L-dopa (3,4-dihydroxyphenylalanine).

    2. Dopa is converted in the cytosol to **dopamine** (3,4-dihydroxyphenylethylamine) by a nonspecific aromatic L-amino acid decarboxylase.

    3. Dopamine enters the chromaffin granule, where it is converted to L-norepinephrine by dopamine-β-hydroxylase, which exists exclusively in the granule.

        a. Norepinephrine is the end product in approximately 20 percent of the chromaffin cells.

        b. In about 80 percent of the chromaffin cells, norepinephrine diffuses back into the chromaffin cytoplasm. There it is N-methylated by phenylethanolamine-N-methyltransferase (PNMT) using S-adenosylmethionine as a methyl donor to form L-epinephrine.

            (1) PNMT selectively is localized in the adrenal medulla, the only site where it exists in significant concentrations.

            (2) PNMT activity is induced by very high local concentrations of glucocorticoids (i.e., cortisol in humans), which are found only in the adrenal portal blood draining the adrenal cortex.

E. CONTROL OF CATECHOLAMINE SECRETION

    1. General Considerations.

        a. ACh provides the major physiologic stimulus for the secretion of the adrenomedullary hormones. In addition, angiotensin II, histamine, and bradykinin stimulate catecholamine secretion.

    (1) Catecholamine release is stimulated by ACh from the preganglionic sympathetic nerve endings innervating the chromaffin cells.

    (2) The final common effector pathway activating the adrenal medulla is the cholinergic preganglionic fibers in the greater splanchnic nerve.

    (3) ACh causes the depolarization of the chromaffin cells followed by the release of catecholamines by **exocytosis**. Ca^{2+} influx secondary to membrane depolarization is the central event in **stimulus-secretion coupling**.

  b. Because catecholamine synthesis is dependent on cortisol, the functional integrity of the adrenal medulla indirectly depends on a functional pituitary gland for ACTH (corticotropin) secretion and a functional median eminence for corticotropin releasing hormone (CRH) secretion. CRH is a hypophysiotropic hormone produced by the parvicellular (tuberal) nuclei of the ventral diencephalon.

2. Physiologic and Psychologic Stimuli for Release. The adrenal medulla constitutes the neuroeffector of the sympathoadrenomedullary axis that is activated during states of emergency. This response to stress is called the **fight-or-flight reaction**. Among the conditions in which the sympathetic nervous system, including the sympathoadrenomedullary axis, is activated are fear, anxiety, pain, trauma, hemorrhage and fluid loss, asphyxia and hypoxia, changes in blood pH, extreme cold or heat, severe exercise, hypoglycemia, and hypotension. During hypoglycemia, the adrenal medulla is activated selectively. In humans, epinephrine and norepinephrine appear to be released independently by specific stimuli.

  a. Anger and active aggressive states or situations, which are challenging and which allow active and appropriate anticipatory behavioral responses to the challenge, are associated with increased norepinephrine secretion.

  b. States of anxiety are associated with increased epinephrine secretion. In addition, epinephrine release is increased by tense but passive emotional displays or threatening situations of an unpredictable nature, in which active coping behavior may be required but has not been achieved.

  c. Increases in norepinephrine release with smaller increases in epinephrine release occur in professional ice hockey players during active competition and in psychiatric patients showing aggressive emotional outbursts. Increases in epinephrine secretion alone are found in hockey players who observe, but do not participate, in games and in psychiatric patients during staff conferences.

3. Regulation of Adrenergic Receptors. A reciprocal relationship exists between catecholamine concentration and the number and function of adrenergic (adrenotropic) receptors.

  a. A sustained decrease in catecholamine secretion is associated with an increased number of adrenergic receptors in target cells and an increased responsiveness to catecholamines, and conversely.

  b. This phenomenon may account in part for the phenomenon of **denervation hypersensitivity**, which is observed in sympathetic neuroeffectors following autonomic fiber denervation.

F. METABOLISM AND INACTIVATION OF CIRCULATING CATECHOLAMINES

1. General Considerations. The plasma half-life of epinephrine and norepinephrine is 1–3 minutes. The biologic effects of circulating catecholamines are terminated rapidly by both nonenzymatic and enzymatic mechanisms.

  a. Neuronal Uptake. Sympathetic nerve endings have the capacity to take up amines actively from the circulation. This uptake of circulating catecholamines leads to nonenzymatic inactivation by intraneuronal storage and to enzymatic inactivation by a mitochondrial enzyme called **monoamine oxidase (MAO)**.

  b. Extraneuronal Uptake. The formation of catecholamine metabolites locally in innervated tissues and systemically in the liver, kidney, lung, and gut implies catecholamine uptake by a variety of cells.

2. Metabolic Pathways for Catecholamine Inactivation.

  a. MAO is found in very high concentrations in the mitochondria of the liver, kidney, stomach, and intestine. MAO catalyzes the oxidative deamination of a number of biogenic amines, including the intraneuronal and circulating catecholamines.

    (1) The combined actions of MAO and **aldehyde oxidase** on epinephrine and norepinephrine produce 3,4-dihydroxymandelic acid by oxidative deamination.

    (2) The combined actions of MAO and aldehyde oxidase on the meta-O-methylated metabolites of epinephrine and norepinephrine (metanephrine and normetanephrine, respectively) produce 3-methoxy-4-hydroxymandelic acid (vanillylmandelic acid or VMA) by oxidative deamination.

b. Catechol-O-methyltransferase (COMT) is found in the soluble fraction of tissue homogenates with the highest levels in liver and kidney. COMT is considered mainly as an extraneuronal enzyme, but it is also found in postsynaptic membranes. COMT metabolizes circulating catecholamines in the kidney and liver and metabolizes locally released norepinephrine in the effector tissue.

(1) COMT requires S-adenosylmethionine as a methyl donor.

(2) The action of COMT produces normetanephrine from norepinephrine, metanephrine from epinephrine, and VMA from 3-4-dihydroxymandelic acid by 3-O-methylation.

3. Significance of Catecholamine Metabolites.

a. Only 2–3 percent of the catecholamines are excreted directly into the urine, mostly in conjugation with sulfuric or glucuronic acid. Most of the catecholamines produced daily are excreted as the deaminated metabolites, VMA and 3-methoxy-4-hydroxyphenylglycol (MOPG). Only a small fraction is excreted unchanged or as metanephrines.

b. Under normal circumstances, epinephrine accounts for a very small proportion of urinary VMA and MOPG. Because the majority is derived from norepinephrine, urinary VMA and MOPG reflect the activity of the nerve terminals of the sympathetic nervous system rather than that of the adrenal medulla.

c. The excretion of unchanged catecholamines (unchanged epinephrine or plasma epinephrine) provides a better index of the physiologic activity of the sympathoadrenomedullary system than does the excretion of catecholamine metabolites, since the latter reflects, to a considerable extent, norepinephrine that is metabolized within nerve endings and the brain and never released at adrenergic synapses in the active form.

G. PHYSIOLOGIC ACTIONS OF CATECHOLAMINES (Table 7-2). The effects of adrenomedullary stimulation and sympathetic nerve stimulation generally are similar. In some tissues, however, epinephrine and norepinephrine produce different effects due to the existence of two types of adrenergic receptors, **alpha** (α) and **beta** (β) receptors, which have different sensitivities for the various catecholamines and, therefore, produce different responses.

1. The α-adrenergic receptors are sensitive to both epinephrine and norepinephrine. These receptors are associated with most of the excitatory functions of the body and with at least one inhibitory function (i.e., inhibition of intestine motility).

2. The β-adrenergic receptors respond to epinephrine and, in general, are relatively insensitive to norepinephrine. These receptors are associated with most of the inhibitory functions of the body and with one important excitatory function (i.e., excitation of the myocardium).

3. Epinephrine is the single most active endogenous amine on both α and β receptors.

H. BIOCHEMICAL EFFECTS OF CATECHOLAMINES. Norepinephrine has little direct effect on carbohydrate metabolism; however, both norepinephrine and epinephrine can inhibit glucose-induced secretion of insulin from the beta cells of the pancreatic islets of Langerhans.

1. Carbohydrate Metabolism. Because hepatic stores of glycogen are limited (about 100 g) and decrease only transiently after epinephrine activation, lactate derived from muscle glycogen (300 g) is the major precursor for hepatic gluconeogenesis, the process that sustains hepatic glucose formation and secretion. Gluconeogenesis mainly accounts for the hyperglycemic action of epinephrine in normal physiologic states. In pathologic states, (e.g., pheochromocytoma), the diabetogenic action of catecholamines is caused by the inhibition of insulin secretion and the gluconeogenic effect of these hormones (usually norepinephrine), which are secreted in excessive amounts. Since propranolol, a β-blocker, attenuates hyperlactacidemia and hyperglycemia, this implies that epinephrine-induced glycogenolysis in muscle and in the liver is mediated by the β and α receptors, respectively (see Table 7-2).

a. Glycogenolysis in the Liver.

(1) Epinephrine stimulates glycogenolysis in the liver via the Ca^{2+}-activated glycogen phosphorylase and the inhibition of glycogen synthetase. The enzyme glucose-6-phosphatase, found mainly in the liver and in lesser amounts in the kidney, forms free glucose, which increases blood glucose.

(2) Since glucagon stimulates and insulin suppresses hepatic glycogenolysis, the effects of epinephrine on insulin secretion (suppression) and glucagon secretion (stimulation) reinforce the breakdown of glycogen and the increase in hepatic glucose secretion.

(3) Epinephrine also increases the hepatic production of glucose from lactate, amino acids, and glycerol, all of which are gluconeogenic substances.

b. Glycogenolysis in Muscle.

(1) Epinephrine stimulates glycogenolysis in muscle by a β-adrenergic receptor mechanism involving the stimulation of adenyl cyclase and cAMP-induced stimulation of glycogen phosphorylase. Concomitantly, glycogen synthetase activity is reduced.

Table 7-2. Some Effects of Catecholamines and Types of Adrenergic Receptors

Effector Organ	Receptor Type	Response
Eye		
Radial muscle	α	Contraction (mydriasis)
Ciliary muscle	β	Relaxation for far vision
Heart		
Sinoatrial node	β	Increase in heart rate (increase in rate of diastolic depolarization and decrease in duration of phase 4 of sinoatrial nodal action potential)
Atrioventricular node	β	Increase in conduction velocity and shortening of functional refractory period
Atria	β	Increase in contractility
Ventricles	β	Increase in contractility and irritability
Blood vessels	α	Constriction (arterioles and veins)
	β	Dilatation (predominates in skeletal muscle)*
Bronchial muscle	β	Relaxation (bronchodilatation)
Gastrointestinal tract		
Stomach	β	Decrease in motility
Intestine	α, β	Decrease in motility
Sphincters	α	Contraction
Urinary bladder		
Detrusor muscle	β	Relaxation
Trigone and sphincter	α	Contraction
Skin		
Pilomotor muscles	α	Piloerection
Sweat glands	α	Selective stimulation (adrenergic sweating)*
Uterus	α	Contraction
	β	Relaxation
Liver	α	Glycogenolysis
Muscle	β	Glycogenolysis
Pancreatic islets	α	Inhibition of insulin secretion
	β	Stimulation of insulin secretion

Note.—Adapted from Morgan HE: Function of the adrenal glands. In *Best and Taylor's Physiological Basis of Medical Practice,* 10th ed. Edited by Brobeck JR. Baltimore, Williams and Wilkins, 1979.
*Mediated via **sympathetic** cholinergic fibers.

(2) Muscle lacks the enzyme glucose-6-phosphatase, and epinephrine-induced glycogenolysis in muscle does not **directly** increase blood glucose. The glucose-6-phosphate is metabolized to lactate or pyruvate, which is converted to glucose by the liver.

(3) The ultimate physiologic effect of epinephrine-stimulated glycogenolysis in muscle is increased hepatic glucose secretion (hyperglycemia) via the hepatic conversion of muscle lactate to glucose (gluconeogenesis).

c. Hyperglycemic Effects of Epinephrine.

(1) The hyperglycemic effects of epinephrine on the liver are important only in conjunction with the effects of epinephrine on glucagon and insulin secretion together with its glycogenolytic effect on muscle in acute emergency situations.

(2) Much higher amounts of epinephrine than glucagon are required to cause hyperglycemia. However, epinephrine has a more pronounced hyperglycemic effect than glucagon for the following important reasons.

(a) Epinephrine inhibits insulin secretion, while glucagon stimulates insulin secretion; therefore, the hyperglycemic effect of glucagon is attenuated.

(b) Epinephrine stimulates glycogenolysis in muscle, thereby providing lactate for hepatic gluconeogenesis.

(c) Epinephrine stimulates glucagon secretion, which amplifies its hyperglycemic effect.

        (d) Epinephrine stimulates ACTH secretion, which then stimulates cortisol secretion. Cortisol also is a potent gluconeogenic hormone via the hepatic conversion of alanine to glucose.

        (e) Circulating catecholamines inhibit muscle glucose uptake, which is in contrast to the effect of glucagon.

    (3) Epinephrine in physiologic concentrations does not have a direct glycogenolytic effect in the liver.

2. Fat Metabolism. In humans, the major site of lipogenesis from glucose is the liver.

    a. A man of average size has fat stores that contain about 15 kg of triglyceride, some of which can be mobilized as free fatty acids.

    b. Epinephrine stimulates lipolysis by activating triglyceride lipase, which is called the intracellular **hormone-sensitive lipase.** The activation of this enzyme is via the β-adrenergic receptor (i.e., cAMP).

    c. Mobilization of free fatty acids from stores in adipose tissue supplies a substrate for ketogenesis in the liver. The ketone bodies acetoacetate and β-hydroxybutyrate are transported from the liver to the peripheral tissues, where they are quantitatively important as energy sources.

        (1) Cardiac muscle and the renal cortex use fatty acids and acetoacetate in preference to glucose, whereas resting skeletal muscle uses fatty acids as the major source of energy.

        (2) During extreme conditions, such as starvation and diabetes, the brain adapts to the use of ketoacids. Ketoacids also are oxidized by skeletal muscle during starvation.

3. Gluconeogenesis refers to the formation of glucose from noncarbohydrate sources.

    a. Gluconeogenic substances include pyruvate, lactate, glycerol, odd-chain fatty acids, and amino acids. However, the major source of endogenous glucose production is protein, with a smaller fraction available from the glycerol contained in fat.

    b. The conversion of even-chain fatty acids is not possible in the mammalian liver because of the absence of the enzymes necessary for the de novo synthesis of the four-carbon dicarboxylic acids from acetyl coenzyme A (acetyl-CoA).

    c. All of the constituent amino acids in protein tissue, with the exception of leucine, can be converted to glucose.

    d. Gluconeogenesis occurs in the liver and the kidney.

I . ENDOCRINOPATHIES

1. Hyposecretion of catecholamines, such as occurs during tuberculosis and malignant destruction of the adrenal glands or following adrenalectomy, probably produces no symptoms or other clinical features.

    a. Catecholamine production from the sympathetic nerve endings appears to satisfy the normal biologic requirements, because the adrenal medulla is not necessary for life.

    b. The functional integrity of the adrenal medulla can be determined experimentally by the administration of 2-deoxy-D-glucose. This nonmetabolizable carbohydrate induces intracellular glycopenia and extracellular hyperglycemia.

    c. Spontaneous deficiency of epinephrine is unknown as a disease of adults, and adrenalectomized patients do not require epinephrine replacement therapy.

2. Hypersecretion of catecholamines from chromaffin cell tumors (pheochromocytomas) produces demonstrable clinical features.

    a. Pheochromocytoma patients have sustained or paroxysmal hypertension.

    b. The hypersecretion of catecholamines is associated with severe headache, sweating (cold or **adrenergic** sweating), palpitations, chest pain, extreme anxiety with a sense of impending death, pallor of the skin caused by vasoconstriction, and blurred vision.

    c. Most pheochromocytomas contain predominantly norepinephrine, and most patients with chromaffin cell tumors secrete predominantly norepinephrine into the bloodstream. Evidence of epinephrine hypersecretion increases the likelihood that the tumor origin is in the adrenal medulla. However, an extra-adrenal site should not be excluded.

        (1) If epinephrine is secreted primarily, the heart rate is increased.

        (2) If norepinephrine is the predominant hormone, the pulse rate decreases reflexly in response to marked hypertension.

    d. Urinary excretion of catecholamines, metanephrines, and VMA is increased.

3. Clinical Tests.

    a. The **adrenolytic test** involves the administration of an α-blocker (phentolamine) to observe the effect of systemic blood pressure. A dramatic fall in blood pressure is pathognomonic of pheochromocytoma because this procedure is without significant effect in normotensive subjects.

    b. The **provocative test** involves the administration of histamine, glucagon, or tyramine, which will cause a rise in blood pressure.

    4. Definitive treatment requires surgical removal of the tumor.

VII. THE ADRENAL CORTEX.

This section describes the physiologic and biochemical aspects of **glucocorticoids**, the most important of which is **cortisol** (also known as **hydrocortisone**). The biologic characteristics of **mineralocorticoids** (specifically, aldosterone) are described in Chapter 4, "Renal Physiology," Section IX.

    A. EMBRYOLOGY. Morphologically and physiologically, the fetal adrenal gland differs strikingly from that of the adult. However, through all stages of life, the function of the adrenal cortex is dependent on ACTH.

        1. The adrenal cortex is a mesodermal derivative. It is axiomatic that all endocrine glands derived from mesoderm synthesize and secrete steroid hormones.

        2. The outer **neocortex**, which is the progenitor of the adult adrenal cortex, comprises about 15 percent of the total volume of this organ, and the inner **fetal zone** constitutes about 85 percent.

        3. The size (volume) of the adrenal gland at birth is greater than it is during adulthood. This is due to the fact that the fetal zone undergoes rapid involution during the first few months of extrauterine life and completely disappears by 3–12 months postpartum.

        4. Near term, the fetal adrenal glands secrete 100–200 mg of steroids daily in the form of sulfoconjugates, the principal one of which is the biologically weak androgen, **dehydroepiandrosterone (DHEA) sulfate**. This 17-ketosteroid is an androgen containing 19 carbon atoms.

    B. MORPHOLOGY

        1. Gross Anatomy.
          a. The adrenal glands are paired structures situated above the kidneys.
          b. Normally, each gland weighs about 5 g, of which the cortex constitutes approximately 80 percent.

        2. Histology and Function. The adrenal cortex consists of three distinct layers or **zones** of cells.
          a. The outermost layer, the **zona glomerulosa**, is the site of **aldosterone** and **corticosterone** synthesis. Aldosterone is the principal mineralocorticoid of the human adrenal cortex.
          b. The wider, middle zone is the **zona fasciculata**, and the innermost layer is the **zona reticularis**. The two inner zones of the adrenal cortex should be considered a functional unit where mainly **cortisol** (and some corticosterone) and DHEA are synthesized.

    C. ADRENOCORTICAL HORMONES: CORTICOSTEROIDS (Tables 7-3 and 7-4).

        1. Secretion.
          a. The human adrenal cortex secretes two glucocorticoids (cortisol and corticosterone*), one mineralocorticoid (aldosterone), biosynthetic precursors of three end products (progesterone, 11-deoxycorticosterone, and 11-deoxycortisol), and androgenic substances (DHEA and its sulfate ester).
          b. The normal human adrenal cortex does not secrete physiologically effective amounts of either testosterone or estrogenic substances.

        2. Transport.
          a. Under physiologic circumstances, about 75 percent of the plasma cortisol is bound to **cortisol-binding globulin** (CBG, or transcortin) which is an α-globulin. (In addition to binding cortisol, transcortin has a high binding affinity for progesterone, deoxycorticosterone, corticosterone, and some synthetic analogs.)
          b. About 15 percent of the plasma cortisol is bound to plasma **albumin**, and about 10 percent is **unbound** and represents the physiologically active steroid.
          c. The 90 percent of the plasma cortisol bound to plasma protein represents the metabolically inactive pool, which serves as a reservoir for free hormone.

        3. Steroidogenesis in the Adrenal Cortex: Corticosteroidogenesis.
          a. Free cholesterol of the plasma is the preferred precursor of the corticosteroids, although the adrenal cortex can form cholesterol from acetyl-CoA. Most of the stored adrenal

---

*At physiologic concentrations, corticosterone has glucocorticoid activity.

Table 7-3. Average 8:00 a.m. Plasma Concentration and Secretion Rate of Corticosteroids in Adult Humans

Corticosteroid	Plasma Concentration (μg/dl)	Secretion Rate (mg/day)
Cortisol	13	15
Corticosterone	1	3
11-Deoxycortisol	0.16	0.40
Deoxycorticosterone	0.07	0.20
Aldosterone	0.009	0.15
18-Hydroxycorticosterone	0.009	0.10
Dehydroepiandrosterone (DHEA)	0.5	15
DHEA sulfate	115	15

Note.—Reprinted with permission from Genuth SM: The adrenal glands. In *Physiology*. Edited by Berne RM and Levy MN. St. Louis, CV Mosby, 1983, p 1046.

cholesterol is esterified with fatty acids, and it is the cholesterol ester content that is reduced following ACTH stimulation of the fasciculata-reticularis complex.

b. The rate-limiting step in the biosynthesis of corticosteroids is the mitochondrial conversion of cholesterol to pregnenolone by 20,22-desmolase.

c. The conversion of pregnenolone to progesterone requires two enzymes—3 β-hydroxysteroid dehydrogenase and Δ⁵-isomerase—which are found in the endoplasmic reticulum (microsomes).

d. Following the formation of pregnenolone and progesterone, a series of hydroxylation reactions occur sequentially.

    (1) The hydroxylation reactions in the biosynthesis of aldosterone from progesterone occur sequentially at the C-21, C-11, and C-18 positions.

    (2) The hydroxylation reactions in the biosynthesis of cortisol occur sequentially at the C-17, C-21, and C-11 positions.

e. Both pregnenolone and progesterone are substrates for the microsomal enzyme, 17α-hydroxylase.

    (1) Pregnenolone is converted to 17α-hydroxypregnenolone, which can either continue along the Δ⁵-pathway to the synthesis of androgens or enter the glucocorticoid pathway.

    (2) Progesterone is converted to 17α-hydroxyprogesterone, which can either proceed along the Δ⁴-pathway to the synthesis of androgens or enter the mineralocorticoid pathway.

    (3) The zona glomerulosa lacks 17α-hydroxylase and does not have the capacity to synthesize 17α-hydroxyprogesterone. For this reason, the zona glomerulosa cannot synthesize cortisol.

        (a) The conversion of pregnenolone to DHEA is in the Δ⁵-pathway.

        (b) The conversion of progesterone to androstenedione is in the Δ⁴-pathway.

f. Both progesterone and 17α-hydroxyprogesterone are substrates for the microsomal enzyme, 21-hydroxylase.

    (1) Progesterone is converted to 11-deoxycorticosterone, which is in the mineralocorticoid pathway.

    (2) 17α-Hydroxyprogesterone is converted to 11-deoxycortisol, which is in the glucocorticoid pathway.

    (3) 11-Deoxycorticosterone and 11-deoxycortisol are not interconvertible.

Table 7-4. Blood Production Rates of Adrenal Androgens

Androgenic Steroid	Plasma Concentration (μg/dl)	Blood Production Rate (mg/day)
Testosterone		
Men	0.8	7.0
Women*	0.034	0.34
Androstenedione		
Men	0.06	1.4
Women	0.14	3.4

Note.—Reprinted with permission from Mulrow PJ: The adrenals. In *Physiology and Biophysics*. Edited by Ruch TC and Patton HD. Philadelphia, WB Saunders, 1973, p 229.

*In women, about 50 percent of the blood testosterone is derived from androstenedione.

g. Both 11-deoxycorticosterone and 11-deoxycortisol are acted upon by mitochondrial 11β-hydroxylase.

(1) 11-Deoxycorticosterone forms corticosterone in the mineralocorticoid pathway.

(2) 11-Deoxycortisol is converted to cortisol, the principal glucocorticoid of the human adrenal cortex.

(3) Corticosterone differs from cortisol only in that the former lacks a 17α-hydroxyl group. Although this structural difference is small, it accounts for a very large difference in the biological activity of these two hormones.

h. Some of the corticosterone serves as the substrate for another mitochondrial enzyme, 18-hydroxylase.

(1) This reaction leads to the formation of 18-hydroxycorticosterone.

(2) 18-Hydroxycorticosterone is converted to aldosterone by the mitochondrial enzyme, 18-hydroxysteroid dehydrogenase, which is found only in the zona glomerulosa.

i. Sex Steroids.

(1) Androgens. The adrenal cortex secretes four androgenic hormones: androstenedione, testosterone, DHEA, and DHEA sulfate. Quantitatively, the most important sex steroids produced by the human adrenal cortex are DHEA and DHEA sulfate. Except for testosterone, adrenocortical androgens are relatively weak and serve as precursors for hepatic conversion to testosterone.

(a) DHEA is derived from 17α-hydroxypregnenolone by the action of 17,20-desmolase.

(b) DHEA in the main is conjugated with sulfuric acid and as such is bound to plasma protein. While in this bound form, DHEA is not readily excreted but circulates in higher concentrations than any other steroid.

(c) DHEA is the principal precursor of urinary 17-ketosteroids; however, the most abundant urinary 17-ketosteroids are androsterone and etiocholanolone.

(d) DHEA and DHEA sulfate have androgenic activity by virtue of their peripheral conversion to testosterone. DHEA sulfate is active as a minor precursor of other 19-carbon steroids formed in the gonads and placenta, which are sites of sulfatase activity.

(i) Normal excretion rates for 17-ketosteroids are 5–14 mg/day in women and 8–20 mg/day in men.

(ii) Normally, adrenocortical precursors represent the bulk of the urinary 17-ketosteroid pool, with a smaller contribution from the gonads.

(iii) The main androgen secreted by premenopausal women is androstenedione, about 60 percent of which is of adrenocortical origin.

(2) The human adrenal cortex can synthesize minute amounts of estrogen. The adrenal cortex makes its major contribution to the body's estrogen (estrone) pool indirectly by supplying androstenedione together with DHEA and its sulfate as substrates for conversion to estrogens by subcutaneous fat, hair follicles, mammary adipose tissue, and other tissues.

4. Metabolism of Corticosteroids.

a. General Considerations.

(1) Corticosteroid inactivation occurs by:

(a) Enzymatic reduction of the Δ^{4-5} double bond in the A-ring to form **dihydrocortisol**

(b) Enzymatic reduction of the ketonic oxygen substituent at the C-3 position to form tetrahydrocortisol

(c) Conjugation with glucuronic acid to form a water-soluble metabolite that is readily excreted by the kidney

(2) The major urinary metabolite of cortisol is **tetrahydrocortisol glucuronide**.

b. Hepatic Conversion.

(1) Cortisol can be enzymatically converted to cortisone by 11β-hydroxysteroid dehydrogenase and excreted as **tetrahydrocortisone glucuronide**.

(2) The ketonic oxygen substituent on the C-20 position of cortisol and other steroids can be enzymatically converted to a hydroxyl group. These steroids can be subjected to A-ring reduction, conjugated, and excreted as glucuronides.

(a) Cortisol is converted to **cortol glucuronide**.

(b) Cortisone is converted to **cortolone glucuronide**.

(c) Progesterone forms **pregnanediol glucuronide**.

(d) 17α-Hydroxyprogesterone forms **pregnanetriol glucuronide**.

(3) Steroids that contain a 17α-hydroxyl group are ketogenic. Cortisol can be enzymatically converted to a 17-ketosteroid by 17,20-demolase. About 5 percent of cortisol appears in the urine as a 17-ketosteroid.

(4) Aldosterone also can undergo A-ring reduction and conjugation to form **tetrahydroaldosterone glucuronide**.

c. Conversion in Other Extra-Adrenal Tissues. Although the liver is the major extra-adrenal site of corticosteroid metabolism, other tissues, including the muscle, skin, fibroblasts, intestine, and lymphocytes, can carry on oxidation-reduction reactions at the C-3, C-11, C-17, and C-20 positions of the corticosteroid molecule.

D. CONTROL OF ADRENOCORTICAL FUNCTION (see Fig. 7-2). Physiologic control of the rate of cortisol secretion occurs by way of a double negative feedback loop, a mechanism that is characteristic of most neuroendocrine control systems.

1. Regulation of ACTH Secretion.
   a. The parvicellular peptidergic neurons of the hypothalamus release **CRH**, which stimulates the secretion of ACTH from the anterior lobe of the pituitary gland via the hypophysial portal system. CRH is a polypeptide consisting of 41 amino acid residues.
   b. CRH secretion from the tuberoinfundibular neurons is regulated by neurotransmitters secreted by the monoaminergic neurons that innervate the peptidergic neurons.
      (1) Hypothalamic secretion of CRH is stimulated by cholinergic neurons. Serotonin also is a stimulatory signal to CRH neurons.
      (2) Adrenergic neuron activity inhibits release of CRH. **Gamma-aminobutyric acid (GABA)** also is a known inhibitor of CRH secretion.

2. Negative Feedback Action of Corticosteroids.
   a. Of the endogenous corticosteroids, only cortisol has ACTH-suppressing activity. The synthetic glucocorticoid, **dexamethasone**, is a potent inhibitor of ACTH secretion and, therefore, of endogenous glucocorticoid secretion.
   b. The negative feedback of cortisol is exerted at the level of the pituitary gland and at the level of the ventral diencephalon.
      (1) If free cortisol levels are supraphysiologic, ACTH secretion is suppressed and the adrenal cortex ceases it secretory activity and undergoes **disuse atrophy**.
      (2) Conversely, if plasma free cortisol levels are subnormal, the anterior lobe is released from inhibition by cortisol, ATCH secretion rises, and the adrenal cortex secretes more cortisol and becomes hypertrophic.

3. Hypophysial-Adrenocortical Rhythm. Normally blood ACTH levels are higher in the morning than in the evening. This accounts for the diurnal rhythms in cortisol secretion, plasma cortisol concentration, and 17-hydroxycorticosteroid excretion. The ACTH-cortisol secretory pattern is discussed in Section II A 1.

4. Hypophysial-Adrenocortical Response to Stress. The normal hypothalmic-hypophysial-adrenocortical control system can be overridden by a variety of challenges, which collectively are referred to as **stress**.
   a. Among the stresses shown to induce increased activity in this system are severe trauma, pyrogens, hypoglycemia, histamine injection, electroconvulsive shock, acute anxiety, burns, hemorrhage, exercise, infections, chemical intoxication, pain, surgery, psychological stress, and cold exposure.
   b. In stress conditions, ACTH secretion is stimulated despite the fact that systemic levels of cortisol are much higher than those required to inhibit ACTH secretion completely in unstressed conditions.

E. PHYSIOLOGIC EFFECTS OF GLUCOCORTICOIDS. Of the naturally occurring steroids, only cortisol, cortisone, corticosterone, and 11-dehydrocorticosterone have appreciable glucocorticoid activity. Full recovery from hypothalamic-hypophysial-adrenocortical suppression may require as long as 1 year following cessation of all steroid therapy.

1. Anti-Inflammatory Effects. Glucocorticoids inhibit inflammatory and allergic reactions in several ways.
   a. They stabilize lysosomal membranes, thereby inhibiting the release of proteolytic enzymes.
   b. They decrease capillary permeability, thereby inhibiting diapedesis of leukocytes.
   c. Glucocorticoids reduce the number of circulating lymphocytes, monocytes, eosinophils, and basophils.
      (1) The decreased number of these formed elements in blood is due primarily to a redistribution of these cells from the vascular compartment into the lymphoid tissue (e.g., the spleen, lymph nodes, and bone marrow). Cellular lysis is not a major mechanism for decreasing the number of these cells in the human circulation.
      (2) The decrease in circulating basophils accounts for the fall in blood histamine levels and the abrogation of the allergic response.
         (a) The migration of inflammatory cells from capillaries is decreased.

(b) Glucocorticoids lessen the formation of edema and thereby reduce the swelling of inflammatory tissue.

(3) Glucocorticoids cause an increase in the number of circulating neutrophils due to the accelerated release from bone marrow and a reduced migration from the circulation. Steroids also inhibit the ability of neutrophils to adhere, or marginate, to the vessel wall.

d. Glucocorticoids cause involution of the lymph nodes, thymus, and spleen, which leads to decreased antibody production.

(1) This lymphocytopenic effect is advantageous in the prevention or reduction of the immune response by the recipient to an organ transplant.

(2) Since recipients pretreated with large doses of glucocorticoids are susceptible to intercurrent infections, antibiotics are a necessary adjunct to the steroid therapy.

(3) Antibody production is not suppressed in humans at conventional steroid doses; however, chronic administration of high doses of glucocorticoids leads to an impairment of host defense mechanisms.

e. It should be noted that the glucocorticoids lead to an increase in the total blood count because of the increased numbers of neutrophils, erythrocytes, and platelets. The increase in circulating erythrocytes (polycythemia) is due to the stimulation of hematopoiesis.

2. Renal Effects.

a. Glucocorticoids restore glomerular filtration rate and renal plasma flow to normal following adrenalectomy. Mineralocorticoids do not have these effects.

b. Glucocorticoids facilitate free-water excretion (clearance) and uric acid excretion.

3. Gastric Effects. Cortisol increases gastric flow and gastric acid secretion, while it decreases gastric mucosal cell proliferation. The latter two effects lead to peptic ulceration following chronic cortisol treatment.

4. Psychoneural effects of glucocorticoids have been noted following chronic high-dose drug therapy with these steroids. Patients may become initially euphoric and then psychotic, paranoid, and depressed.

5. Antigrowth Effects.

a. Large doses of cortisol have been shown to antagonize the effect of active vitamin D metabolites (25-hydroxycholecalciferol and 1,25-dihydroxycholecalciferol) on the absorption of Ca^{2+} from the gut, to inhibit mitosis of fibroblasts, and to cause degradation of collagen. All of these effects lead to osteoporosis, which is a reduction in bone mass per unit volume with a normal ratio of mineral-to-organic matrix.

b. The process of wound healing is delayed because of the reduction of fibroblast proliferation. Connective tissue is reduced in quantity and strength.

c. Chronic supraphysiologic doses of glucocorticoids suppress GH secretion and inhibit somatic growth.

d. Although glucocorticoids increase the ability of muscle to perform work, large doses lead to muscle atrophy and muscular weakness.

6. Vascular Effect. Cortisol in pharmacologic doses enhances the pressor effect of norepinephrine on vascular smooth muscle.

a. In the absence of cortisol, the vasopressor action of catecholamines is diminished, and hypotension ensues. Thus, corticosteroids have a role in the maintenance of normal arterial systemic blood pressure and volume through their support of vascular responsiveness to vasoactive substances.

b. Cortisol enhances catecholamine synthesis via its activation of the epinephrine-forming enzyme, PNMT.

7. Stress Adaptation. Glucocorticoids allow mammals to adapt to various stresses in order to maintain homeostasis.

a. Resistance to stress is not increased by the administration of glucocorticoids.

b. Stress is associated with the activation of the hypothalamic-hypophysial-adrenal (cortex and medulla) axis.

F. METABOLIC EFFECTS OF GLUCOCORTICOIDS

1. Carbohydrate Metabolism. Cortisol is a carbohydrate-sparing hormone, and, therefore, exerts an anti-insulin effect, which leads to hyperglycemia and insulin-resistance.

a. Glucocorticoids maintain blood glucose and the glycogen content of the liver, kidney, and muscle by promoting the conversion of amino acids to carbohydrates (hepatic gluconeogenesis) and the storage of carbohydrate as glycogen.

b. Cortisol is hyperglycemic principally because of its gluconeogenic activity, which is related to its protein catabolic effect on extrahepatic tissues, especially muscle.

   (1) The proteolytic effect of glucocorticoids results in the mobilization of amino acids from muscle and in an increase in plasma amino acid concentration.

   (2) Alanine is quantitatively the major amino acid precursor in the liver. Like acetyl-CoA, alanine inhibits pyruvate kinase activity.

c. Cortisol also exerts an anti-insulin effect by blocking glucose transport in muscle and adipose tissue. This effect accounts for the phenomenon of glucose intolerance or an eventual "steroid" or adrenal "diabetes."

d. Glucocorticoids augment the activity of key gluconeogenic enzymes by the induction of hepatic enzyme synthesis. The gluconeogenic pathway has three steps that differ from those in the glycolytic pathways as a result of their thermodynamic irreversibility. These enzymes together with their substrates are shown as follows:

   (1) Glucose→glucose-6-phosphate (glucose 6-phosphatase)

   (2) Fructose 1,6-diphosphate→fructose-6-phosphate (fructose 1,6-diphosphatase)

   (3) The conversion of pyruvate to phosphoenolpyruvate requires two biochemical steps.

      (a) Pyruvate→oxaloacetate (pyruvate carboxylase)

      (b) Oxaloacetate→phosphoenolpyruvate (phosphoenolpyruvate carboxykinase)

   (4) It should be noted that glucocorticoids are associated with activation of glycogen synthetase by glucose-6-phosphate and of pyruvate carboxylase by acetyl-CoA. They also are associated with indirect inhibition of glycolysis by free fatty acids, resulting in increased glucogenesis and glycogenesis.

e. Cortisol also indirectly inhibits the activities of glycolytic enzymes, which accounts for its anti-insulin effect. Enzymes that are blocked by the effect of glucocorticoids include:

   (1) Glucokinase

   (2) Phosphofructokinase

   (3) Pyruvate kinase

f. Glucocorticoids mobilize fatty acids from adipose tissue to the liver, where the metabolism of free fatty acids may lead to products that inhibit glycolytic enzymes and favor gluconeogenesis.

   (1) The glycerol released from the fat cell with the fatty acids also serves as a secondary substrate for gluconeogenesis.

   (2) The cortisol-inhibited glycolysis in peripheral tissue probably is indirectly blocked via the inhibition of key glycolytic enzymes (see above) by the elevated concentration of plasma free fatty acids.

g. Glucocorticoids are associated with compensatory hyperinsulinemia following hyperglycemia.

2. Protein Metabolism. The most important gluconeogenic substrates are amino acids derived from proteolysis in skeletal muscle.

a. Cortisol enhances the release of amino acids from proteins in skeletal muscle and other extrahepatic tissues, including the protein matrix of bone.

   (1) The amino acids released, especially the glucogenic amino acid **alanine**, are transported to the liver and converted to glucose.

   (2) Increased glucose production by cortisol via gluconeogenesis is associated with increased urea production via the conversion of amino nitrogen to urea. This effect accounts for the increased urinary nitrogen excretion.

   (3) The proteolysis in skeletal muscle brings about a negative nitrogen balance.

b. The amino acids taken up by the liver are used not only to form glucose or glycogen but also to build new protein. This protein anabolic effect at the level of the liver is an important exception to the overall protein catabolic effect of cortisol.

c. Equally important is the ability of glucocorticoids to inhibit the de novo synthesis of protein, probably at the translational level. This is called an **antianabolic effect** of cortisol.

3. Fat Metabolism. Glucocorticoids are lipolytic hormones. The lipolytic effect of glucocorticoids is in part due to the potentiation of the lipolytic actions of other hormones, such as GH, catecholamines, glucagon, and thyroid hormone.

a. Glucocorticoids favor the mobilization of fatty acids from adipose tissue to the liver, where the metabolism of fatty acids inhibits glycolytic enzymes and promotes gluconeogenesis. As a result of increased fatty acid oxidation, glucocorticoids may lead to ketosis, especially in the context of diabetes mellitus.

   (1) The major site of stimulation is the gluconeogenic enzyme, fructose 1,6-diphosphatase, which is activated by fatty acids.

   (2) At the same time, pyruvate kinase and phosphofructokinase both are inhibited by fatty acids. Thus, glycolysis is inhibited while gluconeogenesis proceeds.

b. Glucocorticoids also indirectly stimulate lipolysis by blocking peripheral glucose uptake and utilization. They inhibit reesterification of fatty acids within adipocytes by inhibiting the use of glucose.

    c. Fatty acids synthesis is inhibited in the liver by cortisol, an effect not observed in adipose tissue. The overall effect of glucocorticoids on fat is to induce a redistribution of fat together with an increase in total body fat (i.e., truncal obesity). The increase in body weight is not due to a growth effect which is the accretion of protein. (Recall that cortisol is an antigrowth hormone.)

        (1) There is a characteristic centripetal distribution of fat (i.e., an accumulation of fat in the central axis of the body).

            (a) The deposition of fat in the facies is called "moonface."

            (b) The deposition of fat in the suprascapular region is referred to as "buffalo hump."

            (c) Excessive fat distribution leads to a pendulous abdomen.

        (2) Glucocorticoid-induced obesity reflects increased food intake rather than a change in the rate of lipid metabolism.

    d. Chronic excessive amounts of cortisol lead to hyperlipemia and hypercholesterolemia.

VIII. THE TESTIS

A. EMBRYOLOGY

    1. The Internal Genitalia (Fig. 7-5).

      a. The internal reproductive tract in males and females is derived from one of two pairs of genital ducts. The **wolffian** (mesonephric) ducts give rise to the male internal genitalia, and the **müllerian ducts** give rise to the female internal genitalia. In the male, the internal genitalia (also called **accessory sex organs**) consist of the following ducts and glands for the conveyance of sperm:

        (1) Seminiferous tubules

        (2) Rete testis*

        (3) Ductuli efferentes

        (4) Epididymis

        (5) Vas deferens

        (6) Ejaculatory duct

        (7) Seminal vesicles

        (8) Prostate gland

        (9) Bulbourethral glands (Cowper's glands)

      b. The wolffian ducts are the excretory ducts of the mesonephric kidney and are attached to the primordial gonad; the müllerian duct is formed from the wolffian duct and is not contiguous with the primitive gonad.

      c. At 9–10 weeks gestation, following the appearance of the **Leydig** (interstitial) **cells** in the fetal testis, the two wolffian ducts begin to differentiate and give rise to the following male internal genitalia:

        (1) Epididymis

        (2) Vas deferens

        (3) Seminal vesicles

        (4) Ejaculatory duct

      d. It is the primary function of **androgen** to induce the formation of the male accessory sex organs during fetal life. Specifically, testicular androgen is required for the differentiation of the male genital (wolffian) duct system.

      e. The control over the formation of the male phenotype requires the action of the following three hormones. (Genetic sex determines gonadal sex, and gonadal sex determines phenotypic sex.)

        (1) Müllerian duct inhibiting factor is produced by the Sertoli cells of the fetal testis and induces the regression of the müllerian duct system. Müllerian duct inhibiting factor functions as a paracrine secretion that diffuses to the paired müllerian ducts.

        (2) Testosterone is secreted by the fetal testis and promotes virilization of the urogenital tract in two ways.

            (a) It stimulates the growth and differentiation of the wolffian duct system and the accessory sex organs.

            (b) Testosterone is the precursor for **dihydrotestosterone,** which is necessary for the growth and differentiation of the male external genitalia (see Section VIII A 2 d).

    2. The External Genitalia.

      a. The external genitalia also begin to differentiate between 9 and 10 weeks gestation.

---

*The duct system distal to the rete testis is known as the excurrent (excretory) duct system.

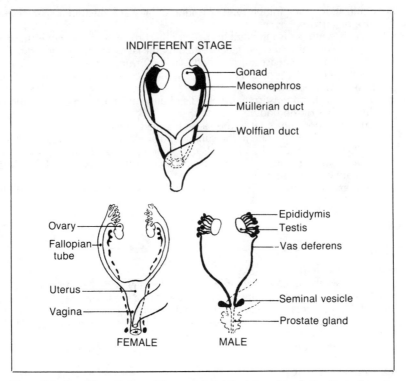

INDIFFERENT STAGE

Gonad
Mesonephros
Müllerian duct
Wolffian duct

Ovary
Fallopian tube
Uterus
Vagina
FEMALE

Epididymis
Testis
Vas deferens
Seminal vesicle
Prostate gland
MALE

Figure 7-5. Sex differentiation of the gonad and the internal genitalia. Up to 6 weeks gestation, the gonad is a bipotential structure (*indifferent stage*), and the urogenital tract in both sexes consists of two pairs of genital ducts (i.e., the wolffian and müllerian ducts) and a mesonephros. In the female, the gonad develops into an ovary and the müllerian ducts become organized into the fallopian tubes, uterus, and upper vagina, while the wolffian ducts remain vestigial. In the male, a testis develops and the wolffian ducts differentiate into the epididymis, vas deferens, and seminal vesicle, while the müllerian ducts regress. (Reprinted with permission from Wilson JD: Embryology of the genital tract. In *Campbell's Urology*, 4th ed. Edited by Harrison JH, et al. Philadelphia, WB Saunders, 1977, p 1473.)

    b. In contrast to the internal genitalia, the external genitalia and urethra in both sexes develop from common anlagen, which are the urogenital sinus, the genital sinus, the genital tubercle, the genital swelling and the genital (urethral) folds. The anlagen of the male external genitalia give rise to the following structures.
      (1) The urogenital sinus forms the male urethra and prostate gland.
      (2) The genital tubercle grows into the glans penis.
      (3) The genital swelling forms the scrotum.
      (4) The urethral folds enlarge to form the outer two-thirds of the penile urethra and corpora spongiosa.
    c. The growth and development of the male external genitalia require dihydrotestosterone, which is formed from the conversion of fetal testicular testosterone within the urogenital sinus and lower urogenital tract.

  3. The Testis (see Fig. 7-5).
    a. The primitive gonads develop midabdominally in association with the mesonephric ridges.
    b. Through 5 weeks gestation, the gonads are undifferentiated and consist of a medulla, a cortex, and primordial germ cells. These germ cells are embedded in a layer of cortical epithelium surrounding a core of medullary mesenchymal tissue.
      (1) At 6 weeks, the seminiferous (spermatogenic) tubules begin to form from the medulla.
      (2) The cortical region of the primitive gonad undergoes regression. (The female gonad develops from the cortical region.)
      (3) The primordial germ cells (gonocytes) arise outside the gonads and migrate from the endodermal yolk sac epithelium to the genital (urogenital) ridge. During migration, the gonocytes undergo continuous mitosis.

    c. At 7 weeks gestation, the **Sertoli (sustentacular) cells** begin to form and secrete the **H-Y** (histocompatibility-Y) **antigen,** which is under control of the Y chromosome. The Sertoli cells are derivatives of the mesenchymal tissue of the urogenital ridge.

    d. At 8 to 9 weeks of fetal life, the Leydig cells are formed and secrete testosterone in response to **chorionic gonadotropin,** which is secreted by the placenta.

        (1) The Leydig cells also are mesenchymal derivatives, which appear in the connective tissue surrounding the seminiferous cords.

        (2) These cells are extensive at birth but virtually disappear within the first 6 months of postnatal life.

        (3) The testes are essentially devoid of Leydig cells throughout childhood until puberty.

    e. At 9 weeks gestation, a definitive testis is present.

    f. At 7 to 9 months gestation, the testes normally descend through the inguinal rings into the scrotum.

B. MORPHOLOGY

  1. Gross Anatomy.

    a. The adult testis is an ovoid gland, which is approximately 4–5 cm long and 2.5–3.0 cm wide and ranges in weight from 10 to 45 g.

    b. The testes normally are situated in the scrotum, where they are maintained at a temperature that is about 2° C lower than normal body temperature. The lower temperature is necessary for normal spermatogenesis.

    c. Each gland is surrounded by a fibrous connective tissue membrane called the **tunica albuginea,** which, in turn, is surrounded by a serous membrane called the **tunica vaginalis.**

    d. By puberty, the testes usually have developed sufficiently to perform the functions of gametogenesis (spermatogenesis) and hormonogenesis (steroidogenesis). Generally, puberty begins between the ages of 12 and 14 years. In the United States, 95 percent of normal boys show signs of puberty by the age of 16 years.

  2. Histology and Physiology of Testicular Parenchyma. The testicular parenchyma is composed of the semiferous tubules and Leydig cells, which comprise about 80 percent and 20 percent of the testicular mass, respectively. These two compartments of tubular and nontubular tissue are separated by boundary tissue, which includes a basement membrane.

    a. Tubular Tissue.

        (1) The spermatogenic tubules are organized into coiled loops, each of which begins and terminates in a single duct called the **tubulus rectus.**

        (2) The tubuli recti join to form the **rete testis** and eventually drain via the **ductuli efferentes** into the **epididymis.**

        (3) The epididymis is the primary storage and final maturation site for spermatozoa.

        (4) From the epididymis, the spermatozoa are transmitted via the **vas deferens** and **ejaculatory duct** into the penile urethra.

    b. Tubular Cells. The basic cellular components of the tubules are the germinal cells and the nongerminal (Sertoli) cells. The spermatogonia and the Sertoli cells are the only tubular cells that lie on the tubular membrane called the **basal lamina.** All of the other germinal cells (spermatocytes) are found between the Sertoli cells.

        (1) The **spermatogonia** are nonmotile stem cells that divide by mitosis to form two cellular pools: a pool of additional stem cells, which undergo continual renewal by mitosis, and a pool of **type A spermatogonia,** which enter the maturation process called **spermatogenesis.**

           (a) Spermatogenesis requires about 70 days in man, and the transport of sperm cells through the epididymis to the ejaculatory duct requires an additional 12–21 days. There are three phases of spermatogenesis.

               (i) The type A spermatogonia divide by mitosis to form **primary spermatocytes,** which are diploid cells.

               (ii) The primary spermatocytes continue the gametogenic process, which leads to the formation of **secondary spermatocytes** and then, through two meiotic divisions, to the formation of **spermatids. Spermatids are haploid cells.**

               (iii) The third phase of spermatogenesis produces mature **spermatozoa** as a result of the metamorphosis of spermatids. This process, called **spermiogenesis,** is characterized by the absence of cell division.

           (b) Because testosterone and FSH act directly on the seminiferous epithelium, these hormones are required for spermatogenesis.

        (2) Sertoli Cells. These nongerminal cells are nonmotile and, in the mature testis, nonproliferating tubular cells that lie on the basal lamina.

(a) The Sertoli cells extend through the entire thickness of the germinal epithelium from the basement membrane to the lumen.

(b) The tight junctions between the bases of the Sertoli cells serve two functions.

　(i) They divide the seminiferous tubular epithelium into two functional pools: a basal compartment containing the spermatogonia and an adluminal compartment containing the spermatogonia and spermatids.

　(ii) The tight junctions form an effective permeability barrier within the seminiferous epithelium, which is defined in man as the blood-testis barrier that limits the transport of many substances from the blood to the seminiferous tubular lumen. This barrier maintains germ cells in an immunologically privileged location, since mature sperm cells are very immunogenic when introduced into the systemic circulation.

(c) Functions of the Sertoli Cells.

　(i) These cells provide mechanical support for the maturing gametes.

　(ii) Sertoli cells have a role in spermatogenesis, which may be attributable to their high glycogen concentration being a potential energy source.

　(iii) In the fetal testis, the Sertoli cells secrete H-Y antigen, which is a product of the testis-organizing genes and is the cell surface glycoprotein responsible for the induction of testicular organogenesis. H-Y antigen has been found in all cell membranes from normal XY males except the cell membranes of immature germ cells.

　(iv) The Sertoli cells secrete müllerian duct inhibiting factor, the glycoprotein that causes regression of the müllerian duct system.

　(v) With the onset of fetal testicular differentiation induced by H-Y antigen and the incorporation of primitive germ cells into the seminiferous tubules, the Sertoli cells secrete a meiosis-inhibiting factor, which suppresses germ cell proliferation and differentiation beyond the spermatogonial stage.

　(vi) Sertoli cells are responsible for the production and synthesis of **inhibin**, which is a protein that exerts a direct negative feedback at the level of the pituitary gland to inhibit FSH secretion.

　(vii) Sertoli cells participate in the release of spermatids by enveloping the residual lobules of spermatid cytoplasm. Thus, the residual bodies are not cast off into the lumen but are retained within the epithelium throughout the spermiation process.

　(viii) Sertoli cells secrete a watery, solute-rich (K^+ and HCO_3^-) fluid into the seminiferous lumen. These cells actively pump ions into the intercellular spaces to create a standing osmotic gradient that moves water from the Sertoli cell base to the free surface of the lumen. This fluid movement provides a driving force for conveying sperm from the testis to the epididymis, where most of this isosmotic fluid is reabsorbed.

　(ix) Sertoli cells produce and secrete an androgen-binding protein in response to FSH stimulation. This macromolecule is secreted into the seminiferous tubular lumen, where it binds testosterone and dihydrotestosterone in the tubular fluid. This protein-androgen complex generates a diffusion potential, resulting in the transport of androgen in close proximity to androgen-dependent germ cells.

　(x) Sertoli cells synthesize estradiol from androgenic precursors.

c. The **Leydig cells** are located between the seminiferous tubules and produce androgenic steroids. These cells also are called **interstitial cells**.

(1) These cells are extensive at birth but virtually disappear within the first 6 months of postnatal life. Their reappearance marks the onset of puberty.

(2) At puberty, the fibroblast-like cells of the testis serve as stem cells that differentiate into Leydig cells.

C. HORMONES OF THE TESTIS: STEROIDS

1. Secretion and Transport.

a. Testosterone is the major hormone produced by the Leydig cells of the testis. Like all naturally occurring androgens, testosterone consists of 19 carbon atoms.

(1) Testosterone is not stored in the testis. Cholesterol esters, the major precursor for testosterone biosynthesis, are stored in the lipid droplets in the Leydig cells.

(2) A normal man secretes 4–9 mg of testosterone daily. More than 97 percent of secreted testosterone is bound to plasma proteins.

(a) Sixty-eight percent is bound to albumin

(b) Thirty percent is bound to testosterone-binding globulin. This β-globulin also is called sex hormone–binding globulin because it binds estradiol as well.

          (c) A very small percentage of the plasma testosterone is unbound.

    b. Androstenedione also is secreted by the testis at a rate of about 2.5 mg/day and is an important steroid precursor for blood estrogens in men.

        (1) Many nonendocrine tissues (e.g., the brain, skin, fat, and liver) have the cytochrome P-450–dependent aromatase, which converts androgens to estrogens.

        (2) Major portions of blood estradiol and estrone in normal men are derived from blood testosterone and androstenedione, respectively.

        (3) In addition, the Sertoli and Leydig cells of the testis secrete small amounts of estradiol.

    c. Dihydrotestosterone is synthesized by the testis probably as a result of the action of **5α-reductase** from the Sertoli cells on testosterone secreted by the Leydig cells.

        (1) Only 20 percent of the plasma dihydrotestosterone is synthesized in the testis. The remainder is derived from the peripheral conversion of testosterone, which serves as a prohormone in the skin, male reproductive tract, prostate gland, and seminal vesicles.

        (2) Dihydrotestosterone is two and one-half times more biologically active than testosterone.

 2. Steroidogenesis in the Testis. The key step in steroidogenesis is the conversion of cholesterol to pregnenolone. This represents the unique and fundamental reaction of all steroid-forming glands.

    a. This reaction is catalyzed by the **20,22-desmolase enzyme**, which is the rate-limiting enzyme for steroid biosynthesis in all steroid-producing tissues (i.e., the testis, ovary, and adrenal cortex).

        (1) LH* activates this desmolase enzyme and, therefore, is the pituitary gonadotropic hormone that regulates testosterone synthesis by the Leydig cells.

        (2) In the developing male fetus, the stimulus for testosterone biosynthesis is chorionic gonadotropin, which is the placental hormone secreted in highest amounts during the first trimester of pregnancy.

        (3) Pregnenolone is a necessary intermediate in the synthesis of all steroid hormones.

    b. The Leydig cells contain 17α-hydroxylase, which hydroxylates pregnenolone at position 17.

        (1) Some pregnenolone may be converted via 3-hydroxysteroid dehydrogenase/Δ^5-isomerase to progesterone, which can be hydroxylated at the C-17 position prior to its conversion to androstenedione along the Δ^4-pathway.

        (2) In the human testis, the Δ^5-pathway is the preferential pathway for the synthesis of testosterone, which is a Δ^4-steroid.

    c. Androstenedione is the common final precursor in the synthesis of testosterone.

 3. Metabolism of Androgenic Hormones.

    a. Testosterone can serve as a prohormone and be metabolized by 5α-reductase to the more active androgen, dihydrotestosterone. The activity of 5α-reductase is high in the skin, prostate gland, seminal vesicles, epididymis, and liver. This enzyme also is present in testicular tissue.

        (1) The principal sites of formation of circulating dihydrotestosterone are thought to be the androgen target tissues. The nuclear membrane and microsomes of androgen-sensitive tissues are the physiologically important sites for conversion of testosterone to dihydrotestosterone. Therefore, dihydrotestosterone not only is secreted by the testis into the circulation but also is synthesized from testosterone that has entered the cells of androgen-dependent tissues.

        (2) The formation of dihydrotestosterone is an irreversible reduction reaction; therefore, dihydrotestosterone cannot serve as a proestrogen.

        (3) The anlagen of the prostate gland and external genitalia are able to form dihydrotestosterone prior to the onset of virilization. Dihydrotestosterone is formed from testosterone prior to the secretion of significant amounts of testosterone from the testis. The wolffian duct derivatives are able to form dihydrotestosterone after the onset of androgen secretion and after differentiation of the male genital system is far advanced.

        (4) Dihydrotestosterone can be metabolized further to 17-ketosteroids and polar derivatives found in the urine.

    b. Testosterone and androstenedione can be converted to estradiol and estrone, respectively, by the action of aromatase. Thus, the estrogens in the male are derived from direct secretion by the testes and from peripheral conversion of circulating androstenedione and testosterone.

        (1) Aromatization of circulating androgens is the major pathway for estrogen formation in the male.

---

*In the male, LH is known as interstitial cell–stimulating hormone **(ICSH)**.

      (2) The peripheral conversion of estrogens from androgens is accomplished by aromatases which are membrane-bound enzymes found in the brain, skin, liver, mammary tissues, and most significantly, the adipose tissue.

   c. Testosterone can be metabolized to less active metabolites that are conjugated in the liver and excreted into the urine as **17-ketosteroids.**

      (1) **Androsterone** and **etiocholanolone** are the major urinary metabolites of testosterone.

      (2) The excretory rate for urinary 17-ketosteroids in normal men is 15–20 mg/day.

         (a) Of this amount, 20–40 percent are of testicular origin. The remainder are adrenocortical secretions, the major one of which is **dehydroepiandrosterone (DHEA).**

         (b) Because the urinary 17-ketosteroid pool reflects mainly adrenocortical activity, a measurement of the 17-ketosteroid excretion is not a good index of testicular function.

      (3) Testosterone and 5α-androstanediol glucuronides are among the other androgenic metabolites measured in urine.

         (a) **Testosterone glucuronide** originates mainly in the liver from testosterone, androstenedione, and dihydrotestosterone. The measurement of plasma testosterone is the mainstay for assessing Leydig cell function.

         (b) **5α-Androstanediol glucuronides** arise from the testosterone metabolites in both the liver and the skin. Because an increased 5α-reductase activity in the skin and other extrahepatic tissues is produced by increased androgen secretion, the measurement of urinary 5α-androstanediol glucuronides has been recommended as an index of clinical androgenicity.

D. HORMONAL CONTROL OF TESTICULAR FUNCTION (see Fig.7-2)

  1. Hypothalamic-Hypophysial-Seminiferous Tubular Axis.

   a. LH (ICSH) and FSH are required for spermatogenesis. Since the effects of LH are mediated by testosterone, this steroid and FSH are two hormones that act directly on the Sertoli cells of the seminferous epithelium to promote gametogenesis.

      (1) Exogenous testosterone alone does not promote spermatogenesis in men lacking Leydig cells. Spermatogenesis requires that a high concentration of testosterone be produced locally by LH action on the Leydig cells.

      (2) The Sertoli cells synthesize androgen-binding protein by an FSH-dependent process. This protein binds testosterone and dihydrotestosterone and functions as a local androgenic pool to support spermatogenesis.

      (3) The Sertoli cells also synthesize **inhibin** in response to FSH secretion. This protein inhibits FSH secretion by direct negative feedback on the pituitary gland. Inhibin is not known to suppress secretion of FSH releasing hormone (FSH-RH), a decapeptide produced by the parvicellular peptidergic neurons (see Section VIII D 2 c). The selective rise in plasma FSH levels in individuals with damaged seminiferous tubules is due to a reduced secretion of inhibin.

   b. Since normal spermatogenesis occurs in men with a 5α-reductase deficiency, dihydrotestosterone is not required form normal sperm development.

   c. Plasma physiologic levels of androgens have little effect in the inhibition of FSH secretion.

   d. Testosterone administration has little effect on FSH secretion, and very large doses are required to suppress FSH in the male.

  2. Hypothalamic-Hypophysial-Leydig Cell Axis.

   a. The rate of testosterone synthesis and secretion by the Leydig cells is stimulated primarily by LH. The secretion of testosterone, in turn, inhibits the secretion of LH. It is free (unbound) testosterone that suppresses LH secretion. It is likely that the major target of negative feedback is the hypothalamus.

   b. Both testosterone and estradiol can inhibit LH secretion; however, since dihydrotestosterone can also suppress LH, androgen conversion to estrogen is not a prerequisite for this inhibitory action on the hypothalamus and pituitary gland.

   c. Some of the neurosecretory neurons of the hypothalamus release a neurosecretory hormone (releasing hormone) into the hypophysial portal system, which then conveys it to the anterior lobe of the pituitary gland.

      (1) This decapeptide, called **gonadotropin releasing hormone (Gn-RH),** stimulates the pituitary basophils.

      (2) Gn-RH also is called LH releasing hormone (LH-RH) and FSH-RH because it elicits the secretion of LH and FSH.

E. PHYSIOLOGIC EFFECTS OF ANDROGENS

  1. Reproductive Function. Androgens are essential for the control of spermatogenesis, the maintenance of the secondary sex characteristics, and the functional competence of the accessory sex organs.

   a. The accessory sex organs consist of excretory ducts and glands that transmit spermatozoa and that secrete seminal fluid necessary for the survival and motility of spermatozoa after ejaculation. The major accessory sex organs are the prostate gland and seminal vesicles.

   b. The secondary sex characteristics are the physiologic characteristics of masculinity (e.g., growth of facial hair, recession of hair at the temples, enlargement of the larynx, and thickening of the vocal cords).

   c. Seminal plasma, the fluid in which spermatozoa normally are ejaculated, originates almost entirely from the prostate gland and seminal vesicles.

   (1) The volume of the human ejaculate is 3–4 ml, most of which is contributed by the seminal vesicles.

   (a) The prostate gland is the origin of citric acid, acid phosphatase, zinc, and spermine.

   (b) The seminal vesicles are the source of prostaglandins, fructose, ascorbic acid, and phosphorylcholine. Metabolism of fructose provides energy for sperm motility.

   (2) Sperm represents 10 percent of the ejaculate volume.

   (a) Euspermia is defined as 40–100 million sperm/ml of ejaculate.

   (b) Oligospermia is defined as 5–20 million sperm/ml of ejaculate.

   (c) Azoospermia is defined as 5 million sperm/ml of ejaculate.

2. Biologic Effects. Androgens stimulate cell division as well as tissue growth and maturation and are classified as protein anabolic hormones. (This anabolic effect in muscle is referred to as the **myotropic effect** of androgens.) Only testosterone and dihydrotestosterone have significant biologic activity. The 17-ketosteroids, androstenedione, DHEA, and etiocholanolone, are weak androgens but important metabolites of testosterone.

   a. In the adolescent, androgens produce linear growth, muscular development, and retention of nitrogen, potassium, and phosphorus. Testosterone also accelerates epiphysial closure (fusion) of the long bones. The skeletal development during puberty, particularly of the shoulder girdle, is pronounced.

   b. Testosterone stimulates differentiation of the wolffian duct system into the epididymis, vas deferens, and seminal vesicles, and dihydrotestosterone stimulates organogenesis of the urogenital sinus and tubercle into the prostate gland, penis, urethra, and scrotum.

   c. Androgens produce a low-pitched voice (as a result of the enlargement of the larynx and thickening of the vocal cords), stimulate growth of chest, axillary, and facial hair, and cause temporal hair recession.

   d. Androgens are responsible for libido.

IX. THE OVARY

A. EMBRYOLOGY

1. The Internal Genitalia (see Fig. 7-5). The primordia of both male and female genital ducts, which are derived from the mesonephros, are present in the fetus at 7 weeks gestation. The paired **müllerian ducts** form from, and parallel to, the paired **wolffian ducts**. In the female fetus, the upper ends of the müllerian ducts are the anlagen of the fallopian tubes (oviducts), whereas the lower ends join to form the uterus, cervix, and upper end of the vagina.

   a. The uterus and fallopian tubes, which develop from the müllerian ducts, do not require the presence of an ovary.

   b. In the absence of a fetal testis, müllerian duct inhibiting factor and testosterone are not secreted by the fetal Sertoli cells and Leydig cells, respectively. Moreover, dihydrotestosterone is not formed from testosterone. Without the presence of these three hormones, at 10–11 weeks gestation, the müllerian ducts begin to differentiate and the wolffian ducts undergo regression.

   c. This process of female genital duct development is completed at 18–20 weeks gestation.

2. The External Genitalia.

   a. The external genitalia of both sexes begin to differentiate at 9–10 weeks gestation. In the female, this process proceeds without any known hormonal influence.

   b. The external genitalia and urethra in both sexes develop from common anlagen, which are the urogenital sinus and the genital tubercle, folds, and swelling.

   (1) The urogenital sinus gives rise to the lower portion of the vagina and to the urethra.

   (2) The genital tubercle is the origin of the clitoris.

   (3) The genital swelling is the primordium of the labia majora.

   (4) The genital folds develop into the labia minora.

3. The Ovary.

a. In the absence of another testicular product, the H-Y antigen, the gonadal primordium develops into an ovary, provided that germ cells are present.

b. The indifferent stage persists in the female fetus weeks after testis organogenesis begins in the male fetus.

  (1) At 9 weeks, a recognizable testis is present and testosterone secretion already is established.

  (2) Ovarian development occurs several weeks later than does testicular differentiation.

c. At 8 weeks, when the testicular secretion of testosterone begins and before the ovarian differentiation is completed, the fetal "ovary" has the capacity to synthesize **estradiol**.

  (1) It is unlikely that the fetal ovary contributes significantly to the circulating estrogens in the fetus. The fetus is exposed to estrogen **(estriol)** of placental origin. The site of estradiol synthesis by the primordial ovary is not known.

  (2) The ovary has no role in sex differentiation of the female genital tract.

d. Also at about 8 weeks gestation, the cortex of the primitive gonad undergoes active mitosis, and epithelial cells infiltrate the gonad as a syncytium of tubules and cords. Primordial germ cells are carried along with this inward migration.

  (1) Proliferation of the cortex ceases at about 6 months.

  (2) The **rete ovarii** secretes a meiosis-inducing factor.

  (3) The germ cells begin to proliferate to form **oogonia**. This mitotic process is maximal between 8 and 20 weeks, after which it diminishes, ceases, and never is resumed.

e. Unlike the fetal testis, the fetal ovary begins gametogenesis. **Oogenesis**, the formation of **primary oocytes** from oogonia, begins at 15 weeks and reaches a peak between 20 and 28 weeks gestation.

  (1) The primary oocytes enter a prolonged prophase of the first meiotic division and remain in this state until ovulation first occurs, between 10 and 50 years later!

  (2) The **diploid primary oocytes** become enveloped by a single layer of flat granulosa cells and in this form are called **primordial follicles**. The formation of primordial follicles reaches a peak between 20 and 25 weeks.

f. Also between 20 and 25 weeks, the gonad has acquired the morphologic appearance of an ovary.

  (1) During this period, the plasma concentration of pituitary FSH reaches a peak.

  (2) During this same period, the first **primary follicles** appear. A primary follicle is an oocyte surrounded by a single layer of cuboidal epithelial cells called **granulosa cells**. The granulosa secrete a protective shell, the **zona pellucida**, which surrounds the oocyte.

B. OOGENESIS

1. Fetal Oogenesis. The period of oogonial proliferation results in a peak population of about 6–7 million germ cells in the two ovaries at 5 months gestation. Included in this group of cells are oogonia, oocytes in various stages of prophase, and degenerating germ cells. This total number of germ cells decreases to 2 million at term. The number of primordial follicles present in the ovary at birth rapidly diminishes thereafter.

2. Postnatal Oogenesis. By 6 months postpartum, all of the oogonia have been converted to primary oocytes, and by the onset of puberty, the number of primary oocytes has decreased to about 400,000.

3. Prepubertal Oogenesis. Between birth and puberty, the primary oocyte is surrounded by the zona pellucida and six to nine layers of granulosa cells. These follicles are in varying stages of development.

4. Pubertal Oogenesis. In contrast to the male who produces spermatogonia and primary spermatocytes continuously throughout life, the female cannot form oogonia beyond 28 weeks gestation and must function with a declining pool of oocytes.

a. Meiosis in the female results in the formation of one viable oocyte **(ootid)** as compared with that in the male, which results in four spermatocytes **(spermatids)**.

b. Oogenesis in the female begins in utero in response to meiosis-stimulating factor, while in the male spermatogenesis is arrested at the spermatogonial stage in response to meiosis-inhibiting factor.

c. Just prior to ovulation, the first polar body is extruded from the primary oocyte, which completes the first meiotic division, and forms a secondary oocyte.

  (1) This haploid cell immediately begins the second meiotic division but remains in metaphase.

  (2) Extrusion of the second polar body **(polocyte)** does not occur until the mature ovum (ootid) is fertilized by a sperm cell. Fertilization normally occurs in the ampulla of the fallopian tube.

C. MORPHOLOGY

1. Gross Anatomy.

a. The ovaries are ovoid glands with a combined weight of 10–20 g during the reproductive years.

b. The ovaries are anchored via the **mesovarium** to the **broad ligament**.

2. Microanatomy.

a. The Ovary.

(1) The ovary consists of three structural divisions: the **cortex**, which is lined by the germinal epithelium and contains all of the oocytes, the **inner medulla**, and the **hilus**, which is the point where the ovary attaches to its mesentery.

(2) The ovary also can be divided into three functional subunits: the **follicle** and **oocyte**, each of which consists of **theca cells** and **granulosa cells**, and the **corpus luteum**. Like the testis, the ovary has two important **functions**.

(a) Gametogenesis in the female denotes **folliculogenesis**, which leads to the formation of a mature ovum.

(i) During the preovulatory phase, the functional unit of the ovary is the follicle.

(ii) During the postovulatory phase, the functional unit of the ovary is the corpus luteum.

(b) Steroidogenesis in the ovary is the synthesis and secretion of **estradiol** and **progesterone**. Although steroidogenesis occurs in three morphologic units (i.e., the follicle, corpus luteum, and stroma), only the follicle and corpus luteum are major steroid-producing units.

b. The Uterus.

(1) Layers. The uterus consists of two major tissue layers.

(a) The outer layer, the **myometrium**, is a thick layer of smooth muscle.

(b) The inner (adluminal) layer of the uterus is the **endometrium**. At the height of its development, during the luteal phase, the endometrium is approximately 5 mm thick. On the basis of blood supply, the endometrium can be divided into two major layers.

(i) The **stratum basalis** is the abluminal (deeper) layer of the endometrium. This layer functions as the regenerative layer in the growth and differentiation of endometrial tissue that is sloughed during menses.

(ii) The **stratum functionalis** is the adluminal (superficial) layer of the endometrium, which is shed during menses.

(2) Blood Supply to the Endometrium.

(a) The stratum basalis receives its vascular supply from the **basal (straight) arteries**, which arise from the uterine radial arteries.

(b) The stratum functionalis is perfused by the **spiral (coiled) arteries**, which also emanate from the radial arteries.

(c) Thus, the arcuate arteries give off radial arteries, and the radial arteries bifurcate to form the basal and coiled arteries.

D. HORMONES OF THE OVARY: STEROIDS. Ovarian hormones include two phenolic steroids, **estradiol** (C-18) and **estrone** (C-18), and the progestogen, **progesterone** (C-21).

1. Secretion and Transport (Table 7-5).

a. Estrogens.

(1) Estradiol is the principal and biologically most active estrogen secreted by the ovary. Ovarian estradiol accounts for more than 90 percent of the circulating estradiol.

(2) Estrone, a weak ovarian estrogen, also is formed by the extraovarian (peripheral) conversion of **androstenedione**.

(a) In premenopausal women, most of the circulating estrone is derived from estradiol by conversion via 17-hydroxysteroid dehydrogenase.

(b) In postmenopausal women, estrone is the dominant plasma estrogen and is derived via the prohormone pathway. Specifically, estrone is derived from the conversion of adrenocortical androstenedione in peripheral tissues (mainly liver). In obese women, there is a significant peripheral conversion of androstenedione to estrone by adipose tissue. This extraovarian synthesis of estrogen is implicated in the higher incidence of endometrial carcinoma in obese women.

(3) Estriol, the weakest of all of the naturally occurring estrogens, is synthesized by the placenta and the liver but not the ovary. In nongravid women, estriol is formed in the liver as a conversion product of estradiol and estrone.

Table 7-5. Types of Steroids and their Systemic Concentrations and Rates of Synthesis in Women

Steroid	Plasma Concentration (ng/dl)	Production Rate (μg/day)
Estrogens (C-18)		
Estradiol		
Early follicular	6	80
Late follicular	50	700
Middle luteal	20	300
Estrone		
Early follicular	5	100
Late follicular	20	500
Middle luteal	10	250
Progestogens (C-21)		
17-Hydroxyprogesterone		
Early follicular	30	600
Late follicular	200	4000
Middle luteal	200	4000
Progesterone		
Follicular	100	2000
Luteal	1000	25,000
Testosterone	40	250
Dihydrotestosterone	20	50
Androstenedione	150	3000
Dehydroepiandrosterone	500	8000

Note.—Reprinted with permission from Lipsett MB: Steroid hormones. In *Reproductive Endocrinology: Physiology, Pathophysiology and Management*. Edited by Yen SSC and Jaffe RB. Philadelphia, WB Saunders, 1978, p 84.

    (4) Over 70 percent of circulating estrogens are bound to a plasma protein called sex hormone–binding globulin, which also binds testosterone. Another 25 percent is bound to plasma albumin.

  b. Progesterone is not bound to sex hormone–binding globulin. The progestogens are bound primarily to transcortin (CBG) and albumin.

2. Steroidogenesis in the Ovary. Two pathways exist for the biosynthesis of ovarian steroids, which have in common the conversion of cholesterol to pregnenolone. This reaction is stimulated by LH and FSH via 20,22-desmolase.

  a. One pathway proceeds by way of the Δ^5-pathway, which involves the synthesis of 17α-hydroxypregnenolone and DHEA via 17α-hydroxylase and 17,20-lyase, respectively.

    (1) DHEA, a 17-ketosteroid,* is converted to another androgenic 17-ketosteroid,* androstenedione, by 3β-hydroxysteroid dehydrogenase and Δ^5-reductase.

    (2) Androstenedione can be reduced by 17-hydroxysteroid dehydrogenase to testosterone, which is a reversible reaction.

    (3) Testosterone is a precursor of estradiol via an aromatase reaction.

  b. The other pathway proceeds via the conversion of pregnenolone to progesterone by 3β-hydroxysteroid dehydrogenase and Δ^5-isomerase. Progesterone is the initial compound in the Δ^4-pathway.

    (1) Progesterone is converted to 17α-hydroxyprogesterone by 17α-hydroxylase.

    (2) 17α-Hydroxyprogesterone is another precursor for androstenedione via another 17,20-lyase step.

    (3) Androstenedione and testosterone are interconvertible with the enzyme 17-hydroxysteroid dehydrogenase.

    (4) Androstenedione and testosterone are converted to estrone and estradiol, respectively, by the action of aromatase. These two 18-carbon steroids (estrogens) also are interconvertible.

3. Metabolism of Ovarian Steroids. The liver is the major site of steroid metabolism.

  a. Catabolism of Progestogens.

*These 17-ketosteroids are steroids that consist of 19 carbon atoms.

    (1) Progesterone is converted to **pregnanediol**.
    (2) 17α-Hydroxyprogesterone is catabolized to **pregnanetriol**.
  b. Catabolism of Estrogens.
    (1) Large quantities of both estradiol and estrone are hydroxylated (primarily in the liver) at the C-16 position to form **estriol**.
    (2) Another major catabolic route for estrogens is hydroxylation at the C-2 and C-4 positions, which yields the **catecholestrogens**.
  c. Estrogens are excreted in the urine in the form of soluble conjugates.
    (1) Estriol, catecholestradiol, and catecholestrone are excreted primarily as glucuronidates.
    (2) Estrone is excreted primarily as a sulfate conjugate.

E. OVARIAN FUNCTION

1. Menstruation (Fig. 7-6).
  a. Normally, the first menstrual cycle begins between the ages of 9 and 12 years and is referred to as **menarche**. Prior to menarche, minimal amounts of estrogen are produced by the peripheral conversion of androgens.
  b. The menstrual cycle usually has a duration of 25–30 days.
    (1) In early adolescence, the cycle is characterized by irregular menses and anovulation.
    (2) However, even in adulthood, a cycle length of 28 days is the exception rather than the rule.
    (3) Between the ages of 45 and 50 years, the menstrual cycles become increasingly infrequent and finally cease at about age 50. Cessation of menses is called **menopause** and is due to a primary hypogonadism (i.e., a cessation of ovarian steroid secretion). Menopause is associated with a rising gonadotropin (predominantly FSH) secretion.

2. Temporal Reference Points for the Menstrual Cycle. The menstrual cycle (see Fig. 7-6) conventionally begins with the first day of menstruation, when the endometrial lining, the stratum functionalis, is shed along with blood and uterine secretions. Days of the menstrual cycle are measured using two different reference points.
  a. One system of measurement designates the first day of menses as **day 0** and the last day of the cycle as **day 28.**
  b. The other system designates the day of the LH peak (ovulation) as **day 0**. With this latter system, preovulatory days, then, are designated by a **minus sign**, and postovulatory days are indicated with a **plus sign**.

3. The Ovarian Cycle.
  a. Preovulatory Phase. This phase is marked by follicular growth and maturation and by endometrial proliferation. The preovulatory phase generally lasts 8–9 days but can be quite variable (10–25 days). During this phase, the stratum basalis regenerates a stratum functionalis, and by the end of this phase, one follicle (rarely more) has reached the final stage of growth.
    (1) A primary follicle begins as an oocyte surrounded by a single layer of cuboidal epithelial cells called granulosa cells. The primary follicle becomes multilaminar by the mitosis of the granulosa cells, which occurs with maturation.
    (2) Upon cavitation of the granulosa, an **antrum** is formed, which is filled with **liquor folliculi** secreted by the granulosa cells. The developing follicle now is called the **secondary follicle** (vesicular follicle or, more commonly, graafian follicle).
      (a) The granulosa cells secrete a protective shell, the **zona pellucida**, which surrounds the oocyte.
      (b) The stroma gives rise to a bilaminar **theca**, which surrounds the granulosa cells but is separated from them by a basal lamina (**lamina propria**).
        (i) The **theca interna** is a well-vascularized layer consisting of steroid-secreting cells that lie on the basal lamina. The blood vessels do not penetrate this membrane and, therefore, the granulosa cells are avascular until after ovulation.
        (ii) The **theca externa** is peripheral to the theca and is composed mainly of fibrous connective tissue. It is less vascular than the theca interna.
    (3) Just prior to ovulation, the primary oocyte of the secondary follicle completes the first meiotic division, which began prior to birth, and forms a secondary oocyte.
  b. Ovulation. The secondary oocyte is released from the secondary follicle by a process called **ovulation**, which usually occurs on **day 15** of the average cycle. (In reference to the LH peak, this midcycle event occurs on **day 1**, that is, one day following the LH surge.)
  c. Postovulatory Phase. The next 13–14 days constitute the postovulatory phase, during which the endometrium is prepared for the possible implantation of the fertilized ovum,

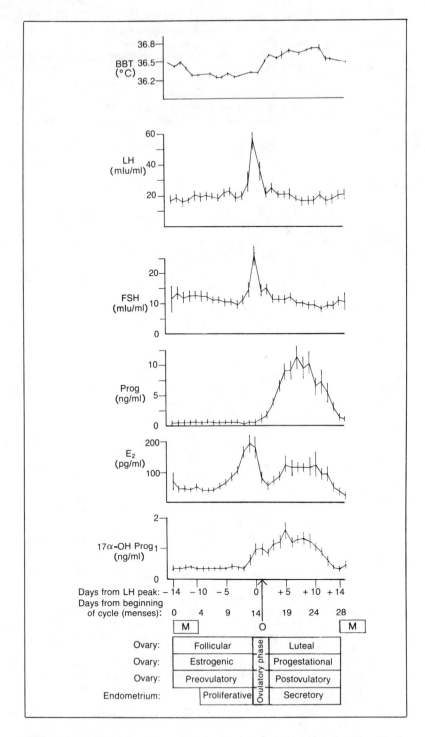

Figure 7-6. Hormonal (ovarian and pituitary), uterine (endometrial), and basal body temperature (*BBT*) correlates of the normal menstrual cycle. Mean plasma concentrations (± SEM) of LH, FSH, progesterone (*Prog*), estradiol (*E₂*) and 17α-hydroxyprogesterone (*17α-OH Prog*) are shown as a function of time. Ovulation occurs on day 15 (day + 1) following the LH surge, which occurs at midcyle on day 14 (day 0). *M* = menses; and *O* = ovulation. (Adapted from Thorneycroft IA, et al: The relation of serum 17-hydroxyprogesterone and estradiol 17-β levels during the human menstrual cycle. *Am J Obstet Gynecol* 111:947–951, 1971.)

which is a blastocyst when it arrives in the uterine cavity. This phase is relatively constant in duration. As a result, the day of ovulation can be estimated by subtracting 14 days from the total length of the menstrual cycle.

(1) If fertilization does not occur, implantation also does not occur because the hormonal maintenance of the endometrial growth and differentiation is withdrawn.

(2) Without conception, ischemia and necrosis of the luminal endometrium result after 14 days, and the ensuing menses marks the beginning of another menstrual cycle.

F. NEUROENDOCRINE CONTROL OF OVARIAN FUNCTION (Fig. 7-2). The ovarian cycle is associated with the secretion of ovarian steroids (estradiol and progesterone), which in turn are regulated by two gonadotropic hormones (FSH and LH) from the pituitary gland. A single hypothalamic hormone (Gn-RH) differentially regulates the secretion of FSH and LH.

1. The Ovarian Cycle.

a. The Preovulatory Phase. Under the influence of FSH and LH, the primary follicle begins to develop. The combined effects of LH on the theca* cells to produce androgenic pro-estrogens and of FSH on the granulosa cells to aromatize these androgens to estrogens (estradiol) result in a slowly increasing blood estradiol concentration. During the preovulatory phase, the dominant gonadotropic hormone is FSH and the dominant steroidal product is estradiol. Therefore, the preovulatory phase of the ovarian cycle also is referred to as the **follicular phase** and the **estrogenic phase**.

(1) Plasma estradiol concentration reaches a peak about 24 hours prior to the surge in LH secretion, or about 48 hours prior to ovulation. The peak in plasma estradiol concentration occurs on day 13 (day − 1).[†]

(a) As the plasma estradiol concentration increases, it exerts a negative feedback on the hypothalamic-hypophysial complex, resulting in a gradual decline in FSH and LH secretion at about day 11 (day − 3).

(b) Inhibin secretion by the granulosa cells also exerts a negative feedback effect on FSH secretion.

(2) The peak in plasma estradiol concentration exerts a positive feedback effect on the hypothalamic-hypophysial axis, causing a reflex release of Gn-RH and a concomitant surge in pituitary LH secretion 24 hours later, on day 14 (day 0).

(a) A lesser increase in plasma FSH secretion occurs on day 14 (day 0).

(b) Increased plasma estradiol levels inhibit FSH secretion both directly and indirectly at the level of the pituitary gland and ventral diencephalon, respectively.

(3) The follicular phase usually has a duration of 14 days, but any variability in the duration of the menstrual cycle usually is attributable to variability in the length of the follicular phase. It should be noted that the follicular phase begins with the first day of menses, while the proliferative phase of the endometrium begins with the last day of menses.

(4) There is a fall in the plasma estradiol level following the estradiol peak. Note on Figure 7-6 that this fall in plasma estradiol precedes ovulation.

(5) During the preovulatory phase, the plasma progesterone concentration remains low.

(a) The bulk of this progesterone is derived from the peripheral conversion of adrenal progestogens; however, large amounts of progesterone exist within the follicular antrum.

(b) The principal progesterone secreted by the granulosa cells during the late follicular phase is 17α-hydroxyprogesterone.

b. The Ovulatory Phase. Ovulation occurs on day 15 (day + 1) in response to the surge of LH secretion that occurred 24 hours earlier.

(1) Ovulation refers to the extrusion of a haploid secondary oocyte into the peritoneal cavity. The oocyte enters the oviduct (fallopian tube) where fertilization occurs.

(2) The rupture of the secondary follicle by the midcycle surge of LH leads to the formation of a new endocrine tissue (i.e., the corpus luteum), which involves proliferation, vascularization, and luteinization of the theca and granulosa cells. LH, therefore, is called the **luteotropic hormone of nonpregnancy.**

c. The Postovulatory Phase. LH also maintains the functional status of the corpus luteum during the postovulatory phase. The hormone secreted in the greatest amounts by the corpus luteum during this phase is progesterone. For these reasons, the postovulatory phase also is called the **luteal phase** and the **progestational phase**.

(1) The decline in estrogen secretion prior to ovulation and at the time of ovulation removes the positive feedback effect of estradiol on gonadotropin secretion.

---

*For the remainder of this chapter, the term theca denotes theca interna.
[†]Also throughout this chapter, the cycle days are numbered with reference to the onset of menses and with reference to the LH peak (in parentheses).

(2) Both FSH and LH levels fall after their midcycle peaks but remain sufficiently high to stimulate the newly formed lutein-theca cells and lutein-granulosa cells to secrete estradiol and progesterone.

(3) About 6 days after ovulation, on day 21 (day + 7), the plasma concentrations of progesterone and 17α-hydroxyprogesterone peak coincidently with the second peak of plasma estradiol concentration. Note in Figure 7-6 that this second peak is lower than the estradiol peak that occurs during the preovulatory phase.

(4) The plasma concentrations of progesterone during the entire ovarian cycle are higher than those of estradiol. It is imperative to note that the units of concentration for progesterone (ng/ml) are 1000 times greater than for estradiol (pg/ml).

(5) The effect of the raised plasma estradiol and progesterone levels is a negative feedback on FSH and LH, respectively. Progesterone acts as an antiestrogen at this time because it inhibits LH secretion when the second estradiol peak occurs.

(6) LH levels continue to decline during the luteal phase, while FSH levels begin to rise progressively during the late luteal phase.

(7) It should be emphasized that the total amount of estradiol secreted during the follicular phase is comparable to that secreted during the luteal phase. This can be appreciated by comparing the areas under the estradiol curve during these two phases, being careful to use day 15 (day + 1) as the dividing line between the follicular and luteal phases.

(8) Unless conception occurs and is followed by implantation of the blastocyst, the corpus luteum undergoes involution following the reduction in gonadotropin secretion.
   (a) The declines in estradiol and progesterone secretion remove the negative feedback effect on the hypothalamic-hypophysial complex.
   (b) The corpus luteum regresses after about 12 days of steroid hormone secretion.

(9) Progesterone is associated with a 0.2–0.5° C rise in basal body temperature, which occurs immediately following ovulation and which persists during most of the luteal phase (see Fig. 7-6).
   (a) The basal body temperature dips during the follicular phase.
   (b) This temperature increment is used clinically as an index of ovulation.

2. The Endometrial Cycle.
   a. Hormonal Effects on the Myometrium. Estradiol and progesterone are antagonistic with respect to their effects on the myometrium.
   (1) Estrogens promote uterine motility.
   (2) Progestogens inhibit myometrial contractility.
   b. Hormonal Effects on the Endometrium. Estradiol and progesterone are synergistic with respect to their effects on the endometrium during the proliferative and secretory phases.
   (1) The proliferative (preovulatory) phase of the menstrual cycle refers to the endometrial changes that occur in response to estradiol.
      (a) Estrogens stimulate mitosis of the stratum basalis, which regenerates the stratum functionalis.
      (b) Estrogens stimulate angiogenesis (neovascularization) in the stratum functionalis as well as stimulate the growth of secretory glands. The blood vessels become the spiral arteries that perfuse the stratum functionalis. The glands contain glycogen but are nonsecretory at this time.
      (c) The cervical epithelium secretes a watery mucus in response to estrogen stimulation.
   (2) The secretory (postovulatory) phase is characterized by the secretion of large amounts of both progesterone and estradiol by the corpus luteum. The endometrium during this secretory phase is hyperemic and has a "lace curtain" or "swiss cheese" appearance.
      (a) Progesterone promotes differentiation of the endometrium, including elongation and coiling of the mucous glands (which secrete a thick viscous fluid containing glycogen) and spiraling of the blood vessels.
      (b) Unless fertilization occurs, hormone secretion by the hypothalamic-hypophysial complex and ovarian steroid secretion decline on about day 25 (day + 11).
         (i) Menses, beginning on day 0 (day – 14) of the following cycle, starts with vasoconstriction of the spiral arteries which causes ischemia and necrosis.
         (ii) The necrotic tissue releases vasodilator substances, causing vasodilation. The necrotic walls of the spiral arteries rupture, causing hemorrhage and shedding of cells over a period of 4–5 days.

G. PHYSIOLOGIC EFFECTS OF OVARIAN STEROIDS

1. Estrogens have important protein anabolic effects.
   a. Estrogens are responsible for the growth and development of the fallopian tubes, uterus,

vagina, and external genitalia as well as the maintenance of these organs in adulthood. These steroids also promote cellular proliferation in the mucosal linings of these structures.
  b. Estrogens stimulate the regeneration of the stratum functionalis during the proliferative phase of the endometrial cycle.
    (1) The water content and blood flow to the endometrium are increased markedly.
    (2) The spiral arteries of the stratum functionalis are especially sensitive to estrogens and grow rapidly under their influence.
  c. Estrogens increase the amount of contractile proteins (i.e., actin and myosin) in the myometrium and, thereby, increase spontaneous muscular contractions. Estrogens also sensitize the myometrium to the action of oxytocin, which promotes uterine contractility.
  d. Under the influence of estrogens, the uterine cervix secretes an abundance of thin, watery mucus.
    (1) This fluid can be drawn into very long threads when placed between two glass slides. This is a clinical index of estrogen activity called **spinnbarkheit**.
    (2) Cervical mucus also demonstrates the phenomenon of crystallization when it is dried on a glass slide. The characteristic **ferning pattern** is due to the accumulation of sodium chloride. This phenomenon also is used diagnostically as an index of endogenous estrogen secretion.
  e. Estrogens promote the development of the tubular duct system of the mammary gland. Estrogens are synergistic with progesterone in stimulating the growth of the lobuloalveolar portions of this gland.
  f. Estrogens, like androgens, exert a dual effect on skeletal growth in that they cause an increase in osteoblastic activity, which results in a growth spurt at puberty.
    (1) Estrogens hasten bone maturation and promote the closure of the epiphysial plates in the long bones more effectively than does testosterone. Therefore, the female skeleton usually is shorter than the male skeleton.
    (2) Estrogens are responsible for the oval or roundish shape of the female pelvic inlet. This inlet in the male is spade-shaped.
    (3) Estrogens, to a lesser degree than testosterone, promote the deposition of bone matrix by causing Ca^{2+} and HPO_4^{2-} retention. In large amounts, estrogens also promote retention of Na^+ and water.
  g. Estrogens stimulate the hepatic synthesis of the transport globulins, including thyroxine-binding globulin and transcortin.
    (1) This results in increased plasma concentrations of thyroxine and cortisol but unchanged amounts of free thyroxine.
    (2) Pregnant women often are in a state of mild hyperadrenocorticism because the elevated placental progesterone competes with cortisol for binding sites on transcortin, thus, increasing plasma free cortisol.

  2. **Progesterone.**
    a. The endometrium, which proliferates under the influence of estrogens, becomes a secretory structure under the influence of progesterone.
      (1) The endometrial glands become elongated and coiled and secrete a glycogen-rich fluid.
      (2) Progesterone accounts for the differentiation of the stratum functionalis.
    b. Under the influence of progesterone, the mucus secreted by the cervical glands is reduced in volume and becomes thick and viscid. This consistency of cervical mucus together with the absence of "ferning" provide presumptive evidence that ovulation and luteinization have occurred.
    c. Progesterone decreases the frequency and amplitude of myometrial contractions.
    d. This steroid also promotes lobuloalveolar growth in the mammary gland.

X. THE ENDOCRINE PANCREAS

A. HISTOLOGY AND FUNCTION OF THE ISLETS OF LANGERHANS (Fig. 7-7).

  1. The endocrine pancreas consists of **islet of Langerhans**, which form less than 2 percent of the pancreatic tissue.

  2. Three cell types, which form the islets of Langerhans, have been identified—the alpha (A), beta (B), and delta (D) cells.
    a. The **A cells** make up about 25 percent of the islet cells and are the source of **glucagon**, which consists of 29 amino acid residues.
    b. The **B cells** constitute about 60 percent of the islet cells and are associated with **insulin synthesis**. This polypeptide consists of two disulfide bridges.

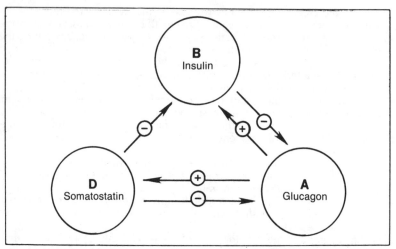

Figure 7-7. The paracrine system of the pancreatic islet cells. The pattern of islet cell hormone secretion represents an integrated response by all of the islet cells to humoral, neural, and paracrine regulation. *Plus signs* indicate stimulation, and *minus signs* indicate inhibition. (Reprinted with permission from Tepperman J: Endocrine function of the pancreas. In *Metabolic and Endocrine Physiology, 4th ed.* Chicago, Year Book, 1980, p 233.)

    (1) The **A chain** contains 21 amino acid residues.
    (2) The **B chain** contains 30 amino acid residues.
  c. The **D cells** comprise about 10 percent of the islet cells and are the source of **somatostatin**, which is a tetradecapeptide.

3. Unmyelinated postganglionic sympathetic and parasympathetic nerve fibers terminate close to the three cell types, and they modulate pancreatic endocrine function via the secretion of **neurotransmitters**.
  a. The liberation of **ACh** causes insulin release only when blood glucose levels are elevated.
  b. **Norepinephrine** secretion due to sympathetic nerve stimulation via the activation of the α-receptor leads to inhibition of insulin release. Despite the dual α- and β-adrenergic receptor system in the B cells, the α-adrenergic action of **epinephrine** predominates so that insulin secretion is inhibited. The release of insulin is mediated via a β-adrenergic receptor.
  c. ACh appears to inhibit somatostatin release, and norepinephrine stimulates somatostatin release.

4. The A, B, and D cells are contiguous with one another and appear to constitute a functional syncytium, which forms a paracrine control system for the coordinated secretion of pancreatic polypeptides (see Fig. 7-7).
  a. Insulin inhibits A cell (glucagon) secretion, which increases peripheral glucose uptake and opposes glucagon-mediated glucose production.
  b. Glucagon stimulates B cell (insulin) secretion and D cell (somatostatin) secretion, which increases hepatic glucose production and opposes hepatic glucose storage.
  c. Somatostatin inhibits A cell (glucagon) and B cell (insulin) secretion, which produces hypoglycemia and inhibition of intestinal glucose absorption. The lowering of blood glucose levels by somatostatin in diabetic patients probably is due both to inhibition of glucagon secretion and to reduced intestinal absorption of glucose.

B. CONTROL OF INSULIN SECRETION

1. Carbohydrates.
  a. Monosaccharides that can be metabolized (e.g., hexose and triose) are more potent stimuli of insulin secretion than carbohydrates that cannot be metabolized (e.g., mannose and 2-deoxy-D-glucose).
  b. The principal stimulus for insulin release is glucose. As the blood glucose level rises above 4.5 mmol/L (80 mg/dl), it stimulates the release and synthesis of insulin.
  c. Substances that inhibit glucose metabolism (e.g., 2-deoxy-D-glucose and D-mannoheptulose) interfere with insulin secretion.
  d. The reduction of glucose to sorbitol may contribute to insulin secretion.
  e. Glucose also stimulates somatostatin release.

2. Gastrointestinal Hormones.

   a. The plasma concentration of insulin is higher after oral administration of glucose than after it has been administered intravenously, even though the arterial blood glucose concentration remains lower. This augmented release of insulin following an oral glucose dose is due to the secretion of gastrointestinal hormones, including:

      (1) Gastric inhibitory peptide (GIP), which appears to be the principal gastrointestinal hormone to potentiate insulin release

      (2) Gastrin

      (3) Secretin

      (4) CCK

   b. Gastrointestinal hormones also augment somatostatin release.

3. Amino Acids.

   a. Amino acids vary in their ability to stimulate the B cells. Among the essential amino acids, in decreasing order of effectiveness, are arginine, lysine, and phenylalanine.

   b. The stimulation of insulin secretion by oral administration of amino acids exceeds that of intravenously administered amino acids. Protein-stimulated secretion of CCK, gastrin, or both may mediate this effect.

   c. The analogs of leucine and arginine that cannot be metabolized also stimulate insulin secretion.

4. Fatty Acids and Ketone Bodies. There is no strong evidence to support the notion that free fatty acids and ketone bodies have an important role in the regulation of insulin secretion in humans. The ingestion of medium-chain triglycerides causes a small increment in insulin levels.

5. Islet Hormones.

   a. As previously mentioned, glucagon stimulates insulin secretion and somatostatin inhibits insulin secretion.

   b. Somatostatin inhibits gastrin and secretin secretion, glucose absorption, and gastrointestinal motility.

6. Other Hormones.

   a. GH induces an elevation in basal insulin levels, which precedes a change in blood glucose levels, suggesting a direct beta-cytotropic effect.

   b. Hyperinsulinemia also has been observed with exogenous and endogenous increments of corticosteroids, estrogens, progestogens, and parathormone. Since blood glucose concentrations are not reduced with these hormones, it is inferred that these hormones have an anti-insulin effect.

7. Obesity. Hyperinsulinemia is observed in obese patients. An increase in body weight in the absence of a disproportionate increase in body fat does not affect insulin levels.

C. PHYSIOLOGIC ACTIONS OF INSULIN (Fig. 7-8)

1. Effects of Insulin on Carbohydrate Metabolism in the Liver.

   a. The liver is freely permeable to glucose, and glucose transport can occur without insulin by simple diffusion.

      (1) Insulin acts on the liver to promote glucose uptake and to inhibit enzymatic processes involved in glucose production and release (glycogenolysis).

      (2) Because the hepatocyte is permeable to glucose, uptake of glucose in the liver is not rate-limiting.

   b. A control point in glucose metabolism occurs when metabolism is initiated by the phosphorylation of glucose to glucose-6-phosphate, which is catalyzed by hexokinase and glucokinase.

      (1) Hexokinase is saturated at normal plasma glucose concentrations and is not regulated by insulin.

      (2) Glucokinase is only half-saturated at blood glucose concentrations between 90 and 100 mg/dl (5–10 mmol/L). Thus, the activity of this enzyme is insulin- and glucose-dependent.

   c. The phosphorylation of fructose-6-phosphate by phosphofructokinase is enhanced by insulin. A decrease in phosphofructokinase activity favors the reversal of glycolysis (i.e., gluconeogenesis and glucose formation).

   d. Insulin diminishes hepatic glucose output by activating glycogen synthetase and by inhibiting gluconeogenesis.

   e. The key intermediary reaction in gluconeogenesis is between pyruvate and phosphoenolpyruvate, which requires the enzymes pyruvate carboxylase and phosphoenolpyruvate carboxykinase. The latter enzyme is inhibited in the presence of glucose and insulin.

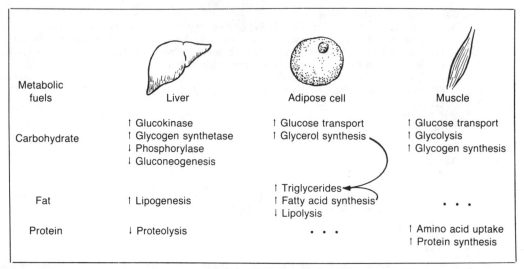

Figure 7-8. The major target sites and metabolic actions of insulin. Insulin is primarily involved in the regulation of metabolic processes, the principal manifestation of which is the control of plasma glucose concentration. (Reprinted with permission from Felig P: Pathophysiology of diabetes mellitus. *Med Clin North Am* 55:821–834, 1971.)

   2. Effects of Insulin on Carbohydrate Metabolism in Muscle.
     a. The **insulin-dependent facilitated diffusion mechanism** for glucose is found in skeletal and cardiac muscle.
     b. Glucose transport across muscle cell membranes requires insulin.
     c. Insulin activates glycogen synthetase and phosphofructokinase, which cause glycogen synthesis and glucose utilization, respectively.
     d. It should be emphasized that glucose uptake in exercising muscle is not dependent on increased insulin secretion. In resting muscle, glucose is a relatively unimportant fuel, with the oxidation of fatty acids supplying most of the energy.

   3. Effects of Insulin on Carbohydrate Metabolism in Adipose Tissue.
     a. The **insulin-facilitated diffusion mechanism** for glucose is found also in adipose tissue.
     b. Insulin acts primarily to stimulate glucose transport.
     c. It activates glycogen synthetase and phosphofructokinase.
     d. The major end products of glucose metabolism in fat cells are fatty acids and α-glycerophosphate. The fat cell depends on glucose as a precursor of α-glycerophosphate, which is important in fat storage because it esterifies with fatty acids to form triglycerides.

   4. Effects of Insulin on Fat Metabolism in the Liver.
     a. Insulin is a lipogenic as well as an antilipolytic hormone.
     b. When insulin and carbohydrate are available, the human liver is quantitatively a more important site of fat synthesis than is adipose tissue.
     c. In the absence of insulin, the liver does not actively synthesize fatty acids, but it is capable of esterifying fatty acids with **glycerol**, which is phosphorylated by glycerokinase.
       (1) Glycerol must be phosphorylated before it can be used in the synthesis of fat.
       (2) In the absence of glycolytic breakdown of glucose to α-glycerophosphate, glycerokinase permits the esterification of fatty acids.
     d. In the absence of insulin, there is an increase in fat oxidation and the production of ketone bodies. Insulin exerts a potent antiketogenic effect.

   5. Effects of Insulin on Fat Metabolism in Adipose Tissue.
     a. Insulin deficiency also decreases the formation of fatty acids in adipose tissue.
     b. The major effect of insulin-stimulated glucose uptake in human fat cells is to provide α-glycerophosphate for esterification of free fatty acids. The absence of α-glycerophosphate formation from glycolysis during insulin deficiency prevents the esterification of free fatty acids, which are constantly released from triglycerides in the adipocytes.
     c. The lipolytic effect in the absence of insulin is due to an increase in the hormone-sensitive lipase known as triglyceride lipase, the activity of which is normally inhibited by insulin.

   6. Effects of Insulin on Amino Acid and Protein Metabolism.
     a. Insulin is an important protein anabolic hormone, and it is necessary for the assimilation of

a protein meal. The protein anabolic effect of insulin is not dependent on increased glucose transport.

b. In diabetic patients, the muscle uptake of amino acids is reduced and elevated postprandial blood levels are observed.

c. During severe insulin deficiency, **hyperaminoacidemia** involving branched-chain amino acids (i.e., valine, leucine, and isoleucine) is present.

d. Insulin increases uptake of most amino acids into muscle and increases the incorporation of amino acids into protein.

e. Insulin increases body protein stores by four mechanisms:
  (1) Increased tissue uptake of amino acids
  (2) Increased protein synthesis
  (3) Decreased protein catabolism
  (4) Decreased oxidation of amino acids

7. Effects of Insulin on Electrolyte Metabolism.
  a. Insulin lowers serum K^+ concentration. This **hypokalemic action of insulin** is due to stimulation of K^+ uptake by muscle and hepatic tissue.
  b. Diabetic patients have a proclivity toward developing hyperkalemia in the absence of acidosis.
  c. Insulin has an antinatriuretic effect.

8. Integration of Insulin Action (Fig. 7-9). The following summary of the actions of insulin, when viewed in the context of a deficiency or absence of insulin, present a concise summary of the metabolic alterations that occur in diabetes mellitus.
  a. Inhibition of hepatic gluconeogenesis decreases the hepatic requirement for amino acids.
  b. The protein anabolic effect of insulin reduces the output of amino acids from muscle, thereby decreasing the availability of glucogenic amino acids for gluconeogenesis.
  c. Glucose uptake by muscle is stimulated, providing an energy source to spare fatty acids, the release of which is inhibited by the antilipolytic action of insulin.
  d. Fat accumulation is enhanced by increased hepatic lipogenesis.
  e. The antilipolytic action of insulin (inhibition of hepatic oxidation of fatty acids) is due to the formation of α-glycerophosphate from glucose in the fat cell.
  f. The antilipolytic action of insulin at the level of the adipose cell reinforces the insulin-mediated inhibition of hepatic ketogenesis and gluconeogenesis by depriving the liver of precursor substrates for ketogenesis and an energy source (fatty acids) and cofactors (acetyl-CoA) necessary for gluconeogenesis.

D. CONTROL OF GLUCAGON SECRETION

1. Metabolic Fuels.
  a. Hypoglycemia stimulates and hyperglycemia inhibits glucagon secretion.
  b. Amino acids (e.g., arginine and alanine) also are stimuli for glucagon release.
  c. Decreasing circulatory levels of fatty acids are associated with glucagon release.

2. Gastrointestinal Hormones.
  a. CCK, gastrin, secretin, and GIP stimulate glucagon secretion.
  b. The potentiation of glucagon secretion by the ingestion of a protein meal is probably mediated via CCK secretion.

3. Fatty acids inhibit glucagon release.

E. PHYSIOLOGIC ACTIONS OF GLUCAGON. The major site of action of glucagon is the liver.

1. Carbohydrate Metabolism.
  a. Glucagon has a hyperglycemic action, resulting primarily from stimulation of hepatic glycogenolysis. It should be emphasized, however, that the hyperglycemic effect of glucagon does not involve inhibition of the peripheral utilization of glucose.

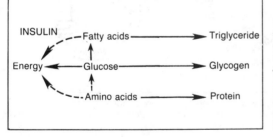

Figure 7-9. Insulin exerts integrated and synergistic actions in the promotion of the storage of body fuels. It enhances the storage of fat and protein, and it promotes both the storage and utilization of carbohydrates. *Solid arrows* denote stimulation, and *dashed arrows* indicate inhibition. (Reprinted with permission from Felig P: Disorders of carbohydrate metabolism. In *Metabolic Control and Disease*, 8th ed. Edited by Bondy PK and Rosenberg LE. Philadelphia, WB Saunders, 1980, p 294.)

b. Glucagon is an important gluconeogenic hormone.

c. The hyperglycemic action of epinephrine is amplified by its stimulation of glucagon secretion and its inhibition of insulin secretion.

d. Suppression of glucagon secretion by glucose is not essential for normal glucose tolerance as long as insulin is available.

2. Fat Metabolism.

a. Glucagon is a lipolytic hormone because of its activation of glucagon-sensitive lipase (triglyceride lipase) in adipose tissue by CAMP.

b. Glucagon causes an elevation in the plasma level of fatty acids and glycerol.

(1) Glycerol is utilized as a glyconeogenic substrate in the liver.

(2) The oxidation of fatty acids as an energy substrate accounts for the glucose-sparing effect of glucagon.

(3) Glucagon is essential for the ketogenesis brought about by the oxidation of fatty acids. In the absence of insulin, glucagon can accelerate ketogenesis, which leads to metabolic acidosis.

3. Protein Metabolism.

a. Glucagon has a net proteolytic effect in the liver.

b. This peptide is gluconeogenic, an effect that leads to increased amino acid oxidation and urea formation.

c. In addition to its protein catabolic effect, glucagon has an antianabolic effect—inhibition of protein synthesis.

XI. THE THYROID GLAND

A. HISTOLOGY AND FUNCTION

1. Thyroid Components. The functional unit of the thyroid gland is the **follicle (acinus)** surrounded by a rich capillary plexus.

a. The follicular (acinar) epithelium consists of a single layer of cuboidal cells.

(1) The cell height of the follicular epithelium varies with the degree of stimulation of TSH.

(2) The glandular epithelium varies with the degree of stimulation, becoming columnar when active and flat when inactive (Fig. 7-10).

b. The lumen of the follicle is filled with a clear, amber proteinaceous fluid called **colloid**, which is the major constituent of the thyroid mass.

c. **Microvilli** extend into the colloid from the apical (adluminal) border, which is the site of the iodination reaction. The initial phase of thyroid hormone secretion (i.e., resorption of the colloid by endocytosis) also occurs at the apical border.

d. The **parafollicular** or **C cells**, which secrete **calcitonin**, do not border on the follicular lumen.

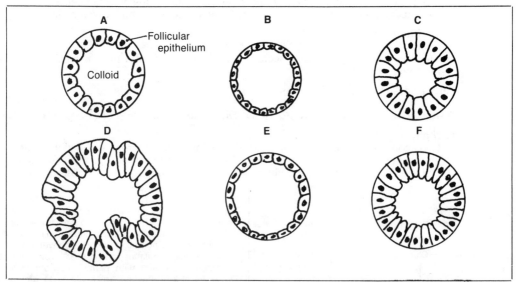

Figure 7-10. The thyroid follicle in various functional states. *A* = normal; *B* = exogenous thyroid hormone treatment (atrophy); *C* = iodide deficiency goiter; *D* = administration of goitrogenic agent; *E* = iodide administration; and *F* = thyrotoxicosis (Graves' disease).

2. Thyroid Products.

a. The thyroid cell performs two parallel functions in the synthesis of thyroid hormone.*

   (1) It synthesizes a protein substrate called **thyroglobulin**.

   (a) This glycoprotein serves as a matrix in which thyroid hormone is formed.

   (b) Thyroglobulin also is the storage form of thyroid hormone.

   (c) Each thyroglobulin molecule contains approximately 120 tyrosyl residues.

   (2) The thyroid cell accumulates inorganic iodide from the plasma.

b. The thyroid gland secretes two hormones, which are **iodothyronines** and, therefore, derivatives of the amino acid **tyrosine**. These hormones also are classified chemically as phenolic derivatives.

   (1) The major secretory product of the thyroid gland is **3,5,3′,5′-tetraiodothyronine (thyroxine)**, which is abbreviated as T_4 to denote the four iodide atoms. The other thyroid hormone is **3,5,3′-triiodothyronine**, which is abbreviated as T_3. T_3 is secreted in small amounts.

   (2) Only these two thyronines have biologic activity.

   (a) The **molar activity** ratio of T_3 to T_4 is 3–5:1

   (b) The **secretory ratio** of T_4 to T_3 is 10–20:1

   (c) The **plasma concentration ratio** of free T_4 to free T_3 is 2:1. Most of the T_3 in the plasma is derived from monodeiodination of T_4 by the action of monodeiodinase found in peripheral tissue.

   (3) **Reverse 3,3′,5′-triiodothyronine (rT$_3$)** is a biologically inactive thyronine also formed by peripheral conversion.

B. DISTRIBUTION OF THYROID IODIDE

1. Iodide Intake.

a. In the United States, the daily dietary iodine intake is about 500 μg.

b. About 1 mg of iodide is required per week to maintain euthyroidism.

c. The thyroid gland stores enough thyroid hormone to maintain a euthyroid state for 3 months without hormone synthesis.

2. Thyroid Iodide.

a. The thyroid gland contains 5–7 mg of iodide.

   (1) Of the total thyroid iodide, 95 percent is in the extracellular space (i.e., stored in the colloid as thyroglobulin).

   (a) Two-thirds of the total iodide content in the colloid is in the form of biologically inactive **iodotyrosines**.

   (b) One-third of the colloid iodide content is in the form of biologically active **thyronines** (i.e., T_4 and T_3).

   (c) The molar storage ratio of T_4 to T_3 is 9:1, and the molar storage ratio of iodotyrosines to iodothyronines is 2:1.

   (2) The remaining 5 percent of the total thyroid iodide is in the intracellular space of the follicular epithelium.

b. The thyroid gland contains the body's largest iodide pool.

C. HORMONE TRANSPORT: EXTRACELLULAR BINDING PROTEINS. It is essential to realize that the extracellular binding proteins for thyroid hormone are in the plasma, while the storage form of thyroid hormone (thyroglobulin) is in the follicular lumen (colloid).

1. Thyroxine Binding Proteins. Virtually all (99.95 percent) of T_4 is bound to plasma proteins, leaving about 0.05 percent unbound (free). This portion of unbound thyroxine represents the biologically active hormone. Thyroxine is mainly associated with two of the three binding proteins.

a. Thyroxine-binding globulin (TBG) binds about 75 percent of the plasma T_4. In normal individuals, less than half of the available binding sites on TBG are saturated with T_4.

b. Thyroxine-binding prealbumin (TBPA) binds about 15–20 percent of the circulating T_4.

c. About 9 percent of the T_4 is bound to albumin.

2. Triiodothyronine Binding Proteins.

a. Almost all (99.5 percent) of T_3 is transported bound to TBG.

b. Very little T_3 is bound to albumin, and practically none is bound to TBPA.

c. About 0.5 percent of the T_3 is unbound. The lower affinity of T_3 for the plasma binding proteins and, thus, the higher concentration of unbound T_3 contribute to the greater biologic activity of T_3.

---

*The term "thyroid hormone" denotes thyroxine (T_4) and triiodothyronine (T_3).

D. BIOSYNTHESIS AND RELEASE OF THYROID HORMONE. The thyroid gland accumulates or "traps" iodide by an active transport mechanism that operates against a concentration and an electric gradient. The normal thyroid iodide to plasma iodide concentration ratio is 25–40:1.

1. **Synthesis.** All of the biosynthetic steps are stimulated by TSH.
   a. **Iodide Uptake.**
      (1) Active iodide uptake occurs at the basal membrane of the follicular cell and is not an essential step in thyroid hormone synthesis.
      (2) Iodide diffuses along an electric gradient into the lumen, where the luminal iodide to follicular cell iodide concentration ratio is 5:1.
   b. **Oxidation** of iodide is mediated by a peroxidase and forms active iodide, which may be in the form of iodinium ion (I^+), a free radical of iodine (IO_3^-), or iodine (I_2).
   c. **Iodination** of active iodide denotes the addition of iodide to the tyrosyl residues of thyroglobulin. The substrate for iodination is thyroglobulin.
      (1) Iodination leads to the formation of the iodotyrosines within the preformed thyroglobulin molecule rather than in free amino acids that are then incorporated into protein. The hormonally-inactive substrates formed by the iodination of thyroglobulin are called **mono-** and **diiodotyrosine**.
      (2) Iodination occurs by the catalytic action of the peroxidase.
   d. **Coupling (condensation)** of iodotyrosines occurs and forms biologically active thyronines (T_3 and T_4).
      (1) The iodothyronines are formed at the apical border of the follicular cell and are held in peptide linkage with thyroglobulin (as are the tyrosines).
      (2) T_4 synthesis requires the fusion of two diiodotyrosine molecules, and T_3 synthesis requires the condensation of a monoiodotyrosine molecule with a diiodotyrosine molecule.
      (3) Peroxidase may also mediate the coupling reaction.

2. **Release.**
   a. Secretion begins with endocytosis of the colloid at the apical border. This brings colloid "droplets" into contact with protease-containing lysosomes.
   b. The release of hormones involves the following reactions:
      (1) Hydrolysis of thyroglobulin by the thyroid protease and by peptidases, which liberate free amino acids
      (2) Secretion of iodothyronines into the blood and deiodination of the iodotyrosines, which form a second iodide pool that can be recycled into hormone synthesis

E. METABOLISM AND EXCRETION OF THYROID HORMONE

1. T_4 is the iodothyronine found in the highest concentration in the plasma and is the only one that arises solely by direct secretion from the thyroid gland.

2. Most of the T_3 present in the plasma is derived from the peripheral conversion of T_4 by monodeiodination.
   a. The extrathyroid deiodination of T_4 accounts for over 80 percent of the circulating T_3.
   b. The liver and kidney deiodinate T_4 to form T_3, but presumably all tissues in the body can do this.

3. Thyroid hormone is metabolized by deiodination, deamination, and by conjugation with glucuronic acid. The conjugate then is secreted via the bile duct into the intestine.

4. In normal individuals, T_4 and T_3 are excreted mainly in the feces, with a small amount appearing in the urine.

F. CONTROL OF THYROID FUNCTION (see Fig. 7-2)

1. **The Hypothalamic-Hypophysial-Thyroid Axis.**
   a. TRH is a tripeptide synthesized by the parvicellular peptidergic neurons in the hypothalamus.
      (1) TRH is transported to the median eminence, where it is stored. From there, TRH is released into the hypophysial portal system and is carried to the anterior lobe of the pituitary gland.
      (2) TRH stimulates some of the basophils **(thyrotrophs)** to secrete TSH.
   b. TSH stimulates the thyroid follicle to secrete thyroid hormone, most of which is bound to plasma protein carriers.
      (1) TSH stimulates the series of chemical reactions that lead to the synthesis of the iodothyronines (see Section XI D).
      (2) It is the circulating free T_3 and T_4 concentrations that influence (regulate) TSH release

by exerting a negative feedback effect at the level of the anterior lobe and probably at the level of the hypothalamus. Both free T_3 and free T_4 are effective inhibitors of TSH secretion when their plasma concentrations are increased.

   (3) The pituitary gland is another tissue that converts T_4 to T_3 by the monodeiodinase enzyme, and this intrapituitary T_3 plays a major role in the negative feedback that occurs at this level.

c. In summary, TSH secretion is influenced by four factors.

   (1) TRH secretion from the median eminence
   (2) The blood level of unbound T_4
   (3) The blood level of unbound T_3 generated by the peripheral conversion of T_4 to T_3
   (4) The peripheral conversion of T_4 to T_3 within the pituitary gland

d. Other Regulators.

   (1) Estrogens enhance TSH secretion.
   (2) Large doses of iodide inhibit thyroid hormone release and, thereby, cause decreases in serum T_4 and T_3 concentrations and an increase in TSH secretion.
   (3) Somatostatin enhances TSH secretion and the response to TRH.
   (4) Dihydroxyphenylethylamine (dopa), dopamine, and bromocriptine decrease the basal secretion of TSH.

2. Thyroid Autoregulation.

a. Thyroid function also is regulated by an intrinsic control system that maintains the constancy of thyroid hormone stores.

   (1) The high concentrations of intrathyroid inorganic iodide lead to the inhibition of thyroid release.
   (2) High concentrations of organic iodide (thyroid hormone) lead to a decrease in iodide uptake (see Fig. 7-10*E*).

b. Both of these effects reduce the fluctuation in thyroid hormone secretion when an acute change occurs in the availability of a requisite substrate (e.g., iodide).

3. Goiter (see Fig. 7-10).

a. Any enlargement of the thyroid gland is called a **goiter**, and antithyroid substances that cause thyroid enlargement are called **goitrogens**.

   (1) Goitrogens are substances that block one or more reactions in the synthesis of thyroid hormone.
   (2) A goiter does not denote the functional state of the thyroid gland.

b. If the goitrogen reduces thyroid hormone synthesis to subnormal levels, TSH secretion is enhanced.

c. Goitrogens lead to the increased synthesis of endogenous TSH, which is responsible for the formation of a hypertropic thyroid gland (goiter).

d. Goitrogenic agents include:

   (1) Perchlorate and thiocyanate, which are monovalent ions that block iodide trapping (see Fig. 7-10*D*)
   (2) Thionamides (propylthiouracil and methimazole), which block the coupling of iodotyrosines (see Fig. 7-10*D*)
   (3) Iodide deficiency (see Fig. 7-10*C*)
   (4) Excess iodide (see Fig 7-10*E*)

G. PHYSIOLOGIC EFFECTS OF THYROID HORMONE

1. Thyroid hormone increases the **basal metabolic rate (BMR)** of most cells in the body.

a. Exceptions to this effect occur in the gonads, brain, lymph nodes, thymus, lung, spleen, dermis, and some accessory sex organs.

b. A correlate of the increase in BMR is an increase in the size and the number of mitochondria together with an increase in the enzymes that regulate oxidative phosphorylation.

c. The increase in BMR also is associated with an increase in Na^+-K^+-ATPase (Na^+ pump) activity. The fluxes of Na^+ (efflux) and K^+ (influx) are estimated to require 10–30 percent of the total energy consumed by cells.

d. The increase in BMR accounts for the thermogenic effect of thyroid hormone.

2. Thyroid hormone is essential for normal bone growth and maturation as well as for the maturation of neurological tissue, especially the brain.

a. In hypothyroidism, there is a marked decrease in the myelination and arborization of neurons in the brain.

b. If hypothyroidism is untreated, mental retardation occurs.

4. Thyroid hormone is necessary for normal lactation.

H. METABOLIC EFFECTS OF THYROID HORMONE

1. Carbohydrate Metabolism.
a. In physiologic amounts, thyroid hormone potentiates the action of insulin and promotes glycogenesis and glucose utilization.
b. In pharmacologic amounts, thyroid hormone is a hyperglycemic agent.
  (1) Thyroid hormone potentiates the glycogenolytic effect of epinephrine, causing glycogen depletion.
  (2) Thyroid hormone is gluconeogenic in that it increases the availability of precursors (lactate and glycerol).
  (3) In large doses, thyroid hormone promotes intestinal glucose absorption.

2. Protein Metabolism.
a. In physiologic amounts, thyroid hormone has a potent protein anabolic effect.
b. In large doses thyroid hormone has a protein catabolic effect.

3. Fat Metabolism.
a. Thyroid hormone stimulates all aspects of lipid metabolism, including synthesis, mobilization, and utilization. On a net basis, the lipolytic effect is greater than the lipogenic effect.
b. There is a general inverse relationship between thyroid hormone levels and plasma lipids.
  (1) Elevated thyroid hormone levels are associated with decreases in blood triglycerides, phospholipids, and cholesterol.
  (2) High levels of thyroid hormones also are associated with increases in plasma, free fatty acids, and glycerol.

4. Vitamin Metabolism.
The metabolism of fat-soluble vitamins is affected by thyroid hormone. For example, thyroid hormone is required for the synthesis of vitamin A from carotene and the conversion of vitamin A to retinene.
a. In hypothyroid states, the serum carotene is elevated and the skin becomes yellow.
b. This skin condition differs from that observed in jaundice in that the sclera of the eye is not yellow.

XII. PARATHORMONE, CALCITONIN, AND VITAMIN D

A. ROLE OF CALCIUM IN PHYSIOLOGIC PROCESSES

1. Hemostasis. Ca^{2+} is necessary for the activation of clotting enzymes in the plasma as well as the enzymes involved in the inflammatory response.

2. Ca^{2+} controls **membrane excitation**, and Ca^{2+} influx occurs during the excitatory process of nerve and muscle.
  a. Excitable membranes contain specific Ca^{2+} channels.
  b. Ca^{2+} entry does not require an active transport process since the concentration gradient across the membrane is larger for Ca^{2+} than for any other ion.
    (1) $[Ca^{2+}]$ in the intracellular fluid (ICF) is approximately 10^{-7} mol/L.
    (2) $[Ca^{2+}]$ in the extracellular fluid (ECF) is about 10^{-3} mol/L (the actual value is 2.5×10^{-3} mol/L).
    (3) The $[Ca^{2+}]$ gradient from outside to inside the cell is in the order of 10,000 to 1!

3. Ca^{2+} is bound to cell surfaces and has a role in the **stabilization of the membrane** and **intercellular adhesion**.

4. Ca^{2+} is necessary for **muscular contraction (excitation-contraction coupling)**.

5. Ca^{2+} is essential in all **excitation-secretion processes**, such as the release of hormone by endocrine cells and the release of other products by exocrine cells. It is also essential for **neurotransmitter release**.

6. Ca^{2+} is necessary for the production of **milk** and the formation of **bone** and **teeth**.

B. CALCIUM DISTRIBUTION

1. Skeletal Storage. More than 99 percent of the total body calcium is stored in the skeleton.
  a. The skeleton of a 70-kg individual contains about 1000 g of calcium compared to about 1 g in the extracellular pool.
  b. The skeleton also serves as a storage depot for phosphorus and contains about 80 percent of the total body phosphorus.
  c. Bone serves as a third-line defense in acid-base regulation by virtue of its PO_4^{3-} content.

2. Plasma. The plasma concentration of total (ionized and nonionized) calcium is about 10 mg/dl, which is equivalent to 5 mEq/L or 2.5 mmol/L.
  a. Calcium is present in the plasma as:
    (1) Ionized or free (45 percent)
    (2) Complexed with HPO_4^{2-}, HCO_3^-, or citrate ion (10 percent)
    (3) Bound to protein [primarily to albumin] (45 percent).
  b. The sum of the ionized and complexed Ca^{2+} consitutes the diffusible fraction (55 percent) of calcium. The protein-bound form constitutes the nondiffusible fraction (45 percent).
  c. Parathormone, calcitonin, and vitamin D regulate the serum-ionized calcium concentration.

3. Calcium Pools. Total body calcium can be conceptualized as two major "pools."
  a. The larger calcium pool, which contains over 99 percent of the total calcium, consists of stable (mature) bone. This represents the calcium pool that is not readily exchangeable, and it is not available for rapid mobilization.
  b. The smaller calcium pool, which contains less than 1 percent of the total body calcium, consists of labile (young) bone. This calcium pool is readily exchangeable because it is in physicochemical equilibrium with the ECF. The pool consists of calcium phosphate salts and provides an immediate buffer to sudden changes in blood [Ca^{2+}].

C. BONE CHEMISTRY

1. Bone calcium is found in the form of **hydroxyapatite crystals**. The empirical chemical formula for this substance is Ca_{10} $(PO_4)_6(OH)_2$ or $[(Ca_3PO_4)_2]_3 \bullet Ca(OH)_2$. Fluoride ion can replace the OH^- group and form **fluorapatite**, or $[Ca_3PO_4)_2]_3 \bullet CaF_2$.
  a. The calcium:phosphorus ratio in bone is about 1.7
  b. A large surface area is provided by the microcrystalline structure of bone; it is estimated to be 100 acres in humans!

2. Dry, fat-free bone consists of two-thirds mineral and one-third organic matrix.
  a. Over 90 percent of the organic matrix is collagen.
  b. The inorganic crystalline structure of bone imparts to it an elastic modulus similar to that of concrete.

D. CALCIUM REGULATION: OVERVIEW

1. General Considerations.
  a. The regulation of calcium involves three **tissues**—the bone, intestine, and kidney.
  b. The regulation of calcium involves three **hormones**—parathormone, calcitonin, and activated vitamin D_3 (activated calciferols).
  c. The regulation of calcium involves three **cell types**—osteoblasts, osteocytes, and osteoclasts.

2. Hormonal Control. The pituitary gland does not play a major role in the regulation of the cells that produce parathormone, calcitonin, and activated vitamin D_3.
  a. Parathormone.
    (1) Parathormone is a polypeptide containing 84 amino acid residues secreted by the chief cells of the four parathyroid glands.
    (2) Parathormone is the **hypercalcemic hormone** of the body and exerts its biologic effect on the bone, intestine, and kidney.
    (3) It regulates only the plasma concentration of the ionic form of calcium. An inverse linear relationship exists between plasma [Ca^{2+}] and parathormone secretion (Fig. 7-11).
      (a) When plasma [Ca^{2+}] falls, the secretion of parathormone increases.
      (b) As plasma [Ca^{2+}] increases, parathormone secretion decreases.
  b. Calcitonin.
    (1) Calcitonin is a 32–amino acid residue polypeptide secreted by the parafollicular cells (C cells) of the thyroid gland.
    (2) Calcitonin is the **hypocalcemic hormone** of the body and exerts a biologic effect on the bone, intestine, and kidney.
    (3) A positive linear correlation exists between plasma [Ca^{2+}] and calcitonin secretion (see Fig. 7-11).
      (a) As the plasma [Ca^{2+}] increases, the secretion of calcitonin also increases.
      (b) When the plasma [Ca^{2+}] decreases, the secretion of calcitonin also decreases.
    (4) Calcitonin release is also stimulated by pentagastrin.
  c. Vitamin D_3 (Cholecalciferol).
    (1) This vitamin hormone is a secosteroid that contains 27 carbons atoms, which makes it the largest of the steroid hormones.

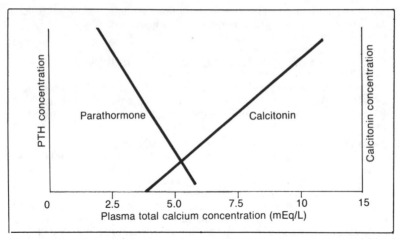

Figure 7-11. Relationship between plasma total calcium concentration and blood levels of parathormone and calcitonin. (Reprinted with permission from Tepperman J: Hormonal regulation of calcium homeostasis. In *Metabolic and Endocrine Physiology, 4th ed.* Chicago, Year Book, 1980, p 297.)

    (2) Only the active metabolites of this hormone exert biologic activity.

        (a) Calcidiol (25-hydroxyvitamin D_3 or 25-hydroxycholecalciferol) is one active metabolite of vitamin D and is the major blood form. This substance is two to five times more effective than vitamin D_3 in preventing rickets.

        (b) Calcitriol (1,25-dihydroxyvitamin D_3 or 1,25-dihydroxycholecalciferol) is another active metabolite of vitamin D_3. On a weight basis it is 100 times more potent than calcidiol.

    (3) The Synthesis of Active Vitamin D_3.

        (a) In the skin 7-dehydrocholesterol is transformed into vitamin D_3 by the action of ultraviolet light.

        (b) Vitamin D_3 is converted to 25-hydroxyvitamin D_3 in the liver by 25-hydroxylase.

        (c) 25-Hydroxyvitamin D_3 is converted in the kidney to 1,25-dihydroxyvitamin D_3 by the action of 1α-hydroxylase. The activity of this enzyme is enhanced by parathormone.

  3. Cellular Aspects of Bone Metabolism (Fig. 7-12).

    a. Osteoblasts are highly differentiated cells that are nonmitotic in their differentiated state. They are the **bone-forming cells** and are located on the bone-forming surface.

      (1) Osteoblasts **synthesize and secrete collagen**.

      (2) They contain abundant alkaline phosphatase activity.

    b. Osteocytes are osteoblasts that have become **buried in bone matrix**.

      (1) Each cell is surrounded by its own lacuna, but an extensive canalicular system connects osteocytes and surface osteoblasts, forming a functional syncytium.

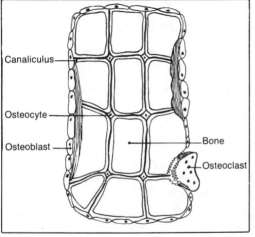

Figure 7-12. Anatomic relationships among the three types of bone cells. The canalicular system provides the structure for a functional syncytium between the osteocytes and osteoblasts. These intercellular connections between these two cell types are disrupted at sites where osteoclasts are found. (Reprinted with permission from Avioli LV, Raisz LG: Bone metabolism and disease. In *Metabolic Control and Disease,* 8th ed. Philadelphia, WB Saunders, 1980, p 1715.)

    (2) In the osteocytic form, these cells no longer synthesize collagen.

    (3) Osteocytes have an **osteolytic activity**, which is stimulated by parathormone. The "osteocytic osteolysis" in the bone matrix provides for the rapid movement of Ca^{2+} from bone into the ECF space.

  c. Osteoclasts are large, multinucleated cells containing numerous lysosomes. They mediate **bone resorption** at bone surfaces.

    (1) These cells contain acid phosphatase.

    (2) Osteoclasts are stimulated by parathormone and form significant amounts of lactic and hyaluronic acids.

    (3) Osteoclasts might cause bone dissolution via an increased local concentration of H^+, which solubilizes bone mineral and increases the activity of enzymes that degrade matrix.

E. PHYSIOLOGIC ACTIONS OF PARATHORMONE

1. Osseous Tissue.

  a. Parathormone action on bone is increased mobilization of calcium and phosphate (i.e., bone dissolution) from the **nonreadily exchangeable Ca^{2+} pool**.

    (1) The long-term effects of parathormone on bone (i.e., the release of Ca^{2+} from bone) may be related to its effects on bone remodeling, which involves bone resorption and accretion.

    (2) Parathormone is known also to stimulate bone synthesis.

    (3) The effects of parathormone on osteogenesis can be both anabolic and catabolic in terms of collagen metabolism.

  b. Parathormone has three important effects on bone that account for its overall osteolytic activity.

    (1) It stimulates osteoclastic and osteocytic activity.

    (2) It stimulates the fusion of progenitor cells to form the multinucleated osteoclastic cells.

    (3) Parathormone causes a transient suppression of osteoblastic activity.

  c. Bone forms from cartilage, which serves as a template for cortical bone by periosteal apposition and for trabecular bone by endochondral ossification.

    (1) Some bones, particularly those in the skull, are formed without a cartilage anlage by intramembranous bone formation.

    (2) In adults, hematopoietic tissue is more abundant in trabecular bone.

  d. Cyclic AMP is a mediator of bone resorption since parathormone stimulates adenyl cyclase in bone cells.

  e. With the dissolution of stable bone, **hydroxyproline** is excreted in the urine. This forms the basis for assessing collagen metabolism and thereby the relative rate of bone resorption.

2. Intestinal Tissue.

  a. Calcium and phosphate are absorbed in the intestine by both active and passive transport, but most of the intestinal absorption of Ca^{2+} occurs via facilitated diffusion.

  b. Parathormone alone does not directly affect the intestinal absorption of Ca^{2+}. Parathormone and 1,25-dihydroxyvitamin D_3 act synergistically to mobilize calcium and phosphate.

    (1) Intestinal absorption of Ca^{2+} does reflect parathyroid status in that hypoparathyroid states are associated with low absorption and hyperparathyroid states are associated with high absorption.

    (2) The increased intestinal absorption of Ca^{2+} by parathormone is mediated indirectly through the increased synthesis of 1,25-dihydroxyvitamin D_3.

    (3) Calcitriol acts on the intestine to promote the transport of calcium and phosphate.

3. Renal Tissue.

  a. Parathormone increases the renal threshold for Ca^{2+} through promoting the active reabsorption of Ca^{2+} by the distal nephron. Parathormone inhibits the proximal tubular reabsorption of Ca^{2+}.

  b. Parathormone inhibits phosphate reabsorption in the proximal tubules [i.e., it lowers the renal threshold for HPO_4^{2-}, which leads to a phosphate diuresis (phosphaturia)]. Thus, parathormone decreases the tubular maximum (Tm) for phosphate.

  c. Both increased parathormone secretion and phosphate depletion stimulate the activation of 1,25-dihydroxyvitamin D_3 via the activation of 1α-hydroxylase.

  d. The phosphaturic effect of parathormone may be mediated by cAMP because parathormone activates adenyl cyclase in the renal cortex.

  e. Parathormone also increases the urinary excretion of Na^+, K^+, and HCO_3^- and decreases the excretion of NH_4^+ and H^+. These effects account for the metabolic acidosis that occurs in hyperparathyroid states.

4. A summary of the unopposed effects of parathormone on bone, intestine, and kidney follows.
  a. Hypercalcemia
  b. Hypophosphatemia
  c. Hypocalciuria, initially due to increased Ca^{2+} reabsorption. (However, in chronic hyperparathyroid states, the hypercalcemia exceeds the renal threshold for Ca^{2+}, and hypercalciuria is observed.)
  d. Hyperphosphaturia.

F. PHYSIOLOGIC ACTIONS OF CALCITONIN

1. Osseous Tissue.
  a. Calcitonin **inhibits osteoclastic activity**, and the antihypercalcemic effect of calcitonin is due principally to the direct inhibition of bone resorption. This effect is not dependent on a functioning kidney, intestine, or parathyroid gland.
  b. Calcitonin also diminishes the osteolytic activity of osteoclasts and osteocytes.
  c. Calcitonin activity is associated with an increase in alkaline phosphatase synthesis from the osteoblasts.

2. Intestinal Tissue.
  a. Calcitonin inhibits gastric motility and gastrin secretion; however, it stimulates intestinal secretion.
  b. Calcitonin inhibits the intestinal (jejunal) absorption of calcium and phosphate.

3. Renal Tissues.
  a. Calcitonin promotes the urinary excretion of phosphate, Ca^{2+}, Na^+, and Cl^-.
  b. It also inhibits renal 1α-hydroxylase activity, which leads to a decrease in the synthesis of calcitriol.

G. PHYSIOLOGIC ACTIONS OF THE BIOLOGICALLY ACTIVE CALCIFEROLS. The actions of 1,25-dihydroxyvitamin D_3 raise the plasma $[Ca^{2+}]$ in concert with parathormone. This vitamin hormone acts directly on the bone, small intestine, and kidney.

1. Osseous Tissue.
  a. Calcitriol, together with parathormone, increases the mobilization of calcium and phosphate from the bone.
    (1) Paradoxically, by raising serum Ca^{2+} and HPO_4^{2-} levels, it also fosters bone deposition.
    (2) This effect of calcitriol on Ca^{2+} and HPO_4^{2-} mobilization is not observed with vitamin D_3 (cholecalciferol) or vitamin D_2 (ergocalciferol).
  b. Calcium-binding protein in bone can be activated by parathormone and calcitriol.
  c. The antirachitic action of vitamin D has traditionally been measured by an increase in bone formation following vitamin D administration to vitamin D–depleted rats.
    (1) It appears that the direct effect of the calciferols on bone is resorption.
    (2) The antirachitic influence of the calciferols appears to be due to an indirect effect on bone through the direct stimulating effect of the calciferols on the intestinal absorption of calcium and phosphate.
  d. In summary, calcitriol acts synergistically with parathormone to cause bone dissolution through the proliferation of osteoclasts. In short, increased osteoclastic activity by parathormone requires 1,25-dihydroxyvitamin D_3.

2. Intestinal Tissue.
  a. Calcitriol is the principal factor in increased Ca^{2+} absorption in intestinal tissue.
    (1) Calcitriol is the mediator hormone for the intestinal actions of parathormone and calcitonin.
    (2) In turn, the action of calcitriol is mediated by the induction of a calcium-binding protein.
  b. Calcitriol causes a lesser increase in intestinal phosphorus absorption.

3. Renal Tissue.
  a. Calcitriol promotes the distal tubular reabsorption of Ca^{2+}.
  b. It promotes proximal tubular reabsorption of HPO_4^{2-}.

4. The renal effects of parathormone and calcitriol are similar in that both promote Ca^{2+} reabsorption; however, the renal effects of these two hormones on phosphate reabsorption are different.
  a. Parathormone promotes phosphate diuresis.
  b. Activated vitamin D_3 (calcitriol) promotes phosphate reabsorption; however, pharmacologic doses of calcitriol do have a phosphaturic effects.

STUDY QUESTIONS

Directions: Each question below contains five suggested answers. Choose the **one best** response to each question.

1. The major glucocorticoid, catecholamine, and sex steroid secreted by the human adrenal gland are

(A) cortisol, norepinephrine, and androstenedione
(B) cortisol, epinephrine, and androstenedione
(C) hydrocortisone, epinephrine, and dehydroepiandrosterone (DHEA)
(D) cortisol, norepinephrine, and DHEA
(E) cortisone, epinephrine, and DHEA

2. Hypersecretion of growth hormone in the adult can lead to

(A) insulin-resistant diabetes
(B) hypogylcemia
(C) decreased amino acid transport
(D) increased urea formation
(E) lipogenesis

3. A goiter always is associated with which of the following conditions?

(A) Hyperthyroidism
(B) Hypothyroidism
(C) Euthyroidism
(D) An enlarged thyroid gland
(E) Iodide deficiency

4. All of the following statements concerning estriol are true EXCEPT

(A) it is produced by the liver
(B) it is synthesized in the placenta
(C) it is quantitatively the major urinary estrogen
(D) it is produced by the theca interna cells of the ovary
(E) it is the least biologically active of the natural estrogens

5. In which of the following triads of terms do all three terms denote the same phase of the menstrual cycle?

(A) Postovulatory, progestational, and luteal
(B) Preovulatory, estrogenic, and luteal
(C) Preovulatory, progestational, and follicular
(D) Postovulatory, estrogenic, and luteal
(E) Postovulatory, progestational, and follicular

6. Pheochromocytoma can produce all of the following clinical features EXCEPT

(A) decreased plasma free fatty acid concentration
(B) decreased release of insulin
(C) hypertension
(D) increased urinary vanillylmandelic acid (VMA)
(E) glycogenolysis

7. Parathormone has a tendency to cause which of the following conditions?

(A) Alkalemia
(B) Hypercalciuria
(C) Hyperphosphaturia
(D) Bone synthesis
(E) Decreases in the size of the labile bone pool

Directions: Each question below contains four suggested answers of which **one or more** is correct. Choose the answer

A   if **1, 2, and 3** are correct
B   if **1 and 3** are correct
C   if **2 and 4** are correct
D   if **4** is correct
E   if **1, 2, 3, and 4** are correct

8. Stimulation of the α- and β-receptors of the intrinsic ocular muscles by catecholamines has several effects on the photomotor reflex including

(1) contraction of the radial muscle of the iris, leading to mydriasis
(2) accommodation for near vision
(3) relaxation of the ciliary muscle
(4) contraction of the pupillary sphincter muscle

9. Testicular function can be estimated by the analysis of seminal fluid for

(1) prostaglandins
(2) citric acid
(3) fructose
(4) acid phosphatase

10. Peptides that are synthesized in nerve cells include

(1) meiosis-inhibiting factor
(2) somatomedin
(3) inhibin
(4) β-endorphin

11. Gluconeogenic substances include which of the following?

(1) Glycerol
(2) Alanine
(3) Lactate
(4) Two-carbon fragments from fatty acid oxidation

12. A description of the activities of estradiol and progesterone would include that these hormones

(1) act synergistically on the endometrium
(2) act synergistically on the myometrium
(3) act synergistically on mammary tissue
(4) promote Na^+ retention

13. Prolactin secretion is different from that of other hormones in that

(1) its secretion is inhibited by stress
(2) its secretion is stimulated by a hormone that also is a monoamine
(3) it is released from pituitary chromophobes
(4) its secretion is under tonic inhibition by the hypothalamus

14. Glucagon has a number of important biochemical and physiologic effects including

(1) stimulation of insulin secretion
(2) lipolysis
(3) glycogenolysis
(4) stimulation of somatostatin secretion

15. Secretory products of the Sertoli cell include

(1) androgen-binding protein
(2) müllerian duct stimulating factor
(3) inhibin
(4) a mucoid fluid

16. The hormone that is produced in many androgen-dependent tissues by the conversion of testosterone

(1) requires the enzyme 5α-reductase
(2) is called dehydroepiandrosterone
(3) is the most active subcellular form of all the 19-carbon steroids
(4) is called dihydroxytestosterone

17. The prohormone produced by the theca interna of the ovary is

(1) a 17-ketosteroid
(2) an estrogen
(3) a 19-carbon steroid
(4) dehydroepiandrosterone

Directions: The groups of questions below consist of lettered choices followed by several numbered items. For each numbered item select the **one** lettered choice with which it is **most** closely associated. Each lettered choice may be used once, more than once, or not at all.

Questions 18–22

Match each of the following phrases with the type of uterine tissue that it describes.

(A) Stratum functionalis
(B) Stratum basalis
(C) Both the stratum functionalis and basalis
(D) Neither the stratum functionalis nor basalis

18. A tissue supplied with blood via the spiral arteries

19. A tissue with a vascular supply from the straight arteries

20. A morphologic component of the myometrium

21. A tissue that regenerates the uterine tissue that is shed during menses

22. Progesterone and estradiol are synergistic with respect to their effect on this tissue

Questions 23–26

Match each of the following phrases with the hormone it describes.

(A) 25-Hydroxycholecalciferol
(B) 1,25-Dihydroxyvitamin D_3
(C) Both of these hormones
(D) Neither of these hormones

23. A substance that is synthesized in the skin from 7-dehydrocholesterol

24. A secosteroid synthesized in the liver from vitamin D_3

25. The immediate precursor substance for the synthesis of calcitriol

26. The most biologically active form of vitamin D_3

ANSWERS AND EXPLANATIONS

1. The answer is C. *[VI B 2 a (1), C 1 a; VII C 3 g (2), i (1)]* The major glucocorticoid secreted by the human adrenal cortex is cortisol, which also is known as hydrocortisone. The principal product of the human adrenal medulla is epinephrine, and the chief sex steroid produced by the adrenal cortex is the 17-ketosteroid, dehydroepiandrosterone (DHEA).

2. The answer is A. *[V D 1 c (2)–(4)]* Growth hormone (GH) has an anti-insulin effect on carbohydrate and fat metabolism. Thus, GH is hyperglycemic primarily because of its lipolytic effect. The increase in fatty acids brought about by fat mobilization blocks glycolysis at the pyruvate kinase and phosphofructokinase steps. Physiologic conditions and hormonal substances that stimulate the breakdown of fat to free fatty acids or that inhibit fatty acid utilization will stimulate gluconeogenesis. The major site of this stimulation is the gluconeogenic enzyme, fructose-1,6-diphosphatase.

The effects of GH are protein anabolic, fat catabolic, and gluconeogenic. GH inhibits glucose utilization and, therefore, produces an insulin-resistant diabetes.

3. The answer is D. *(XI F 3; Fig. 7-10)* The term goiter denotes only a large thyroid gland. A hypertrophied thyroid gland can be caused by iodide deficiency. If compensation to the iodide deficiency occurs, it is possible that the patient would produce adequate amounts of thyroid hormone. Hypothyroidism due to a lesion in the thyroid gland also could cause a goiter. Hyperthyroid patients also can have a diffuse hyperplastic goiter.

4. The answer is D. *[IX D 1 a (3)]* The only organs that produce estriol are the liver and the placenta. Estriol is the major urinary metabolite of all estrogens and the major estrogen produced by the trophoblastic cells. This C-18 steroid is not an ovarian product.

5. The answer is A. *[IX C 2 a (2) (a), F 1 c]* The postovulatory phase is associated with the functioning of the corpus luteum, which secretes large amounts of progesterone and estradiol. The steroid produced in the greater amount is progesterone. Therefore, the postovulatory phase also is known as the progestational phase, during which luteinizing hormone is the dominant gonadotropin. This phase also is associated with the development of the secretory phase of the endometrium.

6. The answer is A. *(VI I 2)* Catecholamines are lipolytic, glycogenolytic, and gluconeogenic hormones. Therefore, hyperglycemia and increased plasma levels of free fatty acids occur in response to elevated circulating catecholamines. Increased systemic arterial pressure also is a finding in this endocrine disease. Catecholamines are diabetogenic mainly because they inhibit insulin secretion. The urinary metabolites, mainly vanillylmandelic acid (VMA), are elevated.

7. The answer is C. *(XII D 2 a, E 3)* Parathormone promotes renal retention of Ca^{2+} and the renal excretion of HPO_4^{2-} (phosphate diuresis). Parathormone also causes HCO_3^- excretion and reabsorption of H^+ and thereby leads to a metabolic acidosis. This hormone has an osteolytic effect on stable (mature) bone.

8. The answer is B (1, 3). *(Table 7-2)* Not all of the intrinsic eye muscles receive a dual autonomic innervation. The pupillary sphincter receives only a parasympathetic innervation, and the radial eye muscle receives only a sympathetic innervation. However, the ciliary muscle does receive an innervation from both outflows of the autonomic nervous system.

Stimulation of the nerves to the radial eye muscle causes contraction of this muscle and pupillary dilation (mydriasis). Sympathetic stimulation of the ciliary muscle causes relaxation of this muscle, which increases the tension of the suspensory ligaments attached to the lens, resulting in a decrease in the thickness of the lens. This decrease in the refractive power of the lens is associated with the accommodation reflex for far vision. Notice that there are no adrenergic receptors associated with the pupillary sphincter muscle.

9. The answer is E (all). *(VIII E 1 c)* The accessory sex organs, which include the seminal vesicles and prostate gland, are androgen target tissues. Therefore, an assessment of the functional status of these accessory sex organs is an indirect measure of testicular function. Seminal fluid is rich in fructose ascorbate and prostaglandins, which are products of the seminal vesicles. The prostate gland contributes acid phosphatase and citrate.

10. The answer is D (4). *[IV B 2 b; V D 1 c (1); VI H 3; VIII B 2 b (2) (c)]* β-Endorphin is produced in nerve cells. It also is a major product of the pituitary gland. Somatomedin is of hepatic origin. The meiosis-inhibiting factor and inhibin are products of the Sertoli cells.

11. The answer is A (1, 2, 3). *(VI H 3)* Glucose can be derived from pyruvate, lactate, glycerol, amino acids, and **odd-chain** fatty acids. It should be noted that in mammalian liver, it is not possible to derive glucose from **even-chain** fatty acids (which account for over 95 percent of the body's total fatty acid content) because there is a lack of the enzymes necessary to synthesize four-carbon dicarboxylic acids from acetyl coenzyme A. With the exception of glycerol, each of the gluconeogenic precursors must be converted to pyruvate and oxaloacetate prior to the formation of glucose.

12. The answer is B (1, 3). *(IX F 2 a, b)* Progesterone and estradiol act synergistically at the levels of the endometrium and mammary tissue. These hormones act antagonistically, however, on the myometrium, where progesterone inhibits uterine motility and estradiol promotes uterine motility. Estradiol has a tendency to cause Na^+ retention, while progesterone induces Na^+ excretion.

13. The answer is D (4). *(V D 2 a–b)* Prolactin secretion usually is under tonic inhibition by the action of the hypothalamic hormone called prolactin inhibiting factor (PIF). This inhibitory hormone is believed to be dopamine. Therefore, dopaminergic substances (e.g., bromocriptine) block prolactin secretion. Prolactin, like growth hormone, is produced by the pituitary acidophils and is released during stress reactions.

14. The answer is E (all). *(VI H 1 c; X E)* Glucagon is a major glycogenolytic hormone at the level of the liver and also is a hyperglycemic hormone. Its hyperglycemic effect is less than that of epinephrine because glucagon stimulates the secretion of two hypoglycemic hormones—insulin and somatostatin—while epinephrine inhibits insulin secretion. Glucagon also has a fat catabolic effect and is one of many lipolytic hormones. It also is gluconeogenic.

15. The answer is B (1, 3). *[VIII B 2 b (2) (c)]* The Sertoli cell produces meiosis-inhibiting factor, androgen-binding protein, estradiol, inhibin, and müllerian duct inhibiting factor. The fluid produced by the Sertoli cell is watery and has a low protein concentration. The Sertoli cell does not produce a müllerian duct stimulating factor. Inhibin is a nonsteroidal hormone that inhibits the secretion of follicle-stimulating hormone.

16. The answer is B (1, 3). *(VIII C 2 c)* Testosterone is converted to dihydrotestosterone by the enzyme 5α-reductase. Dehydroepiandrosterone is not a substrate for this enzyme. Dihydrotestosterone is not a major testicular hormone, but it is produced in androgen-dependent tissues, such as the prostate gland and seminal vesicles. Dihydrotestosterone is the most biologically active androgen.

17. The answer is B (1, 3). *(IX D E, F 1)* The theca interna produces a progestogen which, is utilized by the granulosa cells to form estradiol. This major steroid produced by the theca interna is androstenedione, which is aromatized to estrogens. Androstenedione is a 17-ketosteroid and an androgen. All naturally occurring androgens are 19-carbon steroids.

18–22. The answers are: 18-A, 19-B, 20-D, 21-B, 22-C. *(IX C 2 b, F 2)* The myometrium is composed of smooth muscle. The strata functionalis and basalis are components of the endometrium. The adluminal layer of the endometrium is the stratum functionalis, which is sloughed during menses. The stratum basalis is the abluminal layer of the endometrium and is responsible for regenerating the stratum functionalis. This occurs mainly during the proliferative phase of the endometrial cycle.

The two endometrial layers receive blood from two different sources. The spiral (coiled) arteries constitute the vascular input to the stratum functionalis, and the basal (straight) arteries supply the stratum basalis.

Progesterone and estradiol promote the growth and differentiation of the endometrium. These same two hormones are antagonistic with respect to their actions on the myometrium.

23–26. The answers are: 23-D, 24-A, 25-A, 26-B. *(XII D 2 c)* The prohormone, 7-dehydrocholesterol, found in the skin is converted to cholecalciferol (vitamin D_3). The liver converts cholecalciferol to 25-hydroxycholecalciferol (25-hydroxyvitamin D_3) by the action of hepatic 25-hydroxylase. The kidney forms 1,25-dihydroxycholecalciferol (1,25-dihydroxyvitamin D_3) by the action of renal 1α-hydroxylase. The most biologically active form of vitamin D_3 is 1,25-dihydroxycholecalciferol (calcitriol).

Post-test

QUESTIONS

Directions: Each question below contains five suggested answers. Choose the **one best** response to each question.

1. Which of the following mechanisms increases ventricular end-diastolic volume?

(A) Changing from a reclining to a standing position
(B) Vasodilation of the systemic arterioles
(C) Vasodilation of the systemic venules
(D) Stimulation of cardiac sympathetic neurons
(E) Decreasing heart rate from 150 to 100 beats/min

2. An individual has a ratio of total dead space to tidal volume (V_T) of 0.20, a V_T of 0.6 L, and a respiratory rate of 10 breaths/min. What is the alveolar ventilation?

(A) 1.2 L/min
(B) 2.0 L/min
(C) 3.3 L/min
(D) 4.8 L/min
(E) 6.0 L/min

3. During antidiuresis, the solute found in the highest concentration in the renal medullary interstitium is

(A) urea
(B) Na^+
(C) Cl^-
(D) HPO_4^{2-}
(E) glucose

4. Gastric motility is decreased by all of the following substances EXCEPT

(A) gastrin
(B) cholecystokinin
(C) secretin
(D) atropine
(E) norepinephrine

5. All of the following conditions lead to edema formation EXCEPT

(A) an increase in plasma oncotic pressure
(B) lymphatic obstruction
(C) capillary damage
(D) arteriolar vasodilation
(E) venular constriction

6. The major effect of hypercapnia on ventilation is mediated through which of the following respiratory control mechanisms?

(A) Peripheral chemoreceptors
(B) Vagus nerve
(C) Glossopharyngeal nerve
(D) Chemoreceptors of the fourth ventricle
(E) Carotid bodies

7. An EKG showing wide QRS complexes that occur at a rate of 80/min and that are NOT preceded by P waves is best interpreted as demonstrating an ectopic pacemaker in the

(A) right atrium
(B) left atrium
(C) AV node
(D) common bundle of His
(E) ventricular muscle

8. The two graphs below illustrate how the excretion rate (*graph A*) and the plasma clearance (*graph B*) of a substance relate to the plasma concentration of that substance (P_x). Based on these data, it can be concluded that this substance is

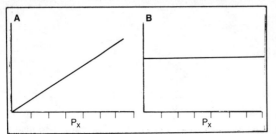

(A) creatinine
(B) PAH
(C) penicillin
(D) inulin
(E) glucose

9. Which of the following changes occur when the pressure in the carotid sinus is increased?

(A) Heart rate decreases and peripheral resistance increases
(B) Both heart rate and peripheral resistance decrease
(C) Heart rate decreases and contractility increases
(D) Both heart rate and contractility increase
(E) Both peripheral resistance and contractility increase

10. One of the few pancreatic enzymes secreted in the active form is

(A) trypsin
(B) amylase
(C) carboxypeptidase A
(D) phospholipase
(E) chymotrypsin

11. O_2 delivery to the tissues is increased immediately after reaching high altitudes. This occurs by all of the following mechanisms EXCEPT

(A) an increase in P_{50}
(B) an increase in cardiac output
(C) erythropoiesis
(D) an increase in 2,3-diphosphoglycerate concentration
(E) an increase in the arteriovenous PO_2 difference

12. What is the cardiac output of a patient who has an O_2 uptake of 360 ml/min and an arteriovenous O_2 difference of 6 ml/dl?

(A) 2160 ml/min
(B) 600 ml/min
(C) 60,000 ml/min
(D) 6000 ml/min
(E) 60 ml/min

13. In the kidney, the largest fraction of filtered Na^+ is reabsorbed in the

(A) proximal tubule
(B) descending limb of the loop of Henle
(C) ascending limb of the loop of Henle
(D) distal tubule
(E) collecting duct

Questions 14 and 15

A 30-year-old woman is admitted to the psychiatric ward, and the following data are reported.

Serum Findings	Arterial Blood Findings
$[Na^+]$ = 140 mEq/L	pH = 7.45
$[K^+]$ = 3.3 mEq/L	PO_2 = 100 mm Hg
$[Cl^-]$ = 106 mEq/L	PCO_2 = 35 mm Hg
CO_2 content = 24 mmol/L	

14. What is the most likely diagnosis of this patient's condition?

(A) Metabolic alkalosis from the unreported use of diuretics
(B) Primary hyperaldosteronemia
(C) Hypokalemic metabolic alkalosis
(D) Renal tubular acidosis
(E) Chronic hyperventilation

15. What is the urine of this patient most likely to show?

(A) Acid pH
(B) Glycosuria
(C) An increased HCO_3^- excretion
(D) An increased Cl^- excretion
(E) A decreased CO_2 content

16. One of the functions of mastication is the mixing of food with saliva. This begins the process of

(A) degradation of polypeptides into amino acids
(B) formation of lipid micelles
(C) breakdown of disaccharides such as sucrose
(D) digestion of polysaccharides
(E) hydrolysis of cellulose

17. Which of the following stimuli causes the appearance of a large quantity of angiotensin II in the blood?

(A) An increase in cardiac output
(B) An increase in blood volume
(C) An increase in sympathetic discharge
(D) A decrease in arterial pressure
(E) Vasoconstriction of arterioles

18. The following pH and blood-gas findings were determined in each of five patients while they were breathing room air. In which of these patients would the presence of chronic obstructive pulmonary disease contraindicate the administration of 100 percent O_2 by mask?

Patient	pH	PCO_2 (mm Hg)	PO_2 (mm Hg)
(A)	7.30	60	38
(B)	7.32	28	90
(C)	7.40	42	84
(D)	7.41	42	37
(E)	7.48	28	50

Questions 19 and 20

The data in the following table were obtained during cardiac catheterization of two patients.

Chamber or Vessel	Patient 1 Pressure (mm Hg)	Oxygen Saturation (%)	Patient 2 Pressure (mm Hg)	Oxygen Saturation (%)
Aorta	140/50	95	95/65	96
Left ventricle	140/14	95	95/9	96
Left atrium	14*	95	31*	96
Pulmonary artery	26/11	75	68/21	74
Right ventricle	26/5	75	68/7	74
Right atrium	5*	75	6*	74

*Denotes mean value

19. Based on the data in the above table, the most likely diagnosis for patient 1 is

(A) aortic insufficiency
(B) aortic stenosis
(C) tricuspid insufficiency
(D) patent ductus arteriosus
(E) mitral stenosis

20. Based on the data in the above table, the most likely diagnosis for patient 2 is

(A) aortic insufficiency
(B) aortic stenosis
(C) tricuspid insufficiency
(D) tricuspid stenosis
(E) mitral stenosis

Directions: Each question below contains four suggested answers of which **one or more** is correct. Choose the answer

A if **1, 2, and 3** are correct
B if **1 and 3** are correct
C if **2 and 4** are correct
D if **4** is correct
E if **1, 2, 3, and 4** are correct

21. Characteristics that distinguish an equilibrium state from a steady state condition include

(1) there is no change in the ionic concentration gradient over time
(2) the Nernst potential for the ion is equal to the membrane potential
(3) the cell is not permeable to the ion
(4) no energy is required to maintain the concentration gradient for the ion

22. Factors leading to reduced pulmonary vascular resistance include

(1) residence at an altitude of 15,000 feet
(2) inspiration to total lung capacity
(3) expiration to residual volume
(4) moderate exercise

23. Signs of atrial fibrillation include

(1) an irregularly irregular rhythm
(2) a pulse deficit
(3) wide fluctuation in the aortic pressure from beat to beat
(4) prolongation of the P-R interval

24. A cell that is permeable only to Na^+ and K^+ will definitely be depolarized by an increase in

(1) K^+ conductance
(2) Na^+ conductance
(3) K^+ transference
(4) Na^+ transference

25. The renal threshold for glucose excretion corresponds to a specific glucose concentration. This concentration can be expressed as

(1) 5.5 mEq/L
(2) 10 mmol/L
(3) 100 mg/dl
(4) 180 mg/dl

26. Transport processes that require a carrier include

(1) osmosis
(2) facilitated diffusion
(3) simple diffusion
(4) active transport

27. Blood flow to various vascular beds is adjusted to need. Mechanisms that regulate this flow include

(1) alteration of cardiac output
(2) autoregulation
(3) local vasodilation
(4) local increases in peripheral resistance

28. Ingestion of a drug that inhibits the active transport of Cl^- will result in

(1) increased water reabsorption by the descending limb of the loop of Henle
(2) decreased Na^+ reabsorption by the ascending limb of the loop of Henle
(3) increased water movement out of the ascending limb of the loop of Henle
(4) decreased urine osmolarity

29. In myelinated fibers, the propagation of action potentials occurs at a velocity that

(1) is directly proportional to fiber diameter
(2) decreases when the action potential invades the nerve terminal
(3) is faster than in unmyelinated fibers
(4) increases as the internodal distance increases

30. In order to achieve maximal cardiac output, the cardiovascular control centers stimulate which of the following responses?

(1) A decrease in vagal tone
(2) An increase in the ventricular filling rate
(3) An increase in myocardial contractility
(4) Venodilation

31. Electrically excitable membrane channels are required for the generation of

(1) an action potential
(2) an inhibitory postsynaptic potential
(3) a local response
(4) an excitatory postsynaptic potential

32. Mechanisms that increase airway resistance include

(1) efferent vagal impulses to the lung
(2) sympathetic impulses to the lung
(3) decreased PCO_2 in the expired air
(4) circulating epinephrine

33. True statements concerning aortic compliance include which of the following?

(1) It increases with advancing age
(2) It is a factor that determines pulse pressure
(3) It is determined by plotting stroke volume against mean arterial pressure
(4) It is essentially linear in the normal range of blood pressures

34. Factors that determine the CO_2 content of the blood include

(1) hemoglobin level
(2) PCO_2
(3) PO_2
(4) pH

35. True statements concerning pulse pressure include which of the following?

(1) It is greater in the aorta than in the pulmonary artery
(2) It is dependent on heart rate
(3) It is dependent on mean arterial pressure
(4) It represents the difference between systolic and diastolic pressures

Questions 36 and 37

A 17-year-old woman is taken to the emergency room complaining of nausea and shortness of breath. She reports fatigability, excessive thirst, and an increased frequency and volume of urine output. Her skin and mucous membranes are dry, and her throat is inflamed. An acetone-like odor is detected on her breath, and the following laboratory data are reported.

Serum Analysis			Urinalysis		
$[Na^+]$	=	130 mEq/L	$[Na^+]$	=	30 mEq/L
$[K^+]$	=	5.8 mEq/L	$[Cl^-]$	=	15 mEq/L
$[Cl^-]$	=	90 mEq/L	Glucose	=	4+
$[HCO_3^-]$	=	10 mEq/L	Ketones	=	4+
Glucose	=	650 mg/dl	pH	=	5.0
PCO_2	=	25 mm Hg			
pH	=	7.2			

36. Likely causes of this patient's condition include

(1) metabolic acidosis
(2) dehydration
(3) ketoacidosis
(4) diarrhea

37. The treatment regimen for this patient's condition is likely to include

(1) intravenous administration of KCl
(2) injection of insulin
(3) infusion of isotonic NaCl
(4) administration of glucocorticoids

Directions: The groups of questions below consist of lettered choices followed by several numbered items. For each numbered item select the **one** lettered choice with which it is **most** closely associated. Each lettered choice may be used once, more than once, or not at all.

Questions 38–41

Match each statement concerning optical defects with the appropriate condition.

(A) Hyperopia
(B) Presbyopia
(C) Both hyperopia and presbyopia
(D) Neither hyperopia nor presbyopia

38. The near point is closer than in emmetropia

39. The lens is inflexible

40. Images of distant objects are focused behind the retina

41. Correction is with glasses that have convex lenses

Questions 42–45

Each of the following phrases describes a characteristic that is unique to one of the nephron components listed below. Match each phase to the nephron region it best describes.

(A) Bowman's capsule
(B) Proximal convoluted tubule
(C) Distal convoluted tubule
(D) Loop of Henle
(E) Collecting duct

42. Site of the highest Na^+ concentration

43. Nephron region where entering tubular fluid always is hypotonic

44. Site where the tubular fluid to plasma (TF/P) glucose concentration ratio is 1

45. Site where the TF/P inulin concentration ratio is between 3 and 4

Questions 46–50

The table below shows the blood acid-base data for five individuals who are designated by the letters A–E. For each of the following descriptions of acid-base status, choose the individual with the appropriate acid-base data.

	pH	PCO_2 (mm Hg)	PO_2 (mm Hg)	$[HCO_3^-]$ (mmol/L)
(A)	7.27	64	64	37
(B)	7.31	30	98	16
(C)	7.34	40	100	24
(D)	7.50	45	83	40
(E)	7.55	20	90	22

46. Normal

47. Metabolic acidosis

48. Respiratory acidosis

49. Metabolic alkalosis

50. Respiratory alkalosis

Questions 51–55

Match each of the following conditions to the appropriate group of blood-gas measurements.

	PO_2 of Inspired Air (mm Hg)	PaO_2 (mm Hg)	$PaCO_2$) (mm Hg)	Arterial pH
(A)	100	60	30	7.45
(B)	150	60	40	7.38
(C)	150	78	60	7.35
(D)	150	97	40	7.40
(E)	150	110	28	7.50

51. Hypoventilation

52. Ventilation:perfusion imbalance

53. Residence at high altitude

54. Hyperventilation

55. Light to moderate exercise

Questions 56–60

Each of the following phrases describes a hormone in terms of its synthesis, secretion, or function. Match each phrase to the appropriate hormone.

(A) Glucagon
(B) Insulin
(C) Somatostatin
(D) Epinephrine
(E) Norepinephrine

56. A hypoglycemic hormone synthesized by the pancreatic delta cells

57. The release of this hormone is stimulated by arginine and glucagon

58. This hormone is protein anabolic and lipogenic and promotes glycogen deposition

59. A stimulator of insulin secretion

60. An equally effective α- and β-agonist that blocks the release of insulin

ANSWERS AND EXPLANATIONS

1. The answer is E. (*Chapter 2 V C 3 c*) In changing from a reclining to a standing position, approximately 500 ml of blood are transferred to dependent vessels, reducing the blood volume in the thoracic vessels and cardiac chambers. Blood also is translocated to the veins during venodilation, as the veins are the major storage site for blood in the cardiovascular system. Arteriolar vasodilation reduces arterial pressure, thereby reducing the ventricular afterload and increasing stroke volume. This sequence decreases end-systolic volume; therefore, end-diastolic volume decreases because the venous return is unchanged. The ventricular end-diastolic volume also is reduced during sympathetic activation of the heart due to the increased contractility augmenting the stroke volume. A decrease in heart rate increases the duration of the diastolic filling, which follows an increase in the ventricular end-diastolic volume.

2. The answer is D. (*Chapter 3 IV B, C*) A dead space to tidal volume (V_T) ratio of 0.2 indicates that the dead space represents 20 percent of the V_T. Dead space ventilation can be determined by multiplying the given values; that is: (0.2 × 0.6 L) × 10 breaths/min = a dead space ventilation of 1.2 L/min. Alveolar ventilation equals minute ventilation (V_T × f = 6 L/min) minus dead space ventilation; that is: 6.0 − 1.2 = 4.8 L/min.

3. The answer is A. (*Chapter 4 VII D 2*) The concentration of urea is much higher than that of any other solute in the medullary interstitium. (It is important to remember that the osmolality of body fluids always is compared to that of normal plasma.)

The medullary interstitial osmolality is higher in antidiuresis than in water diuresis. The difference is largely attributable to urea that is deposited in the medullary interstitium by the process of medullary urea recycling. The osmolality of the interstitium represents the sum of the concentrations of NaCl and urea; urea diffuses into the interstitium along the concentration gradient of urea (from collecting duct to interstitium), and NaCl diffuses into the interstitium along the concentration gradient of Na^+ and Cl^- (from the thin ascending limb of the loop of Henle to the interstitium). Water that has entered the medulla from the descending limb of the loop of Henle is carried away by the ascending vasa recta.

Due to these processes that promote urea reabsorption from the collecting duct, urea constitutes about 40 percent of the total medullary interstitial osmolality during **antidiuresis**, with the balance mainly due to NaCl. During **water diuresis**, urea constitutes less than 10 percent of the medullary osmolality. Thus, the papillary osmolality varies with the secretion of antidiuretic hormone (ADH).

For a medullary interstitial osmolality of 1400 mOsm/kg during maximal antidiuresis, the following medullary solute osmolalities can be estimated:

[Urea]: 40 percent of 1400 mOsm/kg = 560 mOsm/kg

[NaCl]: 60 percent of 1400 mOsm/kg = 840 mOsm/kg

[Na^+]: 420 mOsm/kg

[Cl^-]: 420 mOsm/kg.

4. The answer is A. [*Chapter 6 II A 3 a (1) (2), b (1) (2)*] Gastrin, a polypeptide hormone produced by the G cells of the antral mucosa, increases both intestinal and gastric motility. Cholecystokinin and secretin also are polypeptide hormones; however, they are produced by the small intestine, and their effects include inhibition of gastric motility. Sympathetic nerves have an inhibitory influence on gastric motility, which is mediated by epinephrine and norepinephrine. Since parasympathetic influences predominantly increase gastric motility, an anticholinergic drug such as atropine also depresses motility.

5. The answer is A. (*Chapter 2 I D 4 b; Fig. 2-3*) Edema formation can result from lymphatic obstruction (by decreased removal of tissue fluids), from capillary damage (by increased capillary permeability), and from arteriolar vasodilation or venular constriction (by increased capillary pressure). An increase in plasma oncotic pressure will reduce edema formation by increasing the forces that draw water back into the vascular compartment from the interstitial space.

6. The answer is D. (*Chapter 3 VIII A 3, B*) Approximately 85 percent of the resting ventilation is mediated by the effect of CO_2 on the central chemoreceptors. Although the peripheral chemoreceptors respond to changes in CO_2, the major effect of CO_2 is mediated through the chemoreceptors of the fourth ventricle, which lie near the surface of the medulla.

7. The answer is E. (*Chapter 2 IV B 4 a*) An ectopic pacemaker located in either the right or left atrium would generate a P wave before each QRS complex. These impulses, as well as those from the AV node and bundle of His, also would generate a QRS complex of normal duration. This is because such pacemakers would involve the fast conducting system to the ventricles.

8. The answer is D. (*Chapter 4 IV B; VI B; Fig. 4-6*) In order for the clearance of inulin (C_{in}) to be a true measure of GFR, the C_{in} must be constant and independent of plasma inulin concentration (P_{in}), as shown in **graph B**. This implies that the inulin excretion rate must be directly proportional to (and a linear function of) P_{in}, as shown in **graph A**. The filtered inulin is equal to the product of GFR and P_{in}, and the excreted inulin is equal to the product of urine inulin concentration (U_{in}) and timed urine volume ($\dot{V}$). The relationship between the filtered load of inulin and excreted inulin is expressed in terms of the filtrate-urine mass balance equation:

$$C_{in} \bullet P_{in} = U_{in} \bullet \dot{V}.$$

Therefore, the excretion curve in **graph A** would be congruent with the filtered load curve, with both curves having the same units (i.e., mg/min).

9. The answer is B. [*Chapter 2 VII B 1 b (2)*] A decrease in both heart rate and peripheral resistance leads to a decrease in blood pressure. These events characterize the reflex response to a rise in carotid sinus pressure. A decrease in heart rate coupled with either an increase in peripheral resistance or an increase in contractility does not correct blood pressure. An increase in contractility coupled with either an increase in heart rate or an increase in peripheral resistance leads to a further increase in blood pressure.

10. The answer is B. (*Chapter 6 III B 2 b*) Most pancreatic enzymes are secreted as inactive proenzymes to prevent autodigestion of the pancreas by its own secretions. Trypsin activates the proenzymes chymotrypsinogen, procarboxypeptidases A and B, proelastase, procollagenase, and phospholipases A and B. Trypsinogen is cleaved to its active form by enterokinase or by trypsin itself (autocatalysis). Amylase, however, is secreted by the pancreas in the active form.

11. The answer is C. (*Chapter 3 VII D*) Hypoxic hypoxia (i.e., decreased O_2 supply to the blood) occurs when breathing air at high altitude. Hypoxia reflexly increases cardiac output, which returns to normal after 2 weeks at high altitude. Also, hypoxia increases the concentration of free 2,3-diphosphoglycerate (DPG) in the erythrocytes, which in turn increases the P_{50} of the O_2 dissociation curve and raises tissue PO_2. Erythropoiesis requires several weeks to increase significantly the erythrocyte population. Normally, the arteriovenous PO_2 difference is 60 mm Hg (100 mm Hg − 40 mm Hg). However, because the PaO_2 is reduced at high altitude, the hemoglobin functions on the steep slope of the hemoglobin dissociation curve; the O_2 is delivered to the tissues by a much smaller change in tension between the arterial and venous blood compared to normal (e.g., 60 mm Hg − 30 mm Hg or 30 mm Hg).

12. The answer is D. (*Chapter 2 C 6 a*) The Fick equation can be applied to this proposed case. This equation states that cardiac output equals O_2 consumption divided by the arteriovenous O_2 difference. The calculation can be simplified by converting the arteriovenous O_2 difference to ml O_2/ml blood (i.e., 0.06 ml O_2/ml blood, in this case). The O_2 uptake now can be divided easily (360/0.06) to reveal a cardiac output of 6000 ml/min.

13. The answer is A. (*Chapter 4 VI C 1–2*) Approximately two-thirds of the total filtered Na^+ and water is reabsorbed isosmotically by the proximal tubule. Therefore, the concentration of Na^+ (as opposed to the amount) remains virtually unchanged throughout the proximal tubule. The fraction of filtered Na^+ and water reabsorbed in the proximal tubule decreases during volume expansion and increases during volume depletion.

14. The answer is E. (*Chapter 5 VIII D 2; IX B 2; Fig. 5-13*) The alkaline pH and hypocapnia are diagnostic of respiratory alkalosis due to chronic hyperventilation. The plasma [HCO_3^-] can be determined by subtracting the [dissolved CO_2] from the [total CO_2]. Using the given $PaCO_2$ value, the [dissolved CO_2] is calculated as:

$$35 \text{ mm Hg} \times 0.03 \text{ mmol } CO_2/L/\text{mm Hg} = 1.05 \text{ mm Hg.}$$

To determine the [HCO_3^-], then:

$$[HCO_3^-] = 24 \text{ mmol/L} - 1.05 \text{ mmol/L} = 22.95 \text{ mmol/L.}$$

In respiratory alkalosis, the [H^+] and [HCO_3^-] decrease together in contrast to metabolic alkalosis. Thus, the plasma [HCO_3^-] decline in respiratory alkalosis is the renal compensatory response to hypocapnia. Notice that in alkalotic states, H^+ is released from the ICF buffers into the ECF in an attempt to lower the ECF pH toward normal. To preserve electroneutrality, K^+ and Na^+ enter the cells, resulting in a fall in the plasma [K$^+$]. An inverse relationship exists between plasma [Cl$^-$] and the rate at which HCO$_3$$^-$ is reabsorbed. This inverse relationship serves to maintain a somewhat constant sum of the plasma concentrations of Cl$^-$ and HCO$_3$$^-$. Thus, in a pure respiratory alkalosis, the concentrations of CO$_2$, HCO$_3$$^-$, and H$^+$ are lower than normal.

15. The answer is C. (*Chapter 5 IX B 2 b; Fig. 5-13*) In respiratory alkalosis, H$^+$ excretion is decreased and HCO$_3$$^-$ excretion is increased in the urine. The most important singular determinant of the rate of tubular acid secretion is the PCO$_2$ of the arterial blood. As a result of hyperventilation and the respiratory alkalosis that follows, the patient eliminates CO$_2$ faster than it is produced, thereby lowering the PaCO$_2$ and raising the pH. The decreased PaCO$_2$ reduces the PCO$_2$ within the renal tubular cells, causing a decrease in the formation of H$_2$CO$_3$ and, in turn, a reduced intracellular [H$^+$] and a decline in H$^+$ secretion. Thus, in respiratory alkalosis, the decreased PCO$_2$ reduces tubular H$^+$ secretion so that [HCO$_3$$^-$] reabsorption declines. HCO$_3$$^-$ is excreted, resulting in a decreased plasma [HCO$_3$$^-$].

16. The answer is D. [*Chapter 6 I C 2 d; III C 1 c, E 3 a (3)*] Saliva contains only one digestive enzyme, α-amylase, which initiates digestion of the polysaccharides such as starch and amylose. Disaccharides are not affected by this enzyme; their digestion occurs in the small intestine. Cellulose, although a polysaccharide, is not affected by α-amylase or by any other digestive enzyme. Lipid micelles are formed in the small intestine in the presence of bile salts.

17. The answer is D. (*Chapter 2 VII A 2 b*) A decrease in arterial pressure stimulates the juxtaglomerular cells of the kidney to release renin, which is released into the plasma; here renin cleaves two amino acids from angiotensinogen and forms angiotensin I. As it passes through the pulmonary circulation, angiotensin I is converted into angiotensin II, which acts as a potent vasoconstrictor. An increase in cardiac output, blood volume, or sympathetic discharge and arteriolar vasoconstriction all lead to an increase in arterial pressure and, therefore, an inhibition of renin release.

18. The answer is A. (*Chapter 5 XIII A 2*) In patients whose ventilatory drive depends on hypoxia as a result of CO$_2$ retention and decreased responsiveness to CO$_2$ (CO$_2$ narcosis), the administration of O$_2$-enriched breathing mixtures may seriously decrease ventilation or produce apnea. A high PCO$_2$, which further increases after O$_2$ therapy, is usually seen with chronic respiratory acidosis, and a patient with chronic hypercapnia relies on hypoxemic "drive" to ventilate. The patient designated by (A) can tolerate neither O$_2$ administration nor the PO$_2$ of 38 mm Hg in room air. Therefore, bronchodilators (e.g., aminophylline and epinephrine) may be efficacious in reversing the bronchoconstriction. Corticosteroids and intubation with mechanical ventilation may be necessary to provide adequate oxygenation and to remove mucous plugs.

19. The answer is A. (*Chapter 2 V B 2*) All data for patient 1 are within normal limits except for the low aortic diastolic pressure, the high ventricular diastolic pressure, and the mean left atrial pressure. The lesion is located in the left heart; in order to create these conditions, the lesion must be aortic insufficiency (i.e., regurgitation of the aortic valve). The backflow into the left ventricle during diastole causes a drop in aortic diastolic pressure and an increase in the left-ventricular diastolic pressure. With no systolic gradient between the left ventricle and the aorta, there is no aortic stenosis. Similarly, mitral stenosis does not occur without a significant pressure gradient between the left atrium and the left ventricle.

20. The answer is E. (*Chapter 2 V B 4 a*) In this patient, the aortic pressure is low and the mean left-atrial pressure is abnormally high. The abnormally wide gradient between the left-ventricular (LV) diastolic pressure and the mean left-atrial (LA) pressure could only be created by mitral stenosis. (In order for mitral stenosis to occur, there must be a gradient between LA and LV pressures during diastole.) Mitral insufficiency also raises the mean LA pressure, but the high pressure occurs during systole, when blood flow backs up into the LA. (Phasic pressure pulses that show the diastolic gradient between LA and LV will confirm the diagnosis.) The low aortic pressure is caused by the reduced stroke volume that is secondary to this valvular lesion.

21. The answer is C (2, 4). (*Chapter 1 II B 1–2*) By definition, a system that is in an equilibrium condition is one that does not require energy to keep the concentration gradient from changing. In an equilibrium condition, the membrane potential is equal to the Nernst potential. A system that is described as being in the steady state does require energy to keep the concentration gradient constant. Substances not permeable to the membrane may be actively transported across it, whether the ion is an equilibrium condition, a steady state condition, or neither.

22. The answer is D (4). (*Chapter 3 VI A 1*) Pulmonary vascular resistance is increased at high altitudes due to a hypoxic vasoconstriction, whereas marked increases or decreases in lung volume cause vascular compression and also produce an increased vascular resistance. Vascular resistance is reduced in exercise due to a recruitment of vascular beds in response to the increased pulmonary blood flow.

23. The answer is A (1, 2, 3). (*Chapter 2 IV B 2 d*) Atrial fibrillation is associated with a lack of P waves; therefore, the P-R interval cannot be identified. Atrial fibrillation is characterized by a very irregular ventricular rhythm so that stroke volume varies greatly due to changes in the duration of ventricular filling, contractility, and arterial pressure. Large variations in stroke volume cause large fluctuations in aortic pulse pressure. At times, ventricular contraction may be so weak that stroke volume is zero, and no pulse is transmitted to the periphery; this condition is termed a pulse deficit.

24. The answer is D (4). (*Chapter 1 II B 2*) An increase in Na^+ conductance will not necessarily cause the membrane to depolarize. If K^+ conductance also increases, the relative conductance (transference) of Na^+ may remain the same or even decrease. When the transference of Na^+ increases, the membrane always depolarizes.

25. The answer is C (2,4). [*Chapter 4 VI B 2 b (3); Fig. 4-6A*] The plasma glucose concentration at which glucose begins to appear in the urine is known as the renal threshold concentration for glucose and is equal to 180 mg/dl. It is the plasma concentration of glucose at which a portion of the glucose escapes tubular reabsorption and, therefore, is excreted in the urine (glycosuria). Of course, a glucose concentration of 180 mg/dl is equal to a 1800 mg/L glucose concentration. This converts into mmol/L as follows:

$$mmol/L = \frac{mg/dl \times 10}{molecular\ weight\ (mg)}$$

$$= \frac{180\ (10)}{180}$$

$$= 10\ mmol/L.$$

26. The answer is C (2, 4). (*Chapter 1 1 B 1–3*) Both osmosis and simple diffusion occur without the help of a carrier. The osmotic movement of water occurs through membrane channels and is driven by a hydrostatic pressure created by dissolved particles. In simple diffusion, particles move from an area of high concentration to an area of low concentration as a result of the random motion of particles. In facilitated diffusion, however, a carrier is required to move a particle down its concentration gradient. Active transport also requires the expenditure of energy to move a particle against its concentration gradient.

27. The answer is E (all). (*Chapter 2 I C 2 a; VII A, C*) Blood flow to different tissues, during certain circumstances, is maintained properly by all of the mechanisms listed in the question. Local vasodilation occurs during any increased metabolic activity, such as glandular secretion and exercise. Vasodilation in some vascular beds tends to lower aortic blood pressure, which is compensated by an increase in cardiac output and vascular resistance in other vascular beds. Autoregulation is a variation in peripheral resistance in some vascular beds, which maintains a constant blood flow in response to changes in arterial blood pressure.

28. The answer is C (2, 4). (*Chapter 4 VI C 5; VII B 2 b*) Drugs that block the active transport of Cl^- out of the thick ascending limb of the loop of Henle are called loop diuretics and include furosemide and ethacrynic acid. Since they block the reabsorption of NaCl from this portion of the nephron, loop diuretics cause a marked increase in the urinary excretion of water and NaCl (osmotic diuresis). In the presence of antidiuretic hormone (ADH), these drugs cause a decrease in free-water reabsorption ($T^c_{H_2O}$); in the absence of ADH, they cause a decrease in free-water clearance (C_{H_2O}). During peak diuresis with these drugs, the urine is isosmotic.

29. The answer is E (all). (*Chapter 1 II D*) In myelinated fibers, the propagation velocity in meters per second is equal to about six times the fiber diameter in microns. As fiber diameter increases, the internodal distance and propagation velocity also increase. The fiber diameter becomes smaller as the nerve terminal is approached and, thus, velocity of propagation decreases. Propagation of action potentials is faster in myelinated fibers than it is in unmyelinated fibers, which must generate action potentials on contiguous patches of membrane along the axon. Myelinated fibers are used by the nervous system when high speeds of conduction are needed (e.g., for motor control and tactile discrimination).

30. The answer is A (1, 2, 3). (*Chapter 2 II C 1 b*) Venodilation causes pooling of blood and so reduces venous return and cardiac output. A decreased vagal tone allows heart rate and atrial contractility to increase. Increased contractility and filling rate raise stroke volume and improve cardiac output.

31. The answer is B (1, 3). (*Chapter 1 II C; III C 2*) Both the action potential and local response are initiated by the opening of electrically excitable Na^+ channels. Both the inhibitory and excitatory postsynaptic potentials are generated by the opening of a chemically excitable channel.

32. The answer is B (1, 3). (*Chapter 3 V B 2 d, f*) Efferent vagal impulses and reduced PCO_2 in expired air cause bronchoconstriction and, thus, increase airway resistance. In contrast, sympathetic impulses and circulating epinephrine cause bronchodilation due to the preponderance of β_2-receptors in the bronchi.

33. The answer is C (2, 4). (*Chapter 2 I A 2; VI B 1*) Aortic elastance, not aortic compliance, increases with age. Compliance is plotted using the arterial volume, not the stroke volume. The reciprocal of compliance is elastance, which is calculated using two parameters from the Young's modulus. Increased elastance produces an increase in pulse pressure for any given stroke volume.

34. The answer is E (all). (*Chapter 3 II B 1, D 3*) Hemoglobin acts as a carrier for CO_2 by means of the terminal amino groups. Decreases in PO_2 increase the affinity of the hemoglobin molecules for CO_2 (Haldane effect). An acid blood pH also increases the affinity of the hemoglobin for CO_2. The amount of dissolved CO_2 in plasma depends on the PCO_2 and the solubility constant; this is Henry's law, which is stated as: $[CCO_2] = PCO_2 \times [SCO_2]$.

35. The answer is E (all). (*Chapter 2 VI B*) Pulse pressure is defined as the difference between systolic and diastolic pressures. Heart rate affects pulse pressure by altering the runoff time and by varying the arterial diastolic pressure as well as the stroke volume. Mean arterial pressure affects pulse pressure because it is dependent on the arterial volume. Increases in arterial volume will be associated with smaller pulse pressures due to the relationships expressed in Young's modulus.

36. The answer is A (1, 2, 3). (*Chapter 5 IX B 3; X C; XI D*) Polydipsia and polyuria are two cardinal signs of diabetes mellitus. Acetone, produced by the decarboxylation of acetoacetic acid, has a characteristic fruity odor that may be detected on the breath of a patient with ketoacidosis. Therefore, it is clear that this patient has a metabolic acidosis with a high anion gap, which suggests ketoacidosis. This patient also is dehydrated due to the osmotic (glucose) diuresis.

A patient suffering from diarrhea will lose HCO_3^--containing intestinal secretions. Since this HCO_3^- must be replaced from the ECF, its loss will result in a decreased extracellular $[HCO_3^-]$ and, accordingly, acidosis. Diarrheal fluid is approximately isotonic to plasma, and the fluid loss will not alter either the plasma osmolality or the plasma $[Na^+]$. Since the gastrointestinal secretions contain K^+, diarrhea can lead to K^+ depletion. It is important to emphasize that plasma $[K^+]$ is not always reduced in states of K^+ depletion (diabetes mellitus and diarrhea) because with acidosis, H^+ moves into the ICF in exchange for intracellular K^+ and Na^+. As a result, the plasma $[K^+]$ may be normal or even elevated despite a decrease in body K^+ stores (e.g., due to diabetic ketoacidosis or diarrhea). Thus, diarrhea can be a cause of metabolic acidosis but does not explain the other laboratory findings in this patient.

Metabolic acidosis should reduce urinary K^+ excretion by decreasing intracellular K^+. Increased urinary K^+ excretion can occur in metabolic acidosis, including ketoacidosis, and may be due to a combination of increased distal tubular flow and reabsorption of Na^+ in the presence of a nonreabsorbable anion. In diabetic ketoacidosis, increased quantities of Na^+ and water (due to osmotic diuresis) are presented to the distal tubule with the ketoacid anions β-hydroxybutyrate and acetoacetate. The presentation of Na^+ to the distal tubule with such nonreabsorbable anions results in Na^+ excretion in exchange for K^+ and H^+.

37. The answer is A (1, 2, 3). (*Chapter 5 XIII C*) Treatment for this patient's condition should include immediate intravenous insulin to lower both the plasma glucose concentration (and the plasma osmolality) and the plasma $[K^+]$ and to correct the ketoacidosis. Insulin therapy also enhances ketoacid utilization, resulting in the regeneration of HCO_3^- and the spontaneous correction of the acidosis. Treatment with alkali should be restricted to patients with severe acidosis.

Although patients with diabetic ketoacidosis usually show a depleted total body K^+, the effect of acidosis causes K^+ to move out into the ECF, causing an initial elevation in plasma $[K^+]$ (hyperkalemia). The effect of insulin together with the correction of the acidosis, however, causes K^+ to move intracellularly and reveals the K^+ depletion. Thus, K^+ salts (KCl or K_2HPO_4) should be administered intravenously.

Diabetes mellitus is associated with enhanced water loss relative to solute-induced osmotic (glucose) diuresis. Volume repletion can be achieved with half-isotonic saline ($[Na^+]$ of 77 mEq/L). When severe hypovolemia or shock is present, initial rehydration should begin with isotonic saline ($[Na^+]$ of 154 mEq/L). The average fluid loss with diabetes mellitus is 8–10 L, or approximately 25 percent of the total body water.

Na⁺ is lost without Cl⁻ when it is excreted with excess quantities of organic ions. The loss of extracellular electrolytes is a second cause of dehydration, as well as being a second cause of acidosis. The latter is explained on the basis of overloading the renal mechanisms for acid excretion, which results in a loss of Na⁺ and, consequently, a failure to recover extracellular $NaHCO_3$.

Glucocorticoids are contraindicated because they have a hyperglycemic action and, thus, should not be used in the presence of infection unless therapeutic indications are compelling. The diabetic state may be triggered by infections, which are characteristically of the throat and skin. An important part of the treatment, therefore, is antibiotic therapy.

38–41. The answers are: 38-D, 39-B, 40-A, 41-C. (*Chapter 1 VIII F 1–3*) In hyperopia, the eyeball is too short for the refractive power of the lens, so objects are focused behind the retina. The near point is farther than in emmetropia, which is why this condition is called farsightedness. In presbyopia, the near point recedes further from the eye because the lens loses its flexibility with age and, thus, cannot accommodate for near vision.

42–45. The answers are: 42-D, 43-C, 44-A, 45-B. (*Chapter 4 IV C 2; VI C 2; VII B 4*) In antidiuresis, both the tubular fluid in the loop of Henle and the urine have approximately equimolar concentrations of Na⁺ and urea; however, the relative amounts of Na⁺ and urea in these two sites are inversely related. In the loop of Henle, the [Na⁺] is high and the urea concentration is low, whereas in urine, the opposite is true for these two solutes. In the loop, the [Na⁺] represents about one-third and the urea concentration about one-fifth of the total osmolality.

Because more solute than water is reabsorbed in the water-impermeable ascending limb of the loop of Henle, both the [Na⁺] and the osmolality of the fluid entering the distal tubule are well below those of plasma. The osmolality difference between the loop and the early distal tubule is less than the concentration difference for Na⁺ because of the recycling of urea mainly from the medullary collecting ducts to the interstitium. The urea reenters the thin ascending limb of the loop of Henle, and by the beginning of the distal tubule the fluid is hyposmotic. However, about twice as much urea exists in the distal tubular fluid compared to that in the proximal tubular fluid. This indicates that urea is secreted into the thin ascending limb of the loop of Henle. The hypotonicity of the early distal fluid is due mainly to the reabsorption of Na⁺ and Cl⁻.

The fluid within Bowman's capsule is essentially an ultrafiltrate (i.e., protein-free filtrate) of plasma and contains most of the crystalloids in virtually the same concentrations as in the plasma. It should be noted that the fluid at the end of the proximal tubule is approximately isosmotic to plasma, because the filtrate is reabsorbed isosmotically. However, the concentrations of some solutes (e.g., glucose and HCO_3^-) are low, whereas the concentrations of others (e.g., urea) are higher. In the very early proximal tubule, the tubular fluid to plasma (TF/P) glucose concentration ratio is 1.0 but declines rapidly because glucose is absorbed within the first half of the proximal tubule.

The tubular fluid to plasma concentration ratio for inulin (TF/P_{in}) increases to 3 or 4 by the end of the proximal tubule. Increasing TF/P_{in} concentration ratios are indicative of water reabsorption, because inulin is neither reabsorbed nor secreted. The TF/P_{in} concentration ratios increase progressively along the nephron and range between 15 and 20 by the end of the distal convoluted tubule.

46–50. The answers are: 46-C, 47-B, 48-A, 49-D, 50-E. (*Chapter 5 IX*) (An uncompensated respiratory acidosis is characterized by a low pH (↑[H⁺]), a high PCO_2, and a high CO_2 content. When a respiratory acidosis is compensated by a metabolic alkalosis, the pH tends to return toward normal, but the CO_2 content becomes higher, and the PCO_2 remains unchanged (high). In respiratory acidosis, the [H⁺] and the [HCO_3^-] both increase.

An uncompensated metabolic alkalosis is characterized by a high pH (↓[H⁺]), a high CO_2 content, and a normal PCO_2. When a metabolic alkalosis is compensated by a respiratory acidosis, the pH tends to return toward normal, but the CO_2 content becomes still higher, and the PCO_2 becomes high. In metabolic alkalosis, the [H⁺] decreases and the [HCO_3^-] increases.

An uncompensated respiratory alkalosis is characterized by a high pH (↓[H⁺]), a low CO_2 content, and a low PCO_2. When a respiratory alkalosis is compensated by a metabolic acidosis, the pH tends to return toward normal, but the CO_2 content becomes lower, and the PCO_2 remains unchanged (low). In respiratory alkalosis, the [H⁺] and the [HCO_3^-] both decrease.

An uncompensated metabolic acidosis is characterized by a low pH (↑[H⁺]), a low CO_2 content, and a normal PCO_2. When a metabolic acidosis is compensated by a respiratory alkalosis, the pH tends to return toward normal, the CO_2 content becomes still lower, and the PCO_2 becomes low. In metabolic acidosis, the [H⁺] increases and the [HCO_3^-] decreases.

51–55. The answers are: 51-C, 52-B, 53-A, 54-E, 55-D. (*Chapter 3 IV D–E; VII C 1*) Hypoventilation produces an increase in $PaCO_2$, which is associated with an immediate decrease in arterial pH. Also, PaO_2 is reduced in hypoventilation.

Ventilation:perfusion imbalance usually is associated with arterial hypoxia and a normal $PaCO_2$, although the CO_2 excretion by the lung is impaired. The normal $PaCO_2$ is maintained at the expense of an increase in the minute ventilation.

At high altitudes, inspired air has a reduced PO_2, which stimulates the peripheral chemoreceptors to produce hyperventilation. Hyperventilation reduces $PaCO_2$ and causes respiratory alkalosis. At sea level, hyperventilation also causes an increased PaO_2.

Light to moderate exercise does not alter the arterial blood-gas composition. Vigorous exercise, however, causes lactic acidosis, which produces mild hyperventilation and a lower $PaCO_2$.

56–60. The answers are: 56-C, 57-B, 58-B, 59-A, 60-D. (*Chapter 7 VI G; X A, B*) Somatostatin (growth hormone–inhibiting hormone) is produced in the parvicellular neurosecretory neurons, in the duodenum, and in the pancreatic delta (D) cells. This hormone inhibits both glucagon and insulin secretion and is known to bring about hypoglycemia.

Among the essential amino acids that stimulate the pancreatic beta (B) cells to secrete insulin are arginine, lysine, and phenylalanine, in decreasing order of potency. Glucagon also is a stimulator of insulin release via the paracrine system that exists among the three types of islet cells.

Insulin is one of several protein anabolic hormones (e.g., growth hormone and thyroid hormones) but only one of two hypoglycemic hormones (the other is somatostatin). Insulin also promotes glycogen deposition and glucose utilization.

Of the hormones listed, only glucagon stimulates the pancreatic B cells to release insulin. Somatostatin and epinephrine inhibit the secretion of insulin.

Epinephrine is a hormone and a neurotransmitter that is equally effective as an α- and a β-agonist. The release of insulin is mediated by a β-receptor, while the inhibition of insulin secretion is mediated by an α-receptor. Since epinephrine is both an α- and a β-adrenergic agonist, it activates both types of adrenergic receptors on the pancreatic B cell. When epinephrine is administered with an α-antagonist, insulin secretion occurs. In vivo, without an α-antagonist, epinephrine has a predominant alpha effect and inhibits insulin secretion.

Index

A